THE BUILDING REGULATIONS

EXPLAINED & ILLUSTRATED

ELEVENTH EDITION

Vincent Powell-Smith

LLB, LLM, DLitt, FCIArb

and

M. J. Billington

BSc, ARICS

© 1967,1968,1970,1973,1981,1982 Walter S.Whyte
& Vincent Powell-Smith
© 1986 Seventh Edition by The Estate of Vincent
Powell-Smith & Walter S. Whyte
© 1986 New material The estate of Vincent Powell-
Smith & M.J. Billington
© 1990 Eighth Edition The estate of Vincent Powell-
Smith & M.J. Billington
© 1992 Ninth Edition The estate of Vincent Powell-
Smith & M.J. Billington
© 1995 Tenth Edition The estate of Vincent Powell-
Smith & M.J. Billington
© 1999 Eleventh Edition The estate of Vincent
Powell-Smith & M.J. Billington

Blackwell Science Ltd
Editorial Offices:
Osney Mead, Oxford OX2 0EL
25 John Street, London WC1N 2BL
23 Ainslie Place, Edinburgh EH3 6AJ
350 Main Street, Malden
    MA 02148 5018, USA
54 University Street, Carlton
    Victoria 3053, Australia
10, rue Casimir Delavigne
    75006 Paris, France

Other Editorial Offices:

Blackwell Wissenschafts-Verlag GmbH
Kurfürstendamm 57
10707 Berlin, Germany

Blackwell Science KK
MG Kodenmacho Building
7–10 Kodenmacho Nihombashi
Chuo-ku, Tokyo 104, Japan

First edition published by Crosby Lockwood & Son
Ltd 1967; Second edition 1968; Third edition with
metric supplement 1970; Reprinted 1972; Fourth
edition 1973; Second impression 1974, Third
impression (with supplement) 1976; reprinted 1979;
Fifth edition (amended) 1981 ; Sixth edition 1982;
Reprinted with amendments 1983; Reprint with minor
revisions by Collins Professional Books 1984;
Reprinted with amendments 1985; Seventh edition
1986; Reprinted 1986, 1987; Reprinted by BSP
Professional Books 1987, Reprinted with amendments
1988, 1989 (three times); Eighth edition 1990; Reprinted
with updates 1991; Reprinted 1992; Ninth edition
published by Blackwell Scientific Publications 1992;
Reprinted 1993; Tenth edition 1995; Eleventh edition
published by Blackwell Science Ltd 1999; Reprinted 2000

Set in 10/11pt Times
by DP Photosetting, Aylesbury, Bucks
Printed and bound in Great Britain
by MPG Books Ltd, Bodmin, Cornwall

The Blackwell Science logo is a trade mark of
Blackwell Science Ltd, registered at the
United Kingdom Trade Marks Registry

DISTRIBUTORS

Marston Book Services Ltd
PO Box 269
Abingdon
Oxon OX14 4YN
(Orders: Tel: 01235 465500
         Fax: 01235 465555)

USA
Blackwell Science, Inc.
Commerce Place
350 Main Street
Malden, MA 02148 5018
(Orders: Tel: 800 759 6102
              781 388 8250
         Fax: 781 388 8255)

Canada
Login Brothers Book Company
324 Saulteaux Crescent
Winnipeg, Manitoba R3J 3T2
(Orders: Tel: 204 837-2987
         Fax: 204 837-3116)

Australia
Blackwell Science Pty Ltd
54 University Street
Carlton, Victoria 3053
(Orders: Tel: 03 9347-0300
         Fax: 03 9347-5001)

A catalogue record for this title is available from the
British Library

ISBN 0-632-05069-1

Library of Congress
Cataloging-in-Publication Data
Powell- Smith, Vincent.
   The building regulations: explained & illustrated/
Vincent Powell-Smith and M.J. Billington. --11th ed.
   p.   cm.
   Rev. ed. of: The building regulations explained &
illustrated. 10th ed. 1995.
   Includes bibliographical references and index.
   ISBN 0-632-05069-1
   1. Building laws--Great Britain. I. Billington, M. J.
(Michael J.) II. Powell-Smith, Vincent. Building
regulations explained & illustrated. III. Title.
KD1140.P69 1999
343.41'07869--dc21   99-32997
                                                CIP

For further information on Blackwell Science,
visit our website:
www.blackwell-science.com

Contents

III Appendices

Preface

The last edition of this book was published in 1995 and has since gone through a number of reprints without the need for minor amendments. However a stage has now been reached, following the publication of the Building Regulations (Amendment) Regulations 1997 and 1998, which necessitates a revision of several chapters.

Chapter 15 (formerly entitled *Stairways, ramps and guards*) has been rewritten and retitled *Protection from falling, collision and impact* to reflect new regulations dealing with reducing the risk of injury from collisions with open windows, skylights and ventilators, or when using various types of sliding or powered doors and gates. Similarly, Chapter 18 introduces new provisions governing the safe use and cleaning of glazed elements. These new provisions apply only to workplaces and have been introduced mainly in order to ensure compliance with the Workplace (Health, Safety and Welfare) Regulations 1992.

Chapter 17 (*Access and facilities for disabled people*) has been rewritten to take account of new regulations which come into force on 25 October 1999 whereby accessibility controls are extended to dwellings, and Chapter 8 includes details of the latest amendment to the Approved Document to support regulation 7, covering materials and workmanship.

The early chapters which set out the legal and administrative provisions of the regulations have been completely rewritten to take account of the rapid expansion of the Approved Inspector control system and a number of important amendments to the Building Act 1984. Additionally, a new Chapter 5 has been included which sets out details of 18 Acts of Parliament and statutory instruments which may apply to a building project as well as, or in addition to, the Building Regulations 1991. This includes important changes to the way in which local authorities can charge for dealing with building regulation submissions. These are designed to improve competition with approved inspectors on the setting of charges.

As always, the aim is to provide a convenient and straightforward guide and reference to a complex and constantly evolving subject. It is a guide to the regulations and approved and other documents and is not a substitute for them. The intended readers are all those concerned with building work – architects and other designers, building control officers, approved inspectors, building surveyors and contractors – as well as their potential successors, the current generation of students. This book is designed to be of use to both students and teachers and it is gratifying that successive editions are widely adopted by the various academic institutions and professional bodies.

Finally, it is with sadness that I must record the death in 1997, of my co-author Vincent Powell-Smith. With Walter Whyte, Vincent had the original idea for this book in 1966 and it was already recognised as the authoritative text on the Building Regulations when I took over the technical content from Walter Whyte in 1985. Since then I have collaborated with Vincent on four editions and this will be the first time that he has had no direct involvement in its production. However, past readers will no doubt still recognise his unique style. Accordingly, I am especially grateful to Julia Burden, Deputy Publisher at Blackwell Science and Sue Moore, Senior Editor, for their help, patience, interest and sense of humour during the production of this edition.

The law is stated on the basis of cases reported and other material available to me on 1 March 1999.

Michael J. Billington

Acknowledgements

The tables reproduced from the Approved Documents and Building Research Establishment publications are Crown copyright and are reproduced with the permission of the Controller of HMSO. Packs containing the complete set of Approved Documents and other material can be obtained from HMSO.

I
Legal and Administrative

Chapter 1

Building control: an overview

Introduction

The building control system in England and Wales was radically revised in 1985. After a long period of gestation, building regulations were laid before Parliament and came into general operation on 11 November 1985. They applied to Inner London from 6 January 1986. Subject to specified exemptions, all building work (as defined in the regulations) in England and Wales is governed by building regulations.

The current regulations are the Building Regulations 1991 which came into force on 1 June 1992. The 1991 regulations have been amended five times since then, the latest being the Building Regulation (Amendment) Regulations 1998, and the provisions of all these amendments are reflected in this book.

A separate system of building control applies in Scotland and in Northern Ireland.

The power to make building regulations is vested in the Secretary of State for the Environment by section 1 of the Building Act 1984 which sets out the basic framework. Building regulations may be made for the following broad purposes:

(a) Securing the health, safety, welfare and convenience of people in or about buildings and of others who may be affected by buildings or matters connected with buildings.
(b) Furthering the conservation of fuel and power.
(c) Preventing waste, undue consumption, misuse or contamination of water.

The 1991 Regulations are very short and contain no technical detail. That is found in a series of Approved Documents and certain other non-statutory guidance, all of which refer to other non-statutory documents such as National Standards or Technical Specifications (e.g. British Standards or Agrément Certificates), with the objective of making the system more flexible and easier to use. The 1991 Regulations implement the final conclusions of a major review of both the technical and procedural requirements.

A significant feature of the system is that there are alternative systems of building control – one by local authorities, and the other a private system of certification which relies on 'approved inspectors' operating under a separate set of regulations called The Building (Approved Inspectors, etc.) Regulations

1985. These set out the detailed procedures for operating the system of private certification and came into effect at the same time as the main regulations.

The Building Act 1984

The Building Act 1984 received the Royal Assent on 31 October 1984 and the majority of its provisions came into force on 1 December 1984. It consolidated most, but not all, of the primary legislation relating to building which was formerly scattered in numerous other Acts of Parliament.

Part I of the Building Act 1984 is concerned with building regulations and related matters, while Part II deals with the system of private certification discussed in Chapter 4. Other provisions about buildings are contained in Part III which, amongst other things, covers drainage, the provision of sanitary conveniences, and so on, as well as the local authority's powers in relation to dangerous buildings, defective premises, etc.

The provisions of the 1984 Act are of the greatest importance in practice, and many of them are referred to in this and subsequent chapters.

1984 Act, sec. 121 'Building' is defined in the 1984 Act in very wide terms. A building is 'any permanent or temporary building and, unless the context otherwise requires, it includes any other structure or erection of whatever kind or nature (whether permanent or temporary)'. 'Structure or erection' includes a vehicle, vessel, hovercraft, aircraft or other movable object of any kind in such circumstances as may be prescribed by the Secretary of State. The Secretary of State's opinion is, however, qualified. The circumstances must be those which 'in [his] opinion ... justify treating it ... as a building'.

The result of this definition is that many things which would not otherwise be thought of as a building may fall under the Act – fences, radio towers, silos, air-supported structures and the like. Happily, as will be seen, there is a more restrictive definition of 'building' for the purposes of the 1991 Regulations, but a comprehensive definition is essential for general purposes, e.g. in connection with the local authority's powers to deal with dangerous structures. Hence the statutory definition is necessarily couched in the widest possible terms. In general usage (and at common law) the word 'building' ordinarily means 'a structure of considerable size intended to be permanent or at least to last for a considerable time' (*Stevens* v. *Gourely* (1859) 7 CBNS 99) and considerable practical difficulties arose as to the scope of earlier building regulations which the 1984 definition has removed.

In *Seabrink Residents Association* v. *Robert Walpole Campion and Partners* (1988) (6-CLD-08-13; 6-CLD-08-10; 6-CLD-06-32) for example, the High Court held that walls and bridges on a residential development were not subject to the then Building Regulations 1972 because they were not part of 'a building'. The development was not to be considered as a homogenous whole. The then regulations, said Judge Esyr Lewis QC, were 'concerned with structures which have walls and roofs into which people can go and in which goods can be stored'. Each structure in the development must be looked at separately to see whether the regulations applied. 'Obviously a wall may be part of a building and so, in my view, may be a bridge'.

The linked powers

Local authorities exercise a number of statutory public health functions in conjunction with the process of building control, for example, controls on construction over drains and sewers. These provisions are commonly called 'the linked powers' because their operation is linked with the local authority's building control functions, both in checking deposited plans or considering a building notice, and under the approved inspector system of control. Many of the former linked powers have been brought under the building regulations, but local authorities are responsible for certain functions now found in the 1984 Act. In those cases, the local authority must reject the plans (or building notice) or the approved inspector's initial notice if relevant compliance is not achieved or else must impose suitable safeguards. The relevant provisions are:

(a) Construction over drains and sewers. Section 18, as amended, controls new building on top of drains, sewers, and disposal mains. The local authority must reject plans submitted if they propose building over a sewer or drain shown on the relevant public maps, unless they are satisfied that they can properly give consent either unconditionally or subject to conditions. The most usual condition is that the building owner enters into an access agreement for maintenance purposes. The local authority notify the water authority of the proposal and the water authority may give directions to the local authority as to how their section 18 functions are to be exercised. Disputes under the section are determined by a magistrates' court. **1984 Act, sec. 18**

(b) Provision of drainage. The local authority must similarly reject plans submitted for a new building or extension unless they show that the provision for drainage is satisfactory. They can insist that the drainage connects to a nearby public sewer. Disputes under section 21 are also dealt with by a magistrates' court. **1984 Act, sec. 21**

A related provision is section 98 of the Water Industry Act 1991 under which owners or occupiers of premises can require the water authority to provide a public sewer for domestic purposes in their area, subject to various conditions which can include in an appropriate case the making of a financial contribution. **Water Industry Act 1991, sec. 98**

(c) Provision of water supply. Section 25 requires the local authority to reject plans of a house submitted under the building regulations unless they are satisfied with the proposals for providing the occupants with a sufficient supply of wholesome water for domestic purposes, by pipes or otherwise. Usually, of course, this will be by means of a mains supply provided by the water authority under the Water Industry Act 1991. Disputes are determined by the magistrates' court. **1984 Act, sec. 25**

A related provision is section 37 of the Water Industry Act 1991 which enables a landowner who proposes to erect buildings to require the water authority to lay necessary mains for the supply of water for domestic purposes to a point which will enable the buildings to be connected to the mains at a reasonable cost, a provision which is of considerable use to developers. **Water Industry Act 1991, sec. 37**

Building regulations

The Secretary of State is given power to make comprehensive regulations about the provision of services, fittings and equipment in or in connection with buildings as well as about the design and construction of buildings. A very comprehensive list of the subject matter of building regulations is contained in Schedule 1. The regulations are supported by approved documents, giving 'practical guidance' (see p.2.2ff). **1984 Act, sec. 1**

1984 Act
Sch. 1

Building regulations may include provision as to the deposit of plans of executed, as well as proposed work; for example where work has been done without the deposit of plans or there has been a departure from the approved plans. Broad powers are given to make building regulations about the inspection and testing of work, and the taking of samples.

1984 Act,
secs 3 & 4

Prescribed classes of buildings, services, etc. may be wholly or partially exempted from regulation requirements. Similarly, the Secretary of State may, by direction, exempt any particular building or buildings at a particular location.

1984 Act,
Sch. 1

Schedule 1 of the 1984 Act is a flexible provision and covers the application of the regulations to existing buildings. It enables regulations to be made regarding not only alterations and extensions, but also the provision, alteration or extension of services, fittings and equipment in or in connection with existing buildings. It also enables the regulations to be applied on a *material change of use* as defined in the regulations and, very importantly, makes it possible for the regulations to apply where re-construction is taking place, so that the regulations can deal with the whole of the building concerned and not merely with the new work.

The 1984 Act contains enabling powers for the making of regulations on a number of procedural matters.

The regulations made and currently in force are:

- The Building Regulations 1991
- The Building (Approved Inspectors, etc.) Regulations 1985
- The Building (Local Authority Charges) Regulations 1998
- The Building (Inner London) Regulations 1985

Most of these regulations have been amended, in some cases several times, and care should be taken to ensure that the most recent amendments are being used.

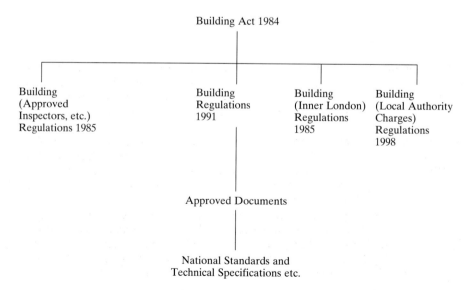

Fig. 1.1 Building control: the legislative scheme.

Building regulations – exemptions

Crown immunity

The Building Regulations do not apply to premises which are occupied by the Crown. The general position regarding Crown exemption is that a statute does not bind the Crown unless it so provides either expressly or by necessary implication. In fact, there is provision in section 44 of the Building Act 1984 to apply the substantive requirements of the Regulations to Crown buildings but this has never been activated.

In recent years a number of premises have lost their Crown immunity, often in response to public opinion. These include National Health Service buildings (see section 60 of the National Health and Community Care Act 1990) and the Metropolitan Police. Additionally, the reorganisation of the Post Office has meant that whilst the Royal Mail is still regarded as Crown property, Post Office Counters is not.

In practice, it is normal for government department building work to be designed and constructed in accordance with the Building Regulations. In some areas the plans and particulars may even be submitted to the local authority for comment, although it is more usual for these to be scrutinised by specialist companies (replacing the service which was formally given by the Property Services Agency) who will also carry out on-site inspections of the works in progress. Even so, such companies have no legal control over the work and cannot take enforcement action in the event of a breach of the Regulations.

Interestingly, Crown premises are not exempt from certification under the Fire Precautions Act 1971. However, they are inspected *not* by the fire authority, but by the Crown Premises Inspection Group within the Home Office Fire Service Inspectorate, a bureaucratic anomaly which has attracted much criticism. Unfortunately, the powers of entry to premises contained in the 1971 Act do not apply to premises occupied by the Crown.

Building Act exemptions

Taken together, the Building Act 1984 and the Building Regulations 1991 (as amended), exempt certain uses of buildings and many categories of work from control. Unfortunately, the situation with regard to exempt building uses is often unclear. Section 4 of the 1984 Act removes control from:

- Any educational establishments which are built to details approved by the Secretary of State (for Education) under section 14 of the Education Act 1980 or section 90 of the Education Reform Act 1988. The education authority or governing body must, however, comply with the Education (School Premises) Regulations 1981 instead, and **1984 Act, sec. 4**
- A building belonging to statutory undertakers, the UK Atomic Energy Authority or the Civil Aviation Authority. The building must be one which is used for the purpose of the undertaking (such as an electricity substation), therefore buildings such as houses or offices are not exempt, unless they are part of a railway station or aerodrome which is owned by the relevant body.

Regarding educational establishments, the situation is far from straightforward since many former polytechnics and colleges of further education are now outside the local authority funding system and are subject to building regulation control. Similarly, many schools have 'opted out' of local education authority control and are self-governing, grant maintained schools responsible for their own management.

Curiously, self-governing schools are required to obtain the approval of the Secretary of State for all building projects and are therefore exempt from Building Regulations whereas local authority schools do not need the Secretary of States' approval for projects costing less than £200,000 and hence, are not exempt from Building Regulations. Independent schools are not exempt since they do not require the approval of the Secretary of State.

Further and higher education establishments are exempt if they remain within the local government sector or are substantially dependent for their maintenance on assistance from education authorities or grants under section 100(1)(b) of the Education Act 1944, otherwise they are subject to control.

Full details of the exempting arrangements for educational establishments were issued on 21 March 1994 in a circular letter from the DOE. Additionally, the DfEE are considering issuing revised guidance to reflect forthcoming implementation of new legislation.

It should be noted that local authority buildings are *not* exempt from either the procedural or substantive requirements of the Building Regulations.

Miscellaneous

There is power to approve the plans of a proposed building by stages. Usually, the initiative will rest with the applicant as to whether to seek approval by stages – subject to the local authority's agreement. However, local authorities may – of their own initiative – give approval by stages; they might, for **1984 Act,** example, await further information. In giving stage approval, local authorities **sec. 16** will be able to impose a condition that certain work will not start until the relevant information has been produced.

Plans may also be approved subject to agreed modifications, e.g. where there is a minor defect in the plans.

1984 Act, Section 19 of the Building Act 1984 deals with the use of short-lived **sec. 19** materials. The provision applies where plans, although conforming with the regulations, include the use of items listed in the regulations for the purpose of section 19. In such circumstances the local authority has a discretion:

(a) To pass the plans;
(b) To reject the plans; *or*
(c) To pass them subject to the imposition of a time limit, whether conditionally or otherwise.

Interestingly, the Building Regulations 1991 contain no specific references to any particular materials, however as will be seen, regulation 7 of the 1991 Regulations requires that building work which must comply with the Schedule 1 requirements must be carried out 'with proper materials which are appropriate for the circumstances in which they are used...', and the supporting approved document deals with the use of short-lived materials.

The local authority may impose a time limit either on the whole of a building or on particular work. Additionally, they may impose conditions as to the use of a building or the particular items concerned. Appeal against the local authority's decision lies to the Secretary of State.

Eventually, section 19 will cease to have effect when section 20, which is wider in scope, is brought into force by the Secretary of State.

Building regulations may impose continuing requirements on the owners and occupiers of buildings, including buildings which were not, at the time of their erection, subject to building regulations. These requirements are of two kinds. Continuing requirements may be imposed *first*, in respect of designated provisions of the regulations to ensure that their purpose is not frustrated, e.g. the keeping clear of fire escapes; and *second*, in respect of services, fittings and equipment, e.g. a requirement for the periodical maintenance and inspection of lifts in flats.

1984 Act, sec. 2

Type relaxations may be granted by the Secretary of State; he may dispense with or relax some regulation requirement generally. A type relaxation can be made subject to conditions or for a limited period only. It should be noted that before granting a type relaxation the Secretary of State must consult such bodies as appear to him to be representative of the interests concerned and he has to publish notice of any relaxations issued.

1984 Act, sec. 11

The Building Act 1984, sections 39 to 43, contains the appeal provisions. The principal appeals to the Secretary of State are:

(a) Appeals against rejection of plans by a local authority.
(b) Appeals against a local authority's refusal to give a direction dispensing with or relaxing a requirement of the regulations or against a condition attached by them to such a direction.

1984 Act secs. 39 to 43

Interestingly, section 38 of the Building Act 1984 is concerned with civil liability but has yet to be activated. Under this section, breach of duty imposed by the regulations will be actionable at civil law, where damage is caused, except where the regulations otherwise provide. 'Damage' is defined as including the death of, or injury to, any person (including any disease or any impairment of a person's physical or mental condition). The regulations themselves may provide for defences to such a civil action and section 38 will not, when operative, prejudice any right which exists at common law.

1984 Act sec. 38 (not yet operative)

Dangerous structures, etc.

Local authorities have power to deal with a building or structure which is in a dangerous condition or is overloaded. The procedure is for the local authority to apply to the magistrates' court for an order requiring the owner to carry out remedial works or, at his option, to demolish the building or structure and remove the resultant rubbish. The court may restrict the use of the building if the danger arises from overloading. If the owner fails to comply with the order within the time limit specified by the court, the local authority may execute the works themselves and recover the expenses incurred from the owner, who is also liable to a fine.

1984 Act, sec. 77

The local authority may take immediate action in an emergency so as to remove the danger, e.g. if a wall is in danger of imminent collapse. Where it is practicable to do so, they must give notice of the proposed action to the owner and occupier. The local authority may recover expenses which they have

1984 Act, sec. 78

reasonably incurred in taking emergency action, unless the magistrates' court considers that they might reasonably have proceeded under section 77. An owner or occupier who suffers damage as a result of action taken under section 78 may in some circumstances be entitled to recover compensation from the local authority.

1984 Act,
sec. 79

1984 Act,
sec. 76

Section 79 of the 1984 Act empowers local authorities to deal with ruinous and dilapidated buildings or structures and neglected sites 'in the interests of amenity', which is a term of wider significance than 'health and safety': *Re Ellis and Ruislip and Northwood UDC* [1920] 1 KB 343. (Section 76 of the Act enables them to deal with defective premises which are 'prejudicial to health or a nuisance'.)

Under section 79, where a building or structure is in such a ruinous or dilapidated condition as to be seriously detrimental to the amenities of the neighbourhood, the local authority may serve notice on the owner requiring him to repair or restore it or, at his option, demolish the building or structure and clear the site.

1984 Act,
sec. 80

Demolition is itself subject to control. Section 80 requires a person who intends to demolish the whole or part of a building to notify the local authority, the occupier of any adjacent building and the gas and electricity authorities of his intention to demolish. He must also comply with any requirements which the local authority may impose by notice under section 82.

The demolition notice procedure does not apply to the demolition of:

- An internal part of an occupied building where it is intended that the building should continue to be occupied.
- A building with a cubic content (ascertained by external measurement) of not more than 1750 cubic feet (50m^3) or a greenhouse, conservatory, shed or prefabricated garage which forms part of a larger building.
- An agricultural building unless it is contiguous to a non-agricultural building or falls within the preceding paragraph.

1984 Act,
secs. 81 &
82

The local authority may by notice require a person undertaking demolition to carry out certain works:

- To shore up any adjacent building.
- To weatherproof any surfaces of an adjacent building exposed by the demolition.
- To repair and make good any damage to any adjacent building caused by the demolition.
- To remove material and rubbish resulting from the demolition and clearance of the site.
- To disconnect and seal and/or remove any sewers or drains in or under the building.
- To make good the ground surface.
- To make arrangements with the gas, electricity and water authorities for the disconnection of supplies.
- To make suitable arrangements with the fire authority (and Health and Safety Executive, if appropriate) with regard to burning of structures or materials on site.
- To take such steps in connection with the demolition as are necessary for the protection of the public and the preservation of public amenity.

Other legislation

Although the Building Act 1984 attempted to rationalise the main controls over buildings, there are in fact a great many pieces of legislation, in addition to the Building Act and the Building Regulations, which affect the building, its site and environment and the safety of working practices on and within the building. A summary of the legislation most commonly encountered is given in Chapter 5.

Chapter 2

The Building Regulations and Approved Documents

Introduction

Although the statutory framework of building control is found in the Building Act 1984, the 1991 Regulations, as amended, contain the detailed rules and procedures. The regulations are comparatively short because the technical requirements have mostly been cast in a functional form.

Each technical requirement is supported by a document approved by the Secretary of State intended to give practical guidance on how to comply with the requirements. The Approved Documents refer to British Standards and other guidance material such as BRE publications and thus give designers and builders a great degree of flexibility.

The 1991 Regulations became effective on 1 June 1992, and have since been amended five times.

Division of the regulations

There are 23 regulations, arranged logically in five parts. The division is as follows:

PART I: GENERAL

Reg. 1. Citation and commencement.
Reg. 2. Interpretation.

PART II: CONTROL OF BUILDING WORK

Reg. 3. Meaning of building work.
Reg. 4. Requirements relating to building work.
Reg. 5. Meaning of material change of use.
Reg. 6. Requirements relating to material change of use.
Reg. 7. Materials and workmanship.
Reg. 8. Limitation on requirements.
Reg. 9. Exempt buildings and work.

PART III: RELAXATION OF REQUIREMENTS

Reg. 10.   Power to dispense with or relax requirements.

PART IV: NOTICES AND PLANS

Reg. 11.   Giving of a building notice or deposit of plans.
Reg. 12.   Particulars and plans where a building notice is given.
Reg. 13.   Full plans.
Reg. 13A.   Unauthorised building work.
Reg. 14.   Notice of commencement and completion of certain stages of work.
Reg. 14A.   Energy rating.
Reg. 15.   Completion certificates.

PART V: MISCELLANEOUS

Reg. 16.   Testing of drains and private sewers.
Reg. 17.   Sampling of material.
Reg. 18.   Supervision of building work otherwise than by local authorities.
Reg. 19.   Revocations.
Reg. 20.   Transitional provisions.
Reg. 21.   Contravention of certain regulations not to be an offence.

There are also three schedules:

SCHEDULE 1 – REQUIREMENTS

This contains technical requirements which are almost all expressed in functional terms and grouped in 13 parts set out in tabular form:

PART A: STRUCTURE – covers loading, ground movement and disproportionate collapse.

PART B: FIRE SAFETY – covers means of escape, internal and external fire spread, and access and facilities for the fire service.

PART C: SITE PREPARATION AND RESISTANCE TO MOISTURE – covers preparation of site, dangerous and offensive substances, subsoil drainage, and resistance to weather and ground moisture.

PART D: TOXIC SUBSTANCES – deals with cavity insulation.

PART E: RESISTANCE TO THE PASSAGE OF SOUND – covers airborne and impact sound.

PART F: VENTILATION – covers means of ventilation and condensation in roofs.

PART G: HYGIENE – deals with bathrooms, hot water storage and sanitary conveniences and washing facilities.

PART H: DRAINAGE AND WASTE DISPOSAL – deals with foul water drainage, cesspools, septic tanks and settlement tanks, rainwater drainage and solid waste storage.

PART J: HEAT PRODUCING APPLIANCES – covers air supply, discharge of combustion products and protection of the building.

PART K: PROTECTION FROM FALLING, COLLISION AND IMPACT – covers stairs, ladders and ramps, protection from falling, vehicle barriers and loading bays, protection from collision with open windows, etc., and protection against impact from and trapping by doors.

PART L: CONSERVATION OF FUEL AND POWER – provides that reasonable provision must be made for the conservation of fuel and power.

PART M: ACCESS AND FACILITIES FOR DISABLED PEOPLE – deals with the provision of facilities for the disabled: access and use of sanitary conveniences, and audience or spectator seating.

PART N: GLAZING – SAFETY IN RELATION TO IMPACT, OPENING AND CLEANING – deals with reducing the risks associated with glazing in critical locations in buildings and covers the safe operation and cleaning of windows, skylights and ventilators, etc.

SCHEDULE 2 – EXEMPT BUILDINGS AND WORK

This lists exempt buildings and work in seven classes, and one of its effects is significantly to reduce the extent of control by giving complete exemptions for certain buildings and extensions.

SCHEDULE 3 – REVOCATION OF REGULATIONS

This lists the former regulations which are revoked, i.e. the Building Regulations 1985, and parts of other relevant regulations.   **Reg. 19**

Approved Documents

There are 14 Approved Documents issued by the Department of the Environment, Transport and the Regions (DETR) intended to give practical guidance on how the technical requirements of Schedule 1 may be complied with. They are written in straightforward technical terms with accompanying diagrams and the intention is that they will be quickly updated as necessary. Additionally, there are two other Approved Documents covering respectively: *Timber intermediate floors for dwellings* published by the Timber Research and Development Association, and *Basements for dwellings* published by the British Cement Association.

The status and use of Approved Documents is prescribed in sections 6 and 7 of the Building Act 1984. Section 6 provides for documents giving 'practical guidance with respect to the requirements of any provision of building regulations' to be approved by the Secretary of State or some body designated by him. The documents so far issued have been approved by the Secretary of State, although they refer to other non-statutory material.

The legal effect of 'Approved Documents' is specified in section 7. Their use is not mandatory, and failure to comply with their recommendations does not involve any civil or criminal liability, but they can be relied upon by either party in any proceedings about an alleged contravention of the requirements of the regulations. If the designer or contractor proves that he has complied with the requirements of an Approved Document, in any proceedings which are brought against him he can rely upon this 'as tending to negative liability'. Conversely, failure to comply with an Approved Document may be relied on by the local authority 'as tending to establish liability'. In other words, the onus will be upon the designer or contractor to establish that he has met the functional requirements in some other way.

The position is illustrated by *Rickards* v. *Kerrier District Council* (1987) CILL 345, 4-CLD-04-26 where it was held that if the local authority proved that the works did not comply with the Approved Document, it was then for the appellant to show compliance with the regulations. If the designer fails to follow an Approved Document, it is for him to prove (if prosecuted) that he used an equally effective method or practice.

All the Approved Documents are in a common format, and their provisions are considered in subsequent chapters. They may be summarised as follows:

A: STRUCTURE – This supports Schedule 1, A1, A2 and A3. Section 1, which deals with houses and other small buildings, contains tables for timber sizes, wall thicknesses, etc., and a lot of technical guidance. Section 2 deals with disproportionate collapse and is relevant to all types of building. It lists Codes and Standards for structural design and construction for all building types and emphasises certain basic principles which must be taken into account if other approaches are adopted.

B: FIRE SAFETY – This supports Schedule B1, B2, B3, B4, and B5 and has separate sections in B2 and B4 which deal with different elements of the building.

C: SITE PREPARATION AND RESISTANCE TO MOISTURE – Read in conjunction with Schedule 1, Part C, it deals with the necessary basic requirements. Section 1 covers site preparation and site drainage and Section 2 deals with contaminants, including the erection of buildings on sites affected by radon gas or the landfill gases, methane and carbon dioxide. Additionally, it covers any substances in the ground which might cause a danger to health, and its provisions effectively replace those of the repealed section 29 of the Building Act 1984. C 4 describes the measures necessary in order to prevent the passage of moisture to the inside of the building.

D: TOXIC SUBSTANCES – This supports Schedule 1, Part D and it is very short. It gives advice on guarding against fumes from urea formaldehyde foam.

E: RESISTANCE TO THE PASSAGE OF SOUND – This supports Schedule 1, part E.

F: VENTILATION – Supporting Part F of Schedule 1 it covers means of ventilation including the precautions to be taken to prevent excessive condensation in the roof voids of dwellings.

G: HYGIENE – Supporting Part G of Schedule 1 it includes the requirements of certain repealed sections (sections 26 to 28) of the Building Act 1984 dealing with water-closets and bathrooms as well as covering unvented hot water systems.

H: DRAINAGE AND WASTE DISPOSAL – This supports Part H of Schedule 1 and covers above and below ground drainage, cesspools and tanks, rainwater drainage and solid waste storage and takes the place of the repealed sub-sections of section 23 of the Building Act 1984.

J: HEAT PRODUCING APPLIANCES – Supporting Part J of Schedule 1, it deals with gas appliances up to 60 kW and solid and oil fuel appliances up to 45 kW.

K: PROTECTION FROM FALLING, COLLISION AND IMPACT – This supports Part K of Schedule 1 and covers the design and construction of stairs, ramps and guarding. It has been extended in the 1998 edition to cover vehicle loading bays, protection from collision with open windows, skylights and ventilators, and protection against impact from and trapping by doors.

L: CONSERVATION OF FUEL AND POWER – Supporting Part L. A new edition was issued in 1995, and Appendix G describes a procedure for calculating the energy rating required by regulation 14A. (See also Chapter 16.)

M: ACCESS AND FACILITIES FOR DISABLED PEOPLE – Supporting Part M of Schedule 1 this gives practical guidance on means of access, use of buildings, sanitary conveniences and audience or spectator seating.

N: GLAZING – SAFETY IN RELATION TO IMPACT, OPENING AND CLEANING – Supporting Part N. The 1998 edition now covers safe operation and access for cleaning windows, etc., in addition to measures designed to reduce the risks of accidents caused by contact with glazing.

There is a further Approved Document – MATERIALS AND WORK-MANSHIP – to support regulation 7 – and it is phrased in very general terms.

Relaxations of the mandatory requirements may be given only by local authorities in appropriate cases, with the possibility of an appeal against refusal to the Secretary of State. An approved inspector cannot grant a relaxation.   **Reg. 10**

Definitions in the regulations

Regulation 2 provides a number of general definitions, but not all of them are equally important or helpful. In this section full definitions are given for purposes of ease of reference, although the various special definitions will be referred to again in later chapters. The definitions are:

Reg. 2(1)   THE ACT – This means the Building Act 1984.

BUILDING – The regulations apply only to buildings as defined. There is a narrow definition of 'building' for the purposes of the regulations:

A building is 'any permanent or temporary building but not any other kind of structure or erection'. When 'a building' is referred to in the regulations this includes a part of a building.

The effect of this definition is to exclude from control under the regulations such things as garden walls, fences, silos, air-supported structures and so forth.

BUILDING NOTICE – A notice in prescribed form given to the local authority under regulations 11(1)(a) and 12 informing the authority of proposed works.

BUILDING WORK – The regulations apply only to building work as defined in regulation 3(1); any work not coming within the definition is not controlled. Building work means:

- The erection or extension of a building;
- The material alteration of a building;
- The provision, extension or material alteration of services or fittings required by Schedule 1, Parts G, H or J (and called 'controlled services or fittings');
- Work required by regulation 6 – which sets out the requirements relating to 'material change of use' (see below);
- The insertion of insulating material into the cavity wall of a building; *or*,
- Work involving the underpinning of a building.

CONSTRUCTION PRODUCTS DIRECTIVE – This refers to the EEC Council Directive (89/106/EEC) on construction products which covers all products 'produced with a view to incorporation in construction works'. If they are shown to be 'fit for their intended use' in that they meet essential requirements laid down in the Directive they are entitled to the EC mark and their use in building work cannot be restricted by member states.

CONTROLLED SERVICE OR FITTING – This means services or fittings required by Parts G, H or J, i.e. bathrooms, hot water storage systems, sanitary conveniences, drainage and waste disposal, and heat producing appliances.

DWELLING – This includes a dwellinghouse and a flat.

DWELLINGHOUSE excludes a flat or building containing a flat.

EUROPEAN TECHNICAL APPROVAL – This means a favourable technical assessment of the fitness for use of a construction product for the purposes of the Construction Products Directive by an authorised body.

FLAT – Separate and self-contained premises (including a maisonette) constructed or adapted for residential purposes and forming part of a building divided horizontally from some other part (see Fig. 2.1).

FLOOR AREA – This means the aggregate area of every floor in a building or extension. The area is to be calculated by reference to the finished internal faces of the enclosed walls or, where there is no enclosing wall, to the outermost edge of the floor (see Fig. 2.2).

FULL PLANS – Plans deposited with a local authority in accordance with regulations 11(1)(b) and 13. The Building Act 1984, section 126, gives a definition of 'plans' as including drawings of any description and specifications or other information in any form.

HARMONIZED STANDARD – A recognised standard under the Construction Products Directive, mandated by the Commission for the European Economic Community and published by the Commission in the *Official Journal* of the European Communities.

HEIGHT – This means the height of a building measured from the mean level of the ground adjoining the outside external walls to a level of half the vertical height of the roof, or to the top of any walls or parapet, whichever is the higher (see Fig. 2.2).

INSTITUTION – This means a hospital, home, school, etc., used as living accommodation for, or for the treatment, care, etc., of people suffering from disabilities due to illness or old age or other physical or mental disability or who are under five years old. Those concerned must sleep on the premises and so day care centres, etc., are not included.

MATERIAL ALTERATION – This is defined in regulation 3(2) and is described fully on page 2.12 below.   **Reg. 3(2)**

MATERIAL CHANGE OF USE – This is defined by reference to regulation 5 and there are seven cases: **Reg. 2(1)**

- Where a building becomes a dwelling when it was not one before.
- Where a building will contain a flat for the first time.
- Where a building becomes a hotel or boarding house, where it previously was not.
- Where a building becomes an institution, where it previously was not.
- Where a building becomes a public building and it was not before.
- Where a building was previously exempt from control (see Schedule 2, below), but is no longer so exempt.
- Where a building containing at least one dwelling is altered so that it provides more or less dwellings than before.

PUBLIC BUILDING – This means a building which consists of or contains: **Reg. 2(2)**

- A theatre, public library, hall or other place of public resort.
- A school or other educational establishment which is not exempt under the 1984 Act, section 4(1)(a).
- A place of public worship.

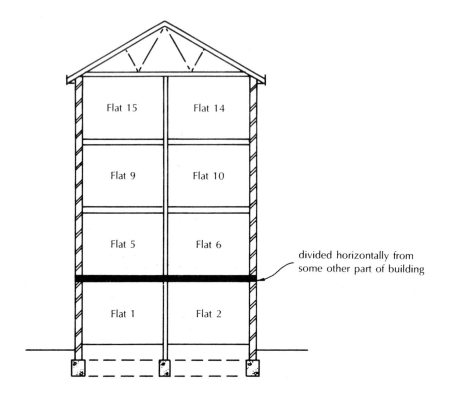

Section x-x

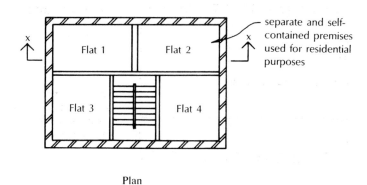

Plan

Fig. 2.1 Flat – regulation 2.

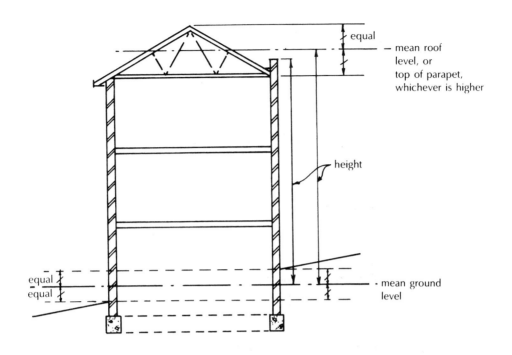

Height

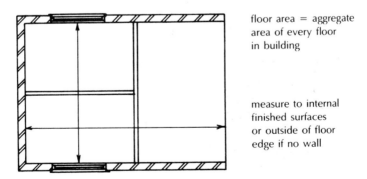

Floor area

Fig. 2.2 Floor area and height – regulation 2.

The definition is restrictive because occasional visits by the public to shops, stores, warehouses or private houses do not make the building a public building.

Reg. 2(1)
SHOP – This includes premises used:
- For sales to the public of food or drink for consumption on or off the premises.
- For retail sales by auction.
- As a barber's or hairdresser's business.
- For the hiring of any item.
- For the treatment or repair of goods.

Exempt buildings and work

Sched. 2
Certain buildings and extensions are granted complete exemption from control. The exempt buildings and work fall into seven classes listed in Schedule 2:

CLASS I – BUILDINGS CONTROLLED UNDER OTHER LEGISLATION
- Buildings subject to the Explosives Acts 1875 and 1923.
- Buildings (other than dwellings, offices or canteens) on a site licensed under the Nuclear Installations Act 1965.
- Buildings scheduled under section 1 of the Ancient Monuments and Archaeological Areas Act 1979.

CLASS II – BUILDINGS NOT FREQUENTED BY PEOPLE
- Detached buildings into which people do not normally go.
- Detached buildings housing fixed plant or machinery, normally visited only intermittently for the purpose of inspecting or maintaining the plant, etc. Such buildings are only exempt where they are at least one-and-a-half times their own height from the boundary of the site or any other building frequented by people.

CLASS III – GREENHOUSES AND AGRICULTURAL BUILDINGS
- A building used as a greenhouse.

A greenhouse is not exempted if the main purpose for which it is used is retailing, packing or exhibiting, e.g. one at a garden centre.

- A building used for agriculture which is:
  Sited at a distance not less than one-and-a-half times its own height from any building containing sleeping accommodation; *and*,
  is provided with a fire exit not more than 30 metres from any point within the building.

The definition of 'agriculture' includes horticulture, fruit growing, seed growing, dairy farming, fish farming and the breeding and keeping of livestock (including any creature kept for the production of food, wool, skins or fur or for the purpose of its use in the farming of land). Agricultural buildings are not exempted if the main purpose for which they are used is retailing, packing or exhibiting.

CLASS IV – TEMPORARY BUILDINGS
- A building intended to remain where it is erected for 28 days or less, e.g. exhibition stands.

CLASS V – ANCILLARY BUILDINGS
- Buildings on a site intended to be used only in connection with the letting or sale of buildings or building plots on that estate.
- Site buildings on all construction and civil engineering sites, provided they contain no sleeping accommodation.
- Buildings, except those containing a dwelling or used as an office or showroom, erected in connection with a mine or quarry.

The exemption does not apply to buildings containing a dwelling or used as an office or showroom.

CLASS VI – SMALL DETACHED BUILDINGS
- Detached single storey buildings of up to $30\,m^2$ floor area, with no sleeping accommodation.

For the exemption to apply, such buildings must either be:
Situated more than one metre from the boundary of their curtilage; *or*;
Constructed substantially of non-combustible material.

- Detached buildings of up to $30\,m^2$ intended to shelter people from the effects of nuclear, chemical or conventional weapons and not used for any other purpose. The excavation for the building must be no closer to any exposed part of another building or structure than a distance equal to the depth of the excavation plus one metre.
- Detached buildings with a floor area not exceeding $15\,m^2$ and which do not contain sleeping accommodation, e.g. garden sheds.

CLASS VII – EXTENSIONS
- Ground level extensions of up to $30\,m^2$ floor area which are greenhouses, conservatories, porches, covered yards or ways or a carport open on at least two sides.

A conservatory or porch which is wholly or partly glazed must satisfy the requirements of Part N.

The regulations do not apply to the erection of any building set out in Classes I **Reg. 9** to VI or to extension work in Class VII. Furthermore, they have no application at all to *any* work done to or in connection with buildings in Classes I to VII provided, of course, that the work does not involve a change of use which takes the building out of exemption, e.g. a barn conversion.

Application of the regulations

The 1991 Regulations apply only to 'building work' or to a 'material change of use', i.e. use for a different purpose. Work or a change of use not coming under these headings is not controlled.

Meaning of 'building work'
The definition of 'building work' means that the regulations apply in six cases:

ERECTION OR EXTENSION OF A BUILDING

Subject to the exemptions set out in the preceding section, the regulations apply to the erection or extension of all buildings. No attempt is made to define what is meant by 'erection of a building', nor is any definition really necessary. There is a good deal of obscure case law under other legislation as to what amounts to 'erection of a building', but none of it is particularly helpful in the light of section 123 of the Building Act 1984.

1984 Act, sec. 123

This gives a relevant statutory definition. For the purposes of Part II of the Act and for building regulation purposes, erection will include related operations 'whether for the reconstruction of a building, [and] the roofing over of an open space between walls or buildings'.

For the purposes of Part III of the 1984 Act (other provisions about buildings) which is also relevant to building control, *certain* building operations are 'deemed to be the erection of a building'. These are:

(a) Re-erection of any building or part of a building when an outer wall has been pulled or burnt down to within ten feet (3 metres) of the surface of the ground adjoining the lowest story of the building.

It follows that the outer wall must have been demolished throughout its length to within ten feet (3 metres) of ground level to constitute re-erection.

(b) The re-erection of any frame building when it has been so far pulled or burnt down that only the framework of the lowest storey remains.

(c) Roofing over any space between walls or buildings. Clearly other operations could be 'the erection of a building'.

PROVISION OR EXTENSION OF CONTROLLED SERVICES AND FITTINGS

Reg. 3(1)

Controlled services and fittings are those required by specified parts of Schedule 1:

- G1 – Sanitary conveniences and washing facilities.
- G2 – Bathrooms in dwellings.
- G3 – Hot water storage systems, except space heating systems, industrial systems, or those with a storage capacity of 15 litres or less.
- H – Drainage and waste disposal systems.
- J – Fixed heat producing appliances burning solid or oil fuel or gas or incinerators.

MATERIAL ALTERATION OF A BUILDING OR OF A CONTROLLED SERVICE OR FITTING

1991 Regs, Reg. 3(2)

The material alteration of an existing building falls within the definition of building work, and is subject to the regulation requirements. Other alterations are not controlled. There are two cases where an alteration is material, namely an alteration to a building or controlled service or fitting, or part of the work involved, which would at any stage result *either*:

Reg. 3(2)

- In the building or controlled service or fitting not complying with the relevant requirements of Schedule 1 where it previously did comply: *or*,
- In the building, which did not comply with such requirements before work started, being made worse in relation to the requirement after the alteration.

The specified requirements (called 'relevant requirements' in the regulations) are: Part A (structure); B1 (means of escape); B3 (internal fire spread – structure); B4 (external fire spread); B5 (access and facilities for the fire service); and Part M (access and facilities for disabled people).

The work done must, of course, comply with all the requirements of Schedule 1. In general, it is not necessary to bring the existing building up to regulation standards, however, it should not be made worse when measured against the standards of the relevant requirements in Schedule 1.

WORK IN CONSEQUENCE OF A MATERIAL CHANGE OF USE

When there is a material change of use, as defined in regulation 5 (see p. 2.7), **Reg. 6** work must be done to make the building comply with some of the regulations, as explained below. Such work is, of course, then subject to control, just as the material change of use is itself controlled. In practical terms, change of use is only subject to control if the change involves the provision of sleeping accommodation or use as a public building or where the building was previously exempt.

'Material change of use' requirements

Material change of use has already been defined (see p. 2.7), and in the six cases falling within that definition, specific technical requirements from Schedule 1 are made to apply in the interests of health and safety, which is the philosophy behind building control. Interestingly, there is no requirement applicable in respect of such things as foul or surface water drainage or stairs, nor is there any definition of 'part' of a building.

The parts of the regulations applicable are set out in Table 2.1.

INSERTION OF INSULATING MATERIAL INTO A CAVITY WALL

When there is the insertion of cavity fill in an existing wall in a building, the work done must comply with certain specific regulation requirements, namely C4 (resistance of walls to the passage of moisture) and D1 (toxic substances).

UNDERPINNING OF A BUILDING

Work involving the underpinning of an existing building is 'building work' for **Reg. 3(1)** the purposes of the regulations and so comes under control.

Regulation requirements

The regulations impose broad general requirements on the builder. Breach of these requirements does not, of itself, involve the builder in any civil liability although such liability may arise, quite independently, at common law.

Compliance with Schedule 1 is mandatory. All building work must be car- **1991 Regs,** ried out so that it complies with the requirements set out in that Schedule. The **Reg. 4(1)** method adopted for compliance must not result in the contravention of another requirement.

The work must also be carried out so that, after completion, an existing **Reg. 4(2)** building or controlled service or fitting to which work has been done continues to comply with the specified requirements if it previously did so comply or, if it did not so comply before in any respect, it must not be more unsatisfactory afterwards.

Table 2.1 Requirements applicable according to material change of use.

Case	Schedule 1 requirements
[A] All cases (dwellings, flats, hotels, boarding houses, and institutions and public buildings, no longer exempt)	B1 (means of escape) B2 and B3 (internal fire spread) B4(2) (external fire spread – roofs) B5 (access etc. for fire services) F1 and F2 (ventilation) G1 (sanitary conveniences & washing facilities) G2 (bathrooms) H4 (solid waste storage) J1 to J3 (heat producing appliances) L1 Conservation of fuel and power
[B] Exempt building to non-exempt, hotel, boarding house, institution, public building	As in **[A]** plus A1 to A3 (structure)
[C] Building more than 15 metres in height	As in **[A]** plus B4(1) (external fire spread – walls)
[D] Building used as a dwelling, where previously it was not	As in **[A]** plus C4 (resistance to weather and ground moisture)
[E] Building used as a dwelling or containing a flat where it did not before and where more or less dwellings are provided than was originally the case	As in **[A]** plus E1 to E3 (resistance to passage of sound)
[F] Change of use of part only of a building	The part itself must comply with the relevant requirements as **[A]**, **[B]**, **[D]** and **[E]**. In **[C]** the whole building must comply with B4(1).

Schedule 1 – Technical requirements

Schedule 1 contains the technical requirements, which are discussed in Chapters 6 to 18 and which are almost all expressed functionally, e.g. C1 dealing with site preparation states that 'the ground to be covered by the building shall be reasonably free from vegetable matter'. These requirements cannot be subject to relaxation.

Which requirements apply depends on the type of building being constructed, but the majority of them is of universal application.

Materials and workmanship

Regulation 7(1) provides that any building work which is required to comply with any relevant requirement of Schedule 1 'shall be carried out (a) with proper materials which are appropriate for the circumstances in which they are used; and (b) in a workmanlike manner'. This is a general statutory obligation imposed on the builder. Guidance on how the obligation may be met is contained in Approved Document to support regulation 7 'Materials and workmanship', although that guidance is of a very general nature.

Reg. 7(1)

This statutory obligation is akin to a building contractor's obligation at common law when, in the absence of a contrary term in the contract, the builder's duty is to do the work in a good and workmanlike manner, to supply good and proper materials and to provide a building reasonably fit for its intended purpose: *Hancock* v. *B.W. Brazier (Anerley) Ltd* (1966) [1966] 1 WLR 1317; [1966] 2 All ER 901, CA. This threefold obligation would normally be implied in any case where a contractor was employed to both design and build, but the third limb of the duty would not arise, for example, where the client employs his own architect (*Lynch* v. *Thorne* (1956) [1956] 1 WLR 303; [1956] 1 All ER 744, CA) although the other two limbs remain.

The principal object of the regulations is to ensure that buildings meet reasonable standards of health and safety, and this is spelled out in regulation 8:

'Parts A to K and N of Schedule 1 shall not require anything to be done except for the purpose of securing *reasonable standards of health and safety* for persons in or about buildings (and any others who may be affected by buildings, or matters connected with buildings).'

Reg. 8

The obligations imposed by the regulations are not therefore absolute obligations, but rather a duty to use reasonable skill and care to secure reasonable standards of health and safety of people using the building and others who may be affected by failure to comply with the requirements of the regulations.

Relaxation of regulation requirements

Section 8 of the Building Act 1984 enables the Secretary of State to dispense with or relax any requirement of the regulations 'if he considers that the operation of [that] requirement would be unreasonable in relation to the particular case'. This power has been delegated to the local authority which may grant a relaxation if, because of special circumstances, the terms of a requirement cannot be fully met.

1984 Act,
sec. 8

1991 Regs
Reg. 10(1)

However, the majority of regulation requirements cannot be relaxed because they require something to be provided at an 'adequate' or 'reasonable' level, and to grant a relaxation would mean acceptance of something that was 'inadequate' or 'unreasonable'.

The application procedure is laid down in sections 9 and 10 of the 1984 Act. There is no prescribed form. Only the local authority (or the Secretary of State on appeal) can grant a relaxation; approved inspectors have no power to do so.

At least 21 days before giving a decision on an application for dispensation or relaxation of any requirement, the local authority must advertise the application in a local newspaper unless the application relates only to internal work. The notice must indicate the situation and nature of the work, and the requirement which it is sought to relax or dispense with. Objections may then be made on grounds of public health or safety. No notice need be published if the effect of the proposal is confined to adjoining premises only, but notice must then be given to the owner and occupier of those premises.

1984 Act, sec. 10

1991 Regs, Reg. 10(2)

Where a local authority refuse an application they must notify the applicant of his right of appeal to the Secretary of State. This must be exercised within one month of the date of refusal. The grounds of the appeal must be set out in writing, and a copy must be sent to the local authority, who must send it to the Secretary of State with a copy of all relevant documents, and any representations they wish to make. The applicant must be informed of the local authority's representations. There is no time limit prescribed for the Secretary of State's decision on the appeal.

1984 Act, sec. 39

Where a local authority fail to give a decision on an application within two months, it is deemed to be refused and the applicant may appeal forthwith.

1984 Act, sec. 36

Neither the Secretary of State nor the local authority may give a direction for any relaxation of the regulations where, before the application is made, the local authority has become statutorily entitled to demolish, remove or alter any work to which the application relates, i.e. as a result of service of a notice under section 36 of the 1984 Act. The same prohibition applies where a court has issued an injunction requiring the work to be demolished, altered or removed.

The procedure may be summarised in tabular form:

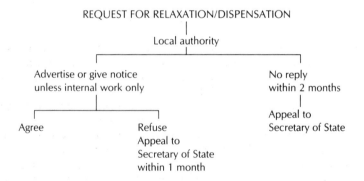

REQUEST FOR RELAXATION/DISPENSATION

Local authority

Advertise or give notice unless internal work only — No reply within 2 months

Agree — Refuse Appeal to Secretary of State within 1 month — Appeal to Secretary of State

Type relaxations

1984 Act, sec. 11

The local authority's power of dispensation and relaxation must be distinguished from that of the Secretary of State to grant a type relaxation, i.e. to dispense with a requirement of the regulations generally. A type relaxation can be made subject to conditions and can be for a limited period only. It can be issued on application to the Secretary of State, e.g. from a manufacturer, in which case a fee may be charged. The Secretary of State may also make a type relaxation of his own accord. Before granting a relaxation the Secretary of State must consult such bodies as appear to him to be representative of the interests concerned and must publish notice of any relaxation issued. No such type relaxations have been granted, under the current legislation.

Continuing requirements

Building regulations can impose continuing requirements on owners and occupiers of buildings. These requirements are of two kinds:

1984 Act, sec. 2

- Continuing requirements in respect of designated provisions of the building regulations, to ensure that the purpose of the provision is not frustrated.

For example, where an item is required to be provided, there could be a requirement that it should continue to be provided or kept in working order. Examples of the possible use of the power are the operation of mechanical ventilation which is necessary for health reasons or the operation of any lifts required to be provided in blocks of flats.

- Requirements with regard to services, fittings and equipment. This enables requirements to be imposed on buildings whenever they were erected and independently of the normal application of building regulations to a building.

This makes it possible for regulations to supersede the continuing requirements of the water byelaws at present made under section 17 of the Water Act 1945. Another possible use of this power would be to require the maintenance and periodic inspection of lifts in flats if they are to be kept in use. This power of continuing requirements has, as yet, not been used.

Testing and sampling

Two regulations empower the local authority to test drains and sewers to ensure compliance with the requirements of Part H of Schedule 1 and to take samples of materials *to be used* in the carrying out of building work. The wording does not appear to cover materials which are already incorporated in the building, but this may prove to be of little importance if the provisions of section 33 of the Building Act 1984, are ever activated.

1991 Regs, Regs 16 & 17

Under that section the local authority may test for compliance with the regulations. They will also be permitted to require a builder or developer to carry out reasonable tests or may carry out such tests themselves and also take samples for the purpose. Section 33(3) sets out the following matters with respect to which tests may be made:

1984 Act, sec. 33 (not yet operative)

(a) Test of the soil or subsoil of the site of any building.
(b) Tests of any material or component or combination of components.
(c) Tests of any service, fitting or equipment.

This is not an exhaustive description of the matters which may be subjected to tests.

The cost of testing is to be borne by the builder or developer, and there will be a right to apply to a magistrates' court regarding the reasonableness of any test required or of any decision of the local authority on meeting the cost of the test. It should be noted that the local authority will have a discretionary power to bear the whole or part of the costs themselves.

In fact the power of testing is given to 'a duly authorised officer of the local

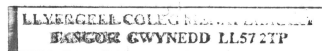

authority'. 'Authorised officer' is defined in section 126 of the Building Act 1984 as:

> '... an officer of the local authority authorised by them in writing, either generally or specially, to act in matters of any special kind, or in any specified matter; or ... by virtue of his appointment and for the purpose of matters within his province, a proper officer of the local authority...'

Section 95 of the 1984 Act confers upon an authorised officer appropriate powers of entry, and penalties for obstructing any person acting in the execution of the regulations are provided by section 112.

A duly authorised officer of the local authority must also be permitted to take samples of the materials used in works or fittings, to see whether they comply with the requirements of the regulations. In practice the authorised officer may ask the builder to have the tests carried out and to submit a report to the local authority. In any event, the builder should be notified of the result of the tests.

1991 Regs,
Reg. 18(2)

It should be noted, however, that regulations 16 and 17 do not apply where the work is supervised by an approved inspector (see Chapter 4) or is done under a public body's notice.

Unauthorised building work

Reg. 13A

Regulation 13A allows local authorities retrospectively to certify unauthorised building work carried out on or after 11 November 1985. The regulation became effective on 1 October 1994 and there are prescribed fees payable. It applies to building work which should have been subject to control, but the person who carried out the work failed to deposit plans with the authority, to give a building notice or to give an initial notice jointly with an approved inspector. The regulation enables the owner of the building (the applicant) to make a written application to the authority for a regularisation certificate.

The applicant's notice should describe the unauthorised work and, if reasonably practicable include a plan of it as well as a plan showing any additional work needed to ensure compliance with the regulations. On receipt of the notice and the accompanying plans, the council may require the applicant to take reasonable steps to enable them to inspect the work, e.g., opening up, testing and sampling. The local authority will then notify the applicant of any work required to ensure compliance, with or without relaxation and when this has been carried out to their satisfaction they may issue a regularisation certificate. This is stated to be evidence (but not conclusive evidence) that the relevant specified requirements have been complied with.

Contravening works

1984 Act,
sec. 36

Where a building is erected, or work is done contrary to the regulations, the local authority may require its removal or alteration by serving notice on the owner of the building. Where work is required to be removed or altered, and the owner fails to comply with the local authority's notice within a period of 28 days, the local authority may remove the contravening work or execute the necessary work themselves so as to ensure compliance with the regulations, recovering their expenses in so doing from the defaulter.

A section 36 notice may not be given after the expiration of twelve months from the date on which the work was completed. A notice cannot be served where the local authority have passed the plans and the work has been carried out in accordance with the deposited plans.

1984 Act, sec. 36(4) & (5)

The recipient of a section 36 notice has a right of appeal to the magistrates' court. The burden of proving non-compliance with the regulations lies on the authority, but if they show that the works do not comply with an approved document (under section 7) then the burden shifts. The appellant against the notice must then prove compliance with the regulations: *Rickards* v. *Kerrier District Council* (1987) CILL 345; 4-CLD-04-26.

Section 37 provides an alternative to the ordinary appeal procedure. Under that section, the owner may notify the local authority of his intention to obtain from 'a suitably qualified person' a written report about the matter to which the section 36 notice relates. Such notices are served where the local authority considers that the technical requirements of the regulations have been infringed.

1984 Act, sec. 37

The expert's report is then submitted to the local authority. In light of it the local authority may withdraw the section 36 notice and *may* pay the owner the expenses which he has reasonably incurred in consequence of the service of the notice, including his expenses in obtaining the report. Adopting this procedure has the effect of extending the time for compliance with the notice or appeal against it from 28 to 70 days.

If the local authority rejects the report, it can then be used as evidence in any appeal under section 40 and section 40(6) provides that

> 'if, on appeal ... there is produced to the court a report that has been submitted to the local authority ... the court, in making an order as to costs, may treat the expenses incurred in obtaining the report as expenses for the purposes of the appeal.'

Thus, in the normal course of events, if the appeal was successful, the owner would recover the cost of obtaining the report as well as his other costs.

The local authority – or anyone else – may also apply to the civil courts for an injunction requiring the removal or alteration of any contravening works. This power is exercisable even in respect of work which has been carried out in accordance with deposited plans, e.g. oversight or mistake on the part of the local authority. In such a case the court might well order the local authority to pay compensation to the owner. The twelve months' time limit does not apply to this procedure which is, however, unusual and rarely invoked in practice. The Attorney-General, as guardian of public rights, may seek an injunction in similar circumstances, and in practice proceedings for an injunction must be taken in his name and with his consent.

1984 Act, sec. 36(6)

Where a person contravenes any provision in the building regulations, he renders himself liable to prosecution by the local authority. The case is dealt with in the magistrates' court. The maximum fine on conviction is £2000, with a continuing penalty of £50 a day.

1984 Act, sec. 35

In *Torridge District Council* v. *Turner* (1991) 9-CLD-07-21, it was held, for reasons which are not entirely clear, that breach of 'do' provisions such as Part A1 which requires that a building 'shall be so constructed' as to meet the specified standards does not constitute a continuing offence which means that the proceedings must be commenced within six months of the commission of the alleged offence. This six month limitation period is specified by section

127(1) of the Magistrates' Courts Act 1980. The Divisional Court held that the person constructing a building commits an offence when the building works are completed in a way not complying with the regulations. He does not commit a continuing offence.

If this decision is correct it makes the enforcement of the regulations a well-nigh impossible task.

Determinations

1984 Act, sec. 16(10)(a) & 50

It is sometimes the case that a local authority rejects a full plans application (or an approved inspector refuses to give a plans certificate – see Chapter 4) on the grounds that the plans show a contravention of the Building Regulations, but the applicant believes that the plans do, in fact, comply. In this case the applicant can apply to the Secretary of State for a determination as to whether or not the work complies with the regulations.

It is possible to apply for a determination at any time after the plans have been submitted to the local authority or an approved inspector has been asked for a plans certificate, but it is usually better to wait until a decision has been made and the plans have been declared unacceptable. Applications for a determination must usually be made before the commencement of work (or before commencement of that part of the work which is the subject of the determination).

The application, in the form of a letter, is made direct to the Department of the Environment, Transport and the Regions if the proposal is in England, or to the Welsh Office if it is in Wales, and it should include the following information:

- The plans of the proposal.
- The precise grounds on which the plans were rejected (or not certified in the case of a plans certificate).
- A statement of the applicant's case.
- The appropriate fee (i.e. half the plan fee, with a minimum of £50 and a maximum of £500).

It is not necessary to obtain the permission of the local authority or approved inspector before making such an application.

Chapter 3

Local authority control

Introduction

Local authorities have exercised control over buildings in England and Wales since 1189, but it was not until 1965 that uniform national building regulations were made applicable throughout the country generally. Inner London retained its own system based on the London Building Acts 1930 to 1978 and byelaws made thereunder until 6 January 1986. Building regulations now apply to Inner London, although many provisions of the London Building Acts continue to apply in modified form. The Building Regulations 1991 introduced a number of substantive changes to the system of local authority control.

Part IV of the Building Regulations 1991, as amended, contains the procedural requirements which must be observed where a person proposes to undertake building work covered by the regulations and opts for local authority control. Although a great deal of building work continues to be under local authority control and supervision, an increasing volume of work is now dealt with under an alternative system – private control and supervision by an approved inspector, as explained in Chapter 4. Until January 1997 the private system was confined to house-builders under the National House Building Council (NHBC) scheme. It is now possible to use an approved inspector for any class of building work and at the date of this edition there were seven corporate and in excess of 20 non-corporate approved inspectors operating in England and Wales. **1991 Regs, Regs. 11 to 15**

Two main procedural options are available under the local authority system of control:

- Control based on service of a building notice.
- Control based on the deposit of full plans.

There is also a case – where the work consists only of the installation of certain gas heating appliances by approved installers – where neither notice nor deposit of plans is required. It is also possible to have an intermediate situation where plans may be passed in stages. **1991 Regs, Reg. 11(3)**

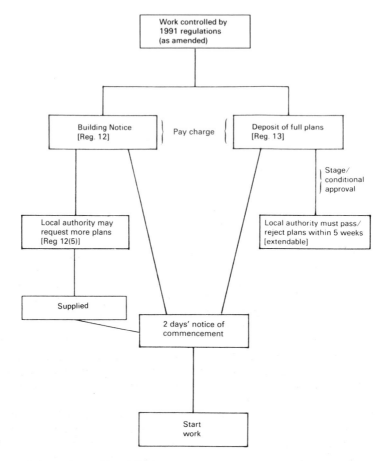

Fig. 3.1 Local authority supervision.

The local authority

1984 Act,
sec. 126(1)
and Sch. 3
The local authority for the purposes of the regulations is the district council, a London borough council, the Common Council of the City of London, the Sub-Treasurer of the Inner Temple, the Under Treasurer of the Inner Temple, and the Council of the Isles of Scilly.

Building notice procedure

The major procedural innovation introduced in 1985, and now to be found in the 1991 Regulations, is based on service of a building notice. There is no approval of plans. Interestingly, this is copied from the former Inner London system.

1991 Regs,
Reg. 11
A person intending to carry out building work or make material change in the use of a building may give a building notice to the local authority unless the building concerned is a workplace subject to the Fire Precautions (Workplace) Regulations 1997 (see Chapter 5), or is one which is designated

under the Fire Precautions Act 1971. The following buildings are designated under the Act:

- *Hotels and boarding houses*

Where there is sleeping accommodation for six or more guests or staff or any number of guests or staff above the first floor or below the ground floor.

- *Factories, offices, shops and railway premises*

Where more than twenty people are employed or more than ten people are employed other than on the ground floor or in factories only where explosive or highly flammable materials are stored or used.

There is no prescribed form of building notice. The notice must be signed by the person intending to carry out the work or on his behalf, and must contain or be accompanied by the following information: **Reg. 12(1)**

- The name and address of the person intending to carry out the work.
- A statement that it is given in accordance with regulation 11(1)(a).
- A description of the proposed building work or material change of use.
- A description of the location of the building to which the proposal relates and the use or intended use of that building.
- If it relates to the erection or extension of a building it must be supported by a plan to a scale of not less than 1:1250, showing size and position in relation to streets and adjoining buildings on the same site, the number of storeys, details of the drainage and the precautions to be taken in building over any drain, sewer or disposal main. Where any local legislation applies, the notice must state how it will be complied with. **Reg. 12(2)**
- Where the building notice involves cavity wall insulation, information must be given about the insulating material to be used and whether or not it has an Agrément Certificate or conforms to British Standards, and whether or not the installer has a BSI Certificate of Registration or has been approved by an Agrément Board. **Reg. 12(3)**
- If the work includes the provision of a hot water storage system covered by Schedule 1, G3 (e.g. an unventilated system with a storage capacity of 16 litres of more) details of the system and whether or not the system and its installer are approved. **Reg. 12(4)**

The local authority are not required to approve or reject the building notice and, indeed, have no power to do so. However, they are entitled to ask for any plans they think are necessary to enable them to discharge their building control functions and may specify a time limit for their provision. They may also require the person giving the notice for information in connection with their linked powers under sections 18 and 21 of the Building Act 1984 (see p.1.3), and if the work involves building over a sewer, they can require an access agreement. **Reg. 12(5)**

The regulations make plain that the building notice and plans shall not be 'treated as having been *deposited* in accordance with the building regulations'. In some ways this is an odd provision because the relevant building control sections of many of the local Acts of Parliament mentioned in Chapter 1 – and which provide for special local requirements – are triggered off by the 'deposit' of plans. At first sight, therefore, this would render such requirements inoperative, but presumably it is thought that compliance will be ensured through **Reg. 12(6)** the requirement that the building notice must contain a statement of the steps to be taken to comply with any local enactment.

Once a building notice has been given, work can be commenced, although there is a requirement (see below) that the local authority be notified at least two days before work commences.

Reg. 12(7)   A building notice remains in effect for a period of three years from the date on which it was given to the local authority. If the work has not been commenced within that period the building notice lapses automatically. This three-year restriction on the validity of building notices was introduced in 1991 and is in line with a full plans submission (see p. 3.5).

Installation of heat-producing gas appliances

Reg. 11(3)   No building notice (or deposit of plans) is required where the building work consists *only* of the installation of a heat-producing gas appliance which is to be installed by a person (or his employee) approved by the British Gas Corporation. This exemption covers gas-fired heaters of various sorts, and does not apply if other building work is involved. The exemption covers installations by installers who have been approved under the Gas Safety (Installation and Use) Regulations 1984, as amended.

Deposit of plans

Reg. 11(1)(b)   This is the traditional system of building control by which full plans are deposited with the appropriate local authority in accordance with section 16 of the Building Act 1984, as supplemented by regulation 13. Section 16 imposes a **1984 Act, sec. 16**   duty on the building control authority to either pass or reject plans deposited for the proposed work.

In *Murphy* v. *Brentwood District Council* (1990) 20 ConLR 1, CA the Court of Appeal held that the duty is imposed on the local authority itself either to pass or reject the deposited plans, and it cannot discharge its duty by delegating performance to outside consultants. If the local authority leaves it to outside consultants to decide whether plans are passed or rejected, the local authority is vicariously responsible if the consultants are negligent, subject to proof of recoverable damage.

However, the Court of Appeal proceeded on the basis that *Anns* v. *London Borough of Merton* [1978] AC 728; [1977] 2 All ER 492, HL; 5 BLR 1 was rightly decided, and in light of the fact that *Anns* was subsequently overruled it is thought that the local authority could only be vicariously liable in these circumstances (if at all) where personal injury was suffered by the occupier or there was damage to other property. Indeed, it is probable that in the current climate of judicial opinion the local authority would be held able to discharge its section 16 duty by reliance on competent outside expertise.

If the plans submitted are not defective the authority has no alternative but to approve them unless, of course, they contravene the linked powers discussed in Chapter 1.

Where the proposed works are subject to the regulations, and it is proposed to deposit full plans, the provisions of section 16 and regulation 13 must be observed. The local authority must give notice of approval or rejection of plans within five weeks unless the period is extended by written agreement. The extended period cannot be later than two months from the deposit of plans, and any extension must be agreed before the five-week period expires. However, the five-week period does not begin to run unless the applicant submits a 'reasonable estimate' of the cost of the works (where applicable) and pays the plan charge at the same time as the plans are deposited.

The approval lapses if the work is not commenced within a period of three years from the date of the deposit of the plans, provided the local authority gives formal notice to this effect. The local authority must pass the plans of any proposed work deposited with them in accordance with the regulations unless the plans are defective, or show that the proposed work would contravene the regulations. The notice of rejection must specify the defects or non-conformity, and the applicant may then ask the Secretary of State to determine the issue. His decision is then final. The Secretary of State may refer questions of law to the High Court and must do so if the High Court so directs.

1984 Act, sec. 32

1984 Act, sec. 16

The local authority may pass plans by stages and, where they do, they must impose conditions as to the deposit of further plans. They may also impose conditions to ensure that the work does not proceed beyond the authorised stage. They have power to approve plans subject to agreed modifications, e.g. where the plans are defective in a minor respect or show a minor contravention. However, it should be noted that local authorities are not obliged to pass plans conditionally or in stages, and the applicant must agree in writing to these procedures.

The 'full plans' required under the deposit method are the same as those required under the building notice procedure, together with such other plans as are necessary to show that the work will comply with the building regulations.

1991 Regs, Regs 13(1), (2) & (4)

Regulation 13 specifies that the plans must be deposited in duplicate; the local authority retains one set of plans and returns the other set to the applicant. They must be accompanied by a statement that they are deposited in accordance with regulation 11 (1)(b) of the 1991 Regulations and if the building is put or is intended to be put to a designated use under the Fire Precautions Act 1971, by a statement to that effect. Two additional copies of the plans must be submitted where Part B (Fire Safety) imposes a requirement in relation to the work and both additional plans may be retained by the local authority, although, it is not necessary to provide additional copies of the plans where the proposed work relates to the erection, extension or material alteration of a dwelling-house or flat.

Reg. 13(2)

Work may be commenced as soon as plans have been deposited – although the local authority must be given notice of commencement at least 2 days before work commences – but it is an unwise practice to commence work before notice of approval is received.

If the applicant wants the authority to issue a completion certificate (see p. 3.7) in due course, a request to that effect should accompany the plans.

Reg. 13(5)

The advantage of the full deposit of plans method of control is that if the work is carried out exactly in conformity with the plans as passed by the local authority, they cannot take any action in respect of an alleged contravention under section 36 of the Building Act 1984.

The deposit of full plans procedure and the possible alternative solutions are shown diagrammatically in Fig. 3.2.

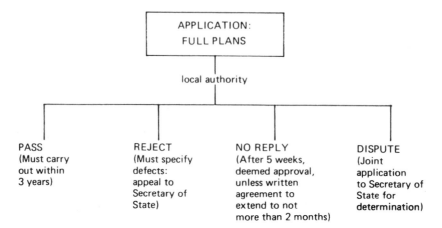

Fig. 3.2

Notice requirements

1991 Regs,
Reg. 14

Wherever the work is to be supervised by the local authority, in addition to the building notice or deposit of plans, the person undertaking the work must pay an inspection charge and give certain notices to the local authority. The notices must be in writing 'or by such other means as [the local authority] may agree', e.g. by telephone, but most building control authorities provide applicants who deposit plans with pre-printed postcards.

Failure to give the required notices is a criminal offence, punishable on summary conviction by a substantial fine. Under the previous regulations, the Court of Appeal has ruled that failure to deposit plans and give notices under the building regulations is not a continuing offence, with the result that magistrates have no power to try informations laid more than six months after the relevant period for compliance. The regulations require the notices to be given by a specified deadline, and once that deadline has passed, the offence has been committed: *Hertsmere Borough Council* v. *Alan Dunn Building Contractors Ltd* (1986) 84 LGR 214; 9-CLD-07-16.

In practice the majority of local authorities do not seek to enforce the penalty but rely on their powers to serve written notice requiring the person concerned within a reasonable time to cut into, lay open or pull down so much of the work as is necessary to enable them to check whether it complies with the regulations.

Where the person carrying out the work is advised in writing by the local authority of contravening works, and has rectified these as required by the local authority, he must give the authority written notice within a reasonable time after the completion of the further work.

'Reasonable time' is not defined in either situation; it is a question of fact in each case. The phrase has been judicially defined as being 'reasonable under ordinary circumstances': *Wright* v. *New Zealand Shipping Co.* [1878] AC 23.

Regulation 14 requires the giving of the following notices to the local authority:

1991 Regs,
Reg. 14(1)

- At least two days' notice of commencement before commencing the work.
- At least one day's notice of:

the covering up of any foundation excavation, foundation, damp-proof course, concrete or other material laid over a site; *or*,
covering up of any drain or private sewer subject to the regulations. **Reg. 14(2)**

These periods of notice commence after the day on which the notice was served. The local authority must be given notice by 'the person carrying out building works' not more than five days after the completion of:

- The laying of any drain or private-sewer, including any haunching, sur-   **Regs 14(3)**
  rounding or trench backfilling. **(4) (5)**
- The erection of a building.
- The completion of any other work.

There is no definition of the term 'person carrying out building works' in the regulations, but in *Blaenau Gwent Borough Council* v. *Khan* (1993) 35 ConLR 65 the High Court held that the owner of a building who authorises a contractor to carry out building works on his behalf fell within the term. The court took the view that those words should not be confined so as to restrict the meaning of the phrase to the person who physically performs the work, 'but includes the owner of the premises on which the works are being performed and who had authorised the work.'

Where a building or part of a building is occupied before completion, the   **1991 Regs,**
local authority must be given notice at least five days before occupation. This   **(1994**
notice is additional to the required notice after completion. Additionally, a   **amendment)**
further notice is required to be given to the local authority where a new   **Reg. 14A**
dwelling is created either by new building work or by a material change of use.

The person carrying out the building work must calculate the energy rating of the dwelling by the government approved Standard Assessment Procedure (SAP) and must give notice of this within five days of completion of the dwelling or at least five days before occupation if this occurs before completion.

If a person fails to comply with the notice requirements of regulation   **Reg. 14(6)**
14(1)(2)(3) the local authority may require him by notice to cut into, lay open or pull down any work so that they may find out whether the regulations have been complied with. The person concerned must comply with the notice within a *reasonable time* which, it is suggested, will normally be short.

Where the local authority has served notice specifying that the work con-   **Reg. 14(7)**
travenes the regulations, on completion of the remedial work, notice must be given to the authority within a reasonable time.

There is now a definition of *day*. It means a period of 24 hours commencing at midnight, i.e. a calendar day, but Saturdays, Sundays and Bank or public   **Reg. 14(8)**
holidays are excluded.

Completion Certificate

The local authority must issue a completion certificate when they are satisfied,   **1991 Regs,**
after having taken all reasonable steps, that the Schedule 1 requirements are   **Reg. 15**
met. Its issue is mandatory in respect of fire safety requirements, i.e. where the building is to be put to a designated use under the Fire Precautions Act 1971, but in other cases the authority need issue the certificate only where they have been requested to do so.

The completion certificate is evidence – but not *conclusive* evidence – that the requirements specified in the certificate have been complied with.

1991 Regs
Reg. 21
It should be noted that the local authority cannot be held liable for a fine if they contravene this regulation by failing to give a completion certificate.

Chapter 4

Private certification

Introduction

One of the Government's aims in reforming the previous system of building control was to provide an opportunity for self-regulation by the construction industry through a scheme of private certification. This is not a complete substitute for local authority control, because local authorities will always remain responsible for taking any enforcement action which may be necessary. Indeed, in certain closely-defined circumstances they may resume their control functions. The local authority also remain responsible for inspecting any connection to, or building over, an existing sewer as well as for work which is covered by local legislation.

The developer is given the option of having the work supervised privately, rather than relying on the local authority control system described in the previous chapter. Essentially, the private certification scheme is based on the proposals set out in a Government White Paper *The Future of Building Control in England and Wales* published by HMSO in February 1981.

The statutory framework of the alternative system is contained in Part II of the Building Act 1984. In broad terms, this provides that the responsibility for ensuring compliance with building regulations may, at the option of the person intending to carry out the work, be given to an approved inspector instead of to the local authority. It also enables approved public bodies to supervise their own work. Various supplementary provisions deal with appeals, offences, and the registration of certain information.

1984 Act, secs 47–53

1984 Act, sec. 54

1984 Act, secs 55–58

The detailed rules and procedures relating to private certification are to be found in the Building (Approved Inspectors, etc.) Regulations 1985, as amended, which also contain prescribed forms which must be used.

It has taken some considerable time for the private certification system to become fully operational even though the first approved inspector, the National House-Building Council (NHBC), was approved on 11 November 1985. Their original approval related only to dwellings of not more than four storeys but this was later extended to include residential buildings up to eight storeys and this was further extended in 1998 to include any buildings.

The approval of further corporate bodies as approved inspectors was held up by a number of factors, but was due mainly to the difficulty posed in obtaining the level of insurance cover which was required by the Department of the Environment, Transport and the Regions. After a period of consultation new proposals for insurance requirements were agreed and these were implemented on 8 July 1996. At the same time the Construction Industry Council (CIC) was designated as the body for approving non-corporate inspectors, although the Secretary of State reserved the right to approve corporate bodies.

Three further corporate bodies were approved by the Secretary of State from 13 January 1997 and there have been four others approved since that date, but at present, the NHBC remains the only body which deals with domestic construction (i.e. self-contained houses, flats and maisonettes). In this context, the DETR has clarified the position with regard to what can be regarded as a dwelling for the purposes of the insurance schemes referred to below. Therefore the following are not treated as dwellings and may be supervised by any approved inspector:

- Dwellings which are ancillary or incidental to one or more non-residential uses of the buildings in which they are situated (such as staff flats in public houses).
- Dwellings which comprise or are contained in institutions (see page 2.7 above).
- Dwellings which comprise or are contained in hotels or boarding houses.
- Dwellings which comprise or which belong to schools, universities or similar establishments and which are used as living accommodation for their staff, pupils or students (such as halls of residence).
- Dwellings which belong to medical establishments and which are used as living accommodation for their staff or students (such as nurses homes).

Furthermore, it has been decided that the CIC will become the designated body for approving both corporate and non-corporate approved inspectors in future (although this will involve a change in the law).

The names and addresses of all approved inspectors may be obtained from Mr K.W.G. Blount, Honorary Secretary, the Association of Corporate Approved Inspectors, c/o TPS Special Services, Waterlinks House, Richard Street, Birmingham, B7 4AA (Tel: 0121 333 2811), or from Ian Jones, Department of the Environment, Transport and the Regions, Zone 3/A1, Eland House, Bressenden Place, London, SW1 5DU.

Insurance requirements

All approved inspectors are required to carry insurance cover in accordance with a scheme approved by the Secretary of State.

Since the NHBC deals mainly with dwellings the insurance cover required is more extensive than that needed for other types of buildings. In fact the NHBC has to provide two different types of insurance policy:

- *Ten year no-fault insurance* against breaches of the Building Regulations relating to site preparation and resistance to moisture, structure, fire, drainage and heat producing appliances. The limit on cover is related to the original cost of the work allowing for inflation during the ten year period up to a maximum of 12% per annum compound.
- *Insurance against the approved inspector's liabilities in negligence* for fifteen years from the issue of the Final Certificate for each dwelling. The limit of

cover is twice the cost of the building work (unless there is a simultaneous claim made under the no-fault policy), together with cover against claims made for personal injury (which is normally £100,000 a dwelling). This is also proof against inflation up to 12% compound per annum and is subject to a minimum of £1 million a site.

For corporate approved inspectors dealing with work other than dwellings it is necessary to provide professional indemnity insurance renewable on an annual basis. Additionally, a 10 year run-off period is required where the approved inspector fails to renew his policy. Indemnity has to extend to any claim reported in writing within 10 years from the acceptance of the final certificate, and claims from the owner of the work, his successors in title, or third parties must be met. Cover must be provided for claims against damage (including injury) resulting from the negligent performance by the approved inspector who issued the Initial Notice.

Approval of inspectors

Section 49 of the Building Act 1984 defines an 'approved inspector' as being a person approved by the Secretary of State or a body designated by him for that purpose. Part II of the Building (Approved Inspectors, etc.) Regulations 1985 (as amended) sets out the detailed arrangements and procedures for the grant and withdrawal of approval. **1984 Act, sec. 49**
AI Regs, Regs 3–7

There are two types of approved inspector:

- Corporate bodies, such as the NHBC or Tarmac Professional Services.
- Individuals, not firms, who must be approved by a designated body.

Approval may limit the description of work in relation to which the person or company concerned is an approved inspector.

Recently, the Construction Industry Council (CIC) became the single designated body for the approval of non-corporate approved inspectors and it has been announced that they will become the designated body for approval of corporate approved inspectors also, although this will necessitate a change in the law.

Approval of an inspector is not automatic. Someone wishing to be a non-corporate approved inspector must have a construction-related qualification and five years post-qualification experience. Applications are made to the CIC registrar, then the application goes to a multi-disciplinary assessment panel who will be looking for knowledge in six areas as follows: **AI Regs, Reg. 3**

- Building Regulations and statutory control.
- Construction technology and materials.
- Law.
- Fire studies.
- Foundation and structural engineering.
- Building services and environmental engineering.

This is followed by a professional interview conducted by three assessors who must be unanimous in their approval of the candidate. Successful candidates are registered and their names are supplied to the DETR. Individuals are

**AI Regs,
Reg. 6**

approved for a period of five years. The CIC can withdraw its approval – for example, if the inspector has contravened any relevant rules of conduct or shown that he or she is unfitted for the work.

More seriously, where an approved inspector is convicted of an offence under section 57 of the 1984 Act (which deals with false or misleading notices and certificates, etc.) the CIC may withdraw their approval. In this case the convicted person's name would be removed from the list for a period of five years. There is no provision for appeals or reinstatement.

**AI Regs,
Reg. 6(3)**

The Secretary of State may himself withdraw his approval of any designated body, thus ensuring that the designated bodies act responsibly in giving approvals. Such action would not necessarily prejudice any approvals given by the designated body but the Secretary of State can, if he so desires, withdraw any approvals given by the designated body.

**AI Regs,
Reg. 7**

Provision is made for the Secretary of State to keep lists of designated bodies and inspectors approved by him, and for their supply to local authorities. He must also keep the lists up-to-date (if there are withdrawals or additions to the list) and must notify local authorities of these changes.

In a similar manner, designated bodies are required to maintain a list of inspectors whom they have approved. There is no express provision for these lists to be open to public inspection, although the designated body is bound to inform the appropriate local authority if it withdraws its approval from any inspector.

**1984 Act,
sec. 49(2)**

In approving any inspector, either the Secretary of State or a designated body may limit the description of work in relation to which the person concerned is approved. Any limitations will be noted in the official lists, as will any date of expiry of approval.

Approved persons and self-certification by competent persons

**1984 Act,
sec. 16(9)**

**AI Regs,
Reg. 27**

The following bodies, together with the Chartered Institution of Building Services Engineers, have been designated to approve private individuals who wish to become approved persons who can certify plans to be deposited with the local authority as complying with the energy conservation requirements:

- The Chartered Institute of Constructors
- The Faculty of Architects and Surveyors
- The Association of Building Engineers
- The Institution of Building Control Officers
- The Institution of Civil Engineers
- The Institution of Structural Engineers
- The Royal Institute of British Architects
- The Royal Institution of Chartered Surveyors

Additionally, the Institution of Civil Engineers and the Institution of Structural Engineers have been designated to approve persons to certify plans as complying with the structural requirements. Approved *persons* under section 16(9) of the Building Act should not be confused with approved *inspectors* under sections 47 to 54. As yet however, no approved persons have been designated in England and Wales although a pilot scheme for structural approvals has been operating in Scotland for some time with limited success.

The DETR is currently consulting on proposals for the approval of competent persons (a new concept which is not currently covered by the Building Act 1984). These people would be able to self-certify work such as drainage system designs, without needing to make an application to the local authority or an approved inspector, in a similar manner to that already in operation for gas appliances.

Independence of approved inspectors

An approved inspector cannot supervise work in which he or she has a professional or financial interest, unless it is 'minor work'. In this context, 'minor work' means:

AI Regs,
Reg. 9

(a) The material alteration or extension of a dwelling-house (not including a flat or a building containing a flat) which has two storeys or less before the work is carried out and which afterwards has no more than three storeys. A basement is not regarded as a storey.
(b) The provision, extension or material alteration of controlled services or fittings (see page 2.6 above for definition of controlled services or fittings).
(c) Work involving the underpinning of a building.

Independence is not required of an inspector supervising minor work but the limitation on the number of storeys should be noted.

There is a broad definition of what is meant by having a professional or financial interest in the work, the effect of which is to debar the following:

● Anyone who is or has been responsible for the design or construction of the work in any capacity, e.g. the architect.
● Anyone who or whose nominee is a member, officer or employee of a company or other body which has a professional or financial interest in the work, e.g. a shareholder in a building company.
● Anyone who is a partner or employee of someone who has a professional or financial interest in the work.

However, involvement in the work as an approved inspector on a fee basis is not a debarring interest!

Approval of public bodies

Public bodies, such as nationalised industries, are able to supervise their own building work by following a special procedure, which is detailed in the regulations.

1984 Act,
sec. 54

Regulation 19 empowers the Secretary of State to approve public bodies for this purpose although, curiously, no criteria have been laid down as to the qualification and experience of the personnel involved. The regulation confers wide discretionary powers on the Secretary of State, but clearly approval will be limited to those bodies which may reasonably be expected to operate responsibly without detailed supervision.

AI Regs,
Reg. 19

Private certification procedure

Initial notice

**1984 Act,
secs 47 &
48**

If the developer decides to employ an approved inspector, whether an individual or a corporate body, the first stage in the process is for the applicant and the approved inspector jointly to give to the local authority an *initial notice* in a prescribed form. This must be accompanied by the details described below and, very importantly, by a declaration that an approved insurance scheme applies to the work. This declaration must be signed by the insurer. The initial notice must contain:

- A description of the work.
- In the case of a new building or extension, a site plan on a scale of not less than 1:1250 showing the boundaries and location of the site, together with relevant documents showing the approximate location of any proposed connection to be made to a sewer or other drainage proposals or the reasons why no drainage is necessary. If it is proposed to build over an existing sewer, the information given must indicate what precautions will be taken to protect it. The local authority need this information in connection with their linked powers. If local legislation is applicable in the area, the necessary information must be provided. For example, some local Acts of Parliament require separate drainage for foul water and rainwater, and others require access for the fire brigade.

**AI Regs,
Reg. 8(3) &
Sch. 3**

It is essential that the initial notice should contain full information, because otherwise it may be rejected. The initial notice must also state whether or not the work is 'minor work'. This is defined in regulation 9(5) (see p. 4.5) and is basically work which consists of alteration or extension to a one- or two-storey house provided that not more than three storeys result.

An undertaking to consult the fire authority must also be included, if appropriate, as well as a statement of awareness of the applicable statutory obligations.

An initial notice submitted by an individual approved inspector must be supported by a copy of the notice of his approval by a designated body.

The initial notice must be on the prescribed form, and it is a contravention of the regulations to start work before the notice has been accepted. Figure 4.1 shows a completed form of initial notice.

**AI Regs,
Reg. 8(4)**

The local authority must accept or reject the initial notice within five working days, which run from the date of its receipt. When sent by post, the presumption is that the notice is received in normal course of post, i.e. assuming first class postage is paid, the notice will be delivered on the working day following posting.

Once accepted, the local authority's powers to enforce the 1991 Regulations are suspended so long as the initial notice remains in force.

**AI Regs,
Sch. 3**

There are only twelve grounds on which the notice may be rejected by the local authority. These are:

- The notice is not in the prescribed form.
- The notice has been served on the wrong local authority.
- The person who signed the notice as an approved inspector is not an approved inspector.

The Building Act 1984, section 47, and the Building (Approved Inspectors, etc.) Regulations 1985 (as amended).

INITIAL NOTICE

To: The Cawsand District Council, Probity House, Cawsand

1. This notice relates to the following work:
   Alteration and extension to dwelling house known as Tragedy House, Cawsand

2. The approved inspector in relation to the work is:
   John Jorrocks, Esq, RIBA, 3 Redcoat Place, Cawsand.
   Tel: 0012-12345

3. The person intending to carry out the work is:
   Facey Romford, Esq, Tragedy House, Cawsand.
   Tel: 0012-45678

4. With this notice are the following documents, which are those relevant to the work described in this notice:
   [a] A copy of the approved inspector's notice of approval.
   [b] A declaration of insurance duly signed on behalf of Structural Failures Insurance Co. PLC.
   [c] A plan (1 : 1250 scale) showing the boundaries and location of the site and details of the proposed sewer connection.

5. The work is not minor work.

6. I, John Jorrocks, declare that I do not, and will not while this notice is in force, have any financial or professional interest in the work described.

7. The approved inspector will be obliged to consult the fire authority.

8. I, John Jorrocks, undertake to consult the fire authority before giving a plans certificate in accordance with section 50 of the Act or a final certificate in accordance with section 51 of the Act in respect of any of the work described above.

9. I, John Jorrocks, am aware of my obligations laid upon me by Part II of the Act and by regulation 10 of the 1985 regulations (as amended).

Signed   Signed

John Jorrocks   *Facey Romford*

Approved Inspector   Person intending to carry out the
                                        work

6 January 1995   6 January 1995

Fig. 4.1 Completed form of initial notice.

- The information supplied is deficient because neither the notice nor plans show the location and description of the work (including the use of any building to which the work relates) or it contains insufficient information about drainage, etc.
- The initial notice is not accompanied by an individual approved inspector's notice of approval.

- Evidence of approved insurance is not supplied.
- The notice does not contain an undertaking to consult the fire authority if this is appropriate.
- The inspector is not independent, i.e. he has a professional or financial interest in the work. (Independence is not required in the case of minor work: see p. 4.5.)
- The local authority is not satisfied about the drainage proposals.
- The work includes the erection or extension of a building over an existing sewer and the local authority is not satisfied that it may properly consent, either conditionally or unconditionally.
- Local legislative requirements will not be complied with.
- There is an overlap with an earlier initial notice which is still effective.

If the local authority do not reject the initial notice within five working days, it is deemed to have been accepted.

1984 Act, secs 18 & 21

The local authority may impose conditions when accepting an initial notice. For example, under section 18 of the Building Act 1984, if the building will be over a sewer, they will usually require the building owner to enter into an access agreement so as to enable the sewer to be properly maintained, while under section 21 of the 1984 Act they can insist that the drainage system is connected to an existing public sewer in defined circumstances.

Once the initial notice is accepted, the local authority's powers to enforce the building regulations are suspended, and supervision of the works is the responsibility of the approved inspector.

1984 Act, secs 48, 52 & 53

AI Regs, Reg. 16

The initial notice generally remains in force during the currency of the works, although as explained below it may be cancelled in certain circumstances. Moreover, after the lapse of certain defined periods of time it will cease to have effect. The time periods depend on the circumstances, but the position may be summarised as follows:

(1) *Where a final certificate is rejected* – four weeks from the date of rejection.
(2) *Occupied buildings and extensions* – if no final certificate is issued, eight weeks from the date of occupation, but in the case of a building designated under the Fire Precautions Act 1971, the period is four weeks.
(3) *Changes of use which have commenced* – eight weeks from the date of occupation if no final certificate is issued.
(4) *All other work* – eight weeks from the date on which the work is 'substantially completed' if no final certificate is issued.

The local authority is given power to extend these time periods.

1984 Act, secs 51A & 51B

AI Regs, Reg. 8A

It is possible to give a final certificate for part of a building or extension if it is needed to be occupied before overall completion of the project. In these circumstances the initial notice is not cancelled but remains in force until final completion of all the work. Additionally, where it is necessary to vary work for which an initial notice has already been served, (e.g. it may be necessary to increase the scope of the work) it is possible to give an amendment notice to the local authority. There is a prescribed form for this, but essentially it contains all the information which was required for the initial notice plus either:

(a) A statement that all plans originally submitted remain unchanged, *or*
(b) Amended plans are submitted with the amendment notice plus a statement that any plans not included remain unchanged.

The procedures for acceptance or rejection of the amendment notice by the local authority are identical to those described for initial notices above.

Cancellation of initial notice

In a number of cases, the approved inspector must cancel the initial notice by issuing to the local authority a cancellation notice in a prescribed form. The grounds on which the initial notice must be cancelled are: **1984 Act, sec. 52; AI Regs, Reg. 17**

- The approved inspector becomes or expects to become unable to carry out (or to continue to carry out) his functions.
- He is of the opinion that because of the way in which the work is being carried out he cannot adequately perform his functions.
- He believes that the requirements of the 1991 Regulations are being contravened and has given notice of contravention to the person carrying out the work and that person has not complied with the notice.

The detailed provisions about contravention are specified in regulation 17. The notice of contravention must inform the person carrying out the work that unless he rectifies the contravention within a period of three months the approved inspector will cancel the initial notice.

If the approved inspector is no longer willing or able to carry out his functions (for example, an individual inspector has died) then the building owner must issue a cancellation notice in the prescribed form and serve it on the local authority and, if practicable, on the approved inspector. Failure by the building owner to issue a cancellation notice is a criminal offence, punishable on conviction by a maximum fine not exceeding level five on the standard scale. **1984 Act, sec. 52(4)**

In both cases, the local authority resume their building control functions and take over the supervision of the work unless a new initial notice is given and accepted. Regulation 18 provides that where the local authority take over they may give a notice to the owner requiring (i) plans of work not covered by a plans certificate to show compliance with the regulations and (ii) where a plans certificate was given and accepted, a copy of the plans as certified. The local authority may also require opening up, etc., of work (not covered by a plans certificate) if this is necessary for them to ensure compliance with the regulations. **AI Regs, Reg. 18**

If the building owner intends to carry out any further work which will be supervised by the local authority, he must deposit plans with the local authority showing that the intended work will comply with the 1991 Regulations. This obligation extends to depositing plans of work already carried out so far as this is necessary to show that future work will comply.

Where the work covered by the initial notice has not been begun within three years from the date on which the initial notice was accepted, the local authority may (not must) cancel the initial notice. **1984 Act, sec. 52(5)**

Functions of approved inspectors

The fees payable to an approved inspector are a matter for negotiation; there is no prescribed scale. The functions which an approved inspector must carry **AI Regs, Reg. 10** out are specified and detailed in the regulations, and his obligation is to 'take such steps as are reasonable to enable him to be satisfied within the limits of professional care and skill that' specified requirements are complied with. An approved inspector is liable for negligence and it is suggested that he *must* inspect the work to ensure compliance, in contrast to local authorities who have a discretion as to whether or not to inspect.

In *NHBC Building Control Services Ltd* v. *Sandwell Borough Council* (1990) 50 BLR 101 the Divisional Court emphasised that regulation 10 does not require a system of individual inspection of every detail covered by the substantive requirements of the regulations. In principle, random sampling is sufficient, although in case of dispute it is for the approved inspector to show that adopting a system of random or selective sampling is a satisfactory way of discharging his duties.

The approach of the court to this important matter was indicated by Lord Justice Leggatt:

'Any system of inspection that is selective involves consideration not only of the importance of a risk against which the inspection is designed to guard, but of the likelihood of its occurrence. In my judgement the justices' conclusion that the [approved inspector's] system is an inadequate precaution is not one that can properly be based solely upon the fact that the risk was obvious and potentially fatal. That amounts to saying that failure in relation to an individual house to detect the absence of rockwool in the gap between the ceiling and wall of its garage could not have occurred unless the system was inadequate or the inspector had shown want of professional skill and care in operating the system. But the liability imposed is not absolute. The system has been impliedly approved by the Secretary of State. In the light of its experience the [inspector] determines the extent and closeness of the inspections to be conducted in respect of the work of any particular builder. Inherent in any selected system is the risk that some defects may escape detection. Except [for] the fact that the defect ... was not spotted, there is no criticism to be made of the system. It follows that the mere fact that an important defect escaped detection in a particular instance cannot ... constitute a proper basis for concluding beyond reasonable doubt that there was any failure to undertake the functions of supervision so as to render false the statement that the [inspector] had performed those functions.'

1984 Act, sec. 49(8)   The approved inspector may arrange for plans or work to be inspected on his behalf by someone else (although only the approved inspector can give plans or final certificates), but delegation does not affect any civil or criminal liability. In particular, the 1984 Act states that:

'an approved inspector is liable for negligence on the part of a person carrying out an inspection on his behalf in like manner as if it were negligence by a servant of his acting in the course of his employment'.

AI Regs, Reg. 10(1)   The approved inspector must be satisfied that:

- The requirements relating to building work and material change of use specified in regulations 4 and 6 of the 1991 Regulations are complied with.
- Satisfactory provision is made for drainage.
- Regulation 10A relating to energy ratings for dwellings is complied with.

This means that the person carrying out the building work must supply a SAP energy rating to the approved inspector not more than 5 days after completion of the dwelling (see also p. 3.7 above).

AI Regs,
Reg. 10A

Where cavity wall insulation is inserted, the approved inspector need not supervise the insulation work, but is required to state in his final certificate whether or not the work has been carried out.

AI Regs,
Reg. 10(2)

As the works progress, if the approved inspector finds that the work is being carried out over a public sewer, he must notify the local authority 'as soon as practicable' of the location of the work unless, of course, the local authority are already aware of the fact from the information provided with the initial notice.

AI Regs,
Reg. 10(3)

Consultation with the fire authority

Where the building is a Workplace Subject to the Fire Precautions (Workplace) Regulations 1997, or will be put to a use designated under section 1 of the Fire Precautions Act 1971 and must be provided with means of escape in case of fire, the approved inspector must, as soon as practicable, consult the fire authority and give them 'sufficient' plans to show that the means of escape comply with the 1991 Regulations, Schedule 1, Paragraph B1. Additionally, he must do this at least fifteen working days before issuing a plans certificate or a final certificate. Thus, the fire authority is given an effective period of three weeks in which to make comment. Silence by the fire authority will imply approval. The approved inspector must 'have regard to any views' expressed by the fire authority.

AI Regs,
Reg. 11

Some local Acts of Parliament impose more extensive consultation requirements than the national legislation. The approved inspector must undertake any consultation required by the local legislation (see Chapter 5).

Plans certificates

A plans certificate is a certificate issued by an approved inspector certifying that the design has been checked and that the plans comply with the 1991 Regulations. Its issue is entirely at the option of the person carrying out the work, and is issued by the approved inspector to the local authority and the building owner.

1984 Act,
sec. 50

If the approved inspector is asked to issue a plans certificate and declines to do so on the grounds that the plans do not comply with the building regulations, the building owner can refer the dispute to the Secretary of State for a determination. A plans certificate can be issued at the same time as the initial notice or at a later stage, provided the work has not been carried out. There are two prescribed forms of plans certificate. There are three preconditions to its issue:

AI Regs,
Regs 12 to
14

- The approved inspector must have inspected the plans specified in the initial notice.
- He must be satisfied that the plans are neither defective nor show any contravention of the regulation requirements.
- He must have complied with any requirements about consultation, etc.

If a plans certificate is issued and accepted and, at a later stage, the initial notice ceases to be effective, the local authority cannot take enforcement action in respect of any work described in the plans certificate if it has been done in accordance with those plans.

The local authority have five working days in which to reject the plans certificate, but may only do so on certain specified grounds:

AI Regs,
Reg. 13;
Sch. 4

- The plans certificate is not in the prescribed form.
- It does not describe the work to which it relates.
- It does not specify the plans to which it relates.
- Unless it is combined with an initial notice, that no initial notice is in force.
- The certificate is not signed by the approved inspector who gave the initial notice or that he is no longer an approved inspector.
- The required declaration of insurance is not given.
- There is no declaration that the fire authority has been consulted (if appropriate).
- There is no declaration of independence (except for minor work).

When combined with an initial notice, the grounds for rejecting an initial notice specified in Schedule 3 (see p. 4.6) also apply.

Plans certificates may be rescinded by a local authority if the work has not been commenced within three years from the date on which the certificate was accepted.

Final certificates

1984 Act,
sec. 51

The final certificate should be issued by the approved inspector when the work is completed, but curiously there are no sanctions against an approved inspector who fails to issue a final certificate. The final certificate need not relate to all the work covered by the initial notice; it can, for example, be given in respect of part of a building which complies with the 1991 Regulations, or one or more of the houses on a development covered by an initial notice. Once given and accepted the initial notice ceases to apply.

It is to be issued, in a prescribed form, where an approved inspector is satisfied that any work specified in an initial notice given by him has been completed and certifies that 'the work described ... has been completed' and that the inspector has performed the functions assigned to him by the

AI Regs,
Reg. 15;
Sch. 5

regulations. If the local authority do not reject the final certificate within ten working days they are deemed to have accepted it. A final certificate can only be rejected on limited grounds. These are:

- The certificate is not in the prescribed form.
- It does not describe the work to which it relates.
- No initial notice relating to the work is in force.
- The certificate is not signed by the approved inspector who gave the notice or he is no longer an approved inspector.
- The required declaration of insurance is not provided.
- There is no declaration of independence (except for minor works).

The flow chart in Fig. 4.2 shows private certification procedure.

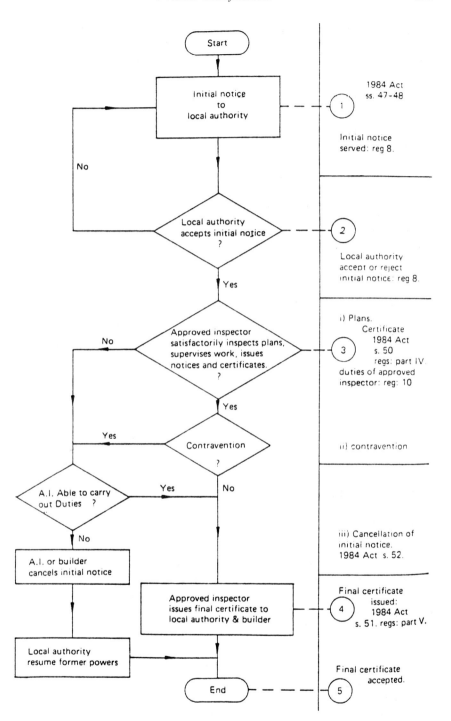

Fig. 4.2 Private certification procedure.

Once the final certificate is accepted by a local authority their powers to take proceedings against a person for contravention of building regulations in relation to the work referred to in the final certificate, are cancelled.

Public body's notices and certificates

1984 Act, sec. 54; AI Regs, Regs 19 to 26

Part VII of the Building (Approved Inspectors, etc.) Regulations 1985 is concerned with public bodies and, read in conjunction with section 54 of the Building Act 1984, its effect is to enable designated public bodies to self-certify their own work.

Public bodies are approved by the Secretary of State, and the regulations, relating to notices, consultation with the fire authority, plans certificates and final certificates mirror those of Part III dealing with approved inspectors. The grounds on which the local authority may reject a public body's notice, etc., mirror those applicable to private certification, except that:

- There is no provision for cancellation of a public body's notice.
- There is no requirement that there should be an approved insurance scheme in force.

Prescribed forms

AI Regs, Sch. 2

Eleven prescribed forms are set out in Schedule 2 of the Building (Approved Inspectors, etc.) Regulations 1985. Regulation 2(1) provides that where the regulations require the use of one of the numbered forms set out in Schedule 2, 'a form substantially to the like effect may be used'. Approved inspectors, public bodies, and local authorities, etc., may therefore have their own forms printed, provided they follow the precedents laid down in Schedule 2.

Chapter 5

Other legislation affecting health and safety

Introduction

Chapters 1 and 2 of this book contain information which should enable a person intending to carry out building work to assess whether or not the Building Regulations apply to that work. It will be seen that there are circumstances where the building or the work itself may be exempt from control. Even if this is the case, there may be legislation, other than the Building Regulations, which does apply.

This chapter provides a brief guide to a range of Acts of Parliament and regulations which might affect a building project, whether or not the Building Regulations also apply to that project. In practice there is a great deal of legislation which affects building development. The Acts and regulations covered in this chapter are those most commonly encountered concerning public health and safety, therefore town planning and conservation area issues, and legislation specific to a particular type of development (such as caravan sites, etc.) are not covered.

Conflicting statutory requirements

Some legislation (such as the Building Regulations) applies to a building when it is being designed and constructed whereas other different legislation will apply to it when it is being used (e.g. the Fire Precautions Act 1971). Therefore, it is possible that two (or more) pieces of legislation, (perhaps enforced by different authorities), might apply to the same building or work. In order to avoid conflicting requirements applying to a building it is often the case that a 'statutory bar' applies whereby one piece of legislation takes precedence over another. Often, conflicts are avoided by the provision in Acts of Parliament or regulations for consultation to take place between the different enforcing authorities (e.g. see page 4.11 above). Therefore, although a project may be exempt from the need to comply with the Building Regulations, it is possible that this may nullify the effect of a statutory bar and hence result in the need to comply with a different piece of legislation (such as the Fire Precautions Act 1971).

The Acts of Parliament and regulations referred to below apply at different stages in the normal life-cycle of the building and are often associated with the age and condition of the building. This is shown in Fig. 5.1.

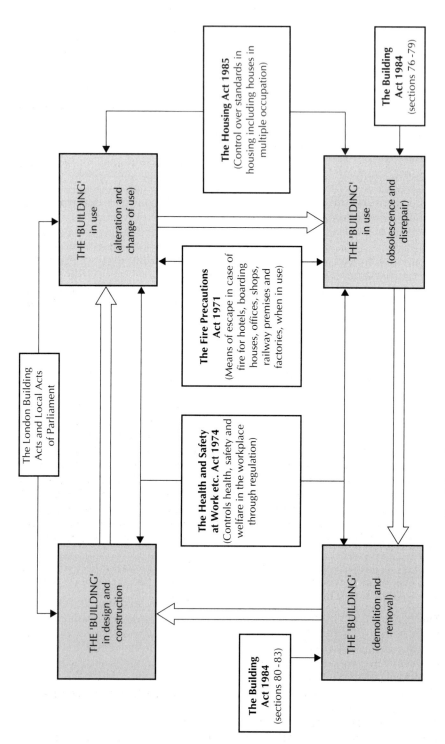

Fig. 5.1 Legislation affecting the Building Life-cycle.

The Fire Precautions Act 1971

This Act is mainly concerned with providing and maintaining safe means of escape from fire in existing buildings when they are being used. Under the Act a fire certificate is required for certain designated premises. These are:

- Hotels or boarding houses where sleeping accommodation is provided for more than six staff or guests (or some sleeping accommodation is provided above the first floor or below the ground floor), *and*
- Factories, offices, shops and railway premises where more than 20 people are employed or more than ten people work other than on the ground floor or in factories only where explosive or highly flammable materials are stored or used.

Since it is possible that an existing building may be put to a designated use where alterations are unnecessary, it may be the case that a fire certificate is needed. Accordingly, application must be made to the fire authority for the area in which the building is located (and which is the body which enforces the Act) in the prescribed form in accordance with the requirements of section 5 of the Act. (Of course, where work is being carried out which must comply with the Building Regulations, the local authority or approved inspector will deal with the application and carry out the necessary consultations with the fire authority. There is no need for a separate application to be made to the fire authority).

The Fire Precautions (Workplace) Regulations 1997

The Fire Precautions (Workplace) Regulations were made to fill a perceived deficiency in current legislation in implementing the general fire safety provisions of the European Framework and Workplace Directives. The Regulations require the provision of minimum fire safety standards in workplaces and impose duties on employers and on others in control of places of work. At present, workplaces covered by other specific fire safety legislation (such as the Fire Precautions Act 1971) are exempt from the requirements of the Regulations (although proposed amendments may result in dual application of the Fire Precautions Act 1971 and the Fire Precautions (Workplace) Regulations 1997), otherwise they apply to any place of work where people are employed.

Usually, employers must undertake a fire risk assessment of the workplace to establish what precautions are necessary to ensure the safety of employees in the event of fire. This could include, for example:

- The provision of fire fighting equipment, fire detectors and alarms.
- Making sure that fire fighting equipment is readily accessible, easy to use and appropriately signed.
- Ensuring that properly trained employees are nominated to put the necessary fire safety measures into practice.
- Making sure that adequate arrangements are made for contacting the emergency services.
- Providing suitable emergency routes and exits.
- Organising a suitable maintenance system so that all equipment is kept in good working order.

Fire authorities are required to enforce the Regulations in their area and they are empowered to inspect premises at any time to ensure compliance with the Regulations.

The Housing Act 1985

This Act allows housing authorities to carry out specified work in order to make houses fit for occupation. For example, houses in multiple occupation (i.e. those in which the occupants do not form part of a single household) are covered by sections 352 to 357 of the Act and among other things, the housing authority can insist that works are carried out which result in the provision of:

- Adequate storage accommodation.
- Food preparation and cooking facilities.
- Adequate toilets, baths, showers and wash hand basins with hot and cold water supplies.
- Suitable means of escape in case of fire and other fire precautions.

The Party Wall Act 1996

The Party Wall Act applies where work is being carried out to a party wall (which may be part of a building), or a boundary wall.
  The Act applies in three main areas:

- It provides a method whereby negotiations may take place over the construction of new party walls.
- It deals with the situation where there is a need to excavate below the level of a neighbouring building or structure which is within 6 m of the boundary.
- It allows owners to carry out work on party walls if appropriate notices have been served.

Any disputes which arise between the parties may be dealt with by means of mechanisms in the Act which allow for the appointment of surveyors to act as adjudicators.

The London Building Acts 1939 to 1982

Originally, the Building Regulations did not apply to Inner London which continued to be dealt with by the Greater London Council under the London Building Acts 1939 to 1982 and the building byelaws made under them. This was altered on 6 January 1986 when the Building (Inner London) Regulations came into operation. Following the abolition of the Greater London Council on 1 April 1986, its building control functions and those of district surveyors under the London Building Acts were transferred to the Common Council of the City of London and the 12 Inner London borough councils.
  Certain transitional provisions were made, but as a result of the new regulations, Inner London building control procedures became essentially the same as elsewhere in England and Wales, since all the London building byelaws and many sections (but not all) of the London Building Acts were repealed, and other sections were amended.

The most important provisions of the London Building Acts which were retained are listed below:

- *Buildings in excess height and cubical content*
  Section 20 of the London Building Acts (Amendment) Act 1939 applies special fire precautions in high buildings, i.e. a building which has a storey at a greater height than 30 m (or 25 m if the area of the building exceeds 930 m^2), or is a large building, or warehouses (over 7100 m^3), by requiring that they be divided up by division walls as defined in the 1939 Act. A wide range of fire protection measures can be required, and there are extra requirements for areas of 'special fire risks' such as boiler rooms.
  Plans must be deposited before any alterations are made to buildings of excess height or cubical content, and the borough council must consult with the London Fire Authority before issuing consent.

- *Uniting of buildings: 1939 Act section 21*
  Local authority consent is required if two buildings are united by making an opening in a party wall, or external wall if access is obtained between the buildings without passing into the external air. This does not apply if the buildings are in one ownership and if united would comply with the London Building Acts.

- *Special and temporary structures: 1939 Act sections 29 to 31*
  These sections deal with the erection and retention of certain temporary buildings which need the consent of the local authority and are not covered by other legislation.

- *Dangerous and neglected structures: 1939 Act sections 60 to 70*
  The procedure includes the service of a dangerous structure notice and is far more efficacious than the procedures in sections 77 and 78 of the Building Act 1984 (which do not apply to Inner London).

Local Acts of Parliament

Although the Building Act 1984 attempted to rationalise the main controls over buildings, there are in fact a great many pieces of local legislation with the result that many local authorities have special powers relevant to building control.

Where a local Act is in force, its provisions must also be complied with, since many of these pieces of legislation were enacted to meet local needs and perceived deficiencies in national legislation. The Building Regulations make it clear that local enactments must be taken into account.

With the growth and development of Building Regulation control over fire precautions in particular, it is likely that most of the current local legislation is now outdated or has been superceded by the Building Regulations. In fact, some local enactments already contain a statutory bar which gives precedence to building regulations.

Local authorities are obliged by section 90 of the Building Act 1984 to keep a copy of any local Act provisions and these must be available for public inspection free of charge at all reasonable times.

A full list of local Acts of Parliament may be found in Appendix 1 to this book where it will be seen that the most common local provisions relating to building control are:

- *Special fire precautions for basement garages or for large garages*
  The usual provision is that if a basement garage for more than three vehicles or a garage for more than twenty vehicles is to be erected, the local authority can impose access, ventilation and safety requirements.

- *Fire precautions in high buildings or for large storage buildings*
  There must be adequate access for the fire brigade in certain high buildings. A high building is one in excess of 18.3 m and the local authority must be satisfied with the fire precautions and may impose conditions, e.g. fire alarm systems, fire brigade access etc. (In many cases these requirements have been superceded by Part B of Schedule 1 to the Building Regulations 1991.) Large storage buildings in excess of 14,000 m^3 are required to be fitted with sprinkler systems by some local Acts. This requirement is likely to be included in the new edition of Approved Document B due out in 1999.

- *Extension of means of escape provisions*
  The Building Act 1984, section 72, is a provision under which the local authority can insist on the provision of means of escape where there is a storey which is more than 20 ft above ground level in certain types of buildings, e.g. hotels, boarding houses, hospitals, etc. Local enactments replace the 20 ft by 4.5 m and make certain other amendments to the national provisions.

- *Drainage systems*
  In some cases, local legislation requires that every building must have separate foul and surface water drainage systems.

- *Safety of stands at sports grounds*
  In many areas, local Acts impose controls over the safety of stands at sports grounds. Again much of this local legislation has been largely superceded by the provisions of the Fire Safety and Safety of Places of Sport Act 1987.

The Highways Act 1980

The Highways Act deals with the creation and control of highways, and the rights and duties of people who use them. Apart from trunk roads which are the responsibility of the Minister of Transport, the controlling authority will be the county council (or borough council in Greater London).

Building work on or immediately adjacent to a highway may present risks to the health and safety of people using it, therefore certain sections control such building work to the extent indicated below:

- *Builder's skips:* Under sections 139 and 140 local authority permission is required to deposit these on the highway and they can impose certain conditions as to lighting and signing.
- *Dangerous land* adjoining a highway and building operations affecting public safety are dealt with under sections 165 and 168 respectively.
- *Building materials* which come into contact with the highway and might damage it, or be a danger to the public, are dealt with under section 170 (mixing of mortar, etc., on the highway), and section 171 (deposit of building materials).

- *Temporary structures* such as scaffolding (section 169) and hoardings (sections 172 and 173) are controlled by a system of licensing administered by the local authority.
- Control is also exercised by the local authority over the construction of buildings:
  (a) Over a highway, sections 176 to 178, and
  (b) Under a highway, sections 179 and 180.

The Water Industry Act 1991

Under the Water Industry Act 1991, water undertakers have responsibilities with regard to the provision of sewers, drains, discharges and drinking water. The following sections illustrate the duties of the water undertakers and the rights of building owners and occupiers under the Act and are of importance to building development:

- Duty to provide a system of public sewers, section 94.
- Owners and occupiers rights to connect to a public sewer, sections 106 to 108.
- Duty to supply water for domestic purposes, section 52.
- Duty to make connections with a water main, section 45.
- The supply of water for non-domestic premises, section 55

The current Water Byelaws are due to be replaced by the Water Regulations made under section 74 of the Water Industry Act, on 1 July 1999.

The Clean Air Acts 1956 and 1993

The Clean Air Act deals, among other things, with the control of atmospheric pollution from furnaces and heating plant. With specific reference to building work, local authorities are empowered to control the height of chimneys for furnaces, and the choice of fuels which may be burnt.

The Environmental Protection Act 1990

The Environmental Protection Act (as amended by the Environment Act 1995) is the principal statute dealing with the collection and disposal of waste. A duty is placed on waste collection authorities, (primarily District Councils and London Boroughs) by virtue of section 45 of the 1990 Act, to collect all household waste in their areas. (Isolated areas, where other adequate disposal arrangements have been made, can be excepted). Additionally, they must collect commercial waste if requested to do so by the occupier of the premises and they may make a charge. They do not have to collect industrial waste but may do so with the consent of the relevant waste disposal authority, and they may charge for this service also.

Under sections 46 and 47, the waste disposal authorities may require the provision of suitable waste receptacles for household, commercial or industrial waste.

The Health and Safety at Work etc. Act 1974

This is the principal Act which deals with securing the health, safety and welfare of people at work and of others whose health and safety may be affected by work activities. The Act sets out certain general principles for health and safety at work, the actual details being contained in regulations which deal with particular situations and practices at work. Those of most relevance to building operations are set out below. Overall, the Act is enforced by the Health and Safety Executive, however many of its duties are carried out by local authorities (i.e. those which apply to work involving non hazardous operations).

The Workplace (Health, Safety and Welfare) Regulations 1992

Made under the Health and Safety at Work etc. Act 1974 the regulations implement provisions of European Workplace Directive 89/654/EEC. A general duty is placed on employers to ensure that workplaces comply with the requirements of the regulations including provisions for:

- Environmental measures such as ventilation, temperature control, lighting levels and adequacy of room dimensions.
- General welfare including cleanliness, sanitary conveniences and washing facilities.
- Safety measures such as the use of windows, doors, stairs, ladders and ramps, etc.

There is a connection with the Building Regulations in that new buildings which follow the guidance in the Approved Documents will, in most cases satisfy the requirements of the Workplace Regulations which, of course, apply to workplaces when they are being used.

The Construction (Design and Management) Regulations 1994

These regulations were made under the Health and Safety at Work etc. Act 1974 and implement provisions of European Directive No. 89/654/EEC. They apply to construction work, in a broad sense, and require the preparation of, and adherence to, a health and safety plan for the project by imposing duties on the client, the planning supervisor, the designer, the main contractor and on other contractors involved in the work.

The Construction (Health, Safety and Welfare) Regulations 1996

Made under the Health and Safety at Work etc. Act 1974, the regulations implement European Directive No. 92/57/EEC on minimum health and safety requirements for temporary or mobile construction sites for building, civil engineering or engineering construction. General duties are imposed on employers, the self-employed and others who control the way in which work is carried out. Employees are also seen to be responsible for their own actions. Enforcement is by the Health and Safety Executive.

The requirements deal with:

- The provision of safe work places (e.g. safe access and egress, sufficient and suitable working space).
- Measures to prevent falling (including through fragile materials).
- Means to ensure the stability of structures.
- Means of ensuring safe methods of demolition or dismantling.
- The safety of excavations.
- Safe traffic routes about the site.
- The provision of safety devices on doors and gates.
- Measures to reduce the risk of fire.
- The provision of fire detection and fire fighting equipment, and emergency escape routes.
- The provision of welfare facilities such as sanitary conveniences, washing facilities, drinking water, rest areas, etc.

The Building (Local Authority Charges) Regulations 1998

These regulations (which replace the Building (Prescribed Fees) Regulations 1994) authorise local authorities to fix and recover charges for the performance of their Building Regulations control functions according to a Scheme governed by principles prescribed in the regulations.

The Charges Regulations (which do not apply to Approved Inspectors) therefore have the effect of making each local authority responsible for fixing the amount of its own Building Regulations control charges for the following functions:

- A plan charge for the passing or rejection of plans of proposed work deposited with them.
- An inspection charge for the inspection of works in progress.
- A building notice charge where the building notice procedure applies. (This is the sum of the plan and inspection charges and usually becomes payable when the notice is given to the local authority, however this may be varied with the consent of the local authorities, see below.)
- A reversion charge where the approved inspector system is used and the initial notice is cancelled so that control reverts to the local authority (see p. 4.9).
- A regularisation charge for unauthorised building work (see regulation 13A, p. 2.18 above).

Local authorities have complete freedom to set the levels of the charges they impose, however the income derived from the charges over a continuous period of three years (commencing on the date on which the LA fixes its charges) must be not less than the cost directly or indirectly (the 'proper costs') incurred in accordance with proper accounting principles. In fact, local authorities are required to estimate the aggregate of their proper costs over the three year period before fixing their charges and they must issue a statement at the end of each financial year which sets out the details of the charging scheme and shows the amount of the income and proper costs. The following points should also be noted:

- Local authorities are not permitted to include any costs in relation to the control of work which is solely for the benefit of disabled people (they are not permitted to levy a charge for such work).

- Before bringing its charges into effect a local authority must give at least seven days' notice, although the charges can be amended whenever necessary provided the due notice is given on each occasion.
- Where the proper costs do not exceed £450,000 over the three year period or where at least 65% of the charges received are connected with work to small domestic buildings such as extensions, garages and car ports, the income derived must cover at least 90% of costs.
- Charges for work may, with the agreement of the local authority, be paid in instalments.
- Reduced charges may be levied for repetitive building work, where either it is part of the same submission (such as the construction of multiple housing units on a single estate), or it is for similar work on different sites but submitted by the same person (such as work to terrace dwellings under a refurbishment scheme).

Charges for work comprising

- The erection of small domestic buildings (up to $300 \, \text{m}^2$ in floor area), *or*
- Small detached garages and car ports with floor area up to $40 \, \text{m}^2$, *or*
- Domestic extensions (including associated access work) with floor areas up to $60 \, \text{m}^2$,

must be applied by reference to the floor area of the building or extension concerned. Where it is intended to erect a number of extensions to a building their floor areas must be aggregated.

All other types of work are subject to charges which must be related to an estimate of the cost of the work and local authorities are permitted to forego their inspection charge if the value of the work is less than £5000; however, a local authority may not charge for building work which comprises:

- The installation of cavity fill material, *or*
- The installation of unvented hot water systems,

where this work forms part of other building work and is being carried out in accordance with Parts D and G respectively of Schedule 1 to the Building Regulations 1991.

Charges are also payable where application is made to the Secretary of State for determination of questions under sections 16 and 50 of the Building Act 1984 (see p.2.20).

The Charges Regulations came into force on 1 April 1999 and prior to that date all submissions were subject to the Building (Prescribed Fees) Regulations 1994. In order to achieve some form of national consistency over levels of charges, the Local Government Association will be publishing a 'model scheme' which it will urge local authorities to follow. At the time of publication this had not been issued.

The Construction Products Regulations 1991

These regulations (as amended) apply to any construction products which are produced for incorporation in a permanent manner in construction works, and which were supplied after 27 December 1991.

The regulations are designed to ensure that when products are used in construction work, the work itself will satisfy any relevant 'essential requirements' of the Construction Products Directive (see p. 8.4). Products bearing the CE marking are presumed to satisfy the requirements of the regulations (see p. 8.6).

The Gas Safety (Installation and Use) Regulations 1994

The Gas Safety Regulations (as amended) are concerned with controlling the risks which arise when using gas from either mains pipes or gas storage vessels. The work must be carried out by a competent person who is a member (or is employed by a member) of the Council for Registered Gas Installers (CORGI).
Additionally:

- Gas appliances and the pipes which supply them, must be maintained in a safe condition if they are situated in workplaces covered by the regulations.
- Safety precautions must be observed for the storage and use of LPG.
- There are restrictions placed on certain kinds of gas appliances used in sleeping accommodation.
- Gas appliances and pipework installed in rented accommodation must be maintained in a safe condition (and maintenance records kept).

The Disability Discrimination Act 1995

The purpose of the Act is to end the discrimination faced by disabled people in the areas of employment, the provision of goods, facilities and services and in the renting or buying of property or land.
From the viewpoint of construction work, after 2004 service providers will have to ensure that there are no physical features which make it unreasonably difficult for disabled people to use their services. For example, this may involve installing accessible entrances, lifts, etc., in existing buildings.

II
Technical

Chapter 6

Structural stability

Introduction

Part A of Schedule 1 to the 1991 Regulations is concerned with the strength, stability and resistance to deformation of the building and its parts. The loads to be allowed for in the design calculations are specified, and recommendations as to construction are given in Approved Document A.

In line with the Government's intention to remove from the regulations those matters which are not directly concerned with public health and safety or the conservation of fuel and power, the previous requirements regarding the ability of a building structure or foundation to resist *damage* due to settlement, etc., have been omitted.

Additionally, control of deflection or deformation of the building structure under normal loading conditions is only relevant if it would impair the stability of another building.

It is conceivable, therefore, that a building constructed under the regulations could be safe and stable but could settle and deflect to such an extent that it would be unusable. In that event, of course, the owner would probably have redress against the designer and/or builder under the general law by way of an action for damages. Insurance cover might be somewhat hard to obtain for such a building.

The section of the regulations dealing with disproportionate collapse has been simplified by the revocation (in the 1994 amendment) of paragraph A4 of Schedule 1. This was concerned with maintaining structural stability in public buildings and shops in the event of roof failures, where roof spans exceeded 9 metres.

The original requirements concerning the failure of long span roof structures were introduced in response to a number of roof collapses which occurred in the 1970s and which led to the banning of high alumina cement in structural work. As this ban still exists the problem of such failures seems largely to have been solved without additional regulatory safeguard.

A further change introduced by the 1991 Regulations concerns the application of the structural requirements to certain buildings subject to a material change of use (see p.2.13 regulations 5 and 6). In these cases it will be necessary to carry out structural appraisals of the existing buildings to see if they are capable of coping with the changed loading conditions necessitated by the change of use. Guidance concerning this may be found in the following documents:

- BRE Digest 366: *Structural Appraisal of Existing Buildings for Change of Use.*
- The Institution of Structural Engineers Report, *Appraisal of Existing Structures*, 1980.

The Institution of Structural Engineers' report contains an item on design checks where a choice of various partial factors should be made to suit the individual circumstances of each case. Since the report was published in 1980 many of the BS Codes and Standards quoted will have been revised. ADA1/2 advises that the latest versions of these documents should be used.

**AD A1/2
sec. 4
4.10**

Loading

**Regs Sch. 1
A1**

Buildings must be constructed so that all dead, imposed and wind loads are sustained and transmitted to the ground:

(a) safely, and,
(b) without causing such settlement of the ground, or such deflection or deformation of the building, as will impair the stability of any other building.

The imposed and wind loads referred to above are those to which the building is likely to be subjected in the normal course of its use and for the purpose for which it is intended.

Ground movement

In addition to the provisions of Paragraph A1 above regarding loading, there are requirements in Paragraph A2 of Schedule 1 that the building shall be so constructed that movement of the ground caused by:

(a) swelling, shrinking or freezing of the subsoil; or,
(b) landslip or subsidence (other than subsidence arising from shrinkage),

will not impair the stability of any part of the building.

**Regs Sch. 1
A2**

It should be noted that the requirement as to landslip and subsidence applies to the extent that the risk can be reasonably foreseen.

Structural safety depends on the successful interrelationship between design and construction, particularly with regard to:

- Degree of loading – dead and imposed loads should be assessed in accordance with BS 6399: Parts 1 and 3, and wind loads to CP3: Chapter V: Part 2.
- The properties of the materials chosen.
- The design analysis used.
- Constructional details.
- Safety factors.
- Standards of workmanship.

It is essential that the numeric values of the safety factors which are used are derived from a consideration of the above factors, since a change in any one of these could disturb the safety of the structure as a whole.

**AD A1/2
0.2**

Additionally, loads used in calculations should take account of possible dynamic, concentrated and peak loads which may arise.

Approved Document A1/2 is arranged in four sections and gives guidance which may be adopted, if relevant, at the discretion of the designer. Where precise guidance is not given, due regard should be paid to the factors listed above.

- Section 1 allows the sizes of certain structural members to be assessed in small buildings of traditional masonry construction.
- Section 2 was added by the 1991 amendment and gives guidance on the fixing and support of external wall cladding.
- Section 3, also added by the 1991 amendment, makes it clear that certain roof re-covering operations may constitute a material alteration to the building and gives guidance to that effect.
- Section 4 lists various codes and standards for structural design and construction and is relevant to all types of buildings. Further information sources are included for the first time regarding landslip and structural appraisal of existing buildings subject to a change of use.

AD A1/2
0.1

Design of structural members in houses and other small buildings

Definitions

The following definitions apply throughout section 1 of AD A1/2.

BUTTRESSING WALL – a wall which provides lateral support, from base to top, to another wall perpendicular to it.

CAVITY WIDTH – the horizontal distance between the leaves in a cavity wall.

COMPARTMENT WALL – see Chapter 7, Fire.

DEAD LOAD – the load due to the weight of all roofs, floors, walls, services, finishes and partitions i.e. all the permanent construction.

IMPOSED LOAD – the load assumed to be produced by the intended occupancy or use, including moveable partitions, distributed, concentrated, impact, inertia and snow loads, but *excluding* wind loads.

PIER – an integral part of a wall which consists of a thickened section occurring at intervals along a wall to which it is bonded or securely tied so as to afford lateral support.

SEPARATING WALL – a wall which is common to two adjoining buildings (see Chapter 7, Fire).

SPACING – the centre to centre distance between two adjacent timbers measured in a plane parallel to the plane of the structure of which they form part.

SPAN – the distance measured along the centreline of a member between centres of adjacent bearings. (However, it should be noted that the spans given in the tables for floor joists, rafters, purlins, ceiling joists, binders and roof joists are *clear spans* i.e. measured between the faces of supports.)

SUPPORTED WALL – a wall which is supported by buttressing walls, piers or chimneys, or floor or roof lateral support arrangements.

WIND LOADS – all loads due to the effect of wind pressure or suction.

Structural stability

The basic stability of a small house of traditional masonry construction is largely dependent on the provision of a braced roof structure which is adequately anchored to walls restrained laterally by buttressing walls, piers or chimneys. If this can be achieved then it should not be necessary to take additional precautions against wind loading.

A traditional fully boarded or hipped roof provides in-built resistance to instability. However, where this is not provided then extra wind bracing may be required.

Trussed rafter roofs have, in the past, been susceptible to collapse during high winds. If this form of construction is used it should be braced in accordance with BS 5268 *Structural use of timber*, Part 3: 1985 *Code of practice for trussed rafter roofs*. The recommendations of this code may also be used for traditional roofs where bracing is inadequate.

Small buildings of masonry construction having walls designed in accordance with Section 1C of ADA1/2 and roofs and floors designed in accordance with Section 1B of ADA1/2 will be satisfactory with regard to structural stability if the roof is braced as mentioned above.

Structural work of timber in single family houses

Section 1B of ADA1/2 provides that if the work concerned is in a floor, ceiling or roof of a single occupancy house of not more than three storeys, that work will be satisfactory if the grades and dimensions of the timbers used are at least equal to those given in Tables A1 to A24 of Appendix ADA1/2 and if the work complies in other respects with BS 5268 *Structural use of timber* Part 2: 1991 *Code of practice for permissible stress design, materials and workmanship.*

In effect this means that for a house of this type it is not necessary to calculate the size of joists, rafters, purlins, etc.; one merely selects the appropriate sizes from the tables in ADA1/2. Unusual load or support conditions might necessitate a check calculation by the recommendations in BS 5268: Part 2: 1991.

Tables A1 to A24 apply to all floor, ceiling and roof timbers in a single occupancy house of three storeys or less.

The timber used for any binder, beam, joist, purlin or rafter must be of a species, origin and grade specified in Table 1 to ADA1/2 (see below) or as given in the more comprehensive tables of BS 5268: Part 2: 1991.

When using Tables A1 to A24 the following points should also be taken into account:

● The imposed load to be sustained by the floor, ceiling or roof of which the member forms part should not exceed:

(a) In the case of a floor: $1.5\,kN/m^2$ (Tables A1 and A2).
(b) In the case of a ceiling: $0.25k\,N/m^2$ and a concentrated load of $0.9\,kN$ acting with the imposed load (Tables A3 and A4).
(c) In the case of a flat roof with access not limited to the purposes of

AD A1/2

Table 1   Common species/grade combinations which satisfy the requirements for the strength classes to which tables A1–A24 in Appendix A relate.

Species	Origin	Grading Rules	Grades to satisfy strength class				
			SC3			SC4	
Redwood or whitewood	Imported	BS 4978	GS	MGS	M50	SS	MSS
Douglas Fir	UK	BS 4978	M50	SS	MSS	—	—
Larch	UK	BS 4978	GS	MGS	M50	SS	MSS
Scotch Pine	UK	BS 4978	GS	MGS	M50	SS	MSS
Corsican Pine	UK	BS 4978		M50		SS	MSS
European Spruce	UK	BS 4978		M75			
Sitka Spruce	UK	BS 4978		M75			
Douglas Fir-Larch Hem-Fir Spruce-Pine-Fir	CANADA	BS 4978	GS	MGS	M50	SS	MSS
Douglas Fir-Larch Hem-Fir Spruce-Pine-Fir	CANADA	NLGA	Joist & Plank Struct. L.F.	No.1 & No.2 No.1 & No.2		Joist & Plank Struct. L.F.	Select Select
Douglas Fir-Larch Hem-Fir Spruce-Pine-Fir	CANADA	MSR		Machine Stress-Rated 1450f-1.3E			Machine Stress-Rated 1650f-1.5E
Douglas Fir-Larch	USA	BS 4978	GS	MGS		SS	MSS
Hem-Fir	USA	BS 4978	GS	MGS	M50	SS	MSS
Western Whitewoods	USA	BS 4978	SS	MSS		—	—
Southern Pine	USA	BS 4978	GS	MGS		SS	MSS
Douglas Fir-Larch	USA	NGRDL	Joist & Plank Struct. L.F.	No.1 & No.2 No.1 & No.2		Joist & Plank Struct. L.F.	Select Select
Hem-Fir	USA	NGRDL	Joist & Plank Struct. L.F.	No.1 & No.2 No.1 & No.2		Joist & Plank Struct. L.F.	Select Select
Western Whitewoods	USA	NGRDL	Joist & Plank Struct. L.F.	Select Select		—	
Southern Pine	USA	NGRDL	Joist & Plank	No.3 Stud grade		Joist & Plank	Select
Douglas Fir-Larch Hem-Fir Southern Pine	USA	MSR		Machine Stress-Rated 1450f-1.3E			Machine Stress-Rated 1650f-1.5E

Notes: The common species/grade combinations given in this table are for particular use with the other tables in Appendix A and for cross section sizes given in those tables. Definitive and more comprehensive tables for assigning species/grade combinations to strength classes are given in BS 5268: Part 2: 1991. The grading rules for American and Canadian Lumber are those approved by the American Lumber Standards (ALS) Board of Review and the Canadian Lumber Standards (CLS) Accreditation Board respectively (see BS 5268: Part 2: 1991). NLGA denotes the National Lumber Grading Association NGRDL denotes the National Grading Rules for Dimension Lumber MSR denotes the North American Export Standard for Machine Stress-Rated Lumber.

maintenance or repair: 1.5 kN/m² or a concentrated load of 0.9 kN (Tables A21 and A22).

(d) In the case of a roof (flat or pitched up to 45°) with access only for maintenance: 0.75 kN/m² or 1.00 kN/m², measured on plan (depending on the location, see below), or a concentrated load of 0.9 kN (Tables A5 to A20 inclusive).

(e) In the case of a roof supporting sheeting or decking pitched at between 10° and 35°: 0.75 kN/m² or 1.00 kN/m² measured on plan (depending on the location, see below), or a concentrated load of 0.9 kN (Tables A23 and A24).

The loading variations on the roofs mentioned in (d) and (e) above are due to

AD A1/2

Diagram 2   Imposed snow roof loading

Site location	Loading
Within hatched area at an altitude of less than 100 m above ordnance datum	1.00 kN/m²
Outside hatched area at an altitude of less than 100 m above ordnance datum	0.75 kN/m²
Outside hatched area at an altitude lying between 100 m and 200 m above ordnance datum	1.00 kN/m²

Note:
For sites at greater altitude reference should be made to BS 6399: Part 3 to determine imposed and snow loading.

the different imposed snow loadings which vary with altitude and location in England and Wales.

Diagram 2 from ADA1/2 is reproduced below and shows how the values of 0.75 kN/m^2 and 1.00 kN/m^2 are chosen. It would seem that designers will now be required to establish the level of their site above ordnance datum when making a building regulation submission.

- Floorboarding is assumed to comply with BS 1297: 1987 *Specification for tongued and grooved softwood flooring* and Table 6.1 below.
- As stated in the footnotes to Tables A1 to A24 in Approved Document A1/ 2, the cross-sectional dimensions given are applicable to either basic sawn or regularised sizes from BS 4471: 1987 Specification for sizes of sawn and processed softwood. For North American timber (CLS/ALS) the tables apply to surface sizes only unless the timber has been resawn to BS 4471 requirements.
- Notches and holes in floor and roof joists should comply with Fig. 6.1. However, no notches or holes should be cut in rafters except for birds-mouths at supports. The rafter may be birdsmouthed to a depth of up to one third the rafter depth. Notches and holes should not be cut in purlins or binders unless checked by a competent person.
- Bearing areas and workmanship should be in accordance with BS 5268: Part 2: 1991 and the following minimum bearing lengths should be provided:

  (a) Floor, ceiling and roof joists – 35 mm.
  (b) Rafters and binders – 35 mm.
  (c) Purlins – 50 mm.

- If the spans of purlins or rafters are unequal it is permissible to choose the section sizes for each span separately or to use the longer span. However, for ceiling joists and binders the longer span only should be used.
- On floor joists no allowances have been made for additional loadings due to baths or partitions. It is recommended that the joists should be duplicated for bath support but no advice is given regarding partition loads. Similarly, when choosing ceiling joist sizes no account has been taken of trimming or of other additional loads such as water tanks.
- Purlins are assumed to be placed perpendicular to the roof slope and adequate connections between the various roof members should be provided as appropriate.
- Tables A1 to A24 are not applicable to trussed rafter roofs.

Example applications of Tables A1 to A24 are given in Fig. 6.2. It should be remembered that all spans, except for floorboards, are measured as the clear dimension between supports, and all spacings are the dimensions between longitudinal centres of members.

AD A1/2 sec. 1B & Appendix A

Table 6.1 Softwood floorboards (tongued and grooved).

Finished thickness of board (mm)	Maximum span of board (centre to centre of joists) (mm) up to
16	500
19	600

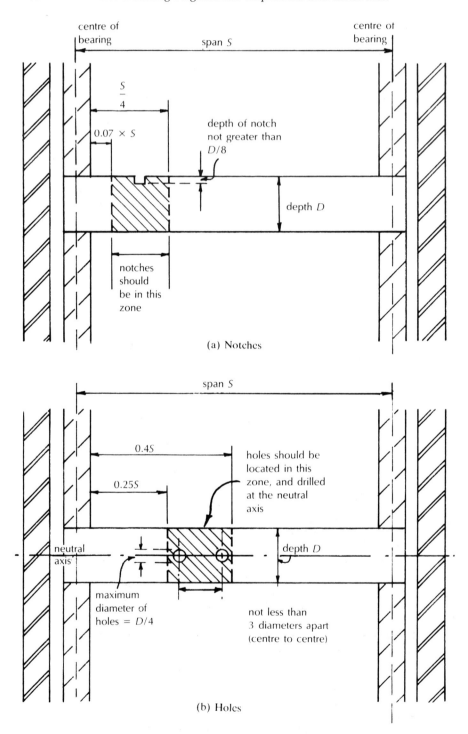

Fig. 6.1 Notches and holes in floor and roof joists.

(a) Floor joists, small house

dead load not more than 0.25 kN/m^2
clear span 4 m, centres 400 mm, timber of strength class SC3

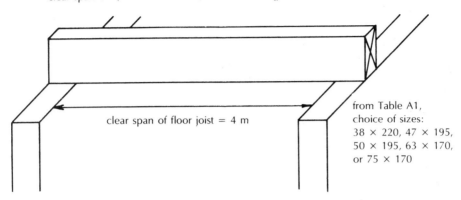

clear span of floor joist = 4 m

from Table A1,
choice of sizes:
38 × 220, 47 × 195,
50 × 195, 63 × 170,
or 75 × 170

(b) Rafter, small house

pitch 20°, dead load not more than 0.50 kN/m^2
clear span 2.90 m, centres 400 mm, timber of strength class SC3
imposed loading 0.75 kN/m^2

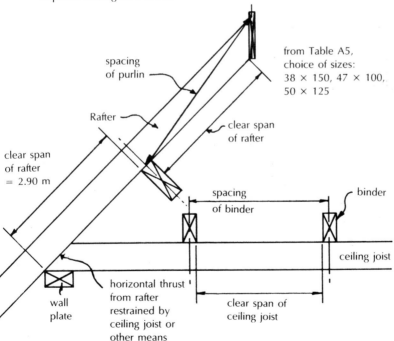

spacing
of purlin

from Table A5,
choice of sizes:
38 × 150, 47 × 100,
50 × 125

Rafter

clear span
of rafter

clear span
of rafter
= 2.90 m

spacing
of binder

binder

ceiling joist

horizontal thrust
from rafter
restrained by
ceiling joist or
other means

wall
plate

clear span of
ceiling joist

Fig. 6.2 Example of application of Tables A1 to A24.

AD A1/2, Appendix A

Table A1   Floor joists

Maximum clear span of joist (m) Timber of strength class SC3 (see Table 1)

Size of joist (mm × mm)	Not more than 0.25			More than 0.25 but not more than 0.50			More than 0.50 but not more than 1.25		
	— Dead Load [kN/m²] excluding the self weight of the joist / Spacing of joists (mm)								
	400	450	600	400	450	600	400	450	600
38 × 97	1.83	1.69	1.30	1.72	1.56	1.21	1.42	1.30	1.04
38 × 122	2.48	2.39	1.93	2.37	2.22	1.76	1.95	1.79	1.45
38 × 147	2.98	2.87	2.51	2.85	2.71	2.33	2.45	2.29	1.87
38 × 170	3.44	3.31	2.87	3.28	3.10	2.69	2.81	2.65	2.27
38 × 195	3.94	3.75	3.26	3.72	3.52	3.06	3.19	3.01	2.61
38 × 220	4.43	4.19	3.65	4.16	3.93	3.42	3.57	3.37	2.92
47 × 97	2.02	1.91	1.58	1.92	1.82	1.46	1.67	1.53	1.23
47 × 122	2.66	2.56	2.30	2.55	2.45	2.09	2.26	2.08	1.70
47 × 147	3.20	3.08	2.79	3.06	2.95	2.61	2.72	2.57	2.17
47 × 170	3.69	3.55	3.19	3.53	3.40	2.99	3.12	2.94	2.55
47 × 195	4.22	4.06	3.62	4.04	3.89	3.39	3.54	3.34	2.90
47 × 220	4.72	4.57	4.04	4.55	4.35	3.79	3.95	3.74	3.24
50 × 97	2.08	1.97	1.67	1.98	1.87	1.54	1.74	1.60	1.29
50 × 122	2.72	2.62	2.37	2.60	2.50	2.19	2.33	2.17	1.77
50 × 147	3.27	3.14	2.86	3.13	3.01	2.69	2.81	2.65	2.27
50 × 170	3.77	3.62	3.29	3.61	3.47	3.08	3.21	3.03	2.63
50 × 195	4.31	4.15	3.73	4.13	3.97	3.50	3.65	3.44	2.99
50 × 220	4.79	4.66	4.17	4.64	4.47	3.91	4.07	3.85	3.35
63 × 97	2.32	2.20	1.92	2.19	2.08	1.82	1.93	1.84	1.53
63 × 122	2.93	2.82	2.57	2.81	2.70	2.45	2.53	2.43	2.09
63 × 147	3.52	3.39	3.08	3.37	3.24	2.95	3.04	2.92	2.58
63 × 170	4.06	3.91	3.56	3.89	3.74	3.40	3.50	3.37	2.95
63 × 195	4.63	4.47	4.07	4.44	4.28	3.90	4.01	3.85	3.35
63 × 220	5.06	4.92	4.58	4.91	4.77	4.37	4.51	4.30	3.75
75 × 122	3.10	2.99	2.72	2.97	2.86	2.60	2.68	2.58	2.33
75 × 147	3.72	3.58	3.27	3.56	3.43	3.13	3.22	3.09	2.81
75 × 170	4.28	4.13	3.77	4.11	3.96	3.61	3.71	3.57	3.21
75 × 195	4.83	4.70	4.31	4.68	4.52	4.13	4.24	4.08	3.65
75 × 220	5.27	5.13	4.79	5.11	4.97	4.64	4.74	4.60	4.07
38 × 140	2.84	2.73	2.40	2.72	2.59	2.17	2.33	2.15	1.75
38 × 184	3.72	3.56	3.09	3.53	3.33	2.90	3.02	2.85	2.47
38 × 235	4.71	4.46	3.89	4.43	4.18	3.64	3.80	3.59	3.11

AD A1/2, Appendix A

Table A2 Floor joists

Maximum clear span of joist (m) Timber of strength class SC4 (see Table 1)

Size of joist (mm × mm)	Dead Load [kN/m²] excluding the self weight of the joist								
	Not more than 0.25			More than 0.25 but not more than 0.50			More than 0.50 but not more than 1.25		
	Spacing of joists (mm)								
	400	450	600	400	450	600	400	450	600
38 × 97	1.94	1.83	1.59	1.84	1.74	1.51	1.64	1.55	1.36
38 × 122	2.58	2.48	2.20	2.47	2.37	2.08	2.18	2.07	1.83
38 × 147	3.10	2.98	2.71	2.97	2.85	2.59	2.67	2.56	2.31
38 × 170	3.58	3.44	3.13	3.43	3.29	2.99	3.08	2.96	2.68
38 × 195	4.10	3.94	3.58	3.92	3.77	3.42	3.53	3.39	3.07
38 × 220	4.61	4.44	4.03	4.41	4.25	3.86	3.97	3.82	3.46
47 × 97	2.14	2.03	1.76	2.03	1.92	1.68	1.80	1.71	1.50
47 × 122	2.77	2.66	2.42	2.65	2.55	2.29	2.38	2.27	2.01
47 × 147	3.33	3.20	2.91	3.19	3.06	2.78	2.87	2.75	2.50
47 × 170	3.84	3.69	3.36	3.67	3.54	3.21	3.31	3.18	2.88
47 × 195	4.39	4.22	3.85	4.20	4.05	3.68	3.79	3.64	3.30
47 × 220	4.86	4.73	4.33	4.71	4.55	4.14	4.26	4.10	3.72
50 × 97	2.20	2.09	1.82	2.08	1.98	1.73	1.84	1.75	1.54
50 × 122	2.83	2.72	2.47	2.71	2.60	2.36	2.43	2.33	2.06
50 × 147	3.39	3.27	2.97	3.25	3.13	2.84	2.93	2.81	2.55
50 × 170	3.91	3.77	3.43	3.75	3.61	3.28	3.38	3.25	2.94
50 × 195	4.47	4.31	3.92	4.29	4.13	3.75	3.86	3.72	3.37
50 × 220	4.93	4.80	4.42	4.78	4.64	4.23	4.35	4.18	3.80
63 × 97	2.43	2.32	2.03	2.31	2.19	1.93	2.03	1.93	1.71
63 × 122	3.05	2.93	2.67	2.92	2.81	2.55	2.63	2.53	2.27
63 × 147	3.67	3.52	3.21	3.50	3.37	3.07	3.16	3.04	2.76
63 × 170	4.21	4.06	3.70	4.04	3.89	3.54	3.64	3.51	3.19
63 × 195	4.77	4.64	4.23	4.61	4.45	4.05	4.17	4.01	3.65
63 × 220	5.20	5.06	4.73	5.05	4.91	4.56	4.68	4.51	4.11
75 × 122	3.22	3.10	2.83	3.09	2.97	2.71	2.78	2.68	2.43
75 × 147	3.86	3.72	3.39	3.70	3.57	3.25	3.34	3.22	2.93
75 × 170	4.45	4.29	3.91	4.27	4.11	3.75	3.86	3.71	3.38
75 × 195	4.97	4.83	4.47	4.82	4.69	4.29	4.41	4.25	3.86
75 × 220	5.42	5.27	4.93	5.25	5.11	4.78	4.88	4.74	4.35
38 × 140	2.96	2.84	2.58	2.83	2.72	2.47	2.54	2.44	2.17
38 × 184	3.87	3.72	3.38	3.70	3.56	3.23	3.33	3.20	2.90
38 × 235	4.85	4.71	4.31	4.70	4.54	4.12	4.24	4.08	3.70

AD A1/2, Appendix A

Table A3   Ceiling joists

Maximum clear span of joist (m) Timber of strength class SC3 and SC4 (see Table 1)

Dead Load [kN/m²] excluding the self weight of the joist

Spacing of joists (mm)

Size of joist (mm × mm)	SC3 — Not more than 0.25			SC3 — More than 0.25 but not more than 0.50			SC4 — Not more than 0.25			SC4 — More than 0.25 but not more than 0.50		
	400	450	600	400	450	600	400	450	600	400	450	600
38 × 72	1.15	1.14	1.11	1.11	1.10	1.06	1.21	1.20	1.17	1.17	1.16	1.12
38 × 97	1.74	1.72	1.67	1.67	1.64	1.58	1.84	1.82	1.76	1.76	1.73	1.66
38 × 122	2.37	2.34	2.25	2.25	2.21	2.11	2.50	2.46	2.37	2.37	2.33	2.22
38 × 147	3.02	2.97	2.85	2.85	2.80	2.66	3.18	3.13	3.00	3.00	2.94	2.79
38 × 170	3.63	3.57	3.41	3.41	3.34	3.16	3.81	3.75	3.58	3.58	3.51	3.32
38 × 195	4.30	4.23	4.02	4.02	3.94	3.72	4.51	4.43	4.22	4.22	4.13	3.89
38 × 220	4.98	4.88	4.64	4.64	4.54	4.27	5.21	5.11	4.86	4.86	4.75	4.47
47 × 72	1.27	1.26	1.23	1.23	1.21	1.17	1.35	1.33	1.30	1.30	1.28	1.24
47 × 97	1.92	1.90	1.84	1.84	1.81	1.73	2.03	2.00	1.93	1.93	1.90	1.83
47 × 122	2.60	2.57	2.47	2.47	2.42	2.31	2.74	2.70	2.60	2.60	2.55	2.43
47 × 147	3.30	3.25	3.11	3.11	3.05	2.90	3.47	3.42	3.27	3.27	3.21	3.04
47 × 170	3.96	3.89	3.72	3.72	3.64	3.44	4.15	4.08	3.89	3.89	3.81	3.61
47 × 195	4.68	4.59	4.37	4.37	4.28	4.04	4.90	4.81	4.57	4.57	4.47	4.22
47 × 220	5.39	5.29	5.03	5.03	4.91	4.63	5.64	5.53	5.25	5.25	5.14	4.84
50 × 72	1.31	1.30	1.27	1.27	1.25	1.21	1.39	1.37	1.34	1.34	1.32	1.28
50 × 97	1.97	1.95	1.89	1.89	1.86	1.78	2.08	2.06	1.99	1.99	1.96	1.88
50 × 122	2.67	2.63	2.53	2.53	2.49	2.37	2.81	2.77	2.66	2.66	2.62	2.49
50 × 147	3.39	3.34	3.19	3.19	3.13	2.97	3.56	3.50	3.35	3.35	3.29	3.12
50 × 170	4.06	3.99	3.81	3.81	3.73	3.53	4.25	4.18	3.99	3.99	3.91	3.69
50 × 195	4.79	4.70	4.48	4.48	4.38	4.13	5.01	4.92	4.68	4.68	4.58	4.32
50 × 220	5.52	5.41	5.14	5.14	5.03	4.73	5.77	5.66	5.37	5.37	5.25	4.95
38 × 89	1.54	1.53	1.48	1.48	1.46	1.41	1.63	1.62	1.57	1.57	1.55	1.49
38 × 140	2.84	2.79	2.68	2.68	2.63	2.50	2.99	2.94	2.82	2.82	2.77	2.63
38 × 184	4.01	3.94	3.75	3.75	3.68	3.47	4.20	4.13	3.94	3.94	3.85	3.64

SC3

SC4

AD A1/2, Appendix A

Table A4   Binders supporting ceiling joists

Maximum clear span of binder (m)   **Timber of strength class SC3 and SC4** (see Table 1)

Size of binder (mm × mm)	Dead Load [kN/m²] excluding the self weight of the binder											
	Not more than 0.25						More than 0.25 but not more than 0.50					
	Spacing of binders (mm)											
	1200	1500	1800	2100	2400	2700	1200	1500	1800	2100	2400	2700
SC3												
47 × 150	2.17	2.05	1.96	1.88	1.81		1.99	1.87				
47 × 175	2.59	2.45	2.33	2.24	2.15	2.08	2.37	2.23	2.11	2.02	1.94	1.87
50 × 150	2.22	2.11	2.01	1.93	1.86		2.04	1.92	1.83			
50 × 175	2.65	2.51	2.39	2.29	2.21	2.13	2.42	2.28	2.16	2.07	1.99	1.91
50 × 200	3.08	2.91	2.77	2.65	2.55	2.47	2.81	2.64	2.50	2.39	2.29	2.21
63 × 125	1.97	1.87					1.82					
63 × 150	2.44	2.31	2.20	2.12	2.04	1.97	2.23	2.11	2.00	1.91	1.84	
63 × 175	2.90	2.74	2.61	2.51	2.41	2.33	2.65	2.49	2.37	2.26	2.17	2.10
63 × 200	3.37	3.18	3.03	2.90	2.79	2.69	3.07	2.88	2.74	2.61	2.51	2.42
63 × 225	3.83	3.61	3.44	3.29	3.16	3.05	3.49	3.27	3.10	2.96	2.84	2.74
75 × 125	2.12	2.01	1.92	1.85			1.95	1.84				
75 × 150	2.61	2.47	2.36	2.26	2.18	2.11	2.39	2.25	2.14	2.05	1.97	1.90
75 × 175	3.10	2.93	2.79	2.68	2.58	2.49	2.83	2.66	2.53	2.42	2.32	2.24
75 × 200	3.59	3.39	3.23	3.09	2.98	2.88	3.27	3.08	2.92	2.79	2.68	2.58
75 × 225	4.08	3.85	3.66	3.51	3.37	3.26	3.71	3.50	3.31	3.16	3.03	2.92
SC4												
47 × 150	2.28	2.16	2.06	1.98	1.90		2.09	1.97	1.87			
47 × 175	2.72	2.57	2.45	2.34	2.26	2.18	2.48	2.34	2.22	2.12	2.03	1.96
50 × 150	2.33	2.21	2.11	2.02	1.95		2.14	2.02	1.92	1.83		
50 × 175	2.78	2.63	2.51	2.40	2.31	2.23	2.54	2.39	2.27	2.17	2.08	2.01
50 × 200	3.23	3.05	2.90	2.78	2.67	2.58	2.95	2.77	2.62	2.51	2.40	2.32
63 × 125	2.07	1.97	1.88	1.81			1.91	1.80				
63 × 150	2.56	2.42	2.31	2.22	2.14	2.07	2.34	2.21	2.10	2.01	1.93	1.86
63 × 175	3.04	2.87	2.74	2.62	2.53	2.44	2.78	2.61	2.48	2.37	2.28	2.20
63 × 200	3.52	3.32	3.16	3.03	2.92	2.82	3.21	3.02	2.86	2.73	2.63	2.53
63 × 225	4.00	3.77	3.59	3.44	3.31	3.19	3.65	3.42	3.24	3.10	2.97	2.86
75 × 125	2.22	2.11	2.01	1.94	1.87	1.81	2.04	1.93	1.84			
75 × 150	2.73	2.59	2.47	2.37	2.28	2.21	2.50	2.36	2.24	2.15	2.06	1.99
75 × 175	3.24	3.07	2.92	2.80	2.70	2.61	2.96	2.79	2.65	2.53	2.43	2.35
75 × 200	3.75	3.54	3.37	3.23	3.11	3.00	3.42	3.22	3.05	2.92	2.80	2.70
75 × 225	4.26	4.02	3.82	3.66	3.52	3.40	3.88	3.65	3.46	3.30	3.17	3.06

AD A1/2, Appendix A

Table A5 Common or jack rafters for roofs having a pitch more than 15° but not more than 22.5° with access only for purpose of maintenance or repair. Imposed loading 0.75 kN/m² (see Diagram 2).

Maximum clear span of rafter (m) Timber of strength class SC3 and SC4 (see Table 1)

				Dead Load [kN/m²] excluding the self weight of the rafter					
	Not more than 0.50			More than 0.50 but not more than 0.75			More than 0.75 but not more than 1.00		
				Spacing of rafters (mm)					
Size of rafter (mm × mm)	400	450	600	400	450	600	400	450	600
SC3									
38 × 100	2.10	2.05	1.93	1.93	1.88	1.75	1.80	1.75	1.61
38 × 125	2.89	2.79	2.53	2.63	2.55	2.34	2.44	2.35	2.15
38 × 150	3.47	3.34	3.03	3.26	3.14	2.78	3.08	2.96	2.57
47 × 100	2.46	2.40	2.18	2.25	2.19	2.03	2.10	2.03	1.87
47 × 125	3.10	2.99	2.72	2.92	2.81	2.56	2.78	2.67	2.41
47 × 150	3.71	3.57	3.25	3.50	3.36	3.06	3.32	3.20	2.86
50 × 100	2.54	2.45	2.23	2.35	2.29	2.09	2.19	2.12	1.95
50 × 125	3.17	3.05	2.78	2.98	2.87	2.61	2.83	2.73	2.48
50 × 150	3.78	3.64	3.32	3.57	3.43	3.12	3.39	3.26	2.94
38 × 89	1.76	1.72	1.63	1.63	1.59	1.49	1.53	1.49	1.38
38 × 140	3.24	3.12	2.83	3.05	2.93	2.61	2.82	2.72	2.41
SC4									
38 × 100	2.42	2.33	2.11	2.28	2.19	1.99	2.16	2.08	1.88
38 × 125	3.01	2.90	2.64	2.83	2.73	2.48	2.69	2.59	2.35
38 × 150	3.60	3.47	3.16	3.39	3.26	2.97	3.22	3.10	2.82
47 × 100	2.59	2.49	2.27	2.44	2.35	2.13	2.32	2.23	2.02
47 × 125	3.22	3.11	2.83	3.04	2.92	2.66	2.89	2.78	2.53
47 × 150	3.85	3.71	3.38	3.63	3.50	3.18	3.45	3.32	3.02
50 × 100	2.64	2.54	2.32	2.49	2.40	2.18	2.37	2.28	2.07
50 × 125	3.29	3.17	2.89	3.10	2.98	2.72	2.95	2.83	2.58
50 × 150	3.93	3.78	3.45	3.70	3.57	3.25	3.52	3.39	3.09
38 × 89	2.16	2.07	1.88	2.03	1.95	1.77	1.92	1.85	1.68
38 × 140	3.37	3.24	2.95	3.17	3.05	2.77	3.01	2.90	2.63

AD A1/2, Appendix A

Table A6   Purlins supporting rafters to which Table A5 refers (Imposed loading 0.75 kN/m²)

Maximum clear span of purlin (m)   Timber of strength class SC3 and SC4 (see Table 1)

Dead Load [kN/m²] excluding the self weight of the purlin — Spacing of purlins (mm)

Class	Size of purlin (mm × mm)	Not more than 0.50						More than 0.50 but not more than 0.75						More than 0.75 but not more than 1.00					
		1500	1800	2100	2400	2700	3000	1500	1800	2100	2400	2700	3000	1500	1800	2100	2400	2700	3000
SC3	50 × 150	1.90																	
	50 × 175	2.22	2.08	1.96	1.87			2.08	1.95	1.84				1.97	1.84				
	50 × 200	2.53	2.37	2.24	2.13	2.02	1.92	2.38	2.22	2.10	1.97	1.85		2.25	2.10	1.95	1.82		
	50 × 225	2.84	2.66	2.52	2.40	2.26	2.14	2.67	2.50	2.35	2.20	2.07	1.96	2.53	2.36	2.18	2.03	1.91	1.81
	63 × 150	2.06	1.94	1.83				1.94	1.82					1.84					
	63 × 175	2.41	2.26	2.13	2.03	1.95	1.87	2.26	2.12	2.00	1.91	1.82		2.14	2.01	1.90	1.80		
	63 × 200	2.75	2.58	2.44	2.32	2.22	2.14	2.58	2.42	2.29	2.18	2.08	1.97	2.45	2.29	2.16	2.05	1.93	1.83
	63 × 225	3.09	2.89	2.74	2.61	2.50	2.40	2.90	2.72	2.57	2.45	2.33	2.20	2.75	2.58	2.43	2.29	2.16	2.04
	75 × 125	1.83																	
	75 × 150	2.19	2.06	1.95	1.86			2.06	1.94	1.83				1.96	1.83				
	75 × 175	2.56	2.40	2.27	2.17	2.08	2.00	2.41	2.26	2.13	2.03	1.95	1.87	2.28	2.14	2.02	1.92	1.84	
	75 × 200	2.92	2.74	2.59	2.47	2.37	2.28	2.75	2.58	2.44	2.32	2.22	2.14	2.61	2.44	2.31	2.20	2.10	2.00
	75 × 225	3.28	3.08	2.91	2.78	2.66	2.56	3.09	2.89	2.74	2.61	2.50	2.40	2.93	2.74	2.60	2.47	2.36	2.23
SC4	50 × 150	1.99	1.86					1.87											
	50 × 175	2.32	2.17	2.05	1.95	1.87		2.18	2.04	1.92	1.83			2.06	1.93	1.82			
	50 × 200	2.64	2.48	2.34	2.23	2.14	2.05	2.49	2.33	2.20	2.09	2.00	1.92	2.36	2.20	2.08	1.98	1.89	
	50 × 225	2.97	2.78	2.63	2.51	2.40	2.31	2.79	2.62	2.47	2.35	2.25	2.16	2.65	2.48	2.34	2.22	2.12	1.94
	63 × 150	2.16	2.02	1.91	1.82			2.03	1.90					1.92	1.80				
	63 × 175	2.51	2.36	2.23	2.13	2.04	1.96	2.36	2.22	2.10	2.00	1.91	1.84	2.24	2.10	1.99	1.89	1.81	
	63 × 200	2.87	2.69	2.55	2.43	2.33	2.24	2.70	2.53	2.39	2.28	2.18	2.10	2.56	2.40	2.27	2.16	2.06	1.98
	63 × 225	3.22	3.02	2.86	2.73	2.61	2.52	3.03	2.84	2.69	2.56	2.45	2.36	2.88	2.70	2.55	2.43	2.32	2.23
	75 × 125	1.91																	
	75 × 150	2.29	2.15	2.04	1.94	1.86		2.16	2.02	1.91	1.82			2.05	1.92	1.82			
	75 × 175	2.67	2.51	2.37	2.26	2.17	2.09	2.51	2.36	2.23	2.13	2.04	1.96	2.39	2.24	2.12	2.02	1.93	1.85
	75 × 200	3.05	2.86	2.71	2.58	2.48	2.39	2.87	2.69	2.55	2.43	2.33	2.24	2.72	2.55	2.42	2.30	2.20	2.12
	75 × 225	3.42	3.21	3.04	2.90	2.78	2.68	3.22	3.02	2.86	2.73	2.62	2.52	3.06	2.87	2.72	2.59	2.48	2.38

AD A1/2, Appendix A

Table A7   Common or jack rafters for roofs having a pitch more than 15° but not more than 22.5° with access only for purposes of maintenance or repair. Imposed loading 1.00 kN/m² (see Diagram 2).

Maximum clear span of rafter (m)   Timber of strength class SC3 and SC4 (see Table 1)

	Size of rafter (mm × mm)	Dead Load [kN/m²] excluding the self weight of the rafter								
		Not more than 0.50			More than 0.50 but not more than 0.75			More than 0.75 but not more than 1.00		
		Spacing of rafters (mm)								
		400	450	600	400	450	600	400	450	600
SC3	38 × 100	2.10	2.05	1.90	1.93	1.88	1.75	1.80	1.75	1.61
	38 × 125	2.73	2.63	2.35	2.59	2.49	2.17	2.44	2.34	2.03
	38 × 150	3.27	3.14	2.79	3.10	2.97	2.58	2.94	2.78	2.41
	47 × 100	2.35	2.26	2.05	2.23	2.15	1.95	2.10	2.03	1.83
	47 × 125	2.93	2.82	2.56	2.78	2.68	2.41	2.66	2.56	2.26
	47 × 150	3.50	3.37	3.07	3.33	3.20	2.86	3.18	3.06	2.68
	50 × 100	2.40	2.31	2.10	2.28	2.19	1.99	2.18	2.09	1.88
	50 × 125	2.99	2.88	2.62	2.84	2.73	2.48	2.71	2.61	2.33
	50 × 150	3.57	3.44	3.13	3.40	3.27	2.95	3.25	3.12	2.76
	38 × 89	1.76	1.72	1.63	1.63	1.59	1.49	1.53	1.49	1.38
	38 × 140	3.05	2.94	2.61	2.90	2.78	2.42	2.76	2.61	2.26
SC4	38 × 100	2.28	2.19	1.99	2.16	2.08	1.89	2.07	1.99	1.80
	38 × 125	2.84	2.73	2.48	2.70	2.59	2.35	2.58	2.48	2.25
	38 × 150	3.40	3.27	2.97	3.23	3.10	2.82	3.09	2.97	2.69
	47 × 100	2.44	2.35	2.14	2.32	2.23	2.03	2.22	2.13	1.94
	47 × 125	3.04	2.93	2.67	2.89	2.78	2.53	2.77	2.66	2.42
	47 × 150	3.64	3.50	3.19	3.46	3.33	3.03	3.31	3.18	2.89
	50 × 100	2.49	2.40	2.18	2.37	2.28	2.07	2.27	2.18	1.98
	50 × 125	3.10	2.99	2.72	2.95	2.84	2.58	2.82	2.72	2.47
	50 × 150	3.71	3.57	3.26	3.46	3.40	3.09	3.38	3.25	2.95
	38 × 89	2.03	1.95	1.77	1.93	1.85	1.68	1.84	1.77	1.60
	38 × 140	3.18	3.06	2.78	3.02	2.90	2.63	2.88	2.77	2.52

AD A1/2, Appendix A

Table A8   Purlins supporting rafters to which Table A7 refers (Imposed loading 1.0 kN/m²)

Maximum clear span of purlin (m)   Timber of strength class SC3 and SC4 (see Table 1)

Dead Load [kN/m²] excluding the self weight of the purlin

	Size of purlin (mm × mm)	Not more than 0.50						More than 0.50 but not more than 0.75						More than 0.75 but not more than 1.00					
		Spacing of purlins (mm)																	
		1500	1800	2100	2400	2700	3000	1500	1800	2100	2400	2700	3000	1500	1800	2100	2400	2700	3000
SC3	50 × 175	2.09	1.95	1.84				1.97	1.85					1.88					
	50 × 200	2.38	2.23	2.10	1.97	1.85		2.26	2.11	1.96	1.87			2.15	1.98	1.83			
	50 × 225	2.68	2.50	2.36	2.20	2.07	1.96	2.54	2.36	2.18	2.04	1.92	1.81	2.42	2.21	2.04	1.90	1.81	
	63 × 150	1.94	1.82					1.84											
	63 × 175	2.27	2.12	2.01	1.91	1.83		2.15	2.01	1.90	1.82			2.05	1.92	1.81			
	63 × 200	2.59	2.42	2.29	2.18	2.09	1.98	2.45	2.30	2.17	2.04	1.95	1.83	2.30	2.19	2.06	1.92		
	63 × 225	2.91	2.72	2.58	2.45	2.33	2.21	2.76	2.58	2.44	2.29	2.16	2.05	2.63	2.46	2.30	2.15	2.02	1.91
	75 × 150	2.07	1.94	1.83				1.96	1.84					1.87					
	75 × 175	2.41	2.26	2.14	2.04	1.95	1.88	2.29	2.14	2.03	1.93			2.18	2.04	1.93	1.84		
	75 × 200	2.75	2.58	2.44	2.33	2.23	2.14	2.61	2.45	2.31	2.21	2.10	2.01	2.49	2.33	2.20	2.10	1.98	1.88
	75 × 225	3.09	2.90	2.74	2.61	2.50	2.41	2.93	2.75	2.60	2.47	2.35	2.24	2.80	2.62	2.48	2.35	2.21	2.09
SC4	50 × 150	1.87																	
	50 × 175	2.18	2.04	1.93	1.83			2.07	1.94	1.82				1.97	1.84				
	50 × 200	2.49	2.33	2.20	2.10	2.00	1.92	2.36	2.21	2.08	1.98			2.25	2.10	1.98	1.88		
	50 × 225	2.80	2.62	2.48	2.36	2.25	2.16	2.65	2.49	2.34	2.23	2.07	1.95	2.53	2.36	2.23	2.12	1.91	
	63 × 150	2.03	1.90	1.80				1.93	1.81										
	63 × 175	2.37	2.22	2.10	2.00	1.91	1.84	2.25	2.11	1.99	1.89			2.14	2.01	1.90	1.80		
	63 × 200	2.70	2.53	2.40	2.28	2.19	2.10	2.57	2.40	2.27	2.16	2.07	1.99	2.45	2.29	2.16	2.06	1.97	1.89
	63 × 225	3.04	2.85	2.70	2.57	2.46	2.36	2.88	2.70	2.55	2.43	2.33	2.23	2.75	2.58	2.43	2.31	2.21	2.12
	75 × 125	1.80																	
	75 × 150	2.16	2.03	1.92	1.83			2.05	1.92	1.82				1.96	1.83				
	75 × 175	2.52	2.36	2.24	2.13	2.04	1.96	2.39	2.24	2.12	2.02	1.93	1.86	2.28	2.14	2.02	1.92	1.84	
	75 × 200	2.87	2.70	2.55	2.43	2.33	2.24	2.73	2.56	2.42	2.31	2.21	2.12	2.61	2.44	2.31	2.20	2.10	2.02
	75 × 225	3.23	3.03	2.87	2.74	2.62	2.52	3.07	2.88	2.72	2.60	2.48	2.39	2.93	2.75	2.60	2.47	2.36	2.27

AD A1/2, Appendix A

Table A9   Common or jack rafters for roofs having a pitch more than 22.5° but not more than 30° with access only for purposes of maintenance or repair. Imposed loading 0.75 kN/m² (see Diagram 2).

Maximum clear span of rafter (m) Timber of strength class SC3 and SC4 (see Table 1)

	Dead Load [kN/m²] excluding the self weight of the rafter								
	Not more than 0.50			More than 0.50 but not more than 0.75			More than 0.75 but not more than 1.00		
	Spacing of rafters (mm)								
Size of rafter (mm × mm)	400	450	600	400	450	600	400	450	600
SC3									
38 × 100	2.18	2.13	2.01	2.01	1.96	1.82	1.88	1.82	1.68
38 × 125	2.97	2.86	2.60	2.74	2.66	2.44	2.54	2.46	2.25
38 × 150	3.55	3.42	3.11	3.34	3.21	2.92	3.17	3.04	2.72
47 × 100	2.55	2.46	2.23	2.35	2.28	2.10	2.18	2.12	1.95
47 × 125	3.18	3.06	2.79	2.99	2.88	2.62	2.84	2.73	2.48
47 × 150	3.80	3.66	3.33	3.57	3.44	3.13	3.39	3.27	2.97
50 × 100	2.60	2.51	2.28	2.45	2.36	2.14	2.28	2.21	2.03
50 × 125	3.24	3.12	2.84	3.05	2.93	2.67	2.89	2.79	2.53
50 × 150	3.87	3.73	3.40	3.65	3.51	3.20	3.46	3.33	3.03
38 × 89	1.82	1.79	1.69	1.69	1.65	1.55	1.59	1.55	1.44
38 × 140	3.32	3.19	2.90	3.12	3.00	2.72	2.94	2.84	2.55
SC4									
38 × 100	2.48	2.38	2.17	2.33	2.24	2.03	2.21	2.12	1.93
38 × 125	3.08	2.97	2.70	2.90	2.79	2.53	2.75	2.65	2.40
38 × 150	3.69	3.55	3.23	3.47	3.34	3.04	3.29	3.17	2.88
47 × 100	2.65	2.55	2.32	2.49	2.40	2.18	2.37	2.28	2.07
47 × 125	3.30	3.18	2.90	3.11	2.99	2.72	2.95	2.84	2.58
47 × 150	3.94	3.80	3.46	3.71	3.58	3.26	3.53	3.40	3.09
50 × 100	2.71	2.61	2.37	2.55	2.45	2.23	2.42	2.32	2.11
50 × 125	3.37	3.24	2.96	3.17	3.05	2.78	3.01	2.90	2.63
50 × 150	4.02	3.87	3.53	3.79	3.65	3.32	3.60	3.46	3.15
38 × 89	2.21	2.12	1.93	2.07	1.99	1.81	1.97	1.89	1.72
38 × 140	3.45	3.32	3.02	3.24	3.12	2.84	3.08	2.96	2.69

AD A1/2, Appendix A

Table A10 Purlins supporting rafters to which Table A9 refers (Imposed loading 0.75 kN/m²)

Maximum clear span of purlin (m) Timber of strength class SC3 and SC4 (see Table 1)

Size of purlin (mm × mm)	Dead Load [kN/m²] excluding the self weight of the purlin																	
	Not more than 0.50						More than 0.50 but not more than 0.75						More than 0.75 but not more than 1.00					
	\Spacing of purlins (mm)\																	
	1500	1800	2100	2400	2700	3000	1500	1800	2100	2400	2700	3000	1500	1800	2100	2400	2700	3000
SC3																		
50 × 150	1.95	1.83					1.83											
50 × 175	2.27	2.12	2.01	1.92	1.83		2.13	1.99	1.88				2.02	1.89				
50 × 200	2.59	2.43	2.30	2.19	2.09	1.99	2.43	2.28	2.15	2.03	1.91	1.81	2.30	2.15	2.01	1.88		
50 × 225	2.92	2.73	2.58	2.46	2.34	2.22	2.74	2.56	2.42	2.27	2.14	2.02	2.59	2.42	2.25	2.10	1.98	1.87
63 × 150	2.12	1.98	1.88				1.99	1.86					1.88					
63 × 175	2.47	2.31	2.19	2.09	2.00	1.92	2.32	2.17	2.05	1.95	1.87		2.19	2.05	1.94	1.85		
63 × 200	2.81	2.64	2.50	2.38	2.28	2.19	2.64	2.48	2.34	2.23	2.13	2.04	2.50	2.35	2.22	2.11	1.99	1.89
63 × 225	3.16	2.97	2.81	2.68	2.56	2.47	2.97	2.78	2.63	2.51	2.40	2.28	2.82	2.64	2.49	2.37	2.23	2.11
75 × 125	1.88																	
75 × 150	2.25	2.11	2.00	1.91	1.83		2.11	1.98	1.87				2.00	1.88				
75 × 175	2.62	2.46	2.33	2.22	2.13	2.05	2.46	2.31	2.19	2.08	1.99	1.92	2.33	2.19	2.07	1.97	1.89	1.81
75 × 200	2.99	2.81	2.66	2.54	2.43	2.34	2.81	2.64	2.50	2.38	2.28	2.19	2.67	2.50	2.36	2.25	2.15	2.07
75 × 225	3.36	3.15	2.99	2.85	2.73	2.63	3.16	2.96	2.80	2.67	2.56	2.46	3.00	2.81	2.66	2.53	2.42	2.31
SC4																		
50 × 150	2.04	1.91	1.81				1.91						1.31					
50 × 175	2.37	2.22	2.10	2.00	1.92	1.84	2.23	2.09	1.97	1.88			2.11	1.97	1.86			
50 × 200	2.71	2.54	2.40	2.29	2.19	2.11	2.54	2.38	2.25	2.14	2.05	2.97	2.41	2.26	2.13	2.02	1.94	1.84
50 × 225	3.05	2.86	2.70	2.57	2.46	2.37	2.86	2.68	2.53	2.41	2.30	2.21	2.71	2.54	2.39	2.28	2.18	2.07
63 × 125	1.84																	
63 × 150	2.21	2.07	1.96	1.87			2.08	1.95	1.84				1.97	1.84				
63 × 175	2.57	2.42	2.29	2.18	2.09	2.01	2.42	2.27	2.15	2.04	1.96	1.88	2.29	2.15	2.03	1.93	1.85	
63 × 200	2.94	2.76	2.61	2.49	2.39	2.30	2.76	2.59	2.45	2.33	2.24	2.15	2.62	2.45	2.32	2.21	2.11	2.03
63 × 225	3.30	3.10	2.93	2.80	2.68	2.58	3.10	2.91	2.75	2.62	2.51	2.42	2.94	2.76	2.61	2.48	2.38	2.28
75 × 125	1.96	1.84					1.84											
75 × 150	2.35	2.20	2.09	1.99	1.91	1.84	2.21	2.07	1.96	1.87			2.09	1.96	1.86			
75 × 175	2.73	2.57	2.43	2.32	2.22	2.14	2.57	2.41	2.28	2.18	2.09	2.01	2.44	2.29	2.16	2.06	1.97	1.90
75 × 200	3.12	2.93	2.78	2.65	2.54	2.45	2.93	2.75	2.61	2.49	2.38	2.29	2.79	2.61	2.47	2.35	2.26	2.17
75 × 225	3.50	3.29	3.12	2.98	2.86	2.75	3.30	3.10	2.93	2.80	2.68	2.58	3.13	2.94	2.78	2.65	2.54	2.44

AD A1/2, Appendix A

Table A11 Common or jack rafters for roofs having a pitch more than **22.5°** but not more than **30°** with access only for purposes of maintenance or repair. Imposed loading **1.00 kN/m²** (see Diagram 2).

Maximum clear span of rafter (m) Timber of strength class SC3 and SC4 (see Table 1)

			Dead Load [kN/m²] excluding the self weight of the rafter						
	Not more than 0.50			More than 0.50 but not more than 0.75			More than 0.75 but not more than 1.00		
	Spacing of rafters (mm)								
Size of rafter (mm × mm)	400	450	600	400	450	600	400	450	600
SC3									
38 × 100	2.18	2.13	1.96	2.01	1.96	1.82	1.88	1.82	1.68
38 × 125	2.80	2.69	2.45	2.65	2.55	2.30	2.53	2.44	2.15
38 × 150	3.35	3.22	2.93	3.18	3.06	2.73	3.03	2.92	2.55
47 × 100	2.41	2.32	2.11	2.28	2.20	2.00	2.18	2.10	1.90
47 × 125	3.00	2.89	2.63	2.85	2.74	2.49	2.72	2.62	2.37
47 × 150	3.59	3.46	3.14	3.41	3.28	2.98	3.25	3.13	2.83
50 × 100	2.46	2.37	2.15	2.33	2.24	2.04	2.23	2.14	1.94
50 × 125	3.06	2.95	2.68	2.91	2.80	2.54	2.78	2.67	2.43
50 × 150	3.66	3.52	3.21	3.48	3.34	3.04	3.32	3.20	2.90
38 × 89	1.82	1.79	1.69	1.69	1.65	1.55	1.59	1.55	1.44
38 × 140	3.13	3.01	2.74	2.97	2.85	2.56	2.83	2.72	2.29
SC4									
38 × 100	2.34	2.25	2.04	2.21	2.13	1.93	2.11	2.03	1.84
38 × 125	2.91	2.80	2.55	2.76	2.66	2.41	2.64	2.53	2.30
38 × 150	3.48	3.35	3.05	3.30	3.18	2.89	3.16	3.04	2.76
47 × 100	2.51	2.41	2.19	2.38	2.29	2.08	2.27	2.18	1.98
47 × 125	3.12	3.00	2.73	2.96	2.85	2.59	2.83	2.72	2.47
47 × 150	3.73	3.59	3.27	3.54	3.41	3.10	3.38	3.26	2.96
50 × 100	2.56	2.46	2.24	2.42	2.33	2.12	2.32	2.23	2.02
50 × 125	3.18	3.06	2.79	3.02	2.91	2.64	2.89	2.78	2.52
50 × 150	3.80	3.66	3.34	3.61	3.48	3.16	3.45	3.32	3.02
38 × 89	2.08	2.00	1.82	1.97	1.90	1.72	1.88	1.81	1.64
38 × 140	3.25	3.13	2.85	3.09	2.97	2.70	2.95	2.84	2.57

AD A1/2, Appendix A

Table A12   Purlins supporting rafters to which Table A11 refers (Imposed loading 1.00 kN/m²)

Maximum clear span of purlin (m) Timber of strength class SC3 and SC4 (see Table 1)

Dead Load [kN/m²] excluding the self weight of the purlin

Spacing of purlins (mm)

Size of purlin (mm × mm)	Not more than 0.50						More than 0.50 but not more than 0.75						More than 0.75 but not more than 1.00					
	1500	1800	2100	2400	2700	3000	1500	1800	2100	2400	2700	3000	1500	1800	2100	2400	2700	3000
SC3																		
50 × 150	1.84																	
50 × 175	2.14	2.00	1.89				2.03	1.89					1.93	1.80				
50 × 200	2.45	2.30	2.16	2.05	1.93	1.82	2.31	2.16	2.03	1.89			2.20	2.05	1.89			
50 × 225	2.75	2.58	2.43	2.29	2.15	2.04	2.60	2.43	2.26	2.11	1.99	1.88	2.48	2.29	2.11	1.97	1.85	
63 × 150	2.00	1.87					1.89						1.88					
63 × 175	2.33	2.18	2.06	1.96	1.88	1.80	2.20	2.06	1.95	1.85			2.10	1.96	1.85			
63 × 200	2.66	2.49	2.35	2.24	2.14	2.05	2.51	2.35	2.22	2.12	2.00	1.90	2.40	2.24	2.12	1.99	1.87	
63 × 225	2.98	2.80	2.65	2.52	2.41	2.29	2.83	2.65	2.50	2.38	2.24	2.12	2.69	2.52	2.38	2.22	2.09	1.98
75 × 150	2.12	1.99	1.88				2.01	1.88					1.92					
75 × 175	2.47	2.32	2.20	2.09	2.00	1.93	2.34	2.20	2.08	1.98	1.89	1.82	2.24	2.09	1.98	1.88	1.80	
75 × 200	2.82	2.65	2.51	2.39	2.29	2.20	2.68	2.51	2.37	2.26	2.16	2.08	2.55	2.39	2.26	2.15	2.05	1.94
75 × 225	3.17	2.98	2.82	2.68	2.57	2.47	3.01	2.82	2.67	2.54	2.43	2.32	2.87	2.69	2.54	2.42	2.29	2.17
SC4																		
50 × 150	1.92						1.82											
50 × 175	2.24	2.10	1.98	1.89	1.80		2.12	1.98	1.87				2.02	1.89				
50 × 200	2.56	2.39	2.26	2.15	2.06	1.98	2.42	2.26	2.14	2.03	1.94	1.86	2.31	2.16	2.03	1.93	1.81	
50 × 225	2.87	2.69	2.54	2.42	2.32	2.22	2.72	2.55	2.40	2.29	2.18	2.09	2.59	2.42	2.29	2.17	2.04	1.83
63 × 150	2.09	1.95	1.85				1.98	1.85					1.88					
63 × 175	2.43	2.28	2.16	2.05	1.97	1.89	2.30	2.16	2.04	1.94	1.86		2.20	2.06	1.94	1.85		
63 × 200	2.77	2.60	2.46	2.35	2.25	2.16	2.63	2.46	2.33	2.22	2.12	2.04	2.51	2.35	2.22	2.11	2.02	1.94
63 × 225	3.12	2.92	2.77	2.64	2.52	2.43	2.95	2.77	2.62	2.49	2.39	2.29	2.82	2.64	2.49	2.37	2.27	2.18
75 × 125	1.85																	
75 × 150	2.22	2.08	1.97	1.88			2.10	1.97	1.86				2.01	1.88				
75 × 175	2.58	2.42	2.29	2.19	2.10	2.02	2.45	2.30	2.17	2.07	1.98	1.91	2.34	2.19	2.07	1.97	1.89	1.81
75 × 200	2.95	2.77	2.62	2.50	2.39	2.30	2.80	2.62	2.48	2.36	2.26	2.18	2.67	2.50	2.37	2.25	2.16	2.07
75 × 225	3.31	3.11	2.94	2.81	2.70	2.59	3.14	2.95	2.79	2.66	2.55	2.45	3.00	2.81	2.66	2.53	2.42	2.33

AD A1/2, Appendix A

Table A13   Common or jack rafters for roofs having a pitch more than 30° but not more than 45° with access only for purposes of maintenance or repair. Imposed loading 0.75 kN/m² (see Diagram 2).

Maximum clear span of rafter (m) Timber of strength class SC3 and SC4 (see Table 1)

				Dead Load [kN/m²] excluding the self weight of the rafter						
		Not more than 0.50			More than 0.50 but not more than 0.75			More than 0.75 but not more than 1.25		
					Spacing of rafters (mm)					
Size of rafter (mm × mm)		400	450	600	400	450	600	400	450	600
SC3	38 × 100	2.28	2.23	2.10	2.10	2.05	1.91	1.96	1.91	1.76
	38 × 125	3.07	2.95	2.69	2.87	2.77	2.52	2.65	2.56	2.35
	38 × 150	3.67	3.53	3.22	3.44	3.31	3.01	3.26	3.14	2.85
	47 × 100	2.64	2.54	2.31	2.45	2.38	2.17	2.28	2.21	2.04
	47 × 125	3.29	3.17	2.88	3.09	2.97	2.70	2.92	2.81	2.56
	47 × 150	3.93	3.78	3.45	3.69	3.55	3.23	3.50	3.37	3.06
	50 × 100	2.69	2.59	2.36	2.53	2.43	2.21	2.38	2.30	2.09
	50 × 125	3.35	3.23	2.94	3.15	3.03	2.76	2.98	2.87	2.61
	50 × 150	4.00	3.86	3.52	3.76	3.62	3.30	3.57	3.44	3.13
	38 × 89	1.91	1.87	1.77	1.77	1.73	1.62	1.67	1.62	1.50
	38 × 140	3.43	3.30	3.01	3.22	3.10	2.82	3.05	2.93	2.66
SC4	38 × 100	2.56	2.47	2.24	2.40	2.31	2.10	2.28	2.19	1.99
	38 × 125	3.19	3.07	2.80	2.99	2.88	2.62	2.84	2.73	2.48
	38 × 150	3.81	3.67	3.35	3.58	3.45	3.14	3.39	3.27	2.97
	47 × 100	2.74	2.64	2.41	2.58	2.48	2.25	2.44	2.35	2.13
	47 × 125	3.41	3.29	3.00	3.21	3.09	2.81	3.04	2.93	2.66
	47 × 150	4.08	3.93	3.59	3.83	3.69	3.36	3.64	3.50	3.19
	50 × 100	2.80	2.70	2.45	2.63	2.53	2.30	2.49	2.40	2.18
	50 × 125	3.48	3.35	3.06	3.27	3.15	2.87	3.10	2.99	2.72
	50 × 150	4.16	4.01	3.66	3.91	3.77	3.43	3.71	3.57	3.25
	38 × 89	2.28	2.20	2.00	2.14	2.06	1.87	2.03	1.95	1.77
	38 × 140	3.56	3.43	3.13	3.35	3.22	2.93	3.17	3.05	2.77

AD A1/2, Appendix A

Table A14 Purlins supporting rafters to which Table A13 refers. (Imposed loading 0.75 kN/m²)

Maximum clear span of purlin (m) Timber of strength class SC3 and SC4 (see Table 1)

Dead Load [kN/m²] excluding the self weight of the purlin

Size of purlin (mm × mm)	Not more than 0.50						More than 0.50 but not more than 0.75						More than 0.75 but not more than 1.00					
	Spacing of purlins (mm)																	
	1500	1800	2100	2400	2700	3000	1500	1800	2100	2400	2700	3000	1500	1800	2100	2400	2700	3000
SC3																		
50 × 150	2.02	1.89					1.89											
50 × 175	2.36	2.21	2.09	1.99	1.90	1.83	2.21	2.06	1.95	1.86			2.08	1.95	1.84			
50 × 200	2.69	2.52	2.38	2.27	2.17	2.09	2.52	2.36	2.23	2.12	2.01	1.90	2.38	2.23	2.10	1.97	1.85	
50 × 225	3.02	2.83	2.68	2.55	2.44	2.34	2.83	2.65	2.50	2.38	2.24	2.12	2.68	2.50	2.36	2.20	2.07	1.96
63 × 125	1.83																	
63 × 150	2.19	2.06	1.95	1.85			2.05	1.93	1.82				1.94	1.82				
63 × 175	2.55	2.40	2.27	2.16	2.07	1.99	2.39	2.24	2.12	2.02	1.94	1.86	2.27	2.12	2.01	1.91	1.83	
63 × 200	2.91	2.74	2.59	2.47	2.37	2.28	2.73	2.56	2.42	2.31	2.21	2.13	2.59	2.42	2.29	2.18	2.09	1.98
63 × 225	3.28	3.07	2.91	2.78	2.66	2.56	3.07	2.88	2.73	2.60	2.49	2.39	2.91	2.72	2.58	2.45	2.33	2.21
75 × 125	1.94	1.82					1.82											
75 × 150	2.33	2.19	2.07	1.97	1.89	1.82	2.18	2.05	1.94	1.85			2.07	1.94	1.83			
75 × 175	2.71	2.55	2.41	2.30	2.21	2.12	2.55	2.39	2.26	2.15	2.06	1.99	2.41	2.26	2.14	2.04	1.95	1.87
75 × 200	3.10	2.91	2.75	2.63	2.52	2.43	2.91	2.73	2.58	2.46	2.36	2.27	2.75	2.58	2.44	2.33	2.23	2.14
75 × 225	3.48	3.27	3.10	2.95	2.83	2.73	3.26	3.06	2.90	2.77	2.65	2.55	3.09	2.90	2.74	2.61	2.50	2.41
SC4																		
50 × 150	2.11	1.98	1.87				1.98	1.85										
50 × 175	2.46	2.31	2.18	2.08	1.99	1.91	2.31	2.16	2.04	1.94	1.86		2.18	2.04	1.93	1.84		
50 × 200	2.81	2.63	2.49	2.37	2.27	2.19	2.63	2.47	2.33	2.22	2.12	2.04	2.49	2.33	2.20	2.09	1.98	1.89
50 × 225	3.16	2.96	2.80	2.67	2.56	2.46	2.96	2.77	2.62	2.50	2.39	2.30	2.80	2.62	2.48	2.36	2.25	2.16
63 × 125	1.91																	
63 × 150	2.29	2.15	2.03	1.94	1.86		2.15	2.01	1.90	1.81			2.03	1.90				
63 × 175	2.67	2.50	2.37	2.26	2.17	2.08	2.50	2.35	2.22	2.12	2.03	1.95	2.37	2.22	2.10	2.00	1.91	
63 × 200	3.04	2.86	2.71	2.58	2.47	2.38	2.86	2.68	2.54	2.42	2.31	2.23	2.71	2.53	2.39	2.28	2.18	2.07
63 × 225	3.42	3.21	3.04	2.90	2.78	2.68	3.21	3.01	2.85	2.72	2.60	2.50	3.04	2.84	2.70	2.56	2.43	2.31
75 × 125	2.03	1.90	1.80				1.90						1.80					
75 × 150	2.43	2.28	2.16	2.06	1.98	1.91	2.28	2.14	2.03	1.93	1.85		2.16	2.03	1.92	1.83		
75 × 175	2.83	2.66	2.52	2.40	2.31	2.22	2.66	2.49	2.36	2.25	2.16	2.08	2.52	2.36	2.24	2.13	2.04	1.96
75 × 200	3.23	3.03	2.88	2.74	2.63	2.54	3.03	2.85	2.70	2.57	2.47	2.37	2.88	2.70	2.55	2.43	2.33	2.24
75 × 225	3.63	3.41	3.23	3.08	2.96	2.85	3.41	3.20	3.03	2.89	2.77	2.67	3.23	3.03	2.87	2.74	2.62	2.52

AD A1/2, Appendix A

Table A15   Common or jack rafters for roofs having a pitch more than 30° but not more than 45° with access only for purposes of maintenance or repair. Imposed loading 1.00 kN/m² (see Diagram 2).

Maximum clear span of rafter (m) Timber of strength class SC3 and SC4 (see Table 1)

		Dead Load [kN/m²] excluding the self weight of the rafter							
		Not more than 0.50			More than 0.50 but not more than 0.75			More than 0.75 but not more than 1.00		
		Spacing of rafters (mm)							
Size of rafter (mm × mm)		400	450	600	400	450	600	400	450	600
SC3	38 × 100	2.28	2.23	2.03	2.10	2.05	1.91	1.96	1.91	1.76
	38 × 125	2.90	2.79	2.54	2.75	2.64	2.40	2.62	2.52	2.26
	38 × 150	3.47	3.34	3.04	3.29	3.16	2.87	3.13	3.01	2.69
	47 × 100	2.50	2.40	2.18	2.36	2.27	2.06	2.25	2.17	1.97
	47 × 125	3.11	2.99	2.72	2.94	2.83	2.58	2.81	2.70	2.45
	47 × 150	3.72	3.58	3.26	3.52	3.39	3.08	3.36	3.23	2.94
	50 × 100	2.55	2.45	2.23	2.41	2.32	2.11	2.30	2.21	2.01
	50 × 125	3.17	3.05	2.78	3.00	2.89	2.63	2.87	2.76	2.51
	50 × 150	3.79	3.65	3.33	3.59	3.46	3.15	3.43	3.30	3.00
	38 × 89	1.91	1.87	1.77	1.77	1.73	1.62	1.62	1.62	1.50
	38 × 140	3.24	3.12	2.84	3.07	2.95	2.68	2.93	2.82	2.52
SC4	38 × 100	2.42	2.33	2.12	2.29	2.20	2.00	2.18	2.10	1.90
	38 × 125	3.02	2.90	2.64	2.86	2.75	2.50	2.72	2.62	2.38
	38 × 150	3.61	3.47	3.16	3.42	3.29	2.99	3.26	3.14	2.85
	47 × 100	2.60	2.50	2.27	2.46	2.36	2.15	2.34	2.25	2.05
	47 × 125	3.23	3.11	2.83	3.06	2.95	2.68	2.92	2.81	2.55
	47 × 150	3.86	3.72	3.39	3.66	3.52	3.21	3.49	3.36	3.06
	50 × 100	2.65	2.55	2.32	2.51	2.41	2.19	2.39	2.30	2.09
	50 × 125	3.30	3.17	2.89	3.12	3.01	2.73	2.98	2.87	2.61
	50 × 150	3.94	3.79	3.46	3.73	3.60	3.27	3.57	3.43	3.12
	38 × 89	2.16	2.08	1.89	2.04	1.96	1.78	1.95	1.87	1.70
	38 × 140	3.37	3.25	2.95	3.19	3.07	2.79	3.05	2.93	2.66

AD A1/2, Appendix A

Table A16 Purlins supporting rafters to which Table A15 refers. (Imposed loading 1.00 kN/m²)

Maximum clear span of purlin (m) Timber of strength class SC3 and SC4 (see Table 1)

Size of purlin (mm × mm)	Dead Load [kN/m²] excluding the self weight of the purlin																	
	Not more than 0.50						More than 0.50 but not more than 0.75						More than 0.75 but not more than 1.00					
	Spacing of purlins (mm)																	
	1500	1800	2100	2400	2700	3000	1500	1800	2100	2400	2700	3000	1500	1800	2100	2400	2700	3000
SC3																		
50 × 150	1.91						1.80											
50 × 175	2.22	2.08	1.97	1.87			2.10	1.96	1.85				2.00	1.87				
50 × 200	2.54	2.38	2.25	2.14	2.03	1.92	2.40	2.24	2.12	1.99	1.87		2.28	2.13	1.99	1.85		
50 × 225	2.85	2.67	2.53	2.40	2.27	2.15	2.70	2.52	2.38	2.22	2.09	1.98	2.56	2.40	2.22	2.07	1.95	1.84
63 × 150	2.07	1.94	1.84				1.96	1.83					1.86					
63 × 175	2.41	2.26	2.14	2.04	1.95	1.88	2.28	2.14	2.02	1.92	1.84		2.17	2.03	1.92	1.83		
63 × 200	2.76	2.58	2.44	2.33	2.23	2.14	2.61	2.44	2.31	2.20	2.10	2.00	2.48	2.32	2.19	2.09	1.97	1.86
63 × 225	3.10	2.90	2.75	2.62	2.51	2.41	2.93	2.74	2.59	2.47	2.36	2.23	2.79	2.61	2.47	2.33	2.20	2.08
75 × 125	1.84																	
75 × 150	2.20	2.07	1.96	1.86			2.08	1.95	1.85				1.98	1.86				
75 × 175	2.57	2.41	2.28	2.17	2.08	2.00	2.43	2.28	2.15	2.05	1.96	1.89	2.31	2.17	2.05	1.95	1.87	
75 × 200	2.93	2.75	2.60	2.48	2.38	2.29	2.77	2.60	2.46	2.34	2.24	2.16	2.64	2.47	2.34	2.23	2.13	2.04
75 × 225	3.29	3.09	2.92	2.79	2.67	2.57	3.12	2.92	2.76	2.63	2.52	2.43	2.97	2.78	2.63	2.50	2.40	2.28
SC4																		
50 × 150	1.99	1.87					1.88											
50 × 175	2.32	2.18	2.06	1.96	1.88	1.80	2.20	2.06	1.94	1.85			2.09	1.96	1.85			
50 × 200	2.65	2.49	2.35	2.24	2.14	2.06	2.51	2.35	2.22	2.11	2.02	1.94	2.39	2.23	2.11	2.00	1.91	
50 × 225	2.98	2.79	2.64	2.52	2.41	2.31	2.82	2.64	2.49	2.37	2.27	2.18	2.68	2.51	2.37	2.25	2.15	2.01
63 × 125	1.81																	
63 × 150	2.16	2.03	1.92	1.83			2.05	1.92	1.81				1.95	1.83				
63 × 175	2.52	2.36	2.24	2.13	2.04	1.97	2.39	2.24	2.11	2.01	1.93	1.85	2.27	2.13	2.01	1.91	1.83	
63 × 200	2.88	2.70	2.56	2.44	2.33	2.24	2.72	2.55	2.41	2.30	2.20	2.12	2.59	2.43	2.30	2.19	2.09	2.01
63 × 225	3.23	3.03	2.87	2.74	2.62	2.52	3.06	2.87	2.71	2.59	2.48	2.38	2.92	2.73	2.58	2.46	2.35	2.26
75 × 125	1.92						1.82											
75 × 150	2.30	2.16	2.04	1.95	1.87		2.18	2.04	1.93	1.84			2.07	1.94	1.84			
75 × 175	2.68	2.51	2.38	2.27	2.18	2.10	2.54	2.38	2.25	2.15	2.06	1.98	2.42	2.27	2.14	2.04	1.96	1.88
75 × 200	3.06	2.87	2.72	2.59	2.49	2.39	2.89	2.72	2.57	2.45	2.35	2.26	2.76	2.59	2.45	2.33	2.23	2.15

AD A1/2, Appendix A

Table A17   Joists for flat roofs with access only for purposes of maintenance or repair. Imposed loading 0.75 kN/m² (see Diagram 2).

Maximum clear span of joist (m)   Timber of strength class SC3 (see Table 1)

Size of joist (mm × mm)	Dead Load [kN/m²] excluding the self weight of the joist								
	Not more than 0.50			More than 0.50 but not more than 0.75			More than 0.75 but not more than 1.00		
	Spacing of joists (mm)								
	400	450	600	400	450	600	400	450	600
38 × 97	1.74	1.72	1.67	1.67	1.64	1.58	1.61	1.58	1.51
38 × 122	2.37	2.34	2.25	2.25	2.21	2.11	2.16	2.11	2.01
38 × 147	3.02	2.97	2.85	2.85	2.80	2.66	2.72	2.66	2.51
38 × 170	3.63	3.57	3.37	3.41	3.34	3.17	3.24	3.17	2.98
38 × 195	4.30	4.23	3.86	4.03	3.94	3.63	3.81	3.72	3.45
38 × 220	4.94	4.76	4.34	4.64	4.49	4.09	4.38	4.27	3.88
47 × 97	1.92	1.90	1.84	1.84	1.81	1.74	1.77	1.74	1.65
47 × 122	2.60	2.57	2.47	2.47	2.43	2.31	2.36	2.31	2.19
47 × 147	3.30	3.25	3.12	3.12	3.06	2.90	2.96	2.90	2.74
47 × 170	3.96	3.89	3.61	3.72	3.64	3.40	3.53	3.44	3.23
47 × 195	4.68	4.53	4.13	4.37	4.28	3.89	4.14	4.04	3.70
47 × 220	5.28	5.09	4.65	4.99	4.81	4.38	4.75	4.58	4.17
50 × 97	1.97	1.95	1.89	1.89	1.86	1.78	1.81	1.78	1.70
50 × 122	2.67	2.64	2.53	2.53	2.49	2.37	2.42	2.37	2.25
50 × 147	3.39	3.34	3.19	3.19	3.13	2.97	3.04	2.97	2.80
50 × 170	4.06	3.99	3.69	3.81	3.73	3.47	3.61	3.53	3.30
50 × 195	4.79	4.62	4.22	4.48	4.36	3.97	4.23	4.13	3.78
50 × 220	5.38	5.19	4.74	5.09	4.90	4.47	4.85	4.67	4.25
63 × 97	2.19	2.16	2.09	2.09	2.06	1.97	2.01	1.97	1.87
63 × 122	2.95	2.91	2.79	2.79	2.74	2.61	2.66	2.61	2.47
63 × 147	3.72	3.66	3.44	3.50	3.43	3.25	3.33	3.26	3.07
63 × 170	4.44	4.35	3.97	4.16	4.07	3.74	3.95	3.85	3.56
63 × 195	5.14	4.96	4.54	4.86	4.69	4.28	4.61	4.47	4.07
63 × 220	5.77	5.57	5.10	5.46	5.27	4.82	5.21	5.02	4.59
75 × 122	3.17	3.12	3.00	3.00	2.94	2.80	2.86	2.80	2.65
75 × 147	3.98	3.92	3.64	3.75	3.67	3.44	3.56	3.48	3.27
75 × 170	4.74	4.58	4.19	4.44	4.33	3.96	4.21	4.11	3.77
75 × 195	5.42	5.23	4.79	5.13	4.95	4.53	4.89	4.72	4.31
75 × 220	6.07	5.87	5.38	5.76	5.56	5.09	5.50	5.30	4.85
38 × 140	2.84	2.79	2.68	2.68	2.63	2.51	2.56	2.51	2.37
38 × 184	4.01	3.94	3.64	3.76	3.68	3.43	3.56	3.48	3.25

AD A1/2, Appendix A

Table A18 Joists for flat roofs with access only for purposes of maintenance or repair. Imposed loading 0.75 kN/m² (see Diagram 2).

Maximum clear span of joist (m) Timber of strength class SC4 (see Table 1)

Size of joist (mm × mm)	Dead Load [kN/m²] excluding the self weight of the joist								
	Not more than 0.50			More than 0.50 but not more than 0.75			More than 0.75 but not more than 1.00		
	Spacing of joists (mm)								
	400	450	600	400	450	600	400	450	600
38 × 97	1.84	1.82	1.76	1.76	1.73	1.66	1.69	1.66	1.59
38 × 122	2.50	2.46	2.37	2.37	2.33	2.22	2.27	2.22	2.11
38 × 147	3.18	3.13	3.00	3.00	2.94	2.79	2.85	2.79	2.64
38 × 170	3.81	3.75	3.50	3.58	3.51	3.30	3.40	3.32	3.12
38 × 195	4.51	4.40	4.01	4.22	4.13	3.78	3.99	3.90	3.59
38 × 220	5.13	4.95	4.51	4.85	4.67	4.25	4.59	4.44	4.04
47 × 97	2.03	2.00	1.94	1.94	1.91	1.83	1.86	1.83	1.74
47 × 122	2.74	2.70	2.60	2.60	2.55	2.43	2.48	2.43	2.30
47 × 147	3.47	3.42	3.26	3.27	3.21	3.04	3.11	3.04	2.87
47 × 170	4.15	4.08	3.76	3.89	3.81	3.54	3.69	3.61	3.36
47 × 195	4.88	4.70	4.29	4.58	4.44	4.05	4.33	4.22	3.85
47 × 220	5.48	5.29	4.83	5.18	5.00	4.56	4.94	4.76	4.33
50 × 97	2.08	2.06	1.99	1.99	1.96	1.88	1.91	1.88	1.79
50 × 122	2.81	2.77	2.66	2.66	2.62	2.49	2.54	2.49	2.36
50 × 147	3.56	3.50	3.32	3.35	3.29	3.12	3.19	3.12	2.94
50 × 170	4.26	4.18	3.83	3.99	3.91	3.61	3.78	3.69	3.43
50 × 195	4.97	4.80	4.38	4.68	4.53	4.13	4.43	4.31	3.93
50 × 220	5.59	5.39	4.93	5.28	5.09	4.65	5.04	4.85	4.42
63 × 97	2.31	2.28	2.20	2.20	2.16	2.07	2.11	2.07	1.97
63 × 122	3.10	3.05	2.93	2.93	2.88	2.74	2.80	2.74	2.59
63 × 147	3.90	3.84	3.58	3.67	3.60	3.38	3.49	3.41	3.21
63 × 170	4.65	4.51	4.12	4.35	4.26	3.89	4.13	4.03	3.70
63 × 195	5.33	5.15	4.71	5.05	4.87	4.45	4.82	4.64	4.24
63 × 220	5.98	5.78	5.30	5.67	5.47	5.00	5.41	5.22	4.76
75 × 122	3.33	3.27	3.14	3.14	3.08	2.93	2.99	2.93	2.77
75 × 147	4.17	4.10	3.78	3.92	3.84	3.57	3.73	3.64	3.40
75 × 170	4.92	4.75	4.35	4.64	4.50	4.11	4.40	4.29	3.92
75 × 195	5.61	5.42	4.97	5.32	5.14	4.70	5.08	4.90	4.48
75 × 220	6.29	6.08	5.59	5.97	5.77	5.28	5.70	5.50	5.04
38 × 140	2.99	2.94	2.82	2.82	2.75	2.63	2.69	2.63	2.49
38 × 184	4.21	4.13	3.79	3.94	3.85	3.57	3.73	3.64	3.39

AD A1/2, Appendix A

Table A19   Joists for flat roofs with access only for purposes of maintenance or repair. Imposed loading 1.0 kN/m² (see Diagram 2).

Maximum clear span of joist (m) Timber of strength class SC3 (see Table 1)

Size of joist (mm × mm)	Dead Load [kN/m²] excluding the self weight of the joist								
	Not more than 0.50			More than 0.50 but not more than 0.75			More than 0.75 but not more than 1.00		
	Spacing of joists (mm)								
	400	450	600	400	450	600	400	450	600
38 × 97	1.74	1.72	1.67	1.67	1.64	1.58	1.61	1.58	1.51
38 × 122	2.37	2.34	2.25	2.25	2.21	2.11	2.16	2.11	2.01
38 × 147	3.02	2.97	2.75	2.85	2.80	2.61	2.72	2.66	2.49
38 × 170	3.62	3.49	3.17	3.41	3.31	3.01	3.24	3.17	2.88
38 × 195	4.15	3.99	3.63	3.94	3.79	3.45	3.77	3.63	3.29
38 × 220	4.67	4.49	4.09	4.44	4.27	3.88	4.25	4.09	3.71
47 × 97	1.92	1.90	1.84	1.84	1.81	1.74	1.77	1.74	1.65
47 × 122	2.60	2.57	2.45	2.47	2.43	2.31	2.36	2.31	2.19
47 × 147	3.30	3.24	2.95	3.12	3.06	2.80	2.96	2.90	2.68
47 × 170	3.88	3.74	3.40	3.69	3.56	3.23	3.53	3.40	3.09
47 × 195	4.44	4.27	3.89	4.23	4.07	3.70	4.05	3.89	3.54
47 × 220	4.99	4.81	4.38	4.75	4.58	4.17	4.55	4.38	3.99
50 × 97	1.97	1.95	1.89	1.89	1.86	1.78	1.81	1.78	1.70
50 × 122	2.67	2.64	2.50	2.53	2.49	2.37	2.42	2.37	2.25
50 × 147	3.39	3.31	3.01	3.19	3.13	2.86	3.04	2.97	2.73
50 × 170	3.96	3.81	3.47	3.77	3.63	3.30	3.61	3.47	3.16
50 × 195	4.53	4.36	3.97	4.31	4.15	3.78	4.13	3.97	3.61
50 × 220	5.09	4.90	4.47	4.85	4.67	4.25	4.65	4.47	4.07
63 × 97	2.19	2.16	2.09	2.09	2.06	1.97	2.01	1.97	1.87
63 × 122	2.95	2.91	2.70	2.79	2.74	2.57	2.66	2.61	2.46
63 × 147	3.70	3.56	3.25	3.50	3.39	3.09	3.33	3.25	2.95
63 × 170	4.26	4.10	3.74	4.06	3.91	3.56	3.89	3.74	3.41
63 × 195	4.86	4.69	4.28	4.64	4.47	4.07	4.45	4.28	3.90
63 × 220	5.46	5.27	4.82	5.21	5.02	4.59	5.00	4.82	4.39
75 × 122	3.17	3.12	2.86	3.00	2.94	2.72	2.86	2.80	2.60
75 × 147	3.90	3.76	3.44	3.72	3.59	3.27	3.56	3.44	3.13
75 × 170	4.49	4.33	3.96	4.29	4.13	3.77	4.11	3.96	3.61
75 × 195	5.13	4.95	4.53	4.89	4.72	4.31	4.70	4.53	4.13
75 × 220	5.76	5.56	5.09	5.50	5.30	4.85	5.28	5.09	4.65
38 × 140	2.84	2.79	2.62	2.68	2.63	2.48	2.56	2.51	2.37
38 × 184	3.92	3.77	3.43	3.73	3.58	3.25	3.56	3.43	3.11

AD A1/2, Appendix A

Table A20 Joists for flat roofs with access only for purposes of maintenance or repair. Imposed loading 1.0 kN/m^2 (see Diagram 2).

Maximum clear span of joist (m) Timber of strength class SC4 (see Table 1)

Size of joist (mm × mm)	Dead Load [kN/m²] excluding the self weight of the joist								
	Not more than 0.50			More than 0.50 but not more than 0.75			More than 0.75 but not more than 1.00		
	Spacing of joists (mm)								
	400	450	600	400	450	600	400	450	600
38 × 97	1.84	1.82	1.76	1.76	1.73	1.66	1.69	1.66	1.59
38 × 122	2.50	2.46	2.37	2.37	2.33	2.22	2.27	2.22	2.11
38 × 147	3.18	3.13	2.86	3.00	2.94	2.71	2.85	2.79	2.59
38 × 170	3.77	3.63	3.30	3.58	3.45	3.13	3.40	3.30	2.99
38 × 195	4.31	4.15	3.78	4.10	3.95	3.59	3.93	3.78	3.43
38 × 220	4.85	4.67	4.25	4.61	4.44	4.04	4.42	4.25	3.86
47 × 97	2.03	2.00	1.94	1.94	1.91	1.83	1.86	1.83	1.74
47 × 122	2.74	2.70	2.55	2.60	2.55	2.42	2.48	2.43	2.30
47 × 147	3.47	3.37	3.07	3.27	3.21	2.91	3.11	3.04	2.79
47 × 170	4.03	3.88	3.54	3.84	3.70	3.36	3.68	3.54	3.22
47 × 195	4.61	4.44	4.05	4.39	4.23	3.85	4.21	4.05	3.68
47 × 220	5.18	5.00	4.56	4.94	4.76	4.33	4.73	4.56	4.15
50 × 97	2.08	2.06	1.99	1.99	1.96	1.88	1.91	1.88	1.79
50 × 122	2.81	2.77	2.60	2.66	2.62	2.47	2.54	2.49	2.36
50 × 147	3.56	3.44	3.13	3.35	3.27	2.97	3.19	3.12	2.85
50 × 170	4.11	3.96	3.61	3.92	3.77	3.43	3.75	3.61	3.28
50 × 195	4.70	4.53	4.13	4.48	4.31	3.93	4.29	4.13	3.76
50 × 220	5.28	5.09	4.65	5.04	4.85	4.42	4.83	4.65	4.23
63 × 97	2.31	2.28	2.20	2.20	2.16	2.07	2.11	2.07	1.97
63 × 122	3.10	3.05	2.81	2.93	2.88	2.67	2.80	2.74	2.56
63 × 147	3.84	3.70	3.38	3.66	3.52	3.21	3.49	3.38	3.07
63 × 170	4.42	4.26	3.89	4.21	4.06	3.70	4.04	3.89	3.54
63 × 195	5.05	4.87	4.45	4.81	4.64	4.24	4.62	4.45	4.06
63 × 220	5.67	5.47	5.00	5.41	5.22	4.76	5.19	5.00	4.56
75 × 122	3.33	3.26	2.97	3.14	3.08	2.83	2.99	2.93	2.71
75 × 147	4.05	3.91	3.57	3.86	3.72	3.40	3.71	3.57	3.25
75 × 170	4.66	4.50	4.11	4.45	4.29	3.92	4.27	4.11	3.75
75 × 195	5.32	5.14	4.70	5.08	4.90	4.48	4.88	4.70	4.29
75 × 220	5.97	5.77	5.28	5.70	5.50	5.04	5.48	5.28	4.83
38 × 140	2.99	2.94	2.72	2.82	2.77	2.59	2.69	2.63	2.47
38 × 184	4.07	3.92	3.57	3.87	3.73	3.39	3.71	3.57	3.24

AD A1/2, Appendix A

Table A21 Joists for flat roofs with access not limited to the purposes of maintenance or repair. Imposed loading 1.50 kN/m².

Maximum clear span of joist (m) Timber of strength class SC3 (see Table 1)

Size of joist (mm × mm)	Dead Load [kN/m²] excluding the self weight of the joist								
	Not more than 0.50			More than 0.50 but not more than 0.75			More than 0.75 but not more than 1.00		
	Spacing of joists (mm)								
	400	450	600	400	450	600	400	450	600
38 × 122	1.80	1.79	1.74	1.74	1.71	1.65	1.68	1.65	1.57
38 × 147	2.35	2.33	2.27	2.27	2.25	2.18	2.21	2.18	2.09
38 × 170	2.88	2.85	2.77	2.77	2.74	2.64	2.68	2.64	2.53
38 × 195	3.47	3.43	3.29	3.33	3.28	3.16	3.21	3.16	3.02
38 × 220	4.08	4.03	3.71	3.90	3.84	3.56	3.75	3.68	3.43
47 × 122	2.00	1.99	1.94	1.94	1.93	1.87	1.89	1.87	1.81
47 × 147	2.60	2.58	2.51	2.51	2.48	2.40	2.44	2.40	2.31
47 × 170	3.18	3.14	3.06	3.06	3.02	2.91	2.95	2.91	2.78
47 × 195	3.82	3.78	3.54	3.66	3.61	3.40	3.52	3.46	3.28
47 × 220	4.48	4.38	3.99	4.27	4.20	3.83	4.10	4.03	3.70
50 × 122	2.06	2.05	2.00	2.00	1.98	1.93	1.95	1.93	1.86
50 × 147	2.68	2.65	2.59	2.59	2.56	2.47	2.51	2.47	2.38
50 × 170	3.27	3.23	3.14	3.14	3.10	2.99	3.04	2.99	2.86
50 × 195	3.93	3.88	3.61	3.76	3.70	3.47	3.62	3.56	3.35
50 × 220	4.60	4.47	4.07	4.38	4.30	3.91	4.21	4.13	3.78
63 × 97	1.67	1.66	1.63	1.63	1.61	1.57	1.59	1.57	1.53
63 × 122	2.31	2.29	2.24	2.24	2.21	2.15	2.17	2.15	2.07
63 × 147	2.98	2.95	2.87	2.87	2.84	2.74	2.78	2.74	2.63
63 × 170	3.62	3.59	3.41	3.48	3.43	3.28	3.36	3.30	3.16
63 × 195	4.34	4.29	3.90	4.15	4.08	3.75	3.99	3.92	3.62
63 × 220	5.00	4.82	4.39	4.82	4.64	4.22	4.62	4.48	4.08
75 × 122	2.50	2.48	2.42	2.42	2.40	2.32	2.35	2.32	2.24
75 × 147	3.23	3.19	3.11	3.11	3.07	2.96	3.00	2.96	2.84
75 × 170	3.91	3.87	3.61	3.75	3.69	3.47	3.61	3.55	3.35
75 × 195	4.66	4.53	4.13	4.45	4.36	3.97	4.28	4.20	3.84
75 × 220	5.28	5.09	4.65	5.09	4.90	4.47	4.92	4.74	4.32
38 × 140	2.19	2.17	2.12	2.12	2.10	2.04	2.07	2.04	1.94
38 × 184	3.21	3.17	3.08	3.08	3.04	2.93	2.98	2.93	2.80

AD A1/2, Appendix A

Table A22 Joists for flat roofs with access not limited to the purposes of maintenance or repair. Imposed loading 1.50 kN/m².

Maximum clear span of joist (m) Timber of strength class SC4 (see Table 1)

Size of joist (mm × mm)	Dead Load [kN/m²] excluding the self weight of the joist								
	Not more than 0.50			More than 0.50 but not more than 0.75			More than 0.75 but not more than 1.00		
	Spacing of joists (mm)								
	400	450	600	400	450	600	400	450	600
38 × 122	1.91	1.90	1.86	1.86	1.84	1.79	1.81	1.79	1.73
38 × 147	2.49	2.46	2.40	2.40	2.38	2.30	2.33	2.30	2.21
38 × 170	3.04	3.01	2.93	2.93	2.89	2.79	2.83	2.79	2.67
38 × 195	3.66	3.62	3.43	3.51	3.46	3.29	3.38	3.33	3.18
38 × 220	4.30	4.25	3.86	4.10	4.04	3.71	3.94	3.87	3.58
47 × 122	2.12	2.10	2.06	2.06	2.04	1.98	2.00	1.98	1.91
47 × 147	2.75	2.73	2.66	2.66	2.62	2.54	2.57	2.54	2.44
47 × 170	3.35	3.32	3.22	3.22	3.18	3.06	3.11	3.06	2.93
47 × 195	4.03	3.98	3.68	3.85	3.80	3.54	3.71	3.64	3.42
47 × 220	4.71	4.56	4.15	4.49	4.39	3.99	4.31	4.23	3.85
50 × 122	2.19	2.17	2.12	2.12	2.10	2.04	2.06	2.04	1.97
50 × 147	2.83	2.81	2.73	2.73	2.70	2.61	2.65	2.61	2.51
50 × 170	3.45	3.41	3.28	3.31	3.27	3.15	3.20	3.15	3.01
50 × 195	4.14	4.09	3.76	3.96	3.90	3.61	3.81	3.74	3.49
50 × 220	4.83	4.65	4.23	4.61	4.47	4.07	4.42	4.32	3.93
63 × 97	1.77	1.75	1.72	1.72	1.71	1.66	1.68	1.66	1.61
63 × 122	2.44	2.42	2.36	2.36	2.34	2.27	2.30	2.27	2.18
63 × 147	3.15	3.12	3.03	3.03	2.99	2.89	2.93	2.89	2.77
63 × 170	3.82	3.78	3.54	3.66	3.61	3.41	3.53	3.47	3.29
63 × 195	4.56	4.45	4.06	4.36	4.29	3.90	4.19	4.11	3.77
63 × 220	5.19	5.00	4.56	5.00	4.82	4.39	4.84	4.66	4.24
75 × 122	2.64	2.62	2.56	2.56	2.53	2.45	2.48	2.45	2.36
75 × 147	3.40	3.36	3.25	3.27	3.23	3.11	3.16	3.11	2.98
75 × 170	4.11	4.07	3.75	3.94	3.88	3.61	3.79	3.73	3.49
75 × 195	4.79	4.70	4.29	4.67	4.53	4.13	4.49	4.38	3.99
75 × 220	5.48	5.28	4.83	5.28	5.09	4.65	5.11	4.93	4.49
38 × 140	2.32	2.30	2.25	2.25	2.22	2.16	2.19	2.16	2.08
38 × 184	3.39	3.35	3.24	3.25	3.21	3.09	3.14	3.09	2.95

AD A1/2, Appendix A

Table A23 Purlins supporting sheeting or decking for roofs having a pitch more than 10° but not more than 35°. Imposed loading 0.75 kN/m².

Maximum clear span of purlin (m) Timber of strength class SC3 and SC4 (see Table 1)

	Dead Load [kN/m²] excluding the self weight of the purlin																	
	Not more than 0.25						More than 0.25 but not more than 0.50						More than 0.50 but not more than 0.75					
	Spacing of purlins (mm)																	
Size of purlin (mm × mm)	900	1200	1500	1800	2100	2400	900	1200	1500	1800	2100	2400	900	1200	1500	1800	2100	2400
SC3																		
50 × 100	1.68	1.63	1.51	1.42	1.34	1.28	1.55	1.48	1.40	1.31	1.24	1.18	1.45	1.37	1.31	1.22	1.16	1.10
50 × 125	2.24	2.03	1.88	1.77	1.67	1.60	2.06	1.88	1.74	1.63	1.54	1.47	1.91	1.77	1.63	1.53	1.44	1.37
50 × 150	2.68	2.44	2.26	2.12	2.01	1.91	2.49	2.26	2.09	1.96	1.85	1.76	2.34	2.12	1.96	1.83	1.73	1.65
50 × 175	3.12	2.84	2.63	2.47	2.34	2.23	2.90	2.63	2.43	2.28	2.16	2.06	2.72	2.47	2.28	2.13	2.02	1.92
50 × 200	3.56	3.24	3.00	2.82	2.67	2.55	3.31	3.00	2.78	2.60	2.46	2.35	3.11	2.81	2.60	2.44	2.30	2.19
50 × 225	4.00	3.63	3.37	3.17	3.00	2.86	3.71	3.37	3.12	2.93	2.77	2.64	3.49	3.16	2.92	2.74	2.59	2.47
63 × 100	1.87	1.77	1.64	1.54	1.46	1.39	1.72	1.64	1.51	1.42	1.34	1.28	1.60	1.52	1.42	1.33	1.26	1.20
63 × 125	2.42	2.20	2.04	1.92	1.82	1.73	2.25	2.04	1.89	1.77	1.68	1.60	2.10	1.91	1.77	1.66	1.57	1.50
63 × 150	2.90	2.63	2.44	2.30	2.18	2.08	2.69	2.44	2.26	2.12	2.01	1.92	2.53	2.29	2.12	2.00	1.88	1.79
63 × 175	3.37	3.07	2.85	2.67	2.54	2.42	3.13	2.84	2.63	2.47	2.34	2.23	2.94	2.67	2.47	2.32	2.19	2.09
63 × 200	3.84	3.50	3.25	3.05	2.89	2.76	3.57	3.24	3.01	2.82	2.67	2.55	3.36	3.05	2.82	2.65	2.51	2.39
63 × 225	4.31	3.92	3.64	3.43	3.25	3.10	4.01	3.64	3.38	3.17	3.01	2.87	3.77	3.42	3.17	2.97	2.82	2.68
SC4																		
50 × 100	1.79	1.71	1.58	1.48	1.40	1.34	1.64	1.57	1.46	1.37	1.30	1.23	1.53	1.45	1.37	1.28	1.21	1.15
50 × 125	2.34	2.13	1.97	1.85	1.75	1.67	2.17	1.97	1.82	1.71	1.62	1.54	2.02	1.85	1.71	1.60	1.51	1.44
50 × 150	2.80	2.55	2.36	2.22	2.10	2.00	2.60	2.36	2.18	2.05	1.94	1.85	2.44	2.21	2.05	1.92	1.81	1.73
50 × 175	3.26	2.97	2.75	2.58	2.45	2.34	3.03	2.75	2.54	2.39	2.26	2.15	2.85	2.58	2.39	2.24	2.12	2.01
50 × 200	3.72	3.38	3.14	2.95	2.79	2.67	3.45	3.13	2.90	2.73	2.58	2.46	3.25	2.94	2.72	2.55	2.42	2.30
50 × 225	4.17	3.80	3.52	3.31	3.14	3.00	3.88	3.52	3.26	3.06	2.90	2.77	3.65	3.31	3.06	2.87	2.72	2.59
63 × 100	1.99	1.84	1.71	1.61	1.52	1.45	1.81	1.71	1.58	1.49	1.41	1.34	1.69	1.60	1.48	1.39	1.32	1.26
63 × 125	2.53	2.30	2.13	2.00	1.90	1.81	2.35	2.13	1.97	1.85	1.76	1.68	2.21	2.00	1.85	1.74	1.65	1.57
63 × 150	3.02	2.75	2.55	2.40	2.28	2.17	2.81	2.55	2.37	2.22	2.10	2.01	2.64	2.40	2.22	2.08	1.97	1.88
63 × 175	3.52	3.20	2.97	2.80	2.65	2.53	3.27	2.97	2.76	2.59	2.45	2.34	3.08	2.79	2.59	2.43	2.30	2.19
63 × 200	4.01	3.65	3.39	3.19	3.03	2.89	3.73	3.39	3.14	2.95	2.80	2.67	3.51	3.19	2.95	2.77	2.62	2.50
63 × 225	4.49	4.10	3.81	3.58	3.40	3.25	4.18	3.80	3.53	3.32	3.15	3.00	3.94	3.58	3.32	3.11	2.95	2.81

AD A1/2, Appendix A

Table A24 Purlins supporting sheeting or decking for roofs having a pitch more than 10° but not more than 35°. Imposed loading 1.0 kN/m².

Maximum clear span of purlin (m) Timber of strength class SC3 and SC4 (see Table 1)

Dead Load [kN/m²] excluding the self weight of the purlin

Size of purlin (mm × mm)	Not more than 0.25						More than 0.25 but not more than 0.50						More than 0.50 but not more than 0.75					
	Spacing of purlins (mm)																	
	900	1200	1500	1800	2100	2400	900	1200	1500	1800	2100	2400	900	1200	1500	1800	2100	2400
SC3																		
50 × 100	1.67	1.51	1.40	1.31	1.24	1.18	1.55	1.42	1.31	1.22	1.16	1.10	1.45	1.34	1.24	1.16	1.09	1.04
50 × 125	2.08	1.88	1.74	1.64	1.55	1.47	1.95	1.77	1.63	1.53	1.45	1.38	1.85	1.67	1.54	1.44	1.36	1.30
50 × 150	2.49	2.26	2.09	1.96	1.85	1.77	2.34	2.12	1.96	1.83	1.73	1.65	2.22	2.00	1.85	1.73	1.64	1.56
50 × 175	2.90	2.63	2.43	2.28	2.16	2.06	2.73	2.47	2.28	2.14	2.02	1.92	2.58	2.34	2.16	2.02	1.91	1.81
50 × 200	3.31	3.00	2.78	2.61	2.47	2.35	3.11	2.82	2.60	2.44	2.31	2.20	2.95	2.67	2.46	2.31	2.18	2.07
50 × 225	3.72	3.37	3.12	2.93	2.77	2.64	3.49	3.16	2.93	2.74	2.59	2.47	3.31	3.00	2.77	2.59	2.45	2.31
63 × 100	1.80	1.64	1.51	1.42	1.35	1.28	1.69	1.54	1.42	1.33	1.26	1.20	1.60	1.45	1.34	1.26	1.19	1.13
63 × 125	2.25	2.04	1.89	1.77	1.68	1.60	2.11	1.92	1.77	1.66	1.57	1.50	2.00	1.81	1.68	1.57	1.49	1.41
63 × 150	2.69	2.44	2.26	2.13	2.01	1.92	2.53	2.29	2.12	1.99	1.88	1.80	2.40	2.17	2.01	1.88	1.78	1.70
63 × 175	3.13	2.85	2.64	2.48	2.35	2.24	2.95	2.67	2.47	2.32	2.20	2.09	2.80	2.53	2.34	2.20	2.08	1.98
63 × 200	3.57	3.25	3.01	2.83	2.68	2.55	3.36	3.05	2.82	2.65	2.51	2.39	3.19	2.89	2.67	2.51	2.37	2.26
63 × 225	4.01	3.65	3.38	3.18	3.01	2.87	3.77	3.43	3.17	2.98	2.82	2.69	3.58	3.25	3.01	2.82	2.67	2.54
SC4																		
50 × 100	1.74	1.58	1.46	1.37	1.30	1.24	1.64	1.48	1.37	1.28	1.21	1.16	1.53	1.40	1.30	1.21	1.15	1.09
50 × 125	2.17	1.97	1.82	1.71	1.62	1.54	2.04	1.85	1.71	1.60	1.52	1.44	1.94	1.75	1.62	1.51	1.43	1.36
50 × 150	2.60	2.36	2.19	2.05	1.94	1.85	2.45	2.22	2.05	1.92	1.82	1.73	2.32	2.10	1.94	1.82	1.72	1.63
50 × 175	3.03	2.75	2.55	2.39	2.26	2.16	2.85	2.58	2.39	2.24	2.12	2.02	2.70	2.45	2.26	2.12	2.00	1.90
50 × 200	3.46	3.14	2.91	2.73	2.58	2.46	3.25	2.95	2.73	2.56	2.42	2.30	3.08	2.79	2.58	2.42	2.28	2.17
50 × 225	3.88	3.52	3.27	3.07	2.90	2.77	3.65	3.31	3.06	2.87	2.72	2.59	3.46	3.14	2.90	2.72	2.57	2.44
63 × 100	1.89	1.71	1.58	1.49	1.41	1.34	1.77	1.61	1.49	1.39	1.32	1.26	1.68	1.52	1.41	1.32	1.25	1.19
63 × 125	2.35	2.13	1.98	1.86	1.76	1.68	2.21	2.00	1.85	1.74	1.65	1.57	2.10	1.90	1.76	1.65	1.56	1.48
63 × 150	2.81	2.55	2.37	2.22	2.11	2.01	2.65	2.40	2.22	2.08	1.97	1.88	2.51	2.27	2.10	1.97	1.87	1.78
63 × 175	3.27	2.97	2.76	2.59	2.46	2.34	3.08	2.79	2.59	2.43	2.30	2.19	2.92	2.65	2.45	2.30	2.18	2.07
63 × 200	3.73	3.39	3.15	2.96	2.80	2.67	3.51	3.19	2.95	2.77	2.63	2.50	3.33	3.02	2.80	2.63	2.48	2.37
63 × 225	4.18	3.81	3.53	3.32	3.15	3.01	3.94	3.58	3.32	3.12	2.95	2.81	3.74	3.30	3.15	2.95	2.79	2.66

Strutting of joists

Where floor joists span more than 2.4m they should be strutted with one or more rows of:

(a) solid timber at least 38 mm wide and 0.75 times the joist depth; *or*,
(b) herringbone strutting in 38 mm × 38 mm timber except where the distance between the joists is greater than three times the joist depth.

In this latter case the alternatives to timber herringbone strutting are not specified and it is not clear if solid timber strutting would be recommended or whether proprietary steel herringbone strutting, for example, could be used.

One row of strutting at mid-span is recommended for joist spans between 2.5 m and 4.5 m. Above 4.5 m, two rows of strutting at the one third positions would be required.

Structural work of bricks, blocks and plain concrete

If a wall of these materials comes within the scope of Section 1C of AD A1/2, it is not necessary to calculate loads or wall thicknesses, provided the wall is built with the thicknesses required by Section 1C and complies with the rules therein and in all other respects complies with BS 5628 *Code of practice for use of masonry*, Part 1: 1978 *Structural use of unreinforced masonry* and Part 3: 1985 *Materials and components, design and workmanship*.

Section 1C may be applied to any wall which is:

(a) an external wall, compartment wall, internal load-bearing wall or separating wall of a residential building of not more than three storeys, *and*,

AD A1/2
sec. 1C
1C1 & 1C2

(b) an external wall or internal load-bearing wall of a small single-storey non-residential building or small annexe to a residential building (such as a garage or outbuilding) *provided that*:

AD A1/2
sec. 1C
1C3 & 1C4

(i) the building design complies with the requirements of Paragraphs 1C14 to 1C17 of Section 1C, *and*,
(ii) the wall construction details comply with the requirements of Paragraphs 1C18 to 1C39 of Section 1C.

Building design requirements (Section 1C, Paragraphs 1C14 to 1C17)

These are concerned with the design wind speed, the imposed load, the building proportions and the plan area of each storey or sub-division.

DESIGN WIND SPEED (V_s). When determined in accordance with CP3: Chapter V *Loading*, Part 2: 1972 *Wind loads*, this should not exceed 44 metres/second.

In order to determine the value of V_s it is necessary to carry out a fairly complicated calculation from CP3 which relates the design wind speed to the basic wind speed adjusted to take account of site conditions and the design of the building. This has been simplifed in AD A1/2, the design wind speed being obtained using the following procedure:

(a) Use Diagram 10 (Map showing basic wind speeds in m/s) to determine the basic wind speed for the building by reference to its location.
(b) Look up the maximum building height permitted in either Table 8 (Maximum height of buildings on normal or slightly sloping sites) or Table 9 (Maximum height of buildings on steeply sloping sites, including hill, cliff and escarpment sites) on the line which corresponds to the basic wind speed.

The height obtained should not be exceeded in the building design.
   Tables 8 and 9, and Diagram 10 are reproduced from AD A1/2.

AD A1/2
sec. 1C
1C17

IMPOSED LOADS. These should not exceed:

(a) on any floor, 2.0 kN/m^2 distributed;
(b) on any ceiling, 0.25 kN/m^2 distributed and 0.9 kN concentrated; *and,*
(c) on any roof, 1.00 kN/m^2 for spans not exceeding 12 m, or 1.5 kN/m^2 for spans not exceeding 6 m.

AD A1/2
sec. 1C
1C16

BUILDING PROPORTIONS. For residential buildings of not more than three storeys:

(a) the height of any part of a wall or roof of the building should not exceed 15 m, as measured from the lowest finished surface of the ground adjoining the building;
(b) the width of the building should not be less than at least half the height of the building;
(c) any wing of the building which projects more than twice its own width from the remainder of the building should have a width at least equal to half its height. ('Height' is measured to the highest part of any roof or wall of the building or wing.)

AD A1/2

Table **8** **Maximum height of buildings on normal or slightly sloping sites**

| | Maximum building height in metres | | | |
| | Location | | | |
Basic wind speed m/s	Unprotected sites, open countryside with no obstructions	Open countryside with scattered windbreaks	Country with many windbreaks, small towns, outskirts of large cities	Protected sites, city centres
36	15	15	15	15
38	15	15	15	15
40	15	15	15	15
42	15	15	15	15
44	15	15	15	15
46	11	15	15	15
48	9	13	15	15

AD A1/2

Diagram **10**   **Map showing basic wind speeds in m/s**

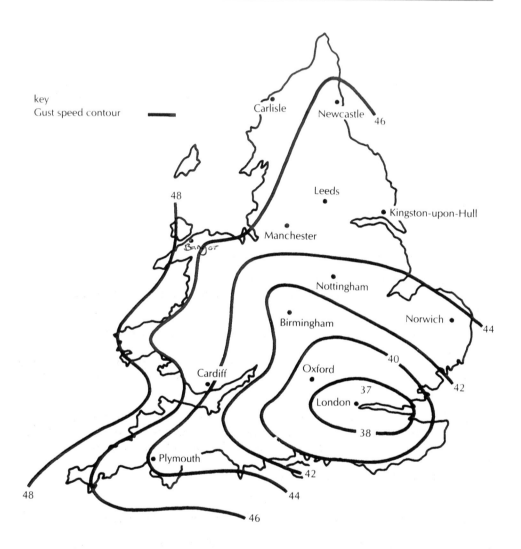

key
Gust speed contour ▬▬

Notes:

1   Maximum gust speed likely to be exceeded on the average only once in 50 years at 10 m above ground in open level country.

2   Contour lines are drawn at 2 m/s intervals.

AD A1/2

Table **9**   **Maximum height of buildings on steeply sloping sites, including hill, cliff and escarpment sites**

Basic wind speed m/s	Maximum building height in metres			
	Location			
	Unprotected sites, open countryside with no obstructions	Open countryside with scattered windbreaks	Country with many windbreaks, small towns, outskirts of large cities	Protected sites, city centres
36	8	11	15	15
38	6	9	15	15
40	4	7.5	14	15
42	3	6	12	15
44	0*	5	10	15
46	0*	4	8	15
48	0*	3	6.5	14

* Section 1C guidance is not applicable.

For small single-storey non-residential buildings:

(a) the height of the building should not exceed 3 m;
(b) the width of the building measured in the direction of the roof span should not exceed 9 m.

For annexes attached to residential buildings the height of any part should not exceed 3 m.

The heights mentioned above may need to be reduced in line with the design wind speed requirement of 44 metres/second maximum. (See above, Tables 8 and 9 and Diagram 10.)

AD A1/2
sec. 1C
1C14

PLAN AREA OF STOREY. The plan area of each storey which is completely bounded by structural walls on all sides should not be more than 70 m². However, if the storey is bounded in this way on all sides but one, the limiting area is 30 m².

AD A1/2
sec. 1C
1C15

These requirements are summarised in Figs. 6.3 and 6.4.

Wall construction requirements (Section 1C, Paragraphs 1C18 to 1C39)

These are concerned with height and length, materials, buttressing, loading conditions, openings and recesses, and lateral support.

Height and length

The height or length of a wall should not be more than 12m and together with storey heights should be measured in accordance with the following rules:

● The height of the ground storey of a building is measured from the base of the wall to the underside of the next floor above.

Residential − not more than 3 storeys excluding basements

design wind speed $V_s > 44$ m/s

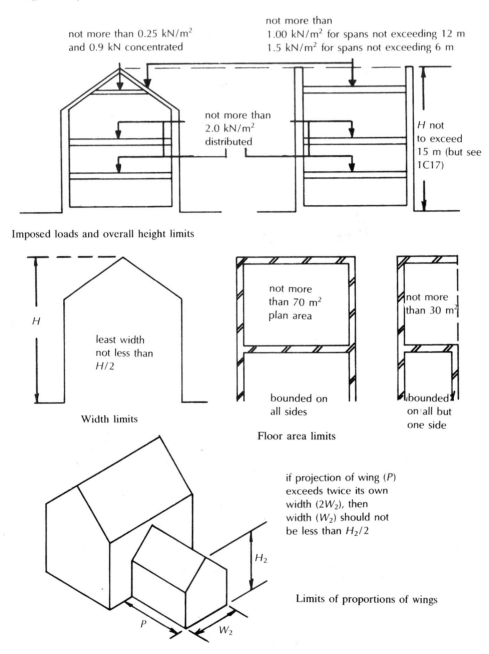

Imposed loads and overall height limits

Width limits

Floor area limits

Limits of proportions of wings

Fig. 6.3 Building design requirements for residential buildings not exceeding three storeys in height.

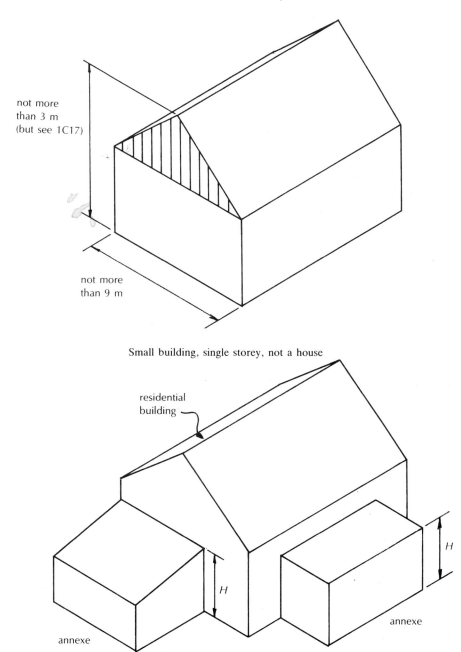

Fig. 6.4 Building design requirements for small non-residential buildings and annexes.

● The height of an upper storey is measured from the level of the underside of the floor of that storey, in each case to the level of the underside of the next floor above.
● For a top storey which comprises a gable wall, measure to a level midway between the gable base and the top of the roof lateral support along the line of the roof slope, but if there is also lateral support at about ceiling level, to the level of that lateral support.
● Where a compartment or separating wall (as defined) comprises a gable, measure the height from its base to the base of the gable.
● Any other gable wall (except a compartment or separating wall) should be measured from its base to half the height of the gable.
● Any wall which is not a gable wall should be measured from its base to its highest part, excluding any parapet not exceeding 1.2 m in height.
AD A1/2
sec. 1C
1C18 &
1C19
● Walls are regarded as being divided into separate lengths by securely tied buttressing walls, piers or chimneys for the purposes of measuring their length. These separate lengths are measured centre to centre of the piers, etc. These special requirements are noted in Fig. 6.5.

Materials and workmanship

BRICKS AND BLOCKS. The wall should be constructed of bricks or blocks, properly bonded and solidly put together with mortar. The materials should comply with the following standards:

● Clay bricks or blocks to BS 3921: 1985 or BS 6649: 1985.
● Calcium silicate bricks to BS 187: 1978 or BS 6649: 1985.
● Concrete bricks or blocks to BS 6073: Part 1: 1981.
● Square dressed natural stone to BS 5390: 1976 (1984).

Additionally, bricks or blocks should have a compressive strength of not less than the following:

(a) when used in any part of a wall with storey heights not exceeding 2.7 m (except in the outer leaf of an external cavity wall of a three-storey building), 5 N/mm^2 for bricks and 2.8 N/mm^2 for blocks.
(b) when used in the inner leaf of a ground storey external cavity wall of a three-storey building, 15 N/mm^2 for bricks and 7 N/mm^2 for blocks.
(c) when used in any circumstances other than those described in (a) or (b) above, 7 N/mm^2.

The above guidance to compressive strengths of blocks or bricks is only applicable where the roof structure is of timber construction. Additionally, the ground storey part of an internal wall in a three-storey building should be at least 140 mm thick if in blockwork or 215 mm thick if in brickwork.
AD A1/2
sec. 1C
1C22
It should be noted that in determining the ground storey height for the purposes of (a) above, the measurement is made from the upper surface of the ground floor and not from the base of the wall.

MORTAR. The mortar used in any wall to which Section 1C of AD A1/2 applies should be at least equal in strength to a 1:1:6 Portland cement/lime/fine aggregate mortar measured by volume of dry materials, or to the proportions

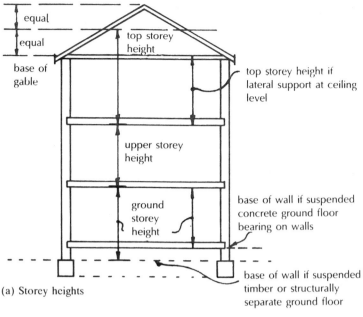

(a) Storey heights

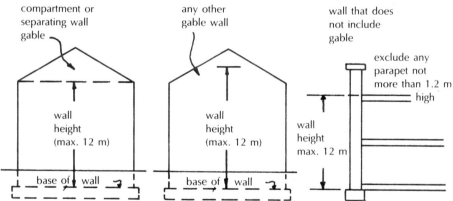

(b) Wall height

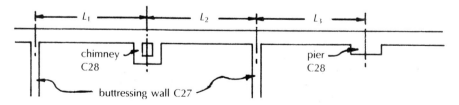

division of wall into separate effective lengths on plan: L_1, L_2, L_3 — each not more than 12 m

(c) Wall length

Fig. 6.5 Rules for measurement, 1C18, 1C19.

given in BS 5628 *Code of practice for use of masonry*, Part 1: 1978 (1985) *Structural use of unreinforced masonry* for mortar designation (iii). The mortar should be compatible with the masonry units and the position of use.

WALL TIES. These should comply with BS 1243: 1978 *Specification for metal ties for cavity wall construction* unless conditions of severe exposure occur. In that case austenitic stainless steel or suitable non-ferrous ties should be used. (Severe exposure is defined in BS 5628: Part 3: 1985.)

Buttressing walls, piers and chimneys

Any load-bearing wall should be bonded or securely tied at each end to a buttressing wall, pier or chimney. (This does not apply to single leaf walls less than 2.5 m in height and length which form part of a small single-storey non-residential building or annexe.) These supporting elements should be of such dimensions as to provide effective lateral support over the full wall height from its base to its top.

If, additionally, such supporting elements are bonded or securely tied to the supporting wall at intermediate points in the length of the wall, then the wall may be regarded as being divided into separate distinct lengths by these buttressing walls, piers or chimneys. Each of the distinct lengths may then be regarded as a supported wall, and the length of any wall is the distance between adjacent supporting elements.

BUTTRESSING WALLS should have:

(a) one end bonded or securely tied to the supported wall;
(b) the other end bonded or securely tied to another buttressing wall, pier or chimney;
(c) no opening or recess greater than 0.1 m in area within a horizontal distance of 550 mm from the junction with the supported wall, and openings and recesses generally disposed so as not to impair the supporting effect of the buttressing wall;
(d) a length of not less than one sixth of the height of the supported wall;
(e) the minimum thickness required by the appropriate rule, according to whether the buttressing wall is actually an external compartment, separating or internal load-bearing wall or a wall of a small building or annexe; but if the wall is none of these, then a thickness, t (see Fig. 6.6), of not less than the greater of:
   (i) half the thickness required of a solid external, compartment or separating wall of the same height and length as the buttressing wall, less 5 mm; *or,*
  (ii) if the buttressing wall is part of a dwellinghouse and the supported wall as a whole is not more than 6 m high and 10 m in length, 75 mm; *or,*
 (iii) in any other case, 90 mm (see Fig. 6.6).

PIERS may project on either or both sides of the supported wall and should:

(a) run from the base of the supported wall to the level of the roof lateral support, or to the top of the wall if there is no roof lateral support;

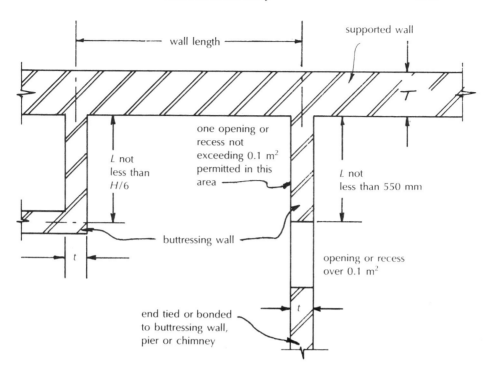

wall length

supported wall

one opening or
recess not
exceeding 0.1 m^2
permitted in this
area

L not
less than
H/6

L not
less than 550 mm

buttressing wall

opening or recess
over 0.1 m^2

t

end tied or bonded
to buttressing wall,
pier or chimney

t

t = thickness from 1C27
of Section 1C of AD/A1/2

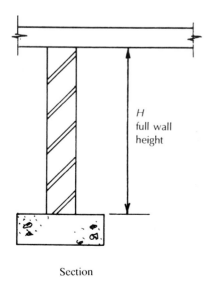

position and shape of
openings or recesses
should not impair lateral
support provided by buttressing
wall

H
full wall
height

Section

Fig. 6.6 Buttressing walls.

(b) have a thickness, measured at right angles to the length of the supported wall and including the thickness of that wall, of at least three times the thickness required of the supported wall; *and,*

(c) measure at least 190 mm in width (the measurement being parallel to the length of the supported wall).

CHIMNEYS should have:

(a) a horizontal cross-section area, excluding any fireplace opening or flue, of not less than the area required of a pier in the same wall; *and,*

(b) a thickness overall of at least twice the thickness required of the supported wall.

The requirements for piers and chimneys are shown in Fig. 6.7.

AD A1/2
sec. 1C
1C28
It should be noted that requirements in respect of plan dimensions of piers do not apply to piers in walls of small buildings and annexes, for which there are special rules (see Fig. 6.18).

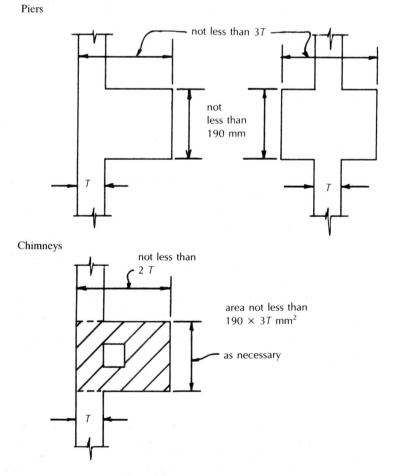

Fig. 6.7 Piers and chimneys.

Loading conditions

FLOOR SPANS. The wall should not support any floor members with a span of more than 6 m. (Span is measured centre to centre of bearings.)

LATERAL THRUST. Where the levels of the ground or oversite concrete on either side of a wall differ, the thickness of the wall as measured at the higher level should not be less than one quarter of the difference in level.

In the case of a cavity wall, the thickness is taken as the sum of the leaf thicknesses. However, if the cavity is filled with fine concrete, the overall thickness may be taken.

The lateral thrust occasioned in these circumstances is the only one which a wall must be expected to sustain, apart from that due to direct wind load and the transmission of wind load.

VERTICAL LOADING. The total dead and imposed load transmitted by a wall at its base should not exceed 70 kN/m. All vertical loads carried by a wall should be properly distributed. This may be assumed for pre-cast concrete floors, concrete floor slabs and timber floors complying with Section 18 of ADA1/2. Distributed loading may also be assumed for lintels with a bearing length of 150 mm or more. Where the clear span of the lintel is 1200 mm or less the bearing length may be reduced to 100 mm.

AD A1/2
sec. 1C
1C24 &
1C25

These requirements are summarised in Fig. 6.8.

Openings and recesses

Openings or recesses in a wall should not be placed in such a manner as to impair the stability of any part of it. Adequate support for the superstructure should be provided over every opening and recess.

As a general rule, any opening or recess in a wall should be flanked on each side by a length of wall equal to at least one sixth of the width of the opening or recess, in order to provide the required stability. Accordingly, the minimum length of wall between two openings or recesses should not be less than one sixth of the *combined* width of the two openings or recesses.

However, where long span roofs or floors bear onto a wall containing openings or recesses it may be necessary to increase the width of the flanking portions of wall. Table 10 of Section 1C of AD A1/2 (see above) contains factors that enable this to be done.

Where several openings and/or recesses are formed in a wall, their total width should, at any level, be not more than two thirds of the length of the wall at that level.

AD A1/2
sec. 1C
1C29 &
1C30

No opening or recess should exceed 3 m in length. These requirements are illustrated in Fig. 6.9.

Chases

The depth of vertical chases should not be more than one third the thickness of the wall, or in a cavity wall, one third the thickness of the leaf concerned. Depth of horizontal chases should be not more than one sixth the thickness of the wall or leaf. Chases should not be placed in such a manner as to impair the stability of the wall, particularly where hollow blocks are used (see Fig. 6.10).

AD A1/2
sec. 1C
1C31

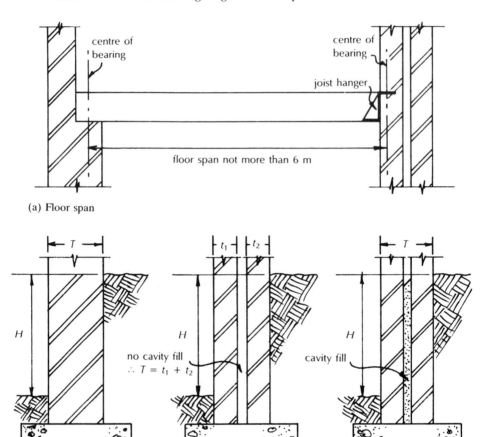

(a) Floor span

(b) Lateral thrust   *T* not less than *H*/4

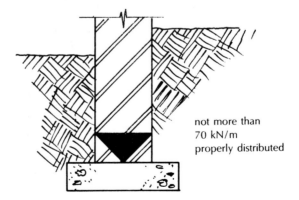

(c) Vertical loading

Fig. 6.8 Loading requirements.

AD A1/2, section 1C

Table **10**   **Value of factor 'X'**

Nature of roof span	Maximum roof span [m]	Minimum thickness of wall inner leaf [mm]	Span of floor is parallel to wall	Span of timber floor into wall		Span of concrete floor into wall	
				max 4.5 m	max 6.0 m	max 4.5 m	max 6.0 m
				Value of factor 'X'			
roof span parallel to wall	not applicable	100	6	6	6	6	6
		90	6	6	6	6	5
timber roof spans into wall	9	100	6	6	5	4	3
		90	6	4	4	3	3

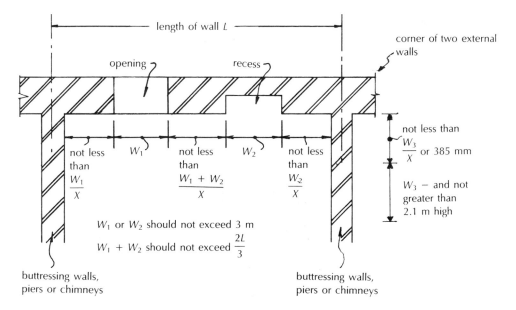

Note: value of X comes from Table 10 of Section 1C of AD A1/2 which is reproduced below OR it may be given the value 6 provided the compressive strength of the blocks or bricks (or cavity wall loaded leaf) is not less than 7 N/mm^2

Fig. 6.9 Openings and recesses.

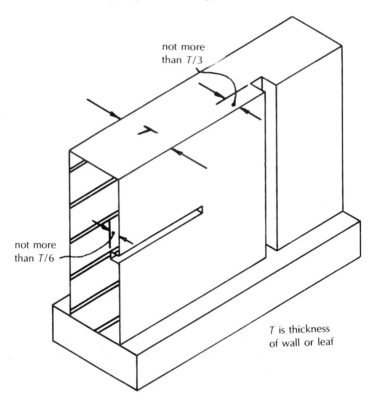

Fig. 6.10 Chases.

Overhanging

AD A1/2
sec. 1C
1C32

Where a wall overhangs a supporting structure beneath it, the amount of the overhang should not be such as to impair the stability of the wall. No limits are specified, but this would generally be interpreted as allowing an overhang of one third the thickness of the wall (see Fig. 6.11).

Lateral support

AD A1/2
sec. 1C
1C33 to
1C35

Floor or roof lateral support is horizontal support or stiffening, intended to stabilise or stiffen a wall by restraining its movement in a direction at right angles to the wall length. The restraint or support is provided by connecting a floor or roof to the wall in such a way that the floor or roof acts as a stiffening frame or diaphragm, transferring the lateral forces to walls, buttressing walls, piers or chimneys.

ROOF LATERAL SUPPORT. This should be provided for all external, compartment, separating and internal load-bearing walls irrespective of their length, at the point of junction between the roof and supported wall (i.e. at eaves level and along the verges).

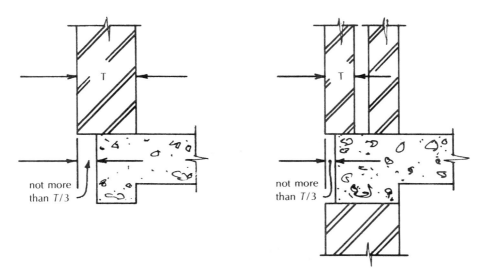

Fig. 6.11 Overhanging.

Walls should be strapped to roofs at not exceeding 2 m centres using galvanised mild steel or other durable metal straps, with a minimum cross section of 30 mm × 5 mm, and a minimum length of 1 m for eaves strapping.

Eaves strapping need not be provided for a roof which:

(a) has a pitch of 15° or more;
(b) is tiled or slated.;
(c) is of a type known by local experience as being resistant to damage by wind gusts;
(d) has main timber members spanning onto the supported wall at intervals of not more than 1.2 m.

Figure 6.12 shows methods of providing satisfactory lateral support at separating or gabled end wall positions.

AD A1/2 sec. 1C 1C37

FLOOR LATERAL SUPPORT. This should be provided for any external, compartment or separating wall which exceeds 3 m in length.

It should also be provided for any internal load-bearing wall (which is not a compartment or separating wall) at the top of each storey, irrespective of its length.

Walls should be strapped to floors above ground level at not exceeding 2 m centres using galvanised mild steel or other durable metal straps, with a minimum cross section of 30 mm × 5 mm.

There are certain cases where, because of the nature of the floor construction, it is not necessary to provide restraint straps:

● where a floor forms part of a house having not more than two storeys and:

(a) has timber members spanning so as to penetrate into the supported wall at intervals of not more than 1.2 m with at least 90 mm bearing directly on the walls or 75 mm bearing onto a timber wall plate; *or,*

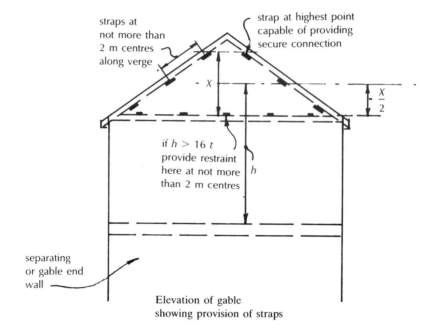

Elevation of gable
showing provision of straps

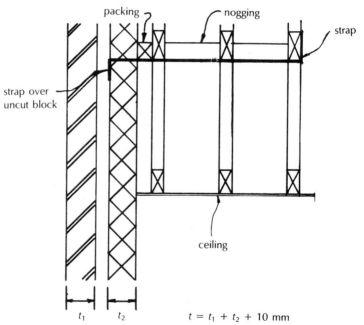

Section through gable at roof level showing method of strapping

Fig. 6.12 Lateral support for roof.

fix each joist to wall plate with
framing anchor or skew nails

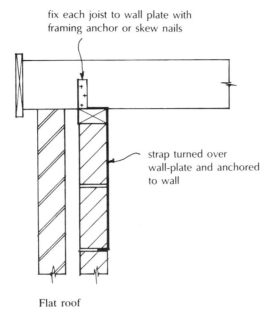

strap turned over
wall-plate and anchored
to wall

Flat roof

fix rafter to wall-plate with
truss clip or framing anchor

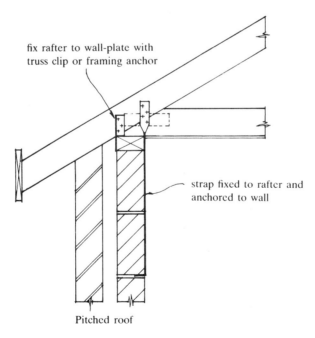

strap fixed to rafter and
anchored to wall

Pitched roof

(b) Section at eaves level showing method of strapping

Fig. 6.12 (Contd).

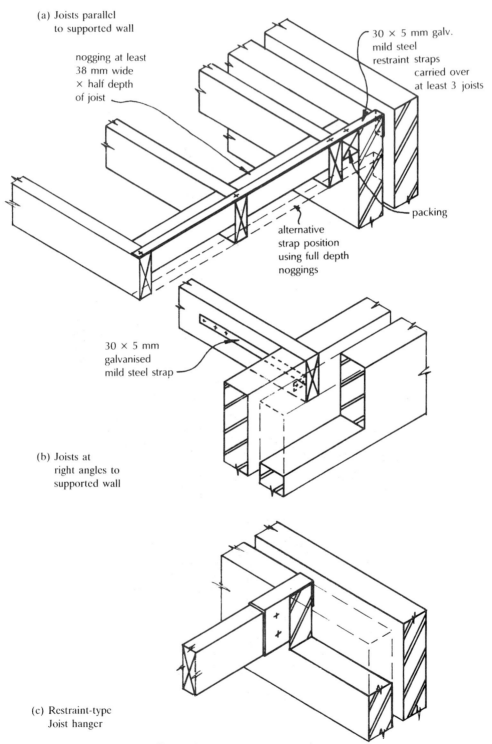

(a) Joists parallel
    to supported wall

nogging at least
38 mm wide
× half depth
of joist

30 × 5 mm galv.
mild steel
restraint straps
carried over
at least 3 joists

packing

alternative
strap position
using full depth
noggings

30 × 5 mm
galvanised
mild steel strap

(b) Joists at
    right angles to
    supported wall

(c) Restraint-type
    Joist hanger

Fig. 6.13 Floor lateral support.

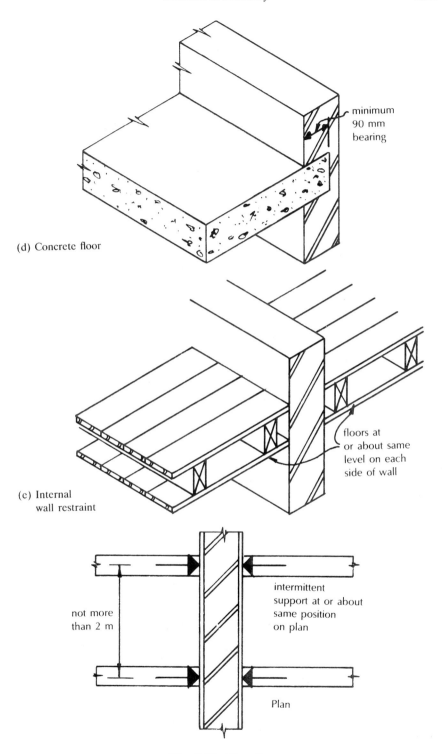

(d) Concrete floor

minimum
90 mm
bearing

floors at
or about same
level on each
side of wall

(e) Internal
wall restraint

not more
than 2 m

intermittent
support at or about
same position
on plan

Plan

Fig. 6.13 (Contd).

(b) the joists are carried on the supported wall by *restraint* type joist hangers, described in BS 5628: Part 1, at not more than 2 m centres.

- where a concrete floor has a bearing onto the supported wall of at least 90 mm.
- where two floors are at or about the same level on either side of a supported wall, contact between floors and wall may be continuous or intermittent. If intermittent, the points of contact should be at or about the same positions on plan at intervals not exceeding 2 m. Figure 6.13 summarises these provisions.

AD A1/2
sec. 1C
1C36

Interruption of lateral support

It is clear that in certain circumstances it may be necessary to interrupt the continuity of lateral support for a wall. This occurs chiefly where a stairway or similar structure adjoins a supported wall and necessitates the formation of an opening in a floor or roof.

This is permitted provided certain precautions are taken:

- The opening extends for a distance not exceeding 3 m measured parallel to the supported wall.
- If the connection between wall and floor or roof is provided by means of anchors, these should be spaced closer than 2 m on either side of the opening so as to result in the same number of anchors being used as if there were no opening.
- Other forms of connection should be provided throughout the length of each part of the wall on either side of the opening.
- There should be no other interruption of lateral support (see Fig. 6.14).

AD A1/2
sec. 1C
1C38

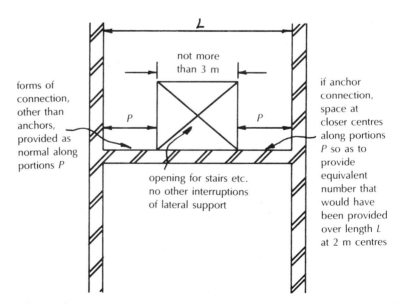

Fig. 6.14 Interruption of lateral support.

Thickness of walls

Provided the building design and wall construction requirements discussed above are satisfied, it is permissible to determine the thickness of a wall without calculation.

The minimum thicknesses required depend upon the wall height and length, and the rules applying to walls of bricks or blocks are set out in Table 5 of section 1C of ADA1/2 (see below) and illustrated in Fig. 6.15.

AD A1/2, section 1C

Table **5**   **Minimum thickness of certain external walls, compartment walls and separating walls**

(1) Height of wall	(2) Length of wall	(3) Minimum thickness of wall
Not exceeding 3.5 m ...	Not exceeding 12 m ...	190 mm for the whole of its height
Exceeding 3.5 m but not exceeding 9 m	Not exceeding 9 m ...	190 mm for the whole of its height
	Exceeding 9 m but not exceeding 12 m	290 mm from the base for the height of one storey, and 190 mm for the rest of its height
Exceeding 9 m but not exceeding 12 m	Not exceeding 9 m ...	290 mm from the base for the height of one storey, and 190 mm for the rest of its height
	Exceeding 9 m but not exceeding 12 m	290 mm from the base for the height of two storeys, and 190 mm for the rest of its height

These thicknesses do not apply to parapet walls, for which there are special rules (see below) or to bays, and gables over bay windows above the level of the lowest window sill. *(AD A1/2 sec. 1C 1C4 & 1C5)*

As a general rule, the thickness of any storey of a brick or block wall should not be less than one sixteenth of the height of that storey. However, walls of stone, flints, clunches of bricks or other burnt or vitrified material should have a thickness of at least $1\frac{1}{3}$) times the thickness required of brick or block walls. *(AD A1/2 sec. 1C 1C6 & 1C7; AD A1/2 sec. 1C 1C9)*

Irrespective of the materials used in construction, no part of a wall should be thinner than any other part of the wall that it supports.

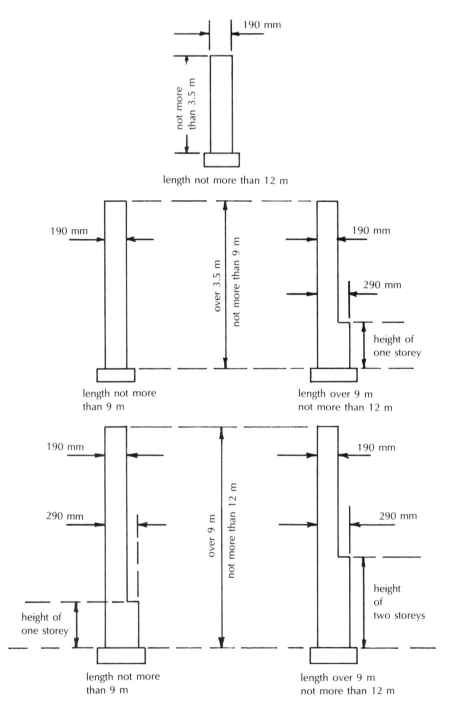

Fig. 6.15 Thickness of solid external, compartment and separating walls (see Table 5).

Solid internal load-bearing walls which are not compartment or separating walls

For these walls the sum of the wall thickness, plus 5 mm, should be equal to at least half the thickness that would be required by Table 5 for an external wall, compartment wall or separating wall of the same height and length.

Where a wall forms the lowest storey of a three-storey building, and it carries loading from both upper storeys, its thickness should not be less than the thickness calculated above or 140 mm, *whichever is greater*. Thus there is an absolute minimum thickness of 140 mm for such walls.

AD A1/2 sec. 1C 1C10

Cavity walls

Any external, compartment or separating wall which is built as a cavity wall should consist of two leaves, each leaf built of bricks or blocks.

The leaves of these walls should be properly tied together with wall ties to BS 1243: 1978, or other not less suitable ties. Ties should be placed at centres 900 mm horizontally and 450 mm vertically, and at any opening at least one tie should be provided for each 300 mm of height within 225 mm of the opening unless the leaves are connected by a bonded jamb.

The cavity should be at least 50 mm, and not more than 75 mm, in width at any level. However, if vertical twist type ties are used, with horizontal spacing reduced to 750 mm, the cavity width may be up to 100 mm. Each leaf should be at least 90 mm thick at any level.

The sum of the thicknesses of the two leaves, plus 10mm, should not be less than the thickness required for a solid wall by Table 5. (See also Fig. 6.16.)

AD A1/2 sec. 1C 1C8

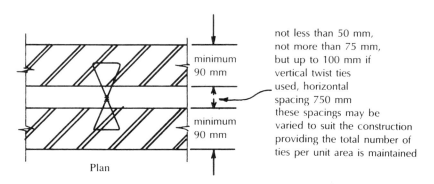

Fig. 6.16 Cavity walls.

Parapets

When referring to Fig. 6.17(a), the minimum thickness, *t*, for a solid parapet wall should not be less than the greater of:

(a) *H*/4; *or*,
(b) 150mm.

For a cavity parapet the minimum thickness, *t*, is related to the maximum parapet height as shown in Fig. 6.17(b).

AD A1/2 sec. 1C 1C11

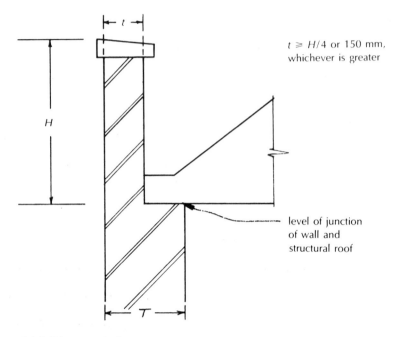

$t \geqslant H/4$ or 150 mm, whichever is greater

level of junction of wall and structural roof

(a) Solid parapet walls

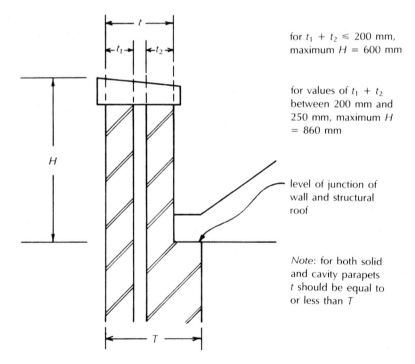

for $t_1 + t_2 \leqslant 200$ mm, maximum $H = 600$ mm

for values of $t_1 + t_2$ between 200 mm and 250 mm, maximum $H = 860$ mm

level of junction of wall and structural roof

Note: for both solid and cavity parapets t should be equal to or less than T

(b) Cavity parapet walls

Fig. 6.17 Height of parapet walls.

Block and brick dimensions

The wall thicknesses specified in Section 1C relate to the *work size* of the materials used. This means the size specified in the relevant British Standard as the size to which the brick or block must conform, account being taken of any permissible deviations or tolerances specified in the British Standard.

Some walls may be constructed of bricks or blocks having modular dimensions derived from BS 6750: 1986 *Specification for modular coordination in building* without the bricks or blocks themselves being covered by a British Standard.

In these cases, the thicknesses prescribed in Section 1C may be reduced by an amount not exceeding that allowed in a British Standard for the same material.

AD A1/2 sec. 1C 1C13

External walls of small buildings and annexes

The external walls of small buildings and annexes have to comply with special rules.

The external walls of such buildings may be not less than 90 mm thick if:

(a) the walls are bonded at each end and intermediately to piers or buttressing walls of not less than 190 mm square in horizontal section (including wall thickness); *and,*

(b) the piers, etc., are positioned so that they divide the wall into distinct lengths and each length is not more than 3 m (this does not apply if the wall is less than 2.5 m high and long); *and,*

(c) the enclosed floor area does not exceed 36 m².

AD A1/2 sec. 1C 1C12 & 1C39

The wall should be built as a solid wall of bricks and blocks and should carry only distributed loading from the roof of the building or annexe, and not be subjected to any lateral thrust from the roof (see Fig. 6.18).

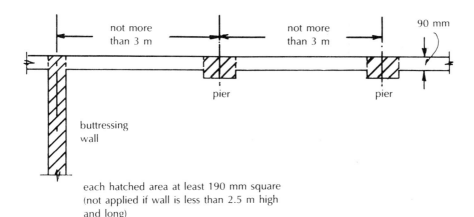

Fig. 6.18 Small buildings and annexes.

Dimensions of chimneys

The wholly external part of a chimney, constructed of masonry and not supported by adequate ties or otherwise stabilised will be deemed satisfactory if the width of the chimney, at the level of the highest point in the line of junction with the roof and at any higher level, is such that its height as measured from that level to the top of the external part of the chimney is not more than $4\frac{1}{2}$ times that width. That height includes any pot or flue terminal on a chimney. Additionally, the masonry should have a density greater than $1500\,\text{kg/m}^3$.

AD A1/2
sec. 1D
1D1

The width of chimney at any level is taken as the smallest width which can be shown on an elevation of the chimney from any direction. This is illustrated in Fig. 6.19.

Foundation recommendations

Section 1E of AD A1/2 provides rules for the construction of strip foundations of plain concrete. It should be remembered that Section 1 applies to certain

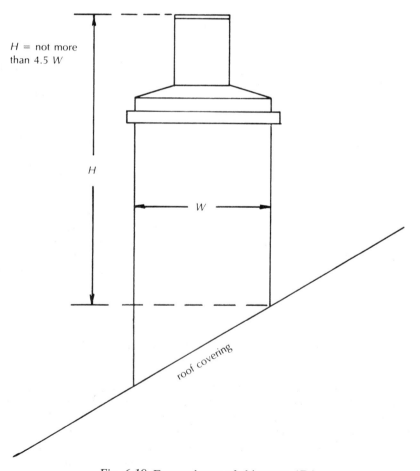

H = not more than 4.5 W

Fig. 6.19 External part of chimneys, 1D1.

residential buildings of not more than three storeys, small single-storey non-residential buildings and annexes. Strictly speaking, the guidance given in Sections 1A to 1E should not be used for any other building types.

Strip foundations of plain concrete placed centrally under the walls will be satisfactory if they comply with the following rules:

(a) There is no made ground or wide soil strength variation in the loaded area or weak soil patches likely to cause foundation failure.
(b) The width of foundation strip is in accordance with Table 12 to Section 1E of ADA1/2 which is reproduced below.
(c) In chemically non-aggressive soils the concrete should be composed of:

   (i) cement to BS 12: 1989 *Specification for Portland cements*; *and,*
   (ii) coarse and fine aggregate to BS 882: 1983 *Specification for aggregates from natural sources for concrete.*

(d) For foundations in chemically aggressive soils, the guidance in BS 5328: Part 1: 1990 *Guide to specifying concrete* should be followed.
(e) The concrete mix is:

   (i) in the proportion 50 kg of cement:0.1 m^3 fine aggregate: 0.2 m^3 coarse aggregate i.e. 1:3:6 or better; *or,*
   (ii) Grade ST1 concrete to BS 5328: Part 2: 1990 *Methods for specifying concrete mixes.*

(f) The concrete strip thickness is equal to or greater than the projection from the wall face, and never less than 150 mm.
(g) The upper level of a stepped foundation overlaps the lower level by twice the height of the step, by the thickness of the foundation or 300 mm, whichever is the greater.
(h) The height of a step is not greater than the thickness of the foundation.
(i) The foundation strip projects beyond the faces of any pier, buttress or chimney forming part of a wall by at least as much as it projects beyond the face of the wall proper.

Table 12 to Section E1 of AD A1/2 specifies seven subsoil types, and the minimum strip widths to use vary according to the calculated load per metre run of the wall at foundation level. The table is reproduced below.

AD A1/2 sec. 1E 1E1 to 1E3

Where a wall load exceeds 70 kN per metre run the foundation will be outside the scope of Section E1 and must be properly designed on structural principles.

These requirements are illustrated in Fig. 6.20.

Design of structural members in buildings of all types

Section 1 of AD A1/2, which is discussed above, deals with a fairly restricted range of building types of traditional masonry construction. If the various parts of Section 1 are complied with it is not necessary to provide design calculations.

Building types falling outside the scope of Section 1 will need full structural calculations and design. Therefore, Sections 2, 3 and 4 of AD A1/2 contain references to British Standards, Codes of Practice and other sources of design information which, if used appropriately, will satisfy the requirements of Paragraphs A1 and A2 of Schedule 1 to the 1991 Regulations.

AD A1/2, section E1

Table 12 Minimum width of strip foundations

(1) Type of subsoil	(2) Condition of subsoil	(3) Field test applicable	(4) Minimum width in millimetres for total load in kilonewtons per lineal metre of load-bearing walling of not more than					
			20 kN/m	30 kN/m	40 kN/m	50 kN/m	60 kN/m	70 kN/m
I Rock	Not inferior to sandstone, limestone or firm chalk	Requires at least a penumatic or other mechanically operated pick for excavation	In each case equal to the width of the wall					
II Gravel Sand	Compact Compact	Requires pick for excavation. Wooden peg 50 mm square in cross-section hard to drive beyond 150 mm.	250	300	400	500	600	650
III Clay Sandy clay	Stiff Stiff	Cannot be moulded with the fingers and requires a pick or pneumatic or other mechanically operated spade for its removal	250	300	400	500	600	650
IV Clay Sandy clay	Firm Firm	Can be moulded by substantial pressure with the fingers and can be excavated with graft or spade	300	350	450	600	750	850
V Sand Silty sand Clayey sand	Loose Loose Loose	Can be excavated with a spade. Wooden peg 50 mm square in cross-section can be easily driven.	400	600	Note: In relation to types V, VI and VII, foundations do not fall within the provisions of this Section if the total load exceeds 30 kN/m			
VI Silt Clay Sandy clay Silty clay	Soft Soft Soft Soft	Fairly easily moulded in the fingers and readily excavated	450	650				
VII Silt Clay Sandy clay Silty clay	Very soft Very soft Very soft Very soft	Natural sample in winter conditions exudes between fingers when squeezed in fist	600	850				

As mentioned above on p. 6.3, Sections 2 and 3 were added to Approved Document A by the 1992 amendment and deal with external wall cladding and re-covering of roofs respectively.

External wall cladding

In recent years a number of accidents have occurred involving heavy concrete cladding panels. Failure of the fixings and deterioration of the concrete has resulted in parts and, in some cases, whole panels becoming detached with the resultant danger to people in the street below.

Guidance is provided in Section 2 of AD A1/2 which relates specifically to heavier forms of cladding. However, some of the guidance is also applicable to curtain walling.

Weather-resistance of wall cladding is not covered by the guidance in AD A1/2; Approved Document C (Site preparation and resistance to moisture) should be consulted for this.

Wall cladding should be:

- capable of safely carrying and transmitting to the structure of the building the combined dead, imposed and wind loads.
- securely fixed to and supported by the structure of the building, the fixing comprising both vertical support and lateral restraint.
- capable of accommodating differential movement between the cladding and the building support structure.
- manufactured of durable materials (including any fixings and associated support components which should also have an anticipated life at least equal to that of the cladding).

In many cases fixings will not be easily accessible for inspection and maintenance. In these cases care is needed in the selection of materials and in the quality of workmanship. Reference should be made to the Approved Document for regulation 7 (Chapter 8).

AD A1/2 sec. 2 2.1 to 2.3

Assessment of loading
Apart from the dead load of the cladding itself the following loads should also be taken into account:

- wind loading – see CP3: Chapter V: *Loading*, Part 2: 1972 *Wind loads* (as amended). Use Class A building size for assessing the ground roughness factor S. Factor S must not be less than one.
- an assessment of the imposed forces from maintenance equipment such as ladders or access cradles which should be based on the actual equipment likely to be used.
- loading from fixtures such as antennae or signboards supported by the cladding.
- lateral loads where the cladding is required to act as pedestrian guarding to stairs, ramps and open wells, or as a vehicle barrier. Refer to Approved Document K (Stairs, ramps and guards) for loading requirements (see Chapter 15).
- lateral pressures from crowds where the wall cladding is required to act as spectator barriers at sports stadia requiring a safety certificate. Recommended design loadings are given in the publication entitled *Guide to Safety at Sports Grounds* (1990) published by the Home Office/Scottish Office.

AD A1/2 sec. 2 2.4 to 2.8

Examples: cavity wall 60 kN/m run in different soil types, to rules of Table 12

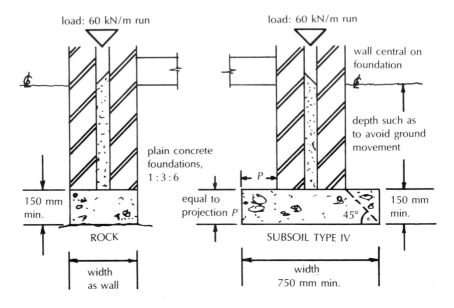

no made ground, no weak patches, no strength variation

(a) Plain strip foundation

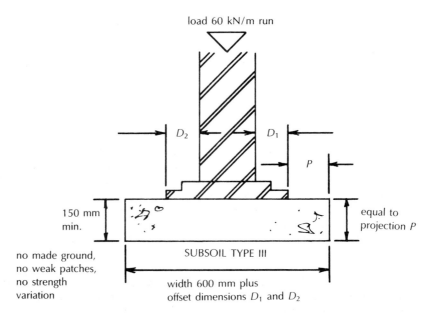

(b) Strip foundation with footing

Fig. 6.20 Strip foundations of plain concrete.

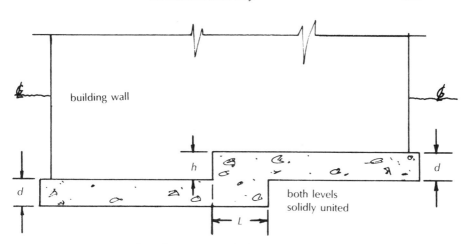

L = 2h or d or 300 mm whichever is greater,
h must not be greater than d

(c) Steps in foundations

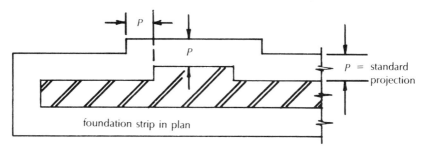

(d) Projections

Fig. 6.20 (Contd).

Design and testing of fixings

The strength of a fixing is a function of the fixing itself and the material into which it is fixed. Therefore, its strength should be derived from tests using materials which are representative of the true in-situ condition. In this way inherent weaknesses in the support structure, such as shrinkage or flexure cracks in concrete and voids in masonry, will be highlighted and may be taken into account in the final design of the fixing.

A number of standards and references exist for assessing the strength of fixings.

- BS 5080: Part 1: 1974 (1982) *Tensile loading* describes a method for testing fixings, such as expanding anchors installed in solid materials either on site or for comparative purposes in a standard material.
- BS 5080: Part 2: 1986 *Method for determination of resistance to loading in shear* gives details of a method for conducting tests under shear force on structural fixings installed in concrete or masonry.

- British Board of Agrément MOAT No. 19: 1981 *The Assessment of Torque Expanded Anchor Bolts When Used in Dense Aggregate Concrete.*

When a load is applied to a fixing from a cladding panel it is possible that the fixing will be subjected to a certain amount of eccentric loading. This can lead to local spalling in the material in which the fixing is anchored and therefore, the fixing should be designed to allow for this by assuming an increase in eccentricity equal to half the diameter of the fixing.

A further problem with fixings and their support components is that there is a possibility of unintended slippage between the parts when in use. Suitable lockable fixings should be used or other means should be sought to mechanically fix the various components together.

Commonly, panel fixings are of the expanding bolt type or are resin bonded into the substrate. When tested in shear and tension the assumed safe working strength of either should not exceed the following:

- 1/3 × (mean shear or tensile failure test load) minus the standard deviation derived from the tests;

unless, for expanding bolt fixings, the mean of the loads which causes a displacement of 0.1 mm under direct tension and 1.0 mm under direct shear is lower, in which case the lower value should be taken.

AD A1/2
sec. 2
2.9 to 2.16

Care should be taken in the design of resin bonded fixings due to their rapid loss of strength at temperatures above 50°C.

Further guidance on fixings may be obtained from BS 8200: 1985, Clause 38 and BS 8298 1989 *Code of practice for design and installation of natural stone cladding and lining*, Clauses 6 and 20, and on the provision of movement joints in BS 5628: Part 3: 1985.

Replacement of roof coverings

It is possible that the re-roofing of a building may result in the existing roof structure having to carry substantially more or less load than it did before the works were carried out. In this case the replacement works would constitute a material alteration under the provisions of Regulation 3(2).

Section 3 of AD A1/2 contains guidance on how the existing roof structure may be assessed to see if it is capable of coping with the changed loading conditions. There are three stages to the assessment procedure:

Stage 1
Compare the proposed and original roof loadings.

Allowance should be made for the increase in loading due to water absorption which may be only 0.3% for oven dry slates but up to 10.5% for plain clay or concrete tiles. These figures are based on the dry mass per unit area of roof coverings.

Stage 2
Carry out a structural inspection on the original roof.

The roof structure must be checked to see if:

- it is capable of sustaining the increased load; *or,*
- it contains sufficient vertical restraints to cope with the wind uplift forces as a result of the lighter roof covering or addition of underlay.

Stage 3
Carry out appropriate strengthening measures.

These may include:

- replacement of defective parts of the roof, such as, structural members, nails or other fixings and vertical restraints.
- provision of additional structural members as necessary to take the increased loads, such as, rafters, purlins, binders or trusses, etc.
- provision of additional restraint straps, ties or fixings to walls as necessary to resist wind uplift forces.

AD A1/2
sec. 3

Guidance on structural design in buildings of all types

Loading

Dead and imposed loads may be assessed by reference to BS 6399 *Loading for buildings*, Part 1: 1984 *Code of practice for dead and imposed loads*. Similarly, imposed roof loads are covered in the same code, Part 3: 1988 *Code of practice for imposed roof loads*.

Wind loads may be assessed by reference to CP 3: Chapter V: Part 2: 1972 *Wind loads*. However, the S factor should never be taken as less than one.

If the actual load is greater than the design load from BS 6399: Part 1: 1984, the actual load should be used. (See also Section 3 of AD A1/2, roof re-covering, p. 6.66 above.)

AD A1/2
sec. 4
4.2

Foundations and ground movement

Foundations should be designed in accordance with BS 8004: 1986 *Code of practice for foundations*.

AD A1/2
sec. 4
4.8

Paragraph A2 of Schedule 1 to the 1991 Regulations requires that ground movement caused by landslip or subsidence must not impair the stability of the building.

Guidance is given in Section 4 of AD A1/2 in the form of reference sources where information may be found regarding the more common forms of ground instability. This includes geological faults, landslides, disused mines or similar unstable strata which may affect the building site or its environs.

The following reviews of various geotechnical conditions have been carried out under the sponsorship of the Minerals and Land Reclamation Division of the Directorate of Planning Services of the DOE and information regarding their availability may be obtained from DPS/2, DOE, Room C15/19, 2 Marsham Street, London SW1P 3EB.

- *Review of research into landsliding in Great Britain.*
- *Review of mining instability in Great Britain.*
- *Review of natural underground cavities in Great Britain.*
- *Review of foundation conditions in Great Britain.*

The reviews are concerned with assessing the general state of knowledge concerning various forms of land instability in order to obtain a general picture of the scale and nature of the problems and how they might be overcome.

The results are presented as reports on a regional basis including 1:250000 scale maps and databases for use by anyone concerned with planning, development or engineering. They cover the nature and causes of instability and the consequent implications for planning and development, investigation methods and remedial preventative measures.

AD A1/2
sec. 4
4.9

Structure above foundations

- Structural work of reinforced, prestressed or plain concrete should comply with BS 8110 *Structural use of concrete*, Part 1: 1985 *Code of practice for design and construction*, Part 2: 1985 *Code of practice for special circumstances* and Part 3: 1985 *Design charts for singly reinforced beams, doubly reinforced beams and rectangular columns*.
- Structural work of aluminium should comply with CP 118: 1969 *The structural use of aluminium*, using one of the alloys listed in Section 1.1 of the code. (Under section 5.3 of the code, the structure should be classified as a safe-life structure.)
- Structural work of masonry should comply with BS 5628 *Code of practice for use of masonry*, Part 1: 1978 (1985) *Structural use of unreinforced masonry*, Part 3: 1985 *Materials and components, design and workmanship*.
- Structural work of timber should comply with BS 5268 *Structural use of timber*, Part 2: 1991 *Code of practice for permissible stress design, materials and workmanship*, Part 3: 1985 *Code of practice for trussed rafter roofs*.
- Structural work of steel should comply with BS 5950 *Structural use of steelwork in building*:

  (a) Part 1: 1990 *Code of practice for design in simple and continuous construction: hot rolled sections*.
  (b) Part 2: 1992 *Specification for materials, fabrication and erection: hot rolled sections*.
  (c) Part 3 *Design in composite construction*, Section 3.1: 1990 *Code of practice for design of simple and continuous composite beams*.
  (d) Part 4: 1982 *Code of practice for design of floors with profiled steel sheeting*.
  (e) Part 5: 1987 *Code of practice for design of cold formed sections*; or,
  (f) BS 449 *Specification for the use of structural steel in building*, Part 2: 1969 *Metric units*.

AD A1/2
sec. 4
4.3 to 4.7

Disproportionate collapse

Buildings of five or more storeys

In May 1968 a gas explosion on the eighteenth floor of a block of flats in London, known as Ronan Point, caused a large portion of the corner of the block to collapse.

Following on the subsequent Tribunal and public inquiry into the disaster, new building regulations were formulated and introduced in 1970 with the express purpose of preventing further similar occurrences of this kind where the extent of the collapse is disproportionate to its cause.

These regulations have been updated and revised in line with current experience and knowledge, the main requirement being stated in Paragraph A3 of Schedule 1 to the 1991 Regulations.

Buildings of five or more storeys (including basement storeys) are required to be constructed so that in the event of an accident the building will not suffer collapse to an extent disproportionate to the cause of that collapse.

Regs Sch. 1
A3

Buildings which come into the five-storey category merely because they have a fifth floor in the roof space are excluded from the provisions if the roof has a pitch of 70° or less to the horizontal.

Approved Document A3 contains guidance on measures designed to reduce the sensitivity of a building to disproportionate collapse in the event of an accident and which may also avoid or reduce the hazards to which the building may be exposed.

Three approaches may be adopted depending on the extent to which it is possible to tie the structural members together.

(a) Provide effective horizontal and vertical ties complying with:

- Clause 2.2.2.2 of BS 8110 *Structural use of concrete*, Part 1: 1985 *Code of practice for design and construction* and Clause 2.6 of Part 2: 1985 *Code of practice for special circumstances*, for structural work of reinforced, prestressed or plain concrete.
- Clause 2.4.5.3 of BS 5950 *Structural use of steelwork in building*, Part 1: 1990 *Code of practice for design in simple and continuous construction: hot rolled sections*, for structural work of steel.
- Clause 37 of BS 5628 *Code of practice for use of masonry*, Part 1: 1978 (1985) *Structural use of unreinforced masonry*, for structural use of masonry.

AD A3
sec. 5
5.2

Compliance with these measures will require no further action to be taken in the structural design.

(b) Provide effective horizontal tying but vertical tying of vertical load-bearing members is not feasible:

- Assume each untied member is removed one at a time, in each storey in turn.
- Check that remainder of structure can bridge over the missing member even if it is in a substantially deformed state.
- In this deformed state certain members, such as cantilevers or simply supported floor panels, will be vulnerable to collapse.
- Check that collapse within the storey and the immediately adjacent storeys would be limited to (i) 15% of the storey area; or, (ii) 70 m², whichever is the less.
- If bridging, as detailed above, is not possible, design the removed member as a protected member (see below).

(c) Effective horizontal and vertical tying of load-bearing members is not feasible:

- Assume each member is removed one at a time, in each storey in turn.
- Check that area at risk of collapse within the storey and the immediately adjacent storeys is limited as in (b) above.

● If the area put at risk cannot be limited as described above when a
member is notionally removed, then design the member as a protected
member (see below).

Design of protected members

Protected members (or key elements) should be designed in accordance with
the codes and standards listed under (a) above. These documents contain
minimum loadings which protected members must be designed to withstand.
For example, in BS 5950: Part 1: 1990, accidental loadings are referred to in
Clause 2.4.5.5. These should be chosen to reflect the importance of the key
element and the consequences of failure, and the key element should be
capable of withstanding a load of at least $34 \, \text{kN/m}^2$ applied in any direction.

AD A3
sec. 5
5.1

Long span roof structures

After the Ronan Point disaster referred to above, a number of long span
buildings suffered roof collapses, this time not caused by accidents. The first of
these occurred in June 1973 at the Camden School for Girls when the roof of
the assembly hall collapsed.

The subsequent investigation by the Building Research Establishment
revealed a number of causes, including loss of strength in the high alumina
cement prestressed concrete roof beams due to conversion of the cement.
Lack of adequate tying at the supports was also indicated as a contributary
cause amongst others.

This failure led to the banning of high alumina cement in structural concrete
work and also to the inclusion of long span roof structures in the disproportionate collapse regulations with the introduction of paragraph A4 of Schedule
1 in the 1991 Regulations.

The 1994 amendment revokes paragraph A4 and therefore the recommendations of Section 6 of AD A4 no longer apply.

Chapter 7

Fire

Introduction

Part B of Schedule 1 to the Building Regulations 1991 is concerned with means of escape from buildings, fire spread within and between buildings, and access for the fire services to fight fires. Since the regulations are made in the interests of public health and safety, they do not attempt to achieve non-combustible buildings. The aims are to ensure the safety of the occupants and others who may be affected by the building, and to provide assistance for fire fighters. The protection of property, including the building itself, is the province of insurers who may require additional measures to provide higher standards before accepting the insurance risk.

Buildings must therefore be constructed so that, in the event of a fire: AD B 0.16

- the occupants are able to reach a place of safety.
- they will resist collapse for a sufficient period of time to allow evacuation of the occupants and prevent further rapid fire spread.
- the spread of fire within and between buildings is kept to a minimum.
- there is satisfactory access for fire appliances and facilities are provided to assist firefighters in the saving of lives.

The first requirement is met by providing an adequate number of exits and protected escape routes. The second is met by setting reasonable standards of fire-resistance for the structural elements – the floors, roofs, load-bearing walls and frames. The third is met by:

(a) dividing large buildings into *compartments* and requiring higher standards of fire-resistance of the walls and floors bounding a compartment;
(b) setting standards of fire-resistance for external walls;
(c) controlling the surface linings of walls and ceilings to inhibit flame spread;
(d) sealing and sub-dividing concealed spaces in the structure or fabric of a building to prevent the spread of unseen fire and smoke; *and*
(e) setting standards of resistance to fire penetration and flame spread for roof coverings.

The fourth is met by:

(a) providing access for fire appliances and firefighting personnel;
(b) providing fire mains within the building; *and*

AD B
0.1

(c) making sure that heat and smoke may be vented from basement
areas.

In some large and complex buildings the provisions contained in AD B may
prove inadequate or difficult to apply. In such buildings, the only viable way to
achieve a satisfactory standard of fire safety may be to adopt a fire safety
engineering approach which takes into account the total fire safety package.
An example of this may be found in BS 5588 *Fire precautions in the design,
construction and use of buildings*, Part 10: 1991 *Code of practice for enclosed
shopping complexes*. (See p. 7.65 below for further reference to enclosed
shopping complexes.)

AD B
0.10, 0.15

Some difficulty may also be encountered when trying to apply the provisions
of AD B to existing buildings, particularly when they are of special historic or
architectural importance. In buildings of this type it may be appropriate to
carry out an assessment of the potential fire hazard or risk to life, and then
incorporate in the design a sufficient number of fire safety features to alleviate
the danger. The risk assessment should take account of:

AD B
0.11

• The likelihood of a fire occurring.
• The anticipated severity of the fire.
• How well the structure of the building is able to resist the spread of smoke
  and flames.
• The consequential danger to persons in or near the building.

AD B
0.12

Fire safety measures which can be incorporated in the design include:

• An assessment of the means to prevent fire.
• The installation of automatic fire detection and warning systems.
• The provision of adequate means of escape.
• The provision of smoke control.
• Design features aimed at controlling the rate of growth of a fire if one does
  occur.
• Improvement of the ability of a structure to resist the effects of fire.
• The extent of fire containment offered by the building.
• The fire separation from other buildings or parts of the same building.
• The standard of firefighting equipment in the building.
• The ease with which the fire service may gain access to fight a potential fire.
• The existence of legislative controls to require staff training in fire safety
  and fire routines (e.g. Fire Precautions Act 1971, or licensing and regis-
  tration controls).
• The existence of continuing control so that fire safety systems can be seen to
  be maintained (as in the certification procedure under the Fire Precautions
  Act).

AD B
0.13

Fire safety in buildings is a complex matter and in recent years a number of
reviews and enquiries have been set up to consider different aspects of fire
safety law. Additionally, matters have been further complicated by the need to
comply with various European Community directives, resulting in the
implementation of the Management of Health and Safety at Work Regulations 1992 and consultation on the revised draft of the Fire Precautions (Places
of Work) Regulations. Currently, the recommendations of the Interdepartmental Review Team on Fire Safety Legislation and Enforcement are out for

consultation and may result in major structural changes to the law concerning fire safety in buildings.

Terminology

Certain terms which apply generally throughout this chapter are defined here. Other terms are defined in the specific section to which they apply.

AD B2/3/4
Appendix E

BASEMENT STOREY – a storey which has some part of the perimeter of its floor more than 1.2m below the highest level of the ground adjoining that part of the floor. A basement storey may be treated as the above ground structure for fire-resistance purposes, if one side is open at ground level for smoke venting or firefighting. (See Fig. 7.1.)

BOUNDARY – when referring to any side of a building or compartment (including any external wall or part), means the usual legal boundary adjacent to that side, being taken up to the centre of any abutting railway, street, canal or river.

CEILING – includes any soffit, roof-light or other part of a building which encloses and is exposed overhead in a room, circulation space or protected

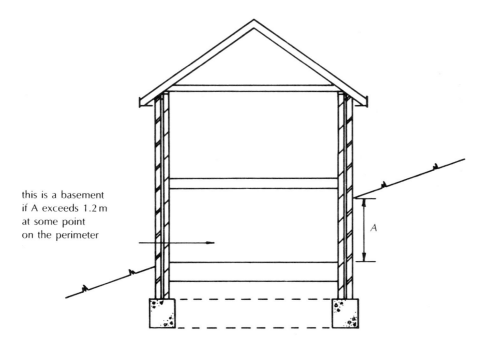

this is a basement
if A exceeds 1.2 m
at some point
on the perimeter

Basements open at ground level
on one side allow smoke venting
or access for fire fighting. Therefore
treat as above ground structure for
fire resistance.

Fig. 7.1 Basement storey.

shaft (but not including the surface of the frame of any roof-light, the upstand of which is considered as part of the wall).

CIRCULATION SPACE – a space (including a protected stairway) used mainly as a means of access between a room and an exit from the compartment or building.

COMPARTMENT – any part of a building (including rooms, spaces or storeys) which is constructed to stop fire spreading to or from another part of the same building, or an adjacent building. If any part of the top storey of a building comes within a compartment, that compartment is taken to include any roof space above that part of the top storey (see Fig. 7.2). See also 'Separated part' below.

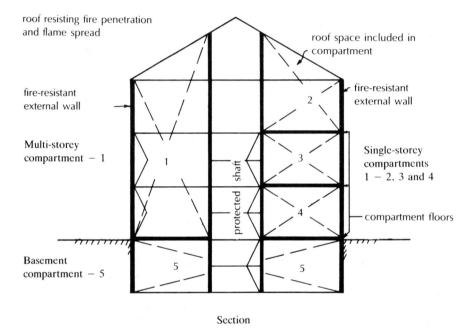

Section

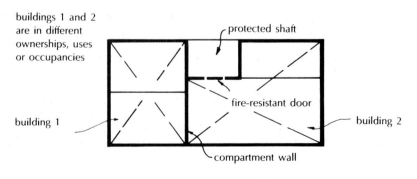

Plan

Fig. 7.2 Division of buildings into compartments.

COMPARTMENT WALL/FLOOR – fire-resisting construction provided to separate one fire compartment from another for the purpose of preventing fire spread.

CONCEALED SPACE (CAVITY) – a space which is concealed by the elements of a building (such as a roof space or the space above a suspended ceiling) or contained within an element (such as the cavity in a wall). This definition does *not* include a room, cupboard, circulation space, protected shaft or space within a flue, chute, duct, pipe or conduit.

CONSERVATORY – a single-storey part of a building in which the walls and roof are substantially glazed with translucent or transparent material.

ELEMENT OF STRUCTURE –

(a) Any member forming part of the structural frame of a building or any other beam or column. (This does not normally include members which form part of a roof structure only unless the roof performs the function of a floor, or the roof structure provides stability for fire-resisting walls.)
(b) A floor (but not the lowest floor in a building, or a platform floor).
(c) An external wall.
(d) A compartment wall (including a wall which is common to two or more buildings).
(e) A load-bearing wall or the load-bearing part of a wall.
(f) A gallery.

These elements are illustrated in Fig. 7.3.

EXTERNAL WALL – includes a portion of a roof sloping at $70°$ or more to the horizontal if it adjoins a space within the building to which persons have access, other than for occasional maintenance and repair (see Fig. 7.4).

FIRE DOOR – includes any shutter, cover or other form of protection to an opening in any fire-resisting wall or floor of a building or in the structure surrounding a protected shaft. The fire door should be able to resist the passage of fire and/or gaseous products of combustion in accordance with specified criteria (see Fire doors on p. 7.93 below). A fire door may have one or more leaves.

GALLERY – a floor (or raised storage area) which projects into another space but has less than half the floor area of that space.

HABITABLE ROOM – a room used for dwelling purposes, including a kitchen (in Part B) but not a bathroom.

NOTIONAL BOUNDARY – a boundary which is assumed to exist between two buildings on the same site where there is no actual boundary. The notional boundary line should be so placed that neither building contravenes any of the requirements of ADB relevant to the external walls facing each other (see Fig. 7.5).

PLACES OF SPECIAL FIRE RISK – oil-filled switchgear and transformer rooms, boiler rooms, stores for fuel or other highly flammable substances, and rooms containing fixed internal combustion engines.

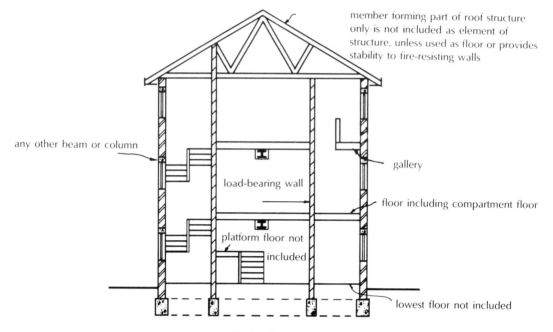

member forming part of roof structure only is not included as element of structure, unless used as floor or provides stability to fire-resisting walls

any other beam or column

gallery

load-bearing wall

floor including compartment floor

platform floor not included

lowest floor not included

Section A−A

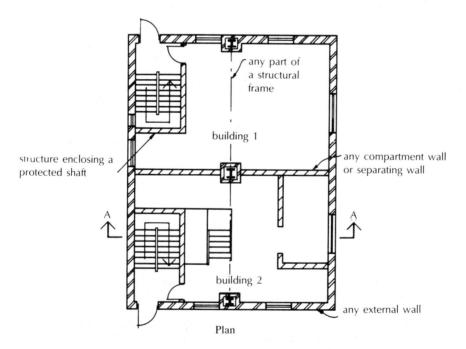

any part of a structural frame

building 1

structure enclosing a protected shaft

any compartment wall or separating wall

building 2

any external wall

Plan

Fig. 7.3 Elements of structure.

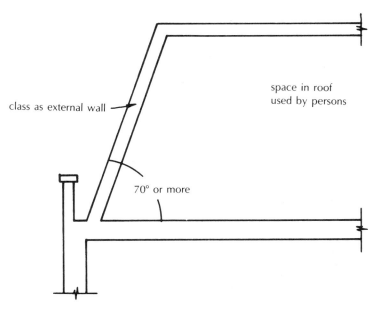

class as external wall

space in roof
used by persons

70° or more

Fig. 7.4 Steeply pitched roofs.

PLATFORM FLOOR – a floor over a concealed space intended to house services, which is supported by a structural floor.

PROTECTED CORRIDOR/LOBBY – a corridor or lobby protected from fire in adjacent accommodation by fire-resisting construction.

PROTECTED SHAFT – a shaft enclosed with fire-resisting construction which enables persons, things or air to pass between different compartments.

PROTECTED STAIRWAY – a stair adequately protected with fire-resisting construction discharging to place of safety through final exit. (Includes passage from foot of stair to final exit.)

RELEVANT BOUNDARY – for a boundary to be considered relevant it should:

(a) be coincident with, or
(b) be parallel to, or
(c) not make an angle of more than 80° with the external wall (see Fig. 7.6).

In certain circumstances a 'notional' boundary, as defined above, will be the relevant boundary. A wall may have more than one relevant boundary.

ROOF-LIGHT – includes any domelight, lanternlight, skylight, ridgelight, glazed barrel vault or other element which is intended to admit daylight.

ROOM – an enclosed space in a building, but not one used solely as a circulation space. (This term would also include cupboards that were not fittings and large rooms such as auditoria.) Excluded are voids such as ducts, roof spaces and ceiling voids.

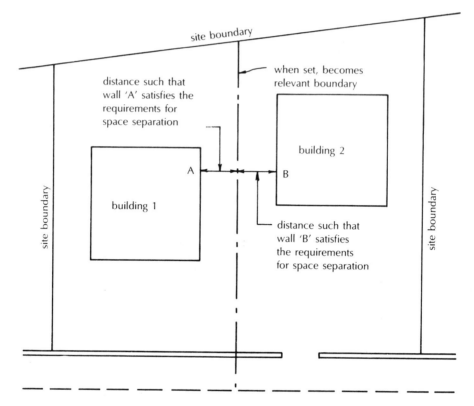

One or both of the buildings new.
One or other of the buildings of residential or assembly and recreation use.
Existing building treated as identical new building but with existing
unprotected area and fire resistance in external wall.

Fig. 7.5 Notional boundary.

SEPARATED PART (OF A BUILDING) – where a compartment wall
completely divides a building from top to bottom and is in one plane, the
divided sections of the building are referred to as separated parts. The height
of each separated part may then be treated individually (see Fig. 7.7 and Rules
for measurement below).

SINGLE-STOREY BUILDING – a building which consists of a ground
storey only. (A separated part consisting of a ground storey only and with a
roof which is accessible only for the purposes of maintenance and repair may
be treated as part of a single-storey building.) Basements are not included
when counting the number of storeys in a building.

STOREY – included in this definition are the following:

(a) any gallery in an assembly building (PG5); *and*
(b) a gallery in any other building if its area exceeds half that of the space into
    which it projects; *and*
(c) a roof, unless used only for maintenance or repair.

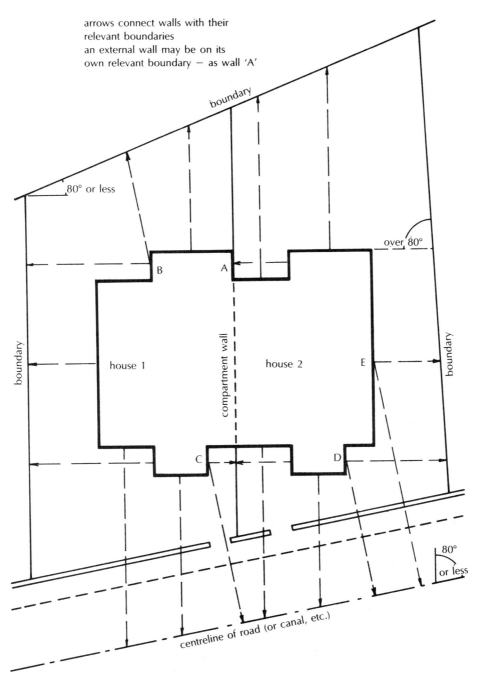

arrows connect walls with their
relevant boundaries
an external wall may be on its
own relevant boundary – as wall 'A'

boundary

80° or less

over 80°

B   A

boundary

house 1   house 2   E   boundary

compartment wall

C   D

80°
or less

centreline of road (or canal, etc.)

an external wall may have more than one relevant boundary,
as walls B, C, D and E

Fig. 7.6 Relevant boundaries.

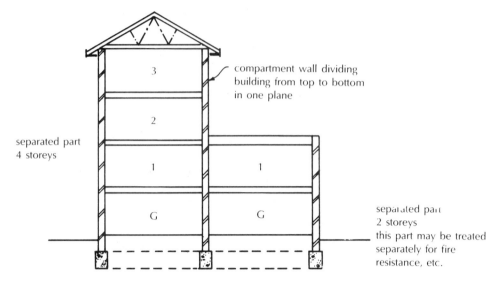

Fig. 7.7 Separated part.

UNPROTECTED AREA – in relation to an external wall or side of a building, means:

(a) a window, door or other opening;
(b) any part of an external wall of fire-resistance less than that required by Section 12 of AD B4;
(c) any part of an external wall with external facing attached or applied, whether as cladding or not, the facing being of combustible material more than 1mm thick (combustible in this context means any material which is not non-combustible or is not a material of limited combustibility) (see Fig. 7.8).

Purpose groups

AD B
Appendix D
D2

The fire hazard presented by a building will, to a large extent, depend on the use to which the building is put. Many provisions concerning means of escape, fire-resistance, compartmentation, etc. are directly related to these use classifications, which in AD B are termed purpose groups (PG).

The purpose groups are set out in Table D1 of Appendix D of AD B and this table is reproduced below. The seven purpose groups can be divided into two main sections:

- Residential – where there is sleeping accommodation (and therefore extra danger in the event of fire).
- Non-residential – where there is no sleeping accommodation.

AD B
Appendix D
D1, D4

Where a building contains compartments or separated parts used for different purposes, each compartment or separated part should be classified separately.

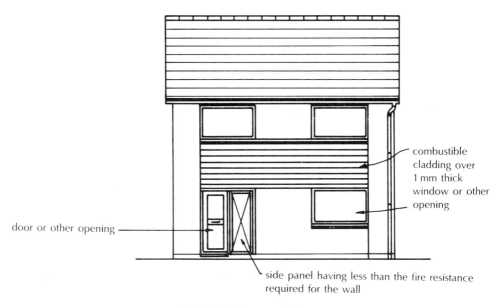

Fig. 7.8 Unprotected areas.

A building or compartment or separated part may be used for several different purposes. Only the main purpose is taken into account in deciding its purpose group classification. However, any ancillary uses may be treated as belonging to a purpose group in their own right in the following cases:

- where a flat or maisonette serves a part of the same building in another purpose group, for example, a flat over a small shop occupied by the shop owner. AD B Appendix D D3
- storage in a shop or commercial building of PG 4 if its area exceeds one-third of the total floor area of the building or compartment.
- any other ancillary use if its area exceeds one-fifth of the total floor area of the building, compartment or separated part.

Some large buildings, such as shopping complexes, involve complicated mixes of purpose groups. In these cases special precautions may need to be taken to reduce any additional risks caused by the interaction of the different purpose groups. AD B Appendix D D5

Rules for measurement

Many of the requirements concerning compartmentation and fire resistance, etc. in ADB are based on the height, area and cubic capacity of the building, compartment or separated part. For consistency, it is necessary to have a standard way of measuring these proportions. Appendix C of AD B indicates diagrammatically how the various forms of measurement should be made. These rules can be summarised as follows: AD B Appendix C

AD B, Appendix D

Table **D1** **Classification of purpose groups**

Title	Group	Purpose for which the building or compartment of a building is intended to be used.
Residential*	1(a)	Flat or maisonette.
(dwellings)	1(b)	Dwellinghouse which contains a habitable storey with a floor level which is more than 4.5 m above ground level.
	1(c)	Dwellinghouse which does not contain a habitable storey with a floor level which is more than 4.5 m above ground level.
Residential (Institutional)	2(a)	Hospital, nursing home, home for old people or for children, school or other similar establishment used as living accommodation or for the treatment, care or maintenance of people suffering from illness or mental or physical disability or handicap, place of detention, where such people sleep on the premises.
(Other)	2(b)	Hotel, boarding house, residential college, hall of residence, hostel, and any other residential purpose not described above.
Office	3	Offices or premises used for the purpose of administration, clerical work (including writing, book keeping, sorting papers, filing, typing, duplicating, machine calculating, drawing and the editorial preparation of matter for publication, police and fire service work), handling money (including banking and building society work), and communications (including postal, telegraph and radio communications) or radio, television, film, audio or video recording, or performance [not open to the public] and their control.
Shop and Commercial	4	Shops or premises used for a retail trade or business (including the sale to members of the public of food or drink for immediate consumption and retail by auction, self-selection and over-the-counter wholesale trading, the business of lending books or periodicals for gain and the business of a barber or hairdresser) and premises to which the public is invited to deliver or collect goods in connection with their hire repair or other treatment, or (except in the case of repair of motor vehicles) where they themselves may carry out such repairs or other treatments.
Assembly and Recreation	5	Place of assembly, entertainment or recreation; including bingo halls, broadcasting, recording and film studios open to the public, casinos, dance halls; entertainment, conference, exhibition and leisure centres; funfairs and amusement arcades; museums and art galleries; non-residential clubs, theatres, cinemas and concert halls; educational establishments, dancing schools, gymnasia, swimming pool buildings, riding schools, skating rinks, sports pavilions, sports stadia; law courts; churches and other buildings of worship, crematoria; libraries open to the public, non-residential day centres, clinics, health centres and surgeries; passenger stations and termini for air, rail, road or sea travel; public toilets; zoos and menageries.
Industrial	6	Factories and other premises used for manufacturing, altering, repairing, cleaning, washing, breaking-up, adapting or processing any article; generating power or slaughtering livestock.
Storage and other non-residential†	7(a)	Place for the storage or deposit of goods or materials [other than described under 7(b)] and any building not within any of the purpose groups 1 to 6.
	7(b)	Car parks designed to admit and accommodate only cars, motorcycles and passenger or light goods vehicles weighing no more than 2500 kg gross.

Notes

* Includes any surgeries, consulting rooms, offices or other accommodation, not exceeding 50 m² in total, forming part of a dwelling and used by an occupant of the dwelling in a professional or business capacity.

† A detached garage not more than 40 m² in area is included in purpose group 1(c); as is a detached open carport of not more than 40 m², or a detached building which consists of a garage and open carport where neither the garage nor open carport exceeds 40 m² in area.

- HEIGHT – the height of the building or part is measured from the mean level of the ground adjoining the outside of the building's external walls to the level of half the vertical height of the roof, or to the top of the walls or parapet, whichever is the higher. This rule applies to double pitch, mono-pitch, flat and mansard type roofs (see Fig. 7.9(a)).

- AREA – the area of any storey of a building, compartment or separated part should be calculated as the total area in that storey within the finished inner surfaces of the enclosing walls. If there is no enclosing wall, the area is measured to the outermost edge of the floor on that side. The area should include any internal walls or partitions.

  The area of a room, garage, conservatory or outbuilding is calculated by measuring to the inner surface of the enclosing walls.

  The area of any part of a roof should be calculated as the actual visible area of that part, as measured on a plane parallel to the roof slope. For a lean-to roof, the measurement should be taken from the wall face to the outer edge of the roof slope. For a hipped roof, the measurement should be to the outer point of the roof as a base area (see Fig. 7.9(a)).

- CUBIC CAPACITY – the cubic capacity of a building, compartment or separated part should be calculated as the volume of the space between the finished surfaces of the enclosing walls, the upper surface of its lowest floor and the under surface of the roof or ceiling surface as appropriate. If there is no enclosing wall, the measurement should be taken to the outermost edge of the floor on that side. The cubic capacity should again include space occupied by other walls, shafts, ducts or structures within the measured space (see Fig. 7.9(b)).

- NUMBER OF STOREYS – the number of storeys in a building or separated part should be calculated at the position which gives the maximum number. Basement storeys are not counted. In most purpose group buildings galleries are also not counted as storeys. However, in assembly buildings, a gallery is included as a storey unless it is a fly gallery, loading gallery, stage grid, lighting bridge or other similar gallery, or is for maintenance and repair purposes. (The common factor here is that these excluded galleries are not generally accessible to the public.) (See Fig. 7.9(b).)

- HEIGHT OF TOP STOREY – This is measured from ground level on the lowest side of the building to the upper surface of the top floor. Roof-top plant areas are excluded from this measurement.

Means of escape in case of fire

The mandatory requirement

Buildings must be designed and constructed so that there are means of escape in case of fire capable of being used safely and effectively at all material times. (No definition is given of material times.) The mandatory requirement goes on to state that the means of escape must be to a place of safety outside the building; however, it will be seen below that for certain classes of buildings the

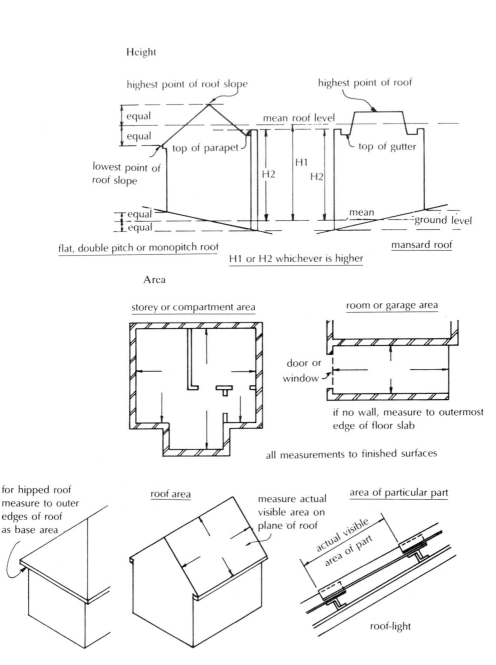

Fig. 7.9(a) Rules for measurement.

Cubic capacity

cubic capacity of compartment or building

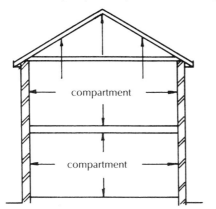
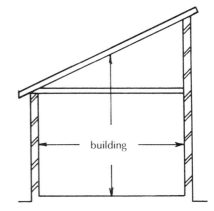

Number of storeys

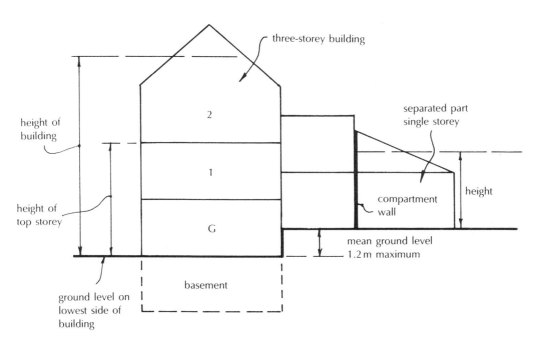

Fig. 7.9(b) Rules for measurement.

Regs Sch. 1
B1

place of safety may be within the building itself. This requirement applies to all buildings except prisons provided under section 33 of the Prisons Act 1952.

Standard of performance to meet the mandatory requirement

The mandatory requirement may be met by:

AD B1

- providing suitably located routes of sufficient number and size to enable the occupants to escape to a place of safety in the event of fire;
- providing escape routes with adequate lighting;
- ensuring exits are suitably signed;
- providing appropriate facilities to limit the ingress of smoke to the escape route; *or*
- taking suitable measures to restrict the fire and remove smoke;

to an extent necessary depending on the size, height and use of the building.

Approved Document B1, Means of escape, contains details of practical ways to satisfy the mandatory requirement and the performance standard referred to above. It does this in two ways:

- By giving actual recommendations in the text (described in detail below); *and*
- By reference to other sources of guidance including:

    (a) BS 5588 *Fire precautions in the design, construction and use of buildings*, Parts 1–10.
    (b) *Fire Prevention Guide No. 1, Fire precautions in town centre redevelopment.* Home Office, HMSO, 1972.
    (c) *Building Bulletin 7, Fire and the design of educational buildings.* Department of Education & Science, HMSO, 1988.
    (d) *Firecode. Health Technical Memorandum 81, Fire precautions in new hospitals.* Department of Health & Social Security, HMSO, 1987.
    (e) *Guide to safety at sports grounds*, Home Office/Scottish Office. HMSO, 1990.

Indeed, the Approved Document recommends that these other sources should be used in the case of shopping complexes, health care buildings, schools and sports grounds instead of the more generalised guidance contained in AD B1. In other buildings, where a choice exists between the use of the relevant Part of BS 5588 and the Approved Document, it would appear to be permissible to choose those parts of each document which are most beneficial to the designer. However, this approach is not permitted in the case of offices or large shops. For these building types all the relevant provisions of BS 5588 should be used, if this document is chosen as the basis for design, and not a mixture of the code and AD B1.

Means of escape – links with other legislation

In addition to building regulations there is a large amount of other legislation covering means of escape in case of fire. Reference to some other provisions will be found in Chapter 1 (Building control: an overview). However, the purpose of this book is not to review all the legislation on this subject, but to show how it may interact with and affect building regulations. Some of the most important provisions are mentioned below.

Fire Precautions Act 1971 and the Fire Precautions (Workplace) Regulations 1997

Together, these probably represent the most significant pieces of legislation which are linked with the Building Regulations. They are enforced by fire authorities. The Fire Precautions Act designates certain uses of premises for which a fire certificate is required (see page 3.3 and Chapter 5) and imposes a statutory duty on the occupiers of certain smaller premises to provide a reasonable means of escape. Additionally, the Fire Precautions (Workplace) Regulations 1997 impose minimum fire safety standards on all workplaces where more than one person is employed (see Chapter 5). A duty is placed on the employer to carry out a fire risk assessment of the workplace to establish what precautions are necessary to ensure the safety of employees in the event of fire.

AD B1
0.20

In the case of a workplace to which the Fire Precautions (Workplace) Regulations do not apply (e.g. a small family-run designated hotel which does not employ any staff) fire authorities cannot, as a condition of issuing a fire certificate, make requirements for structural or other alterations to means of escape if the building plans comply with the building regulations. This does not stop the fire authority from requiring compliance with provisions which were not required to be supplied to the local authority or approved inspector under the building regulations if they consider the means of escape to be inadequate, and in practice, this gives fire authorities considerable leeway to demand additional measures which are not covered by building regulations. Additionally, some things may be required by the Fire Precautions Act which are outside the scope of building regulations, e.g. the provision of first aid fire fighting equipment for use by the building occupants.

Health and Safety at Work etc. Act 1974

In the case of some highly specialised industrial or storage premises where hazardous processes are carried out or materials are kept the Health and Safety Executive may have certification responsibilities similar to those of the fire authority.

AD B1
0.21

Housing Act 1985

Local authorities are obliged to require means of escape from houses in multiple occupation, i.e. houses which are occupied by persons who do not form part of a single household. (For guidance on the interpretation of this definition see DOE/Home Office/Welsh Office Circular *Memorandum on overcrowding and houses in multiple occupation.*) Many local authorities use the services of the fire authority in deciding on the adequacy of an existing means of escape in such buildings. If new dwellinghouses, flats and maisonettes are designed in accordance with AD B1, they should be acceptable for use as 'houses' in multiple occupation.

AD B1
0.22

Interpretation of AD B1

AD B
Appendix E

A large number of terms are used in Approved Document B1 which relate specifically to means of escape:

ACCOMMODATION STAIR – a stair which is provided for the convenience of the occupants of a building and is additional to those required for means of escape. (When calculating the number of stairs required in a building, accommodation stairs are ignored.)

ALTERNATIVE ESCAPE ROUTES – routes which are sufficiently separated from one another by fire-resisting construction or space and direction so that one route will still be available even if the other is affected by fire. In most buildings this will mean alternative protected corridors and stairs; however, in dwellings an alternative escape route could be via a balcony or flat roof.

ALTERNATIVE EXIT – one of two or more exits, each of which is separate from the other. (There are rules in AD B1 for determining whether exits are sufficiently far enough apart to be considered as alternatives and these are illustrated on p. 7.45 below.)

ATRIUM – a vertical space in a building which openly connects three or more floors and is enclosed at the top by a roof or floor. (Shafts containing exclusively stairs, escalators, lifts or services are not included in this definition.)

COMMON BALCONY – an escape route from more than one flat or maisonette which is formed by a walkway open to the air on one or more sides.

COMMON STAIR – an escape stair which serves more than one maisonette or flat.

CORRIDOR ACCESS – a common horizontal internal access or circulation space which serves each dwelling in a building containing flats. It may include a common entrance hall.

DEAD END – An area from which it is only possible to escape in one direction.

DIRECT DISTANCE – the shortest distance which can be measured from within the floor area of the inside of the building, to the storey exit. All internal walls, partitions and fittings are ignored when measuring the direct distance, except the walls enclosing the protected stairway.

DWELLING – a unit of residential accommodation which is occupied by:

- a single person or family; *or*
- not more than six residents living together as a single household, including a household where the residents receive care.

The dwelling need not be the sole or main residence of the occupants.
   (This is an important definition since it shows the difference, for the purposes of means of escape, between a dwelling and a house in multiple

occupation. It also appears to indicate that certain buildings, where people receive care, may be regarded as dwellings and not institutional buildings, again, for the purposes of means of escape.)

EMERGENCY LIGHTING – lighting which is provided to be used when the normal lighting fails.

ESCAPE LIGHTING – part of the emergency lighting which is provided specifically to light escape routes.

ESCAPE ROUTE – that part of the means of escape from any point in the building to the final exit.

EVACUATION LIFT – a lift used to evacuate disabled people in the event of fire.

FINAL EXIT – the termination of an escape route from a building sited so that people may be able rapidly to get clear of any danger from smoke or fire in the vicinity of the building. It should give direct access to a street, passageway, walkway or open space.

INNER ROOM – a room contained within another room (termed the access room). Escape is only possible by passing through the access room.

MEANS OF ESCAPE – structural means whereby a safe route or routes is or are provided for persons to travel to a place of safety from any point in the building, in the event of fire.

OPEN SPATIAL PLANNING – the internal arrangements of a building whereby a number of floors are contained within one undivided space, i.e. split-level floors. AD B1 makes a distinction between atria and open spatial planning; however, since there are no recommendations for the use of either in the document, it would appear to be a somewhat academic distinction.

PRESSURISATION – a technique for preventing the ingress of smoke to protected escape routes by maintaining a pressure differential between the protected area and the adjoining accommodation.

PROTECTED CIRCUITS – an electrical circuit which is protected against fire.

PROTECTED ENTRANCE HALL/LANDING – a hall or circulation space in a dwelling which is protected by fire-resisting construction (other than any part of the wall which is external).

STOREY EXIT – a doorway giving direct access to:

- A protected stairway (defined on p. 7.7).
- A firefighting lobby (defined on p. 7.145).
- An external escape route.

Also the following:

- A final exit (see above for definition).
- A door in a compartment wall in a hospital.

TRAVEL DISTANCE – the actual distance travelled by a person from any point in the floor area to the nearest storey exit. In this case the layout of the floor in terms of walls, partitions and fittings *is* taken into account. (Cf Direct distance above.)

General requirements for means of escape

AD B1
0.25

There are certain basic principles which govern the design of means of escape in buildings and which apply to all building types. In general the design should take account of:

- The activities of the users.
- The form of the building.
- The degree to which it is likely that a fire will occur.
- The potential fire sources, and
- The potential for fire spread throughout the building.

In assessing the above, the following assumptions are made in AD B1 in order that a safe and economical design may be achieved:

(a) In general, when a fire occurs the occupants should be able to escape safely, without external assistance or rescue from the fire service or anyone else. Obviously, there are some institutional buildings where it is not practical to expect the occupants to escape unaided and special arrangements are necessary in these cases. Similar considerations apply to disabled people. Aided escape is also permitted in certain low-rise dwellings.

AD B1
0.26

(b) Fires do not normally break out in two different parts of a building at the same time.

(c) Fires are most likely to occur in the furnishings and fittings of a building or in other items which are not controlled by the building regulations.

(d) Fires are less likely to originate in the building structure and accidental fires in circulation spaces, corridors and stairways are unlikely due to the restriction on the use of combustible materials in these areas.

(e) When a fire breaks out the initial hazard is to the immediate area in which it occurs and it is unlikely that a large area will be affected at this stage. When fire spread does occur it is usually along circulation routes.

AD B1
0.27

(f) The primary danger in the early stages of a fire is the production of smoke and noxious gases. These obscure the way to escape routes and exits and are responsible for the most casualties. Therefore, limiting the spread of smoke and fumes is vital in the design of a safe means of escape.

AD B1
0.24

(g) Buildings covered by AD B1 are assumed to be properly managed. Where there is a failure of management responsibility, the building owner or occupier may be prosecuted under the Fire Precautions Act or the Health and Safety at Work etc. Act, which may result in prohibition of the use of the building.

Alternative escape routes

AD B1
0.30

When a fire occurs it should be possible for people to turn their backs on it and travel away from it to either a final exit or a protected escape route leading to a place of safety. This means that an alternative escape route should be provided in most situations as follows:

- The first part of the escape route will be within the accommodation or circulation areas and will usually be unprotected. It should be of limited length so that people are not exposed to fire and smoke for any length of time. Where the horizontal escape route is protected it should still be of limited length since there is always the risk of premature failure. AD B1
0.28, 0.31
- The second part of the escape route will usually be in a protected stairway designed to be virtually 'fire sterile'. Once inside it should be possible to proceed direct to a place of safety without rushing. Therefore, flames, smoke and gases must be excluded from these routes by fire-resisting construction or adequate smoke control measures or by both these methods. AD B1
0.32
- The protected stairway should lead direct to a place of safety or it may do this via a protected corridor. The ultimate place of safety is open air clear of the effects of fire; however in certain large and complex buildings reasonable safety may be provided within the building if suitable planning and protection measures can be included in the design. AD B1
0.28

Dead ends

Escape in one direction only (a dead end), is acceptable under certain conditions depending on: AD B1
0.30

- The use of the building.
- Its associated fire risk.
- Its size and height.
- The length of the dead end.
- The number of people accommodated in the dead end.

Unacceptable means of escape

Certain paths of travel are not acceptable as means of escape, including: AD B1
0.29

- Lifts, unless designed and installed as evacuation lifts for disabled people in the event of fire.
- Portable or throw-out ladders.
- Manipulative apparatus and appliances such as fold-down ladders.
- Escalators. These should not be counted as additional escape routes due to the uneven nature of the top and bottom steps; however it is likely that people would use them in the event of a fire. Mechanised walkways could be acceptable if they were properly assessed for capacity in the static mode.
- Accommodation stairs. Normally, these would be ignored when assessing the adequacy of a means of escape in a building.

Security

It is possible that security measures intended to prevent unauthorised access to a building may hinder the entry of the fire services when they need access to AD B1
0.35

fight a fire or rescue trapped occupants. Advice may be sought from architectural liaison officers attached to most police forces so that possible conflicts between security and access may be solved at the design stage.

Rules for measurement

AD B1
0.37

In addition to the rules for measurement described on p. 7.11 above, which apply to all parts of AD B, there are certain rules which relate only to AD B1. These are concerned with assessing the length and capacity of escape routes which may, in turn, have a bearing on the quantity of routes and the number of stairways that it is necessary to provide.

Occupant capacity

AD B1
0.38

This is the total number of people that a building, storey or room is designed to contain. If this number is not known it may be calculated for an individual room or storey by dividing its area by the floor space factor in square metres (m^2) per person obtained from Table 1 of AD B1 which is reproduced below. For storey occupant capacities it is permissible to ignore stair enclosures, lifts and sanitary accommodation when calculating the floor area. Having obtained the occupant capacity for each storey in this way, the building occupant capacity may be obtained by totalling the individual floor capacities.

For example, an exhibition hall has a floor area (excluding stair enclosures, lifts and sanitary accommodation) of $2250\,m^2$. From Table 1, item 5, the floor space factor is $1.5\,m^2$ per person. Therefore, the occupant capacity is 2250 divided by 1.5 i.e. 1500 people.

Travel distance

AD B1
0.39

This is measured along the shortest, most direct route. Where there is fixed seating or other fixed obstructions, the travel distance is measured along the centreline of the seating or gangways. If a stair (such as an accommodation stair) is included in the escape route, the travel distance is measured along the pitch line on the centreline of travel.

Width

AD B1
0.40

Usually, the narrowest part of an escape route will be at door openings. These are measured by taking the width of the door leaf (or the sum of both door leaves in the case of double doors) even though this may give a width which is greater than the actual door clear opening width.

Escape route widths are measured at a height of 1.5 m above floor or stair pitch line if the route is defined by walls. Handrails fixed to walls may be ignored. Where the width is not so defined then the narrowest distance between fixed obstructions is taken.

Stair widths are measured clear between walls and balustrades. Stringers which do not project more than 30 mm may be ignored as may handrails with less than 100 mm projections.

AD B1

Table 1   **Floor space factors**

Type of accommodation(1)(6)	Floor space factor m²/person
1. Standing spectator areas	0.3
2. Amusement arcade, Assembly hall (including a general purpose place of assembly), Bar (including a lounge bar), Bingo hall, Dance floor or hall, Club, Crush hall, Venue for pop concert and similar events, Queuing area	0.5
3. Concourse or shopping mall (2)	0.75
4. Committee room, Common room, Conference room, Dining room, Licensed betting office (public area), Lounge (other than a lounge bar), Meeting room, Reading room, Restaurant, Staff room, Waiting room (3)	1.0
5. Exhibition hall	1.5
6. Shop sales area (4), Skating rink	2.0
7. Art gallery, Dormitory, Factory production area, Office (open-plan exceeding 60 m²), Workshop	5.0
8. Kitchen, Library, Office (other than in 7 above), Shop sales area (5)	7.0
9. Bedroom or Study-bedroom	8.0
10. Bed-sitting room, Billiards room	10.0
11. Storage and warehousing	30.0
12. Car park	two persons per parking space

Notes:

1. Where accommodation is not directly covered by the descriptions given, a reasonable value based on a similar use may be selected.

2. Refer to section 4 of BS 5588: Part 10 for detailed guidance on the calculation of occupancy in common public areas in shopping complexes.

3. Alternatively the occupant capacity may be taken as the number of fixed seats provided, if the occupants will normally be seated.

4. Shops excluding those under item 8, but including supermarkets and department stores (all sales areas), shops for personal services such as hairdressing and shops for the delivery or collection of goods for cleaning, repair or other treatment or for members of the public themselves carrying out such cleaning, repair or other treatment.

5. Shops (excluding those in covered shopping complexes, and excluding department stores) trading predominantly in furniture, floor coverings, cycles, prams, large domestic appliances or other bulky goods, or trading on a wholesale self-selection basis (cash and carry).

6. If there is to be mixed use, the most onerous factor(s) should be applied.

Means of escape from dwellinghouses

Introduction

Approved Document B1 deals with means of escape from dwellinghouses according to the height of the top storey above ground level. (It should be remembered that this is the ground level on the lowest side of the building.) This is probably a sensible approach since storey heights can vary and means

AD B1
sec. 1
1.1

of escape through upper windows become more hazardous with increasing height. Thus, the divisions chosen are:

- Houses with all floors less than 4.5 m above ground (i.e. ground and first floor only).
- Houses with one floor more than 4.5 m above ground (i.e. ground floor, first floor and second floor).
- Houses with two or more floors more than 4.5 m above ground (i.e. ground floor and three or more upper floors).

Therefore, as the height of the top floor increases above ground level, the means of escape provisions become more complex and these are dealt with under separate sections below. Certain recommendations however, are common to all dwellings and these include:

- The installation of smoke alarms.
- Special provisions to deal with basements and inner rooms.
- Windows and external doors used for escape purposes.
- Balconies and flat roofs.

Smoke alarms

AD B1
sec. 1
1.4, 1.12

The minimum recommendation is that all dwellings should be fitted with self-contained smoke alarms permanently wired to a separately fused circuit at the distribution board. Low-voltage units operating through a mains transformer are permitted. The wiring installation should conform to the IEE Wiring Regulations although it is not necessary to use special fire-resistant cables. Mains units with a secondary power supply are acceptable but units operated by primary batteries alone are not.

Smoke alarms should be:

AD B1
sec. 1
1.5

- Designed and manufactured in accordance with BS 5446 *Specification for components of automatic fire alarm systems for residential premises*, Part 1: 1990 *Point-type detectors.*

AD B1
sec. 1
1.9

- Located in circulation areas within 7 m of rooms where fires are likely to start (kitchens and living rooms) and within 3 m of bedroom doors (all distances measured horizontally).

AD B1
sec. 1
1.13

- At least 300 mm from any wall or light fitting if fixed to the ceiling and preferably mounted in the centre.
- Between 150 mm and 300 mm below the ceiling if designed to be wall mounted.

AD B1
sec. 1
1.14

- Fixed in positions that allow for routine maintenance, testing and cleaning (i.e. not over a stairwell or other floor opening).

AD B1
sec. 1
1.15

- Sited away from areas where steam, condensation or fumes could give false alarms (this would include heaters, air-conditioning outlets, bathrooms, showers, cooking areas or garages, etc.).
- Sited away from areas that get very hot or very cold. They should not be fitted to surfaces which are either much hotter or much colder than the rest of the room since air currents might be created which would carry smoke away from the unit.

AD B1
sec. 1
1.11

The number of alarms which are installed will depend on the size and complexity of the layout of the dwelling, but the following minimum provisions should be observed:

- At least one alarm should be installed in each storey of the dwelling.
- Corridors which are more than 15 m long should contain more than one smoke alarm.
- Where more than one smoke alarm is installed they should be interconnected so that the detection of smoke in any unit will activate the alarm signal in all of them.

It should be noted that maintenance of the system in use is of utmost importance. Since this cannot be made a condition of the passing of plans for building regulation purposes it is important to ensure that occupiers are provided with full details of the use of the equipment and its maintenance. BS 5839: Part 1: 1988 *Code of practice for system design, installation and servicing* recommends that this be done.

AD B1
sec. 1
1.15

Alternative approach to the provision of smoke alarms

It is permissible to provide dwellings with an automatic fire detection and alarm system in accordance with the relevant recommendations of BS 5839 *Fire detection and alarm systems for buildings*, Part 1: 1988 *Code of practice for system design, installation and servicing*. This should be to at least the L3 standard specified in the code. (This is a system intended only to protect escape routes, although it may involve the installation of smoke detectors in rooms immediately adjacent to the escape route.) For large dwellinghouses (where the distance from any part of one room to the most distant part of any other room exceeds 30 m), self-contained smoke alarms are unlikely to be adequate; in these cases an automatic fire detection and alarm system complying with BS 5839 should be installed.

AD B1
sec. 1
1.5

AD B1
sec. 1
1.6

Basements and inner rooms

An inner room will be put at risk by a fire occurring in the access room since escape is only possible by passing through that access room. Therefore, an inner room should only be used as:

AD B1
sec. 1
1.16

- a kitchen, laundry or utility room;
- a dressing room;
- a bathroom, shower room or WC; *or*
- any other room which has a suitable escape window or door (see below for details), provided the room is in the basement, or is on the ground or first floor.

Escape from a basement fire may be particularly hazardous if an internal stair has to be used, since it will be necessary to pass through a layer of smoke and hot gases. Therefore, any bedroom or inner habitable room in a basement should have an alternative escape route via a suitable door or window. In dwellings where all the floors are less than 4.5 m above ground, it is permissible for the basement accommodation to connect directly with the rest of the dwelling. In this case the rooms on the other floors should be treated as inner rooms and should contain suitable escape windows and doors.

AD B1
sec. 1
1.17

Windows and external doors used for escape purposes

AD B1
sec. 1
1.18
To be suitable for escape purposes, windows and external doors should conform to the dimensions given in Fig. 7.10. Dormer windows and roof windows situated above the ground storey should comply with Fig. 7.13(b). Escape should be to a place of safety free from the effects of fire. Where this is to an enclosed back garden or yard its length should be at least equivalent to the height of the dwelling (see Fig. 7.11).

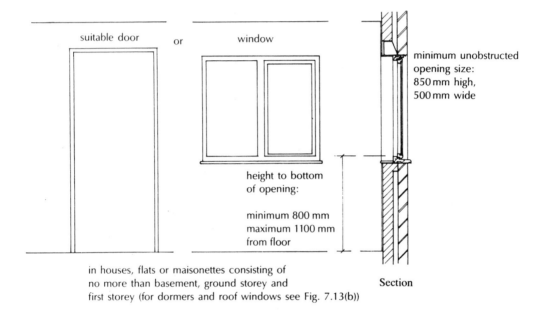

suitable door   or   window

minimum unobstructed opening size: 850 mm high, 500 mm wide

height to bottom of opening:

minimum 800 mm
maximum 1100 mm
from floor

in houses, flats or maisonettes consisting of no more than basement, ground storey and first storey (for dormers and roof windows see Fig. 7.13(b))

Section

Fig. 7.10 Windows and doors for escape purposes.

Balconies and flat roofs

AD B1
sec. 1
1.19
If used for escape routes, balconies and flat roofs should be guarded in accordance with the provisions of ADK (see Chapter 15).

Houses with all floors less than 4.5 m above ground

The provisions outlined above are suitable for dwellings with all floors less than 4.5 m above ground and no further measures for means of escape are necessary.

Houses with one floor more than 4.5 m above ground

AD B1
sec. 1
1.20
Houses with one floor more than 4.5 m above ground are likely to consist of ground, first and second floors. Two alternative solutions are recommended in AD B1.

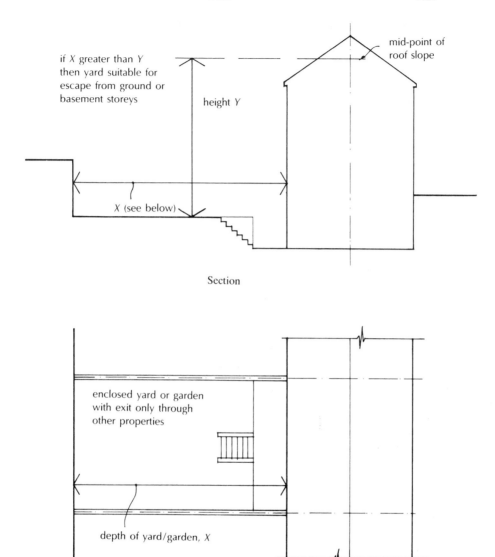

if X greater than Y
then yard suitable for
escape from ground or
basement storeys

mid-point of
roof slope

height Y

X (see below)

Section

enclosed yard or garden
with exit only through
other properties

depth of yard/garden, X

Plan

Fig. 7.11 Enclosed yard or garden suitable for escape purposes –
dwellinghouses.

(a) Provide the first and second floors with a protected stairway which should
either:

(i) Terminate directly in a final exit, or

(ii) Give access to a minimum of two escape routes, separated from each
other by self-closing fire doors and fire-resisting construction and
each leading to a final exit.

These alternatives are illustrated in Fig. 7.12.

(b) Separate the top floor from the other floors by fire-resisting construction and provide it with an alternative means of escape leading to its own final exit. (A variation on this solution can be used where it is intended to convert the roof space of a two-storey dwelling, effectively making it three-storey. Further details of this are described below.)

AD B1
sec. 1
1.22
Where ducted warm air heating systems are provided in houses in this category, additional precautions may be needed to prevent smoke or fire from spreading into the protected stairway. See BS 5588: Part 1, clause 6, for details of guidance on this.

Houses with two or more floors more than 4.5 m above ground

AD B1
sec. 1
1.21
Normally, this would include houses with three or more storeys above the ground floor. No practical guidance is given in AD B1 for houses of this size and the reader is referred to clause 4.4 of BS 5588 *Fire precautions in the design and construction of buildings*, Part 1: 1990 *Code of practice for residential buildings*. In most circumstances it will be necessary to provide a separate means of escape for all floors which are more than 7.5 m above ground.

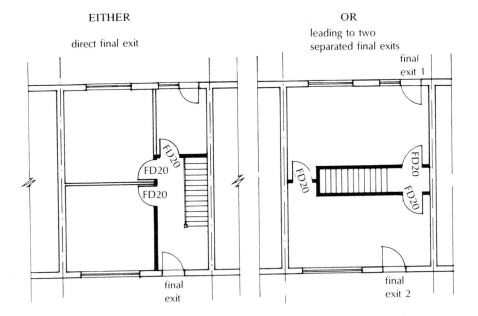

Fig. 7.12 Alternative final exit arrangements.

Conversions to provide rooms in a roof space

The following provisions apply when it is proposed to convert the roof space of a two-storey house to provide living accommodation. The additional floor provided is most likely to be more than 4.5 m above ground and will therefore require additional protection to the means of escape or an alternative escape route. The measures described do not apply if:

AD B1 sec. 1 1.23–1.31

- It is necessary to raise the ridge line; *or*
- The new floor exceeds 50 m² in area; *or*
- It is proposed to provide more than two habitable rooms in the new storey.

The recommendations for means of escape are illustrated in Figs. 7.13(a) and 7.13(b). The general principles being:

- Doors to habitable rooms in the existing stair enclosure should be self-closing to prevent the movement of smoke into the means of escape.
- New doors to habitable rooms should be fire-resisting and self-closing.
- The stairway in the ground and first floors should be adequately protected with fire-resisting construction and should terminate as shown in Fig. 7.12 above.
- Glazing in the existing stair enclosure should be fire-resisting. This includes glazing in all doors except those to bathrooms or WCs.
- The new storey should be served by a stair which is located as shown in Figs. 7.13(a) and 7.13(b) and complies with AD K (see Chapter 15). This could be an alternating tread stair.
- The new storey should be separated from the remainder of the dwelling by fire-resisting construction. Any openings should be protected by self-closing fire doors situated as shown in Figs. 7.13(a) and 7.13(b).
- Windows in the new third storey should be large enough and suitably positioned to allow escape by means of a ladder to the ground.
- There should be suitable means of pedestrian access to the place at which the ladder would be set.

Means of escape from flats and maisonettes

Introduction

Approved Document B1 deals with a limited range of common designs for flats and maisonettes. Where less common arrangements are desired reference should be made to BS 5588: Part 1: 1990, clauses 9 and 10.

AD B1 sec. 1 2.1, 2.2

The means of escape recommendations listed above for dwellinghouses which consist of basement, ground and first floors only, apply equally to flats and maisonettes situated at these levels. At higher levels, escape through upper windows becomes more hazardous and more complex provisions are necessary especially in maisonettes where internal stairs will need protection. In addition to the general assumptions stated above for all building types, the following assumptions are made when considering the means of escape from flats and maisonettes:

AD B1 sec. 2 2.3

- Fires generally originate in the dwelling.
- Rescue by ladders is not considered suitable.

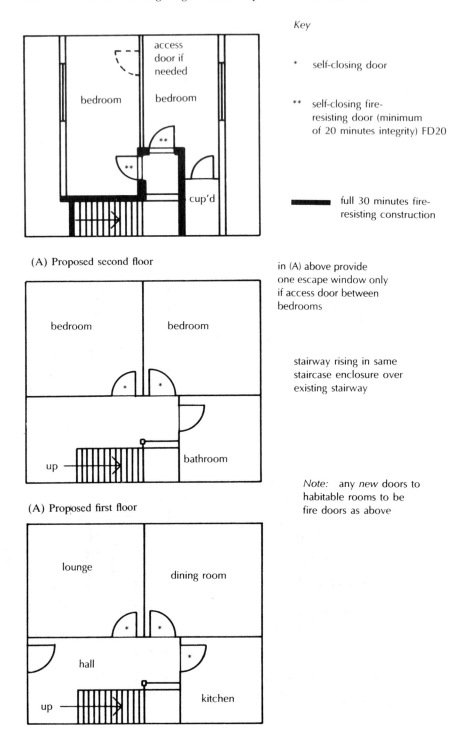

Key

*   self-closing door

**   self-closing fire-
      resisting door (minimum
      of 20 minutes integrity) FD20

━━━━━ full 30 minutes fire-
      resisting construction

(A) Proposed second floor

in (A) above provide
one escape window only
if access door between
bedrooms

stairway rising in same
staircase enclosure over
existing stairway

(A) Proposed first floor

Note: any *new* doors to
habitable rooms to be
fire doors as above

Existing Ground Floor

Fig. 7.13(a) Conversion to provide rooms in roof.

Window or roof-light
minimum 850 mm high
× 500 mm wide when
open and located so as
to be accessible from
ground by ladder

Note: window could also be in end or gable
wall of house

dormer window
bottom of opening   roof-light

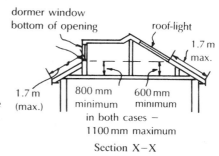

1.7 m
max.

1.7 m   800 mm   600 mm
(max.)   minimum   minimum
        in both cases –
        1100 mm maximum

Section X–X

doors provided at top of stairway

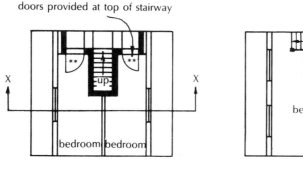

X   X

(B) Proposed second floor

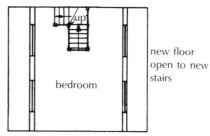

new floor
open to new
stairs

bedroom

(C) Proposed second floor

full 30 minutes fire-resisting construction including
underside of staircase if exposed in bedroom

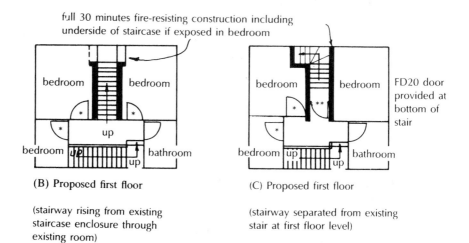

bedroom   bedroom

up

bedroom   bathroom

(B) Proposed first floor

(stairway rising from existing
staircase enclosure through
existing room)

bedroom   bedroom

FD20 door
provided at
bottom of
stair

bedroom   bathroom

(C) Proposed first floor

(stairway separated from existing
stair at first floor level)

Fig. 7.13(b) Conversion to provide rooms in roof.

- Fire spread beyond the dwelling of origin is unlikely due to the compartmentation recommendations in AD B3, therefore simultaneous evacuation of the building should be unnecessary.
- Fires which occur in common areas are unlikely to spread beyond the immediate vicinity of the outbreak due to the materials and construction used there.

AD B1
sec. 2
2.4

There are two main components to the means of escape from flats and maisonettes:

- Escape from within the dwelling itself; *and*
- Escape from the dwelling to the final exit from the building.

The following sections consider these two components.

Means of escape from within flats and maisonettes

AD B1
sec. 2
2.7

The provisions relating to automatic smoke detection, inner rooms, basements, balconies and flat roofs in dwellings also apply to the inner parts of flats and maisonettes, with the following qualifications:

- The provisions for automatic smoke detection and alarm are not required to be extended to common areas in blocks of flats and do not include interconnection of the installations in separate flats.
- Maisonettes should be treated in the same way as multi-storey houses for the provision of automatic smoke detection and alarm.

AD B1
sec. 2
2.8

- Where the floor of a flat is not more than 4.5 m above ground, the flat should be planned so that any inner habitable room is provided with a suitable escape window or door, as shown in Fig. 7.10 above.

AD B1
sec. 2
2.17

- Where ducted warm air heating systems are provided in flats and maisonettes with any floor more than 4.5 m above ground, additional precautions may be needed to prevent smoke or fire from spreading into protected entrance halls or landings. See BS 5588: Part 1, clause 15, for details of guidance on this.

Flats with floors more than 4.5 m above ground

AD B1
sec. 2
2.12

Provided that the restrictions on inner rooms are observed, three possible solutions to the internal planning of flats are given in AD B1:

(a) All the habitable rooms in the flat are arranged to have direct access to a protected entrance hall. The maximum distance from the entrance door to the door of a habitable room should not exceed 9 m. (See Fig. 7.14(a).)

(b) In flats where a protected entrance hall is not provided, the 9 m distance referred to in (a) should be taken as the furthest distance from any point in the flat to the entrance door. Cooking facilities should be remote from the entrance door and positioned so that they do not prejudice the escape route from any point in the flat. (See Fig. 7.14(b).)

(c) Provide an alternative exit from the flat. In this case the internal planning will be more flexible. An example of a typical flat plan where an alternative exit is provided, but not all the habitable rooms have direct access to the entrance hall, is shown in Fig. 7.14(c).

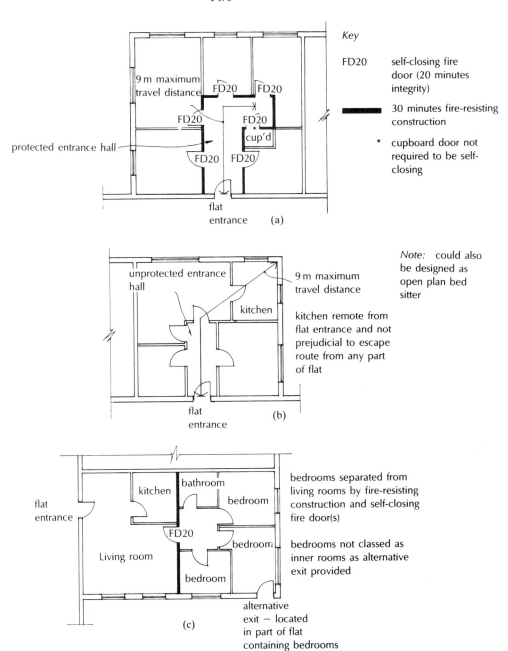

Fig. 7.14 Alternative internal layout to flats.
(a) Flat with protected entrance hall.
(b) Flat with unprotected entrance hall and limited travel distance.
(c) Flat with alternative exit but no common entrance hall.

Maisonette with independent external entrance at ground level

AD B1
sec. 2
2.14

A maisonette of this type is similar to a dwellinghouse and should have a means of escape which complies with the recommendations for dwellings described on p. 7.24 above, depending on the height of the top storey above ground.

Maisonette with floor more than 4.5 m above ground and no external entrance at ground level

AD B1
sec. 2
2.15

Two internal planning arrangements are described in AD B1 for maisonettes of this type. These are illustrated in Fig. 7.15, as follows:

(a) Provide each habitable room which is not on the entrance floor with an alternative exit, or
(b) Provide a protected entrance hall or landing entered directly from all the habitable rooms on that floor. Additionally, one alternative exit should be provided on each floor other than the entrance floor.

Alternative exits from flats and maisonettes

AD B1
sec. 2
2.16

Reference has been made above, to the provision of alternative exits in certain planning arrangements for flats and maisonettes. Alternative exits will only be effective if they are remote from the main entrance door to the dwelling and lead to a common stair or final exit by means of:

- A door to a common balcony or access corridor.
- An internal private stairway leading to a common balcony or access corridor on another level.
- A door to an external stair.
- A door to an escape route over a flat roof.

Means of escape from flats and maisonettes to the final exit from the building

AD B1
sec. 2
2.18

In general, flats and maisonettes should have access to an alternative means of escape from their entrance doors. In this way it will be possible to escape from a fire in a neighbouring flat by walking away from it. It is not always possible to provide alternative escape routes (which means providing two or more staircases) in all buildings containing flats and maisonettes, therefore single staircase buildings are permissible in certain circumstances. Typical examples of single and multi-stair buildings are described below, although more of these provisions apply where the top floor is not more than 4.5 m above ground.

Flats and maisonettes with single common stairs

AD B1
sec. 2
2.18, 2.19

In larger buildings it will be necessary to separate the entrance to each dwelling from the common stair by a protected lobby or common corridor.

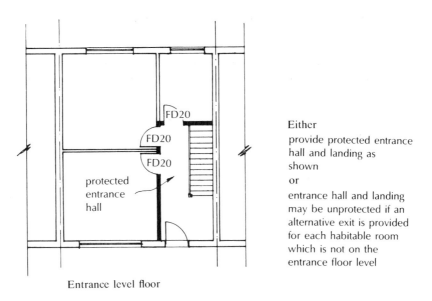

Either

provide protected entrance hall and landing as shown

or

entrance hall and landing may be unprotected if an alternative exit is provided for each habitable room which is not on the entrance floor level

Entrance level floor

these provisions apply where at least one storey is more than 4.5 m above ground level

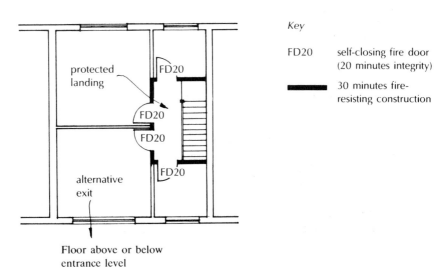

Key

FD20   self-closing fire door (20 minutes integrity)

▬▬▬   30 minutes fire-resisting construction

Floor above or below entrance level

Fig. 7.15 Maisonette with no independent external entrance at ground level.

The maximum distance from any entrance door to the common stair or protected lobby should not exceed 7.5 m. (See Figs. 7.16(a) and 7.16(b).)

These recommendations may be modified for smaller buildings where:

- The building consists of a ground storey and no more than three other storeys.
- The top floor does not exceed 11m above ground level.
- The stair does not connect to a covered car park unless it is open-sided.

The modified recommendations are illustrated in Fig. 7.16(c). The maximum distance from the dwelling entrance door to the stair entrance should be reduced to 4.5 m. If the intervening lobby is provided with an automatic opening vent this distance may be increased to 7.5 m. Where the building contains only two flats per floor further simplifications as shown in Fig. 7.16(d) are possible.

Flats and maisonettes with more than one common stair

AD B1
sec. 2
2.21

Where escape is possible in two directions from the dwelling entrance door, the maximum escape distance to a storey exit may be increased to 30 m. Furthermore, if all the dwellings on a storey have independent alternative means of escape, the maximum travel distance of 30 m does not apply. (There is still, however, the need to comply with the fire service access recommendations in Approved Document B5, where all parts of the building should be within 60 m of a fire main.)

In buildings of this type it is possible to have a dead end situation where the stairs are not located at the extremities of each storey. This is permissible provided that the dead end portions of the corridor are fitted with automatic opening vents and the dwelling entrance doors are within 7.5 m of the common stair entrance.

Typical details of flats and maisonettes with more than one common stair are shown in Fig. 7.17.

Additional provisions for common escape routes

Common escape routes should be planned so that they are not put at risk by a fire in any of the dwellings or in any stores or ancillary accommodation. The following recommendations are designed to provide additional protection to these routes:

AD B1
sec. 2
2.21

- It should not be necessary to pass through one stairway enclosure to reach another. Where this is unavoidable a protected lobby should be provided to the stairway. This lobby may be passed through in order to reach the other stair.

AD B1
sec. 2
2.22

- Common corridors should be designed as protected corridors and should be constructed to be fire-resisting.
- The wall between each dwelling and the common corridor should be a compartment wall.

AD B1
sec. 2
2.24

- A common corridor connecting two or more storey exits should be sub-divided with a self-closing fire door and/or fire-resistant screen, positioned so that smoke will not affect access to more than one storey exit.

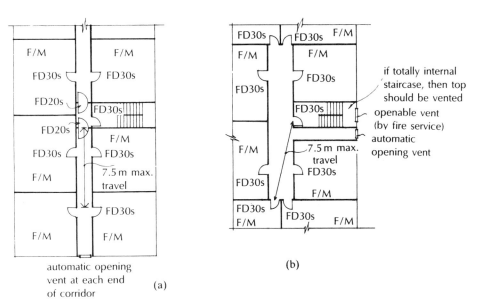

Note: automatic opening vents to
have min. free area of 1.5 m²

maximum travel
distance may be
increased to 7.5 m
if automatic
opening vent
provided in lobby

openable vent
(by fire service)

door free from
security fastenings
(lobby may be
omitted if flats/maisonettes
have protected entrance
halls)

openable vent
(by fire service)

maximum 2 dwellings
per floor

single stair access in
small buildings shown in
(c) and (d) permitted if:

● maximum 5 storeys
● too floor not greater than
  11 m above ground level
● escape route does not connect
  to covered car park at ground
  level unless open sided

Key

FD30s   self-closing fire door
(30 minutes integrity and
restricted smoke
leakage)

▬▬▬   fire-resisting
construction

F/M   flat or maisonette

Fig. 7.16 Flats and maisonettes with single common stairs. (a) Corridor access.
(b) Lobby access. (c) and (d) Single stair access in small buildings.

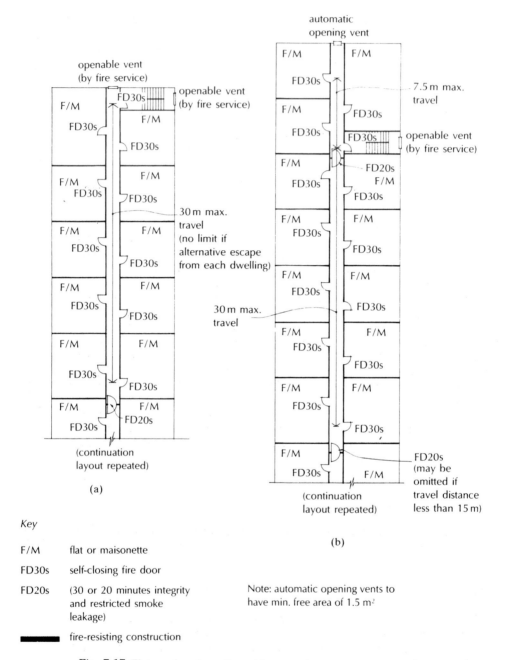

Key

F/M   flat or maisonette

FD30s   self-closing fire door

FD20s   (30 or 20 minutes integrity
           and restricted smoke
           leakage)

▬▬▬   fire-resisting construction

Note: automatic opening vents to
have min. free area of 1.5 m²

Fig. 7.17 Flats and maisonettes with more than one common stair.
(a) Corridor access. (b) Corridor access with dead ends.

- A dead end section of a common corridor should be separated in a similar manner from the rest of the corridor. **AD B1** sec. 2 2.25
- Protected lobbies and corridors should not contain any stores, refuse chutes, refuse storage areas or other ancillary accommodation. **AD B1** sec. 2 2.27

Ventilation of common escape routes

Although precautions can be taken to prevent the ingress of smoke onto common corridors and lobbies, it is almost inevitable that there will be some leakage since a flat entrance door must be opened in order that the occupants can escape. Provisions for ventilation of the common areas are therefore vital and may be summarised as follows: **AD B1** sec. 2 2.23

- Subject to the variations shown in Figs. 7.16(a) and 7.16(b), common corridors or lobbies in larger, single-stair buildings should be provided with automatic opening ventilators, triggered by automatic smoke detection. These should be positioned as shown in the diagram, should have a free area of at least $1.5\,\text{m}^2$ and be fitted with a manual override.
- Small single-stair buildings should conform to the guidance shown in Figs. 7.16(c) and 7.16(d).
- Common corridors in multi-stair buildings should extend at both ends to the external face of the building where openable ventilators, or automatic opening ventilators, should be fitted for fire service use. They should have a free area of $1.5\,\text{m}^2$ at each end of the corridor.
- It is possible to protect escape stairways, corridors and lobbies by means of smoke control systems employing pressurisation. These systems should comply with BS 5588: Part 4: 1978 *Code of practice for smoke control in protected escape routes using pressurization.* Where these are provided the cross corridor fire doors and automatic opening ventilators referred to above should be omitted. **AD B1** sec. 2 2.26

Escape routes across flat roofs

Where more than one escape route exists from a storey or part of a building, one of those routes may be across a flat roof if the following conditions are observed: **AD B1** sec. 2 2.28

- The flat roof should be part of the same building.
- The escape route over the flat roof should lead to a storey exit.
- The roof and its structure forming the escape route should be fire-resisting.
- Any opening within 3 m of the route should be fire-resisting.
- The route should be adequately defined and guarded in accordance with AD K (see Chapter 15).

Provision of common stairs in flats and maisonettes

Stairs which are used for escape purposes should provide a reasonable degree of safety during evacuation of a building. Since they may also form a potential route for fire spread from floor to floor there are recommendations contained **AD B1** sec. 2 2.31, 2.32

in AD B3 which are designed to prevent this. Stairs may also be used for fire fighting purposes. In this case reference should be made to the recommendations contained in AD B5. Both these Approved Documents are described in detail below. The following recommendations are specifically for means of escape purposes:

AD B1
sec. 2
2.33,
2.34

- Each common stair should be situated in a fire-resisting enclosure with the appropriate level of fire resistance taken from Tables A1 and A2 of Appendix A of AD B (see pp. 7.80 and 7.83 below).

AD B1
sec. 2
2.35

- Each protected stair should discharge either:

  (a) direct to a final exit, or
  (b) by means of a protected exit passageway to a final exit.

AD B1
sec. 2
2.36

- If two protected stairways or protected exit passageways are adjacent, they should be separated by an imperforate enclosure.

AD B1
sec. 2
2.37

- A protected stairway should contain no other accommodation apart from a lift well.

AD B1
sec. 2
2.38

- Openings in the external walls of protected stairways should be protected from fire in other parts of the building if they are situated where they might be at risk. (See p. 7.55 below for details of protection measures.)

AD B1
sec. 2
2.30

- A common stair should have a minimum width of 1 m, and if used as a firefighting stair this should be increased to 1.1 m.

AD B1
sec. 2
2.40–2.42

- Basement stairs will need to comply with special measures (see p. 7.55 below).

AD B1
sec. 2
2.39

- Gas service pipes and meters should only be installed in protected stairways if the installation complies with the requirements for installation and connection set out in the Gas Safety Regulations 1972 (SI 1972/1178) and the Gas Safety (Installation and use) (Amendment) Regulations 1990.

AD B1
sec. 2
2.43,
2.44

- A common stair which forms part of the only escape route from a flat or maisonette should not also serve any fire risk area such as a covered car park, boiler room, fuel storage space or other similar ancillary accommodation.

- Where more than one common stair is provided as an escape from a dwelling it is permitted to serve ancillary accommodation.

External escape stairs

AD B1
sec. 2
2.45

External escape stairs are stairs which are not part of the structure of the building and are not, therefore, subject to the requirements of Part K of Schedule 1 to the 1991 Regulations. They are permitted to be used as an alternative means of escape where there is more than one escape route available from a storey or part of a building if the following provisions can be met:

- The stair should not serve any floors which are more than 6 m above the ground or a roof or podium. The roof or podium should itself be served by an independent protected stair.
- All the doors which lead onto the stair should be fire-resisting and self-closing. This does not apply to the only exit door to the landing at the head of a stair which leads downward.
- External walls which are within 1.8 m of and 9 m vertically below the stair,

should be of fire-resisting construction. This 1.8 m dimension may be reduced to 1.1 m above the top landing level provided that this is not the top of a stair up from basement level to ground.

- Any part of the building which is within 3 m of the escape route from the stair to a place of safety should be protected with fire-resisting construction.

These recommendations are shown in Fig. 7.18.

Stairs to flats in mixed use buildings

Many buildings consist of a mix of flats and other uses. Sometimes the flats are ancillary to the main use (such as a caretaker's flat in an office block), and sometimes they form a distinct separate use (as in the case of shops with flats

AD B1
sec. 2
2.46, 2.47

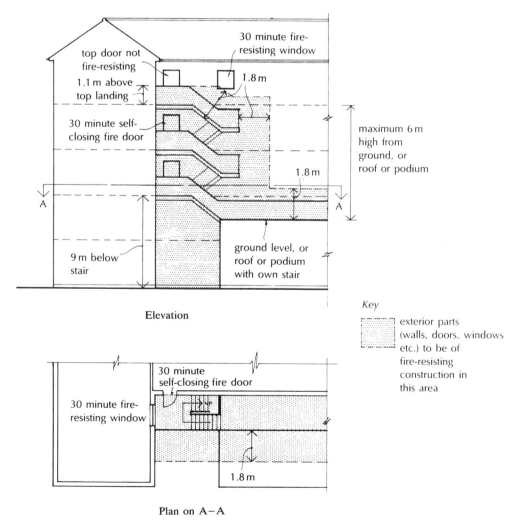

Fig. 7.18 External escape stairs.

over). Clearly, the degree of separation of the uses for means of escape purposes will depend on the height of the building and the extent to which the uses are interdependent.

Where a building has no more than three storeys above the ground storey, the stairs may serve both non-residential and dwelling uses, with the proviso that each occupancy is separated from the stairs by protected lobbies at all levels.

In larger buildings where there are more than three storeys above the ground storey, the flats should have their own separate stairs and these should not communicate with any other part of the building. However, a flat which is ancillary to the main use of the building is permitted to share a stair with the rest of the building if:

- The stair is separated from other parts of the building at lower storey levels by protected lobbies at those levels.
- The flat is provided with an independent alternative escape route.
- Any automatic fire detection and alarm system fitted in the main part of the building is extended to the flat.

Means of escape from buildings other than dwellinghouses, flats and maisonettes

Introduction

So far we have discussed the provisions contained in AD B1 for means of escape in buildings of purpose groups 1(a), (b) and (c) i.e. dwellinghouses, flats and maisonettes. Although only a few of the more common arrangements for flats and maisonettes are covered in the Approved Document, the recommendations which are given are quite detailed and should form the basis for sound design guidance.

All other building types are grouped together in AD B1 and design recommendations are given for horizontal escape in Section 3, and vertical escape in Section 4. Of necessity, the guidance given is general in nature and designers may well find that it is better to use other relevant design documents where more comprehensive guidance may be given.

Horizontal escape routes – buildings other than dwellings

AD B1
sec. 3
3.1

Section 3 of AD B1 deals with the provision of means of escape from any point in the floor of a building to the storey exit of that floor. It covers all buildings apart from dwellinghouses, flats and maisonettes.

AD B1
sec. 3
3.2

The main decision that needs to be taken when designing the means of escape from a building is the number of escape routes and exits that are required. This will depend on:

- the maximum travel distance which is permitted to the nearest exit; *and*
- the number of occupants in the room, tier or storey under consideration.

Maximum travel distances and alternative escape routes in buildings other than dwellings

AD B1
sec. 3
3.3, 3.4

Ideally, there should be alternative escape routes provided from every part of the building. This is especially important in multi-storey buildings and in buildings where a mixture of purpose groups are present. In fact, if a mixed use building also contains residential, or assembly and recreation purpose groups, these should be served by their own independent means of escape. (But see the exceptions to this for flats, described on p. 7.41 above.)

Where alternative escape routes are provided, escape will be possible in more than one direction. AD B1 places limits on the travel distance from any part of a room, tier or storey to a storey exit and these are shown in Table 3 from AD B1 which is reproduced below. It is important to read the notes which accompany this table since the distances given are actual maximum travel distances. Where the layout of a room or storey is not known at the design stage, the direct distance measured in a straight line should be taken. Direct distances are two-thirds of the travel distance.

It will be observed from Table 3 that where escape is possible in one direction only (i.e. a dead end), the travel distances are much reduced. Where these one-directional travel distances are adhered to, single direction escape routes are permitted in the following circumstances:

AD B1 sec. 3 3.5

- where a storey has an occupant capacity of not more than 50 people (except storeys used for in-patient care in hospitals); *and*
- where a room in a storey has an occupant capacity of not more than 50 people, or not more than 30 people if the building is in the institutional purpose group.

Similarly, it is often the case that there will not be alternative escape routes, especially at the beginning of an escape route. A room may have only one exit onto a corridor from where it may be possible to escape in two directions. This is permissible provided that:

AD B1 sec. 3 3.6, 3.8

- the overall distance from the furthest point in the room to the storey exit complies with the two-directional travel distance from Table 3; *and*
- the single direction part of the route (in this case, in the room) complies with the travel distance in 'one direction only' specified in Table 3.

Although a choice of escape routes may be provided from a room or storey, it is possible that they may be so located, relative to one another, that a fire might disable them both. In order to consider them as true alternatives they should be positioned as shown in Fig. 7.19(a) i.e. the angle which is formed between the exits and any point in the space should be at least 45°. Where this angle cannot be achieved:

- the alternative escape routes should be separated from each other by fire-resisting construction; *or*
- the maximum travel distance for escape in one direction will apply.

Figure 7.19 illustrates these rules covering alternative escape routes.

Number of exits related to the number of occupants in buildings other than dwellings

The number of occupants in a room, tier or storey also affects the number of exits and escape routes which need to be provided from that area. Table 4 of

AD B1 sec. 3 3.7

AD B1

Table 3 Limitations on travel distance

Purpose group	Use of the premises or part of the premises	Maximum travel distance[1] where travel is possible in:	
		one direction only (m)	more than one direction (m)
2(a)	Institutional[2]	9	18
2(b)	Other residential:		
	(a) in bedrooms[3]	9	18
	(b) in bedroom corridors	9	35
	(c) elsewhere	18	35
3	Office	18	45
4	Shop and Commercial[4]	18	45
5	Assembly and Recreation:		
	(a) buildings primarily for the handicapped except schools	9	18
	(b) elsewhere	15	32
6	Industrial[5]	25	45
7	Storage and other non-residential[5]	18	45
2–7	Place of special fire risk[6]	9[3]	18[3]
2–7	Plant room or rooftop plant:		
	(a) distance within the room	9	35
	(b) escape route not in open air (overall travel distance)	18	45
	(c) escape route in open air (overall travel distance)	60	100

Notes:
1. The dimensions in the table are travel distances. If the internal layout of partitions, fittings, etc. is not known when plans are deposited, direct distances may be used for assessment. The direct distance is taken as $\frac{2}{3}$rds of the travel distance.
2. If provision for means of escape is being made in a hospital or other health care building by following the detailed guidance in the relevant part of the Department of Health 'Firecode', the recommendations about travel distances in the appropriate 'Firecode' document should be followed.
3.. Maximum part of travel distance within the room.
4. Maximum travel distances within shopping malls are given in BS 5588: Part 10: 1991. Guidance on associated smoke control measures is given in a BRE report *Design principles for smoke ventilation in enclosed shopping centres BR 186.*

5. In industrial buildings the appropriate travel distance depends on the level of fire risk associated with the processes and materials being used. Control over the use of industrial buildings is exercised through the Fire Precautions Act. Attention is drawn to the guidance issued by the Home Office *Guide to fire precautions in existing places of work that require a fire certificate Factories Offices Shops and Railway Premises.* The dimensions given above assume that the premises will be of 'normal' fire risk, as described in the Home Office guidance. If the building is high risk, as assessed against the criteria in the Home Office guidance, then lesser distances of 12 m in one direction and 25 m in more than one direction, would apply.
6. Places of special fire risk are listed in the definitions in Appendix E of AD B.

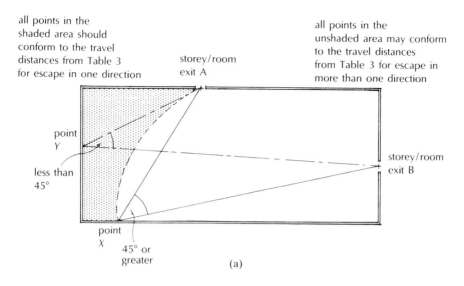

all points in the shaded area should conform to the travel distances from Table 3 for escape in one direction

all points in the unshaded area may conform to the travel distances from Table 3 for escape in more than one direction

storey/room exit A

point Y

less than 45°

storey/room exit B

point X

45° or greater

(a)

OR: if 45° angle cannot be achieved, separate alternative escape routes from each other with fire-resisting construction

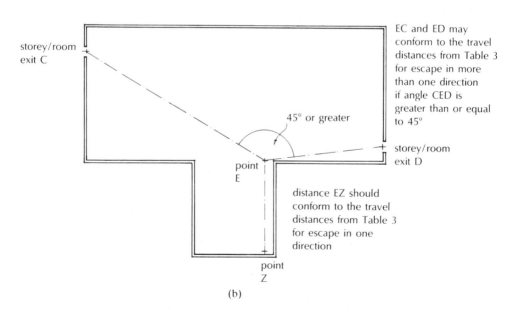

storey/room exit C

EC and ED may conform to the travel distances from Table 3 for escape in more than one direction if angle CED is greater than or equal to 45°

45° or greater

storey/room exit D

point E

distance EZ should conform to the travel distances from Table 3 for escape in one direction

point Z

(b)

Fig. 7.19 Alternative escape routes.

AD B1

Table **4**   **Minimum number of escape routes and exits from a room, tier or storey**

Maximum number of persons	Minimum number of escape routes/exits
500	2[1]
1000	3
2000	4
4000	5
7000	6
11000	7
16000	8
more than 16000	8[2]

Notes:
1. See paragraph 3.5 [p. 7.43] about the circumstances in which single exits and escape routes are acceptable.
2. Plus 1 per 5000 persons (or part thereof) over 16 000.

AD B1 lists the minimum number of escape routes which should be provided related to the maximum numbers of occupants, and is reproduced below. The figures for the maximum number of occupants should be based on the design information. Where this is not known the numbers should be calculated using the floor space factors from Table 1 of AD B1 (see p.7.23 above).

Width and height of escape routes and exits

AD B1
sec. 3
3.14 to 3.17

The width of an escape route will also be related to the number of people who will need to use it. The numbers should be assessed as described above if they are not known at the design stage and then the width can be chosen from Table 5 of AD B1 (see below).

Where a storey has two or more exits it is assumed that one of them will be disabled by a fire. Therefore the remaining exits should have sufficient width to take the occupants safely and quickly. This means that the widest exit should be discounted and the remainder should be designed to take the occupants of the storey. Since stairs need to be as wide as the exit leading onto them, this recommendation for exit width may influence the width of the stairways. (Stairways may also need to be discounted and this is discussed on p. 7.51 below.)

Except in doorways, all escape routes should have a clear headroom of at least 2 m.

Horizontal escape routes – factors affecting internal planning

The efficacy of an escape route in a storey may be affected by a number of internal planning considerations, such as:

AD B1

Table **5**   **Widths of escape routes and exits**[1]

Maximum number of persons	Minimum width (mm)[2]
50	800[3]
110	900
220	1100
more than 220	5 per person

Notes:
1. Refer to p. 7.22 for methods of measurement of width.
2. Refer to the guidance in the Approved Document to part M on minimum widths for areas accessible to disabled people.
3. May be reduced to 530 mm for gangways between fixed storage racking, other than in public areas of Purpose Group 4 (shop and commercial).

- The need for inner rooms.
- The relationship between circulation routes and stairways.
- The need for different occupancies in a building to use the same escape route.
- The design and layout of means of escape corridors.

These are considered in more detail in the following paragraphs.

Provision of inner rooms

The rules governing the provision of inner rooms are more stringent than those for dwellings. Inner rooms are only acceptable under the following conditions:

AD B1
sec. 3
3.9

- The occupant capacity should be as for a room with only one direction of escape (see p. 7.43 above).
- It should not be a bedroom.
- Only one access room should be passed through when escaping from the inner room.
- The maximum travel distance from the furthest point in the inner room to the exit from the access room should not exceed the appropriate limit given in Table 3.
- The access room should be in the control of the same occupier as the inner room.
- The access room should not be a place of special fire risk (e.g. a kitchen).

Where these conditions are met the inner room should be designed to conform to one of the following arrangements:

(a) The walls or partitions of the inner room should stop at least 500 mm from the ceiling, or

(b) A vision panel, which need not be more than 0.1 m² in area, should be situated in the walls or door of the inner room (this is to enable the occupiers to see if a fire has started in the access room), or

(c) A suitable automatic fire detection and alarm system should be fitted in the access room which will give warning of fire in that room to the occupiers of the inner room.

Horizontal escape routes and stairways

Care must be taken in the design of horizontal escape routes since they also form part of the normal circulation in a building and may jeopardise access to stairways unless the following points are considered:

AD B1
sec. 3
3.11

- It should not be necessary to pass through one stairway to reach another. Where this is unavoidable a protected lobby should be provided to the stairway. This lobby may be passed through in order to reach the other stair.

AD B1
sec. 3
3.12

- As part of the normal circulation in a building, it should not be necessary to pass through a stairway enclosure in order to reach another part of the building on the same level. (Such circulation patterns should be avoided since familiarity breeds contempt, and fire doors may become ineffective due to excessive use or misuse.)

AD B1
sec. 3
3.10

- Where buildings are planned with more than one exit round a central core, these exits should be remote from each other and should not be approached from the same lift hall, common lobby or undivided corridor, or linked together by any of these.

Use of common escape routes by different occupancies

AD B1
sec. 3
3.13

It is common in mixed-use or multi-tenanted buildings for common escape routes to be used by all the occupants. There are restrictions on this and these have been referred to above on pp. 7.34 and 7.41. Where common escape routes are permitted the following rules should be observed:

- The means of escape from one occupancy should not pass through another.
- Common corridors should either be fire-protected or fitted with an automatic fire detection and alarm system which extends throughout the storey.
- The rules on compartmentation referred to below should be observed.

Means of escape corridors – design factors

AD B1
sec. 3
3.19

The following means of escape corridors should be fire-protected:

- All corridors in residential accommodation.
- All dead end corridors.
- Those corridors which are common to two or more different occupancies (but see also above).

AD B1
sec. 3
3.20

The enclosures to means of escape corridors which are not required to be fire-protected will still provide some defence against the spread of smoke in the early stages of a fire. This defence against smoke spread may be maintained if the partitions are carried up to the soffit of the floor above or to a suspended ceiling. Openings into rooms should be fitted with doors (which do not need to be fire-resistant or self-closing).

Corridors which give access to alternative escape routes may become blocked by smoke before all the occupants of a building have escaped and may block the access to any permitted dead end corridors. Therefore, corridors connecting two or more storey exits should be sub-divided by means of self-closing fire doors (and screens, if necessary) if they exceed 12 m in length. The doors should be positioned so that:

AD B1
sec. 3
3.21

- No undivided corridor is common to two storey exits; *and*
- The route is protected from smoke, having regard to any adjacent fire risks and the layout of the corridor.

Dead end corridors exceeding 4.5 m in length, which give access to a point from which alternative escape routes are available, should also be provided with fire doors positioned so that they are separated:

AD B1
sec. 3
3.22

- from any corridor which provides two directions of escape; *or*
- from any corridor which continues past one storey exit to another.

These provisions for means of escape corridors are summarised in Fig. 7.20.

Escape routes across flat roofs

With the exception of escape routes serving institutional buildings or any part of a building intended for use by members of the public, the recommendations given above for escape over flat roofs from flats and maisonettes (p. 7.39) also apply to all other building types.

AD B1
sec. 3
3.25

Vertical escape routes – buildings other than dwellings

Section 4 of AD B1 deals with the provisions for vertical escape, by means of a sufficient number of adequately sized and protected escape stairs, for all buildings apart from dwellinghouses, flats and maisonettes.

AD B1
sec. 4
4.1

The main decision to be taken when designing the vertical means of escape in a building is the number of stairways that need to be provided. It has already been shown above that alternative means of escape are required:

- in mixed use buildings which also contain residential, or assembly and recreation purpose groups, where separate, independent means of escape are required; *and*
- where horizontal constraints are imposed by travel distances, exit widths and numbers of occupants.

AD B1
sec. 4
4.2

Section 4 provides additional recommendations for assessing the number of stairways that are needed with regard to:

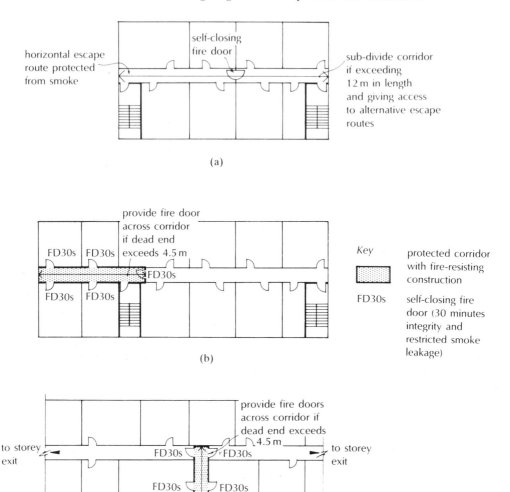

Fig. 7.20 Escape corridors – buildings other than dwellings.
(a) Sub-division of corridors. (b) Dead end separation from continuation corridor.
(c) Dead end separation from alternative escape routes.

- the acceptability of single stairs for means of escape in a building; *and*
- the influence that adequate width of stairs may have on their provision.

AD B1
sec. 4
4.3

One further influence on the provision of stairs is the necessity to provide firefighting stairs in some large buildings. This is covered by AD B5 and may mean that extra stairways are required beyond those needed merely for means of escape purposes.

The provision of single stairs in buildings other than dwellings

Assuming that a building is not excluded from having a single escape route by virtue of the recommendations listed above, the situations where a building may be served by a single escape stair are:

AD B1
sec. 4
4.5

- From a basement.
- From a building which has no floor more than 11 m above ground.
- From an office building which complies with the recommendations of clause 8 of BS 5588: Part 3: 1983 *Code of practice for office buildings.* (These are small offices with single direction travel restrictions and limitations on the height of the top storey above ground.)

Width of stairs in buildings other than dwellings

Stairs for escape purposes need to be sufficiently wide to allow the total number of people in the building or storey affected by fire, to escape. Research has shown that people prefer to stay close to a handrail when making a long descent. Therefore the centre of a very wide stairway would be little used and might, in fact, be hazardous. For this reason, AD B1 puts a limit of 1400 mm on the width of a stairway where its vertical extent is more than 30 m. If a stairway wider than this is dictated by the design of the building it should be at least 1800 mm wide and contain a central handrail. In this case, the stair width on either side of the central handrail will need to be considered separately when assessing stair capacity.

AD B1
sec. 4
4.6, 4.7,
4.8

If the points raised above are taken into consideration, the width of an escape stairway should:

- Be at least as wide as any exits serving it.
- Comply with the minimum width given by Table 6 of AD B1 (see below).
- Not exceed 1400 mm if more than 30 m in vertical extent.
- Be at least 1800 mm if centrally divided by a handrail.
- Not reduce as it approaches the final exit.

If the exit route from a stair also picks up occupants of the ground and basement storeys, it may need to be increased in width accordingly.

Method of assessing minimum stair width

Since the width of a stair is related to the number of people that it can carry in an emergency, it is possible to calculate the minimum stair width required for means of escape. Before this calculation can be made however, it is necessary to consider the way in which the building is likely to be evacuated in a fire.

AD B1
sec. 4
4.9

Where more than one stairway is provided, it is possible that one of the stairs may be unavailable due to fire or smoke unless special precautions are taken. Hence, it is necessary to discount the largest stair in order to check that the remaining stairways are capable of coping with the demand. This discounting is unnecessary if:

AD B1
sec. 4
4.11 to 4.14

- The escape stairs are approached through a protected lobby at each floor level (except for the top floor).

AD B1

Table 6   Minimum width of escape stairs

Situation of stair	Max number of people served[1]	Minimum stair width (mm)
1. In an institutional building (unless the stair will only be used by staff)	150	1000
2. In an assembly building and serving an area used for assembly purposes (unless the area is less than 100 m²)	220	1100
3. In any other building and serving an area with an occupancy of more than 50	over 220	note[2]
4. Any stair not described above	50	800

Notes:
1. Assessed as likely to use the stair in a fire emergency.
2. See Table 7 for sizing stairs for total evacuation, and Table 8 for phased evacuation.

• The stairs are protected by a pressurisation smoke control system designed in accordance with BS 5588: Part 4: 1978.

AD B1
sec. 4
4.15 to 4.18

It is usual to assume that all the occupants would be evacuated. This is termed 'total evacuation' and should be the design approach for:

• all stairs which serve basements,
• all stairs which serve buildings with open spatial planning; *and*
• all stairs which serve assembly and recreation buildings (PG 5) or other residential buildings (PG 2b).

Using this approach the escape stairs should be wide enough to allow all the floors to be evacuated simultaneously. The calculations take into account the number of people temporarily housed in the stairways during evacuation. The figures for the maximum number of occupants should be based on the design information. Where this is not known the numbers should be calculated using the floor space factors from Table 1 of AD B1.

The simplest way of assessing the escape stair width is to use Table 7 from AD B1, which is reproduced below. This covers the capacity of stairs with widths from 1000 mm to 1800 mm for buildings up to ten storeys high. Where the building has more than ten floors the following formula may be used:

$$P = 200w + 50(w - 0.3)(n - 1)$$

where P is the number of people that can be accommodated,
    w is the width of the stair in metres, and
    n is the number of storeys in the building.

AD B1

Table 7   **Capacity of a stair for basements and for total evacuation of the building**

No. of floors served					Maximum number of persons served by a stair of width:				
	1000 mm	1100 mm	1200 mm	1300 mm	1400 mm	1500 mm	1600 mm	1700 mm	1800 mm
1.	150	220	240	260	280	300	320	340	360
2.	190	260	285	310	335	360	385	410	435
3.	230	300	330	360	390	420	450	480	510
4.	270	340	375	410	445	480	515	550	585
5.	310	380	420	460	500	540	580	620	660
6.	350	420	465	510	555	600	645	690	735
7.	390	460	510	560	610	660	710	760	810
8.	430	500	555	610	665	720	775	830	885
9.	470	540	600	660	720	780	840	900	960
10.	510	580	645	710	775	840	905	970	1035

Note:
The capacity of stairs serving more than 10 storeys may be obtained by using the formula in paragraph 4.18 [p. 7.52].

(*Note:* the working of the example given in AD B1 is incorrect although the final answer is correct. It should also be noted that the final comment in the answer suggests that one stair should be discounted, whereas in a building of 12 storeys, lobbies would have to be provided and therefore discounting would be unnecessary. Hence, only two stairs would be needed, each 1.1 m wide.)

In certain buildings it may be more advantageous to design stairs on the basis of 'phased evacuation'. Indeed, in high buildings it may be impractical or unnecessary to evacuate the building totally, especially if the recommendations regarding fire resistance, compartmentation and the installation of sprinklers are adhered to.

AD B1
sec. 4
4.19 to 4.23

In phased evacuation it is usual to evacuate people with reduced mobility and those most immediately affected by the fire first (i.e. those people on the floor of fire origin and the one above it). After that, if the need arises, floors can be evacuated two at a time.

Unless buildings are of the types listed above as needing total evacuation, phased evacuation should be used for buildings over 30 m high and it may be used for buildings less than 30 m high.

Where a building is designed for phased evacuation the following conditions should be met:

- The stairs should be approached through a protected lobby or protected corridor at each floor level. (This does not apply to a top storey containing only plant rooms.)
- Each floor should be a compartment floor.
- If the building has a floor which is more than 30 m above ground, it should be protected throughout by an automatic sprinkler system which complies with BS 5306 *Fire extinguishing installations and equipment on premises*, Part 2: 1990 *Specification for sprinkler systems*. (This does not apply to flats in a mixed use building.)
- An appropriate fire warning system should be fitted which complies with BS 5839: Part 1, to at least the L3 standard.

- A telephone, intercom system or similar should be provided so that conversation is possible between a fire warden at every floor level and a control point at the fire service access level.

When phased evacuation is used as the basis for design, the minimum stair width needed may be taken from Table 8 of AD B1, assuming phased evacuation of not more than two floors at a time (see below). It will be seen that this method allows narrower stairs to be used and causes less disruption in large buildings than total evacuation.

AD B1

Table **8**   **Minimum aggregate width of stairs designed for phased evacuation**

Maximum number of people in any storey	Stair width[1] (mm)
100	1000
120	1100
130	1200
140	1300
150	1400
160	1500
170	1600
180	1700
190	1800

Notes:
1. Stairs with a rise of more than 30 m should not be wider than 1400 mm unless provided with a central handrail (see para 4.6 [p. 7.51]).
2. As an alternative to using this table, provided that the minimum width of a stair is at least 1000 mm, the width may be calculated from: $[(P \times 10) - 100]$ mm where P = the number of people on the most heavily occupied storey.

Provision and protection of stairs

AD B1
sec. 4
4.24

To be effective as an area of relative safety during a fire, escape stairs need to have an adequate standard of fire protection. This relates not only to the presence of fire-resisting enclosures but also to the provision of protected lobbies, corridors and final exits.

Fire-resisting enclosures

AD B1
sec. 4
4.25

Each internal escape stair should be a protected stair situated in a fire-resisting enclosure.

Accommodation stairs cannot normally be used for escape purposes unless the number of people and the travel distance involved are very limited. (BS 5588: Part 2: 1985 *Code of practice for shops* contains provisions in clause 9 for the use of unprotected stairs as a means of escape in small shops. The storeys must not exceed $90\,m^2$ in area and the building must contain no more than a basement, ground floor and first floor.)

Additional measures may be necessary for a stairway which is also a protected shaft or a firefighting shaft.

Protected lobbies and corridors

Protected lobbies and corridors should be provided at all levels including basements (but not at the top storey) where:

AD B1
sec. 4
4.26

- The building has a single stair and there is more than one floor above or below the ground storey.
- The building has a floor which is more than $20\,m$ above ground.
- The building is designed for phased evacuation.
- The stairway serves an area of special risk (such as a covered car park). In this case, the lobby should be ventilated by permanent vents with an area of at least $0.4\,m^2$, or should be protected by a mechanical smoke control system.

AD B1
sec. 4
4.27

Final exits from protected stairs

- Each protected stair should discharge either:

AD B1
sec. 4
4.28, 4.29,
4.30

  (a) direct to a final exit; *or*
  (b) by means of a protected exit passageway to a final exit.

- Each exit should be as wide as the stairway leading to it.
- If two protected stairways or protected exit passageways are adjacent, they should be separated by an imperforate enclosure.

Restrictions on the use of space in protected stairways

Since a protected stairway is considered to be a place of relative safety, it should be free of potential sources of fire. Therefore, the space in protected stairways is restricted to the provision of:

AD B1
sec. 4
4.31

- Washrooms or sanitary accommodation provided that the accommodation is not used as a cloakroom. The only gas appliances that may be installed are water heaters or sanitary towel incinerators.
- A lift well, on condition that the stairway is not a firefighting stair.
- An enquiry office or reception desk area of not more than $10\,m^2$, provided that there is more than one stair serving the building.
- Fire-protected cupboards, provided that there is more than one stair serving the building.
- Gas service pipes and meters, but only if they comply with the Gas Safety Regulations referred to on p. 7.40 above.

Protection of the external walls of protected stairways

AD B1
sec. 2
2.38
& sec. 4
4.32

If a protected stairway is situated on the external wall of a building it is not necessary for the external part of the enclosure to be fire-protected and in many cases it may be fully glazed. This is because fires are unlikely to start in protected stairways. Therefore these areas will not contribute to the radiant heat from a building fire which might put at risk another building.

In some building designs the stairway may be situated at an internal angle in the building façade and may be jeopardised by smoke and flames coming from windows in the facing walls. This may also be the case if the stair projects from the face of the building or is recessed into it.

In these cases any windows or other unprotected areas in the face of the building and in the stairway should be separated by at least 1.8 m (see Fig. 7.21). This recommendation also applies to flats and maisonettes.

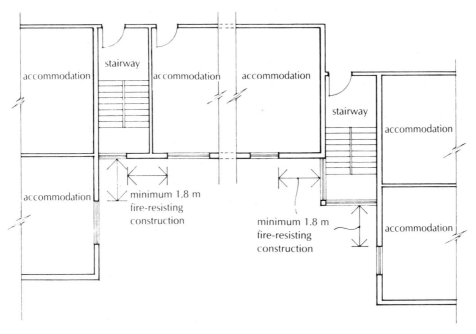

Fig. 7.21 Protected stairways – external protection.

Basement stairs

AD B1
sec. 4
4.34
& sec. 2
2.40 to 2.42
Basement fires are particularly serious since combustion products tend to rise and find their way into stairways unless other smoke venting measures are taken. (See AD B5, p. 7.154 below.)

Therefore, it is necessary to take additional precautions to prevent a basement fire endangering upper storeys in a building as follows:

- In buildings with only one escape stair serving the upper storeys, this stair should not continue down to the basement i.e. the basement should be served by a separate stair.
- In buildings containing more than one escape stair, at least one of the stairs should terminate at ground level and not continue down to the basement. The other stairs may terminate at basement level on condition that the

basement accommodation is separated from the stair(s) by a ventilated protected lobby or corridor at basement level.

These recommendations also apply to flats and maisonettes.

External escape stairs

An external escape stair is permitted to be used as an alternative means of escape where there is more than one escape route available from a storey, or part of a building, on condition that:

AD B1
sec. 4
4.35 & 4.36

- It is not intended for use by members of the public if installed in assembly and recreation buildings (PG 5).
- It serves only office or residential staff accommodation if installed in an institutional building (PG 2a).
- It is sufficiently protected from the weather, although full enclosure is not necessary. The stair may be located so that protection may be obtained from the building itself.
- It is adequately protected from a fire in the building.
- Any part of an external wall of the building which is within 3 m of the horizontal part of the escape route should be protected with fire-resisting construction up to 1.1 m from ground or paving level.

If these conditions can be met, the stair may be provided in accordance with the recommendations for external escape stairs from flats and maisonettes (see p. 7.40 above). The 6 m height restriction may be ignored for buildings other than dwellings if the stair is protected from the effects of ice and snow.

General recommendations common to all buildings except dwellinghouses

Section 5 of AD B1 gives general guidance on a number of features of escape routes which apply to all buildings, except dwellinghouses, concerning:

- Construction and protection.
- The provision of doors.
- The construction of escape stairs.
- The position and design of final exits.
- Lighting and signing.
- Mechanical services including lifts.
- Refuse chutes and storage.

These recommendations should be read in conjunction with the provisions described above for flats and maisonettes, and other buildings except dwellinghouses.

AD B1
sec. 5
5.1

Construction and protection of escape routes – general provisions

Those parts of a means of escape which are required to be fire-resisting should comply with the recommendations given in AD B3 or AD B5 in

AD B1
sec. 5
5.2 to 5.4

addition to AD B1, and also with the following general constructional provisions:

AD B1
sec. 5
5.7 to 5.9

- Glazed elements in fire-resisting enclosures and doors which are only able to meet the requirements for integrity in the event of a fire, will be limited in area to the amounts shown in Table A4 of Appendix A of AD B. (This is reproduced on p. 7.96 below.)
- There are no limitations on the use of glazed elements, which can also meet the insulation requirements, in AD B1. However, there may be some restrictions on the use of glass in firefighting stairs and lobbies in BS 5588: Part 5: 1991 *Code of practice for firefighting stairways and lifts*. This is referred to in AD B5 and is mentioned below.
- Glazed elements may also need to comply with AD N (see Chapter 18).

AD B1
sec. 5
5.24 to 5.28

- All escape routes should have a minimum headroom of 2 m; the only projections allowed below this are door frames.
- The floors of escape routes, including the surfaces of steps and ramps, should not be unduly slippery when wet.
- Ramps on escape routes should not be steeper than 1 in 12.
- Sloping floors or tiers should not have a pitch greater than 35° to the horizontal.
- Further guidance on the provision of ramps, stairs, aisles and gangways may be found in AD K (see Chapter 15) and AD M (see Chapter 17).

The provision of doors

AD B1
sec. 5
5.5, 5.6

Doors on escape routes should be readily openable if undue delay in escaping from a building is to be avoided. They should also comply with the following general provisions:

- If they are required to be fire-resisting they should comply with the relevant test criteria and performance standards set out in Appendix B of AD B and in Table B1 (see p. 7.95 below).

AD B1
sec. 5
5.10 to 5.18

- Escape doors should open in the direction of the means of escape where it is reasonably practicable to do so and should always do so if more than 50 people are likely to use the door in an emergency.
- Ideally, the doors should not be fitted with fastenings unless these are simple to use and can be operated from the side of the door which is approached by people escaping. Any fastenings should be able to be operated without a key and without having to operate more than one mechanism.
- Where security of final exit doors is important, as in assembly and recreation (PG 5) and shop and commercial (PG 4) buildings, panic bolts may be used.
- Recommendations for self-closers and hold-open devices for fire doors are contained in Appendix B of AD B.
- Doors on escape routes should swing through at least 90° to open, and should not reduce the effective width of any escape route across a landing. The swing should be away from any changes in floor level, although a single step or threshold on the line of a door opening is permitted.
- Any door that opens towards a corridor or stairway should be recessed so that it does not encroach on or reduce the effective width of the corridor or stairway.

- Doors on escape routes which sub-divide corridors, or are hung to swing in two directions, should contain vision panels. (See also AD M and Chapter 17 for vision panels in doors across accessible corridors.)
- If revolving or automatic doors, or turnstiles, are placed across an escape route, in the event of an emergency they should:

  (a) fail safely in the open position, or
  (b) be easily openable, or
  (c) have swing doors adjacent to them which can provide an alternative exit.

The construction of escape stairs

Escape stairs and their associated landings in certain high risk situations or buildings require the extra safeguard of being constructed in materials of limited combustibility. This recommendation applies in the following cases:

AD B1
sec. 5
5.19 to 5.21

- Where a building has only one stair serving it. (This does not apply to two- and three-storey flats and maisonettes.)
- Where a stair is located in a basement storey (except if it is a private stair in a maisonette).
- To any stair serving a storey in a building which is more than 20 m above ground level.
- To any external stair (except where it connects the ground floor or paving level to a floor or flat roof which is not more than 6 m above ground).
- If the stair is a firefighting stair.

In all the above, except for the firefighting stair, it is permissible to add combustible materials to the upper surface of the stair.

Where possible, single steps should be avoided on escape routes unless prominently marked, since they can cause falls. It is permissible though, to have a single step on the line of a doorway.

Special stairs and ladders

Spiral and helical stairs, and fixed ladders are not as inherently safe as conventional stairs and are subject to extra controls over their use.

AD B1
sec. 5
5.22, 5.23

Spiral and helical stairs should comply with BS 5395 *Stairs, ladders and walkways*, Part 2: 1984 *Code of practice for the design of helical and spiral stairs*. If they are intended for use by members of the public they should be a type E (public) stair from the Code.

Fixed ladders are not suitable as a means of escape for members of the public. They should only be used to access areas which are not normally occupied, such as plant rooms, where it is not practical to provide a conventional stair. They should be constructed of non-combustible materials.

The position and design of final exits

Final exits should not be narrower than the escape routes they serve and should be positioned to facilitate evacuation of people out of and away from the building. This means that they should be:

AD B1
sec. 5
5.29 to 5.32

- Positioned so that rapid dispersal of people is facilitated to a street, passageway, walkway or open space clear of the effects of fire and smoke. The route from the building should be well defined and guarded if necessary.
- Clearly apparent to users. This is very important where stairs continue up or down past the final exit level in a building.
- Sited so that they are clear of the effects of fire from risk areas in buildings such as basements, and openings to transformer chambers, refuse chambers, boiler rooms and other similar areas.

Lighting and signing

AD B1
sec. 5
5.33 to 5.35

All escape routes should have adequate artificial lighting. In certain cases emergency escape lighting which illuminates the route if the mains supply fails, should also be provided. These are listed in Table 9 to AD B1 which is reproduced below.

The lighting to escape stairs will also need to be on a separate circuit from that which supplies any other part of the escape route.

Protected power circuits are provided in situations where it is critical that the circuit should continue to function during a fire. The cable used in a protected power circuit for operation of equipment in the event of fire should:

- Meet the requirements for classification as CWZ in accordance with BS 6387: 1983 *Specification for performance requirements for cables required to maintain circuit integrity under fire conditions.*
- Follow a route which passes through parts of the building in which there is negligible fire risk.
- Be separate from circuits which are provided for other purposes.

Escape lighting systems may comply with standards for installation such as those given in BS 5266 *Emergency lighting*, Part 1: 1988 *Code of practice for the emergency lighting of premises other than cinemas and certain other specified premises used for entertainment*, or CP 1007: 1955 *Maintained lighting for cinemas.*

Except in dwellinghouses, flats and maisonettes, exit signs should be provided to every doorway or other exit giving access to a means of escape. It is not necessary to sign exits which are in ordinary, daily use. The exit should be distinctively and conspicuously marked by a sign with letters of adequate size in accordance with BS 5499 *Fire safety signs, notices and graphic symbols*, Part 1: 1990 *Specification for fire safety signs*. In some buildings other legislation may require additional signs.

Lift installations

AD B1
sec. 5
5.36 to 5.42

Lifts are not normally used for means of escape since there is always the danger that they may become immobilised due to power failure and may trap the occupants. It is possible to provide lifts as part of a management plan for evacuating disabled people if the lift installation is appropriately sited and protected. It should also contain sufficient safety devices to ensure that it

AD B1

Table 9 Provisions for escape lighting

Purpose group of the building or part of the building	Areas requiring escape lighting
Residential	All common escape routes
Office, Shop and Commercial[1], Industrial, Storage, Other non-residential	a. Underground or windowless accommodation b. Stairways in a central core or serving storey(s) more than 20 m above ground level c. Internal corridors more than 30 m long d. Open-plan office areas of more than 60 m²
Shop and Commercial[2] and car parks to which the public are admitted	All escape routes (except in shops of 3 or fewer storeys with no sales floor more than 280 m² provided that the shop is not a restaurant or bar)
Assembly and Recreation	All escape routes and accommodation except for: a. accommodation open on one side to view sport or entertainment during normal daylight hours b. toilet accommodation having a gross floor area not more than 8 m²
Any purpose group	a. electricity generator rooms b. switch room/battery room for emergency lighting system c. emergency control room

Notes:
1. Those parts of the premises to which the public are not admitted.
2. Those parts of the premises to which the public are admitted.

remains usable during a fire. Further details may be found in BS 5588: Part 8: 1988 *Code of practice for means of escape for disabled people.*

A further problem with lifts is that they connect floors and may act as a vertical conduit for smoke or flames. This may be prevented if the following recommendations are observed:

- Lift wells should be enclosed throughout their height with fire-resisting construction if their siting would prejudice an escape route. Alternatively, they should be contained within the enclosure of a protected stairway.
- Any lift well which connects different compartments should be constructed as a protected shaft.
- Unless the lift is in a protected stairway enclosure, it should be approached through a protected lobby or corridor if it is situated in a basement or enclosed car park. This also applies where the lift delivers directly into

corridors serving sleeping accommodation if any of the storeys also contain high fire risk areas such as kitchens, lounges or stores.
- A lift should not continue down to serve a basement if there is only one escape stairway in the building or if it is in an enclosure to a stairway which terminates at ground level.
- Lift machine rooms should be located over the lift shaft wherever possible. Where a lift is within the only protected stairway serving a building and the machine room cannot be located over the lift shaft, then it should be sited outside the protected stairway. This is to prevent smoke from a fire in the machine room from blocking the stair.
- Wall-climber and feature lifts which do not have a conventional well may be at risk if they pass through a smoke reservoir, such as an atrium or open mall. Care will be needed in the design in order to prevent spread of smoke from the reservoir and to protect the occupants of the lift.

Mechanical ventilation and air-conditioning services

AD B1
sec. 5
5.43 to 5.46

Mechanical ventilation systems should be designed so that in a fire:

- Air is drawn away from protected escape routes and exits, or
- the system (or the appropriate part of it) is closed down.

Systems which recirculate air should comply with BS 5588: Part 9: 1989 *Code of practice for ventilation and air conditioning ductwork* for operation under fire conditions. Recommendations for the use of mechanical ventilation in a place of assembly are given in BS 5588: Part 6: 1991 *Code of practice for assembly buildings*. Guidance on the design and installation of mechanical ventilation and air-conditioning plant is given in BS 5720: 1979 *Code of practice for mechanical ventilation and air conditioning in buildings*.

Where pressurisation systems are installed they should be compatible with any ventilation or air-conditioning systems in the building, when operating under fire conditions.

Refuse chutes and storage

AD B1
sec. 5
5.47 to 5.50

Fires in refuse chute installations are extremely common and they are required to be built of non-combustible materials in AD B3. So that escape routes are not jeopardised, refuse chutes and rooms for refuse storage should:

- be separated, by fire-resisting construction, from the rest of the building; *and*
- not be located in protected lobbies or stairways.

Rooms which store refuse or contain refuse chutes should:

- be approached directly from the open air; *or*
- be approached from a protected lobby with at least $0.2\,m^2$ of permanent ventilation.

Refuse storage chamber access points should be sited away from escape routes, final exits and windows to dwellings.

Refuse storage chambers, chutes and hoppers should be sited and constructed in accordance with BS 5906: 1980 *Code of practice for storage and on-site treatment of solid waste from buildings.*

Alternative approach to the provision of means of escape in selected premises

Reference has been made throughout the text above to the use of design guides, other than AD B1, in the provision of means of escape. There are certain specialised types of premises where it is recommended that this other guidance be used in preference to the more general guidance in AD B1. Additionally, whilst AD M covers access and facilities for disabled people, there are no specific recommendations for means of escape for disabled people in AD B1. It may not be necessary to provide special structural measures to aid the escape of disabled people other than suitable management arrangements to cater for emergencies. Where it is felt that special provisions for means of escape are desirable, reference should be made to BS 5588: Part 8: 1988 *Code of practice for means of escape for disabled people.* Advice is given in the Code on refuges, evacuation lifts and the need for efficient management of escape.

AD B1
0.33

Means of escape in health care premises

Health care premises such as hospitals, nursing homes and homes for the elderly differ from other premises in that they contain people who are bed-ridden or who have very restricted mobility. In such buildings it is unrealistic to expect that the patients will be able to leave without assistance, or that total evacuation of the building is feasible.

AD B1
0.34,
& sec. 3
3.26

Hence, the approach to the design of means of escape in these premises will demand a very different approach to that embodied in much of AD B1 and the Department of Health has prepared a set of guidance documents under the general title of *Firecode* for use in health care buildings. These documents are also applicable to non-National Health Service premises and are as follows:

- Means of escape in new hospitals –
  Firecode. Nucleus fire precautions recommendations. Department of Health, HMSO, 1989.
  Firecode. Health Technical Memorandum 81, Fire precautions in new hospitals. Department of Health & Social Security, HMSO, 1987.

If an existing house of not more than two storeys is converted for use as an unsupervised Group Home for not more than seven mentally handicapped or mentally ill people, it may be regarded as a dwellinghouse (PG 1c) if it has means of escape designed in accordance with *Firecode. Health Technical Memorandum 88.*

Other guidance documents include:

- Means of escape in existing hospitals –
  Draft guide to fire precautions in existing hospitals. Home Office, 1982.
- Means of escape in existing residential care premises –
  Draft guide to fire precautions in existing residential care premises. Home Office/Scottish Home and Health Department, 1983.

Progressive horizontal evacuation

AD B1
sec. 3
3.28 to 3.30
Since total evacuation of health care premises is inappropriate in most cases, the *Firecode* documents contain guidance on progressive horizontal evacuation of premises. In-patients are evacuated, in the event of fire, to adjoining compartments or sub-divisions of compartments, the object being to provide a place of relative safety within a short distance. If necessary, further evacuation can be made from these safe places but under less pressure of time.

Section 3 of AD B1 gives a limited amount of guidance on progressive horizontal evacuation in some other residential buildings to which the *Firecode* documents do not apply, such as residential rest homes. When storeys are being planned for progressive horizontal evacuation, the following conditions should be considered when they are being divided into compartments:

- The compartment into which the evacuation is to take place should be large enough to take the occupants from the adjoining compartment and its own occupants. The design occupancy figures should be used to assess the total number of people involved.
- Each compartment should have an alternative escape route which is independent of the adjoining compartment. This may be through another compartment which also has an independent escape route.
- All upper storeys used for in-patient care should be divided into at least two compartments complying with the above rules for evacuation.

The *Firecode* documents also contain management and other safety provisions which are outside the scope of building regulations.

Sheltered housing

AD B1
sec. 1
1.7
sec. 2
2.6
Sheltered housing schemes which consist of specially adapted groups of houses, bungalows or two-storey flats with warden assistance and few communal facilities need not be treated differently from other one- or two-storey houses or flats. The automatic smoke detection equipment should have a connection to a central monitoring point so that the supervisor (or central alarm relay station) is aware that a fire has occurred and knows its location. This would not apply to communal facilities such as lounges.

Sheltered accommodation in the institutional or other residential purpose groups would need to comply with the provisions for other buildings listed above. Additional guidance may be found in BS 5588: Part 1: 1990, clause 17.

Assembly buildings

AD B1
sec. 3
3.31
A principal problem with assembly buildings is the difficulty in escaping from fixed seating. This occurs in theatres, concert halls, conference centres and at sports events. Specific guidance on this may be obtained from:

- BS 5588: Part 6: 1991 *Code of practice for assembly buildings*, where guidance on the spacing of fixed seating may be found; *and*
- *Guide to safety at sports grounds*, Home Office/Scottish Office. HMSO, 1990, for sports stadia, etc.

Schools and other educational buildings

Means of escape in these buildings should be in accordance with DES *Building Bulletin 7, Fire and the design of educational buildings*, HMSO, 1988.

AD B1
sec. 3
3.32

Any specialised buildings found on school or other educational premises which are outside the scope of *Building Bulletin 7* may have means of escape designed in accordance with the relevant British Standard for that building type or may use AD B1.

Shops and shopping complexes

British Standard BS 5588: Part 2: 1985 *Code of practice for shops* may be used instead of AD B1. The relevant sections are 2, 3, 4 and 5 and these should be used exclusively if it is decided to use BS 5588 (i.e. it is not recommended that a mixture of AD B1 and the Code be used).

AD B1
sec. 3
3.34, 3.35

Small shops (those having no more than two storeys plus a basement storey and no storey larger than 280 m^2) may be designed in accordance with clause 9 of BS 5588: Part 2 instead of AD B1.

Shopping complexes are not covered adequately by AD B1 and should be designed in accordance with BS 5588: Part 10: 1991 *Code of practice for enclosed shopping complexes.*

Offices

Sections 2, 3, 4 and 5 of BS 5588: Part 3: 1983 *Code of practice for office buildings* may be used in preference to AD B1. Again, the Code should be used exclusively and not with provisions from AD B1.

AD B1
sec. 3
3.36

Internal fire spread (linings)

Introduction

Although the linings of walls and ceilings are unlikely to be the materials first ignited in a fire (this is more likely to occur in furniture and fittings), they can significantly affect the spread of fire and its rate of growth. This is especially true in circulation areas where rapid fire spread may prevent occupants from escaping. Part B2 of Schedule 1 to the 1991 Regulations seeks to control these surface linings.

AD B2
0.41, 0.44

Control of wall and ceiling linings

The spread of fire within a building may be inhibited by paying attention to the lining materials used on walls, ceilings, partitions and other internal structures. These linings must:

Regs. Sch. 1
B2

(a) offer adequate resistance to spread of flame over their surfaces, and
(b) if ignited, have a rate of heat release which is reasonable in the circum-
    stances.

It should be noted that floors and stairs are not covered by the requirements since they are not usually involved in a fire until it is well established. Consequently, they will not contribute to fire spread in the early stages of a fire.

Methods of test

In order to meet the requirements of AD B2 it is necessary for materials or products to meet certain levels of performance under surface spread of flame and/or fire propagation tests.

The surface spread of flame characteristics of a material may be determined by testing it in accordance with the method specified in BS 476 *Fire tests on building materials and structures*, Part 7: 1971 *Surface spread of flame tests for material*, or Part 7: 1987 *Method for classification of the surface spread of flame of products*.

A strip of the material under test is placed with one end resting against a furnace and the rate at which flames spread along the material is measured.

Materials or products are thus placed in Classes 1, 2, 3 or 4, Class 1 representing a surface of very low flame spread. Class 4 (a surface of rapid flame spread) is not acceptable under the provisions of the approved document.

In the event of fire, some materials have a higher rate of heat release or ignite more easily than others. They are therefore more hazardous and this may mean a reduced time to flashover. The way in which the rate of heat release may be assessed is contained in BS 476: Part 6: 1981 and 1989 *Method of test for fire propagation for products*.

The material or product is tested for a certain period of time in a furnace and is given two numerical indices related to its performance. The sub-index (i') is derived from the first three minutes of the test whilst the overall test performance is denoted by the index of performance (I).

Class 0 materials

In order to establish a high product performance classification for lining materials in high risk areas (such as circulation spaces), AD B2 recommends that materials in these areas should conform to the Class 0 standard. This is not a classification found in any British Standard test as such; however it is evident that it draws on BS 476 test results as the following definition shows.

The Class 0 standard will be achieved by any material or the surface of a composite product which:

(a) is composed of materials of limited combustibility (see below) throughout;
    or
(b) is a material of Class 1 which has an index of performance (I) of not more than 12 and a sub-index (i_1) of not more than 6.

Interpretation

The following terms are common to all parts of AD B but occur most frequently in AD B2, 3 and 4:

NON-COMBUSTIBLE MATERIAL – a material which has the highest level of reaction to fire performance as follows:

AD B
Appendix A
A7

- When tested to BS 476: Part 11, the material does not flame and there is no rise in temperature on either the centre (specimen) or furnace thermo-couples; *or*
- A totally inorganic material, such as concrete, fired clay, masonry, etc., containing not more than 1% by weight or volume of organic material, including concrete bricks or blocks to BS 6073: Part 1: 1981; *or*
- Products classified as non-combustible when tested in accordance with BS 476: Part 4: 1970 *Non-combustibility test for materials.*

Table A6 from Appendix A of AD B is reproduced below and gives details of where non-combustible materials should be used.

AD B, Appendix A

Table **A6**   **Use of non-combustible materials**

Use	Non-combustible materials
1. Ladders referred to in the guidance to B1, paragraph 5.22.	a. Any material which when tested to BS 476: Part 11 does not flame nor cause any rise in temperature on either the centre (specimen) or furnace thermocouples.
2. Refuse chutes meeting the provisions in the guidance to B3, paragraph 8.28.	
3. Suspended ceilings and their supports where there is provision in the guidance to B3, paragraph 9.13, for them to be constructed of non-combustible materials.	b. Totally inorganic materials such as concrete, fired clay, ceramics, metals, plaster and masonry containing not more than 1 per cent by weight or volume of organic material. (Use in buildings of combustible metals such as magnesium/ aluminium alloys should be assessed in each individual case.)
4. Pipes meeting the provisions in the guidance to B3, Table 15.	
5. Flue walls meeting the provisions in the guidance to B3, Diagrams 35.	
6. Construction forming car parks referred to in the guidance to B3, paragraph 11.3.	c. Concrete bricks or blocks meeting BS 6073: Part 1: 1981.
	d. Products classified as non-combustible under BS 476: Part 4: 1970.

MATERIALS OF LIMITED COMBUSTIBILITY – materials in this group, whilst not regarded as non-combustible, would contribute little heat energy to a fire. Therefore, they can be used in situations where control of fire spread is essential, such as stairs to basements. They are defined in AD B, Table A7 (see below) as follows:

AD B
Appendix A
A8

- Any material with a density of $300\,kg/m^3$ or more, which when tested to BS 476: Part 11, does not flame and the rise in temperature on the furnace thermocouple is not more than $20°C$.

AD B, Appendix A

Table **A7**   Use of materials of limited combustibility

Use	Materials of limited combustibility
1. Stairs where there is provision in the guidance to B1 for them to be constructed of materials of limited combustibility (see 5.19). 2. Materials above a suspended ceiling meeting the provisions in the guidance to B3, paragraph 9.13. 3. Reinforcement/support for fire-stopping referred to in the guidance to B3, see 10.13. 4. Roof coverings meeting the provisions: a. in the guidance to B3, paragraph 9.11, or b. in the guidance to B4, Table 17, or c. in the guidance to B4, Diagram 42. 5. Roof deck meeting the provisions of the guidance of B3, Diagram 24a. 6. Class 0 materials meeting the provisions of Appendix A, paragraph A12(a). 7. Ceiling tiles or panels of any fire protecting suspended ceiling (Type D) in Table A3. 8. Compartment walls and compartment floors in hospitals where required to have fire resistance of 60 minutes or more.	a. Any non-combustible material listed in Table A6. b. Any material of density 300 kg/m^3 or more, which when tested to BS 476: Part 11, does not flame and the rise in temperature on the furnace thermocouple is not more than 20°C. c. Any material with a non-combustible core at least 8 mm thick having combustible facings (on one or both sides) not more than 0.5 mm thick. (Where a flame spread rating is specified, these materials must also meet the appropriate test requirements.)
9. Insulation material in external wall construction referred to in paragraph 12.7. 10. Insulation above any fire-protecting suspended ceiling (Type D) in Table A3.	Any of the materials (a), (b), or (c) above, or: d. Any material of density less than 300 kg/m^3, which when tested to BS 476: Part 11, does not flame for more than 10 seconds and the rise in temperature on the centre (specimen) thermocouple is not more than 35°C and on the furnace thermocouple is not more than 25°C.

- Any material with a non-combustible core at least 8 mm thick with combustible facings not more than 0.5 mm thick on one or both sides (e.g. plasterboard). However, such a material should also meet the flame spread requirements mentioned above.

Table A7 also gives details of where these materials should be used. It should be noted that certain insulating materials in group (d) of the table are of a lower standard than is shown above and may only be used in the situations shown in items 9 and 10 of the table.

It is of course permissible to use non-combustible materials whenever a recommendation for materials of limited combustibility is specified.

Materials for surface linings

As a guide to the materials which may be used for wall and ceiling linings, AD B lists in Table A8 (see below) the typical performance ratings for some generic materials and products. Test results for proprietary materials may be obtained from manufacturers and trade associations. However, small differences in detail (e.g. thickness, colour, fixings, adhesives, etc.) can significantly affect the rating. Therefore, the reference used to substantiate the spread of flame rating should be carefully checked to ensure that it is suitable, adequate and applicable to the construction to be used.

AD B
Appendix A
A14, A15

Thermoplastic materials

A thermoplastic material is a synthetic polymer which has a softening point below 200°C when tested in accordance with BS 2782 *Methods of testing plastics*, Part 1: 1976: Method 120A. If the thickness of the product to be tested is less than 2.5 mm then specimens for the test may be fabricated from the original polymer.

Some thermoplastic materials can be tested under BS 476: Parts 6 and 7, and can be used in accordance with their ratings as described above. Alternatively, they may be classified as TP(a) rigid, TP(a) flexible, or TP(b), as described below, but their use would be restricted to roof-lights, lighting diffusers, suspended ceiling panels and external window glazing (except in circulation areas). These uses are described more fully below.

AD B
Appendix A
A16 to A19

TP(a) rigid means:

AD B
Appendix A
A19

- Rigid solid PVC sheet.
- Solid (i.e. not double- or multi-skin) polycarbonate sheet at least 3 mm thick.
- Multi-skinned rigid sheet made from uPVC or polycarbonate with a BS 476: Part 7 rating of Class 1.
- Any other rigid thermoplastic product which, when tested in accordance with BS 2782: Part 5: 1970 (1974): Method 508A, performs so that the flame extinguishes before reaching the first mark, and the duration of the flame or after-glow after removal of the burner does not exceed five seconds.

AD B, Appendix A

Table **A8** **Typical performance ratings of some generic materials and products**

Rating	Material or product
Class 0	1. Any non-combustible material or material of limited combustibility. (Composite products listed in Table A7 must meet the test requirements given in paragraph A12(b).)
	2. Brickwork, blockwork, concrete and ceramic tiles.
	3. Plasterboard (painted or not, or with a PVC facing not more than 0.5 mm thick) with or without an air gap or fibrous or cellular insulating material behind.
	4. Woodwool cement slabs.
	5. Mineral fibre tiles or sheets with cement or resin binding.
Class 3	6. Timber or plywood with a density more than 400 kg/m^3, painted or unpainted.
	7. Wood particle board or hardboard, either treated or painted.
	8. Standard glass reinforced polyesters.

Notes:

1. Materials and products listed under Class 0 also meet Class 1.

2. Timber products listed under Class 3 can be brought up to Class 1 with appropriate proprietary treatments.

3.. The following materials and products may achieve the ratings listed below. However, as the properties of different products with the same generic description vary, the ratings of these materials/products should be substantiated by test evidence.

Class 0 aluminium faced fibre insulating board, flame retardant decorative laminates on a calcium silicate board, thick polycarbonate sheet, phenolic sheet and UPVC;

Class 1 phenolic or melamine laminates on a calcium silicate substrate and flame retardant decorative laminates on a combustible substrate.

TP(a) flexible means:

- Flexible products not more than 1mm thick complying with the Type C requirements of BS 5867 *Specification for fabrics for curtains and drapes*, Part 2: 1980 *Flammability requirements*, when tested to BS 5438: 1976, Test 2 1989. In the BS 5438 test, the flame should be applied to the specimens for 5, 15, 20 and 30 seconds respectively, although it is not necessary to include the cleansing procedure described in the British Standard.

TP(b) means:

- Rigid solid polycarbonate sheet products less than 3mm thick, or multi-skin polycarbonate sheet products which do not qualify as TP(a) by test.
- Any other product, a specimen of which between 1.5 mm and 3 mm thick, when tested in accordance with BS 2782: Part 5: 1970 (1974): Method 508A, has a rate of burning not exceeding 50mm/minute.

If it is not possible to cut or machine a 3mm thick test specimen from the product, then it is permissible to mould one from the original material used for the product.

Specific recommendations for wall and ceiling linings

Table 10 from Section 6 of AD B2 (shown below) gives the recommended flame spread classifications for the surfaces of walls and ceilings in any room or circulation space, for all building types.   **AD B2 6.1**

Different standards are set for 'small rooms', which are totally enclosed rooms with floor area of not more than $4\,m^2$ in residential buildings or $30\,m^2$ in non-residential buildings, and for other rooms and circulation spaces.

When considering the performance of wall linings, window glazing and ceilings which slope at more than $70°$ to the horizontal are treated as a wall. Certain vertical surfaces are excluded from this definition, such as: **AD B2 6.2**

- Doors, door frames and glazing in doors.
- Window frames and other frames containing glazing.
- Narrow members, such as architraves, cover moulds, picture rails and skirtings.
- Fireplace surrounds, mantelshelves and fitted furniture.

Similarly, ceiling surfaces include glazing and walls which slope at $70°$ or less to the horizontal, but exclude: **AD B2 6.3**

- Trap doors and frames.
- Window frames, roof-light frames and other frames in which glazing is fitted.
- Narrow members, such as architraves, cover moulds and picture rails.

AD B2, Section 6

Table **10**   **Classification of linings**

Location	Class
Small rooms of area not more t̶̶̶̶̶̶ ̶ ̶ ̶ ̶ ̶building and 30 m² in a non-residential building	3
Other rooms	
	1
Circulation spaces within dwellings	
Other circulation spaces, including the common areas of flats and maisonettes	0

AD B2
6.4

Approved Document B2 permits certain variations from the strict lining classifications shown in Table 10 provided that no lining is lower than Class 3, as follows:

(a) Wall linings in rooms may be lower than Class 1 if their area does not exceed half the floor area of the room subject to the following maxima:

  (i) In residential buildings – $20\,m^2$.
  (ii) In non-residential buildings – $60\,m^2$.

AD B2
6.7

(b) Plastic roof-lights, and lighting diffusers fitted in suspended ceilings, may be lower than Class 0 or Class 1 (but not lower than Class 3) if they comply with recommendations for thermoplastic roof-lights or diffusers shown below. Roof-lights of other materials should comply with the Table 10 recommendations.

AD B2
6.9

(c) External windows to rooms may be glazed with a TP(a) rigid thermoplastic product, but not in circulation areas. However, internal glazing should follow the recommendations of Table 10.

AD B2
6.8, 6.14

(d) Suspended or stretched-skin ceilings made from thermoplastic material with a TP(a) flexible classification are permitted provided they do not form part of a fire-resisting ceiling. Each panel should be supported on all its sides and should not exceed $5\,m^2$ in area.

Roof-lights and lighting diffusers

AD B2
6.10 to 6.13

Roof-lights, and lighting diffusers which form an integral part of a ceiling (i.e. not attached to the soffit or suspended beneath the ceiling), may:

- comply with the classification contained in Table 10; *or*
- consist of thermoplastic materials with a TP(a) rigid classification (unless used in a protected stairway); *or*
- consist of any plastics materials with at least a Class 3 rating (or a thermoplastic material with a TP(b) classification) if used under the conditions described in Table 11 from AD B2 (see below), and limited in extent and layout as illustrated in Fig. 7.22.

The space in the ceiling above a lighting diffuser should comply with the recommendations in Table 10 for flame spread, according to the type of space below the ceiling. Lighting diffusers should only be used in fire-resisting or fire-protecting ceilings if they have been satisfactorily tested as part of the ceiling system being used to provide the appropriate fire protection.

AD B4
14.5

Additionally, the external surfaces of roof-lights may need to follow the recommendations for roofs contained in AD B4. Table 18 of AD B4 is reproduced below and gives details of the limitations on the use of plastic roof-lights in roofs. The main recommendations of the table are summarised below:

- TP(a) rigid thermoplastic material may be used for a roof-light over any space except a protected stairway. Unless located in any of the buildings listed in (a) and (b) immediately below, the roof-light should be at least 6 m from any point on a boundary.
- Roof-lights serving:

AD B2, Section 6

Table **11**   **Limitations applied to thermoplastic roof-lights and lighting diffusers in suspended ceiling and Class 3 plastics roof-lights**

Minimum classification of lower surface	Use of space below the diffusers or roof-light	Maximum area of each diffuser panel or roof-light [1]	Max. total area of diffuser panels and roof-lights as percentage of floor area of the space in which the ceiling is located	Minimum separation distance between diffuser panels or roof-lights [1]
TP(a)	Any except protected stairway	No limit [2]	No limit	No limit
Class 3 [3] or TP(b)	Rooms	5 m²	50	3 m
	Circulation spaces except protected stairways	5 m²	15	3 m

Note:

1. Smaller panels can be grouped together provided that the overall size of the group and the space between one group and any others satisfies the dimensions shown in Fig. 7.22.

2. Lighting diffusers of TP(a) flexible rating should be restricted to panels of not more than 5 m² each, see paragraph 6.14 of AD B2.

3. There are no limitations on Class 3 material in small rooms.

(a) balconies, verandas, carports, covered ways or loading bays with one longer side permanently open, or detached swimming pools; *or*

(b) garages, conservatories or outbuildings with a floor area not exceeding 40 m²;

may consist of plastics materials with a lower surface of not less than Class 3 surface spread of flame or TP(b). The roof-light should be 6 m from any point on the boundary if the external surface is designated AD, BD, CA, CB, CC, CD or TP(b). (For details of designatory letters see pp. 7.137 and 7.138.) If the external surface is designated DA, DB, DC or DD the roof-light should be 20 m from any point on the boundary.

- The internal and external surface limits specified in (b) immediately above also apply to roof-lights in all other types of buildings. However, there are additional limitations on the area and spacing of the roof-lights.
- Individual roof-lights serving circulation spaces or rooms should not exceed 5 m² in area. They should be separated by roof covering materials of limited combustibility at least 3 m wide. If the roof-light is not thermoplastic it should consist of a single-skin material.

These recommendations for plastic roof-lights and lighting diffusers are illustrated in Fig. 7.23. It should be noted that none of the above designations for plastics roof-lights are suitable for use in protected stairways.

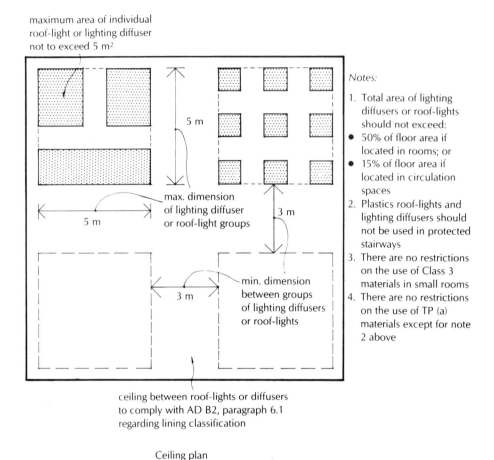

maximum area of individual
roof-light or lighting diffuser
not to exceed 5 m²

5 m

max. dimension
of lighting diffuser
or roof-light groups

5 m

3 m

min. dimension
between groups
of lighting diffusers
or roof-lights

3 m

Notes:

1. Total area of lighting diffusers or roof-lights should not exceed:
   - 50% of floor area if located in rooms; or
   - 15% of floor area if located in circulation spaces
2. Plastics roof-lights and lighting diffusers should not be used in protected stairways
3. There are no restrictions on the use of Class 3 materials in small rooms
4. There are no restrictions on the use of TP (a) materials except for note 2 above

ceiling between roof-lights or diffusers
to comply with AD B2, paragraph 6.1
regarding lining classification

Ceiling plan

Fig. 7.22 Limitations on use of Class 3 plastic roof-lights, TP(b) roof-lights and TP(b) lighting diffuers.

Further requirements for ceilings

AD B2
6.5, 6.6

Fire-protecting suspended ceilings may be used to contribute towards the fire-resistance of a floor if they meet certain criteria. These are contained in Table A3 of Appendix A of AD B which is reproduced below. The highest grade of suspended ceiling is type D. This should be constructed of materials of limited combustibility (including any insulating material) and should not contain easily openable access panels. Type D ceilings may be used anywhere under any conditions and should be used where the fire-resistance of the total floor/ceiling assembly exceeds 60 minutes. All the ceiling types need to comply with certain surface spread of flame requirements for the upper surface facing into the cavity, in addition to the Table 10 recommendations for the lower surface.

Similarly, fire-resisting ceilings may be used to reduce the need for cavity barriers in concealed void spaces in some floors and roofs. These are discussed on p. 7.104 below, in the section on cavity barriers.

AD B4, Section 14

Table **18** Plastics roof-lights: limitations on use and boundary distance

Classification on lower surface [1]	Space which roof-light can serve	Minimum distance from any point on relevant boundary to roof-light with an external surface classification [2] of:		
		TP(a)	AD BD CA CB CC CD OR TP(b)	DA DB DC DD
1. TP(a) rigid	Any space except a protected stairway	6 m [3]	6 m [5]	20 m
2. Class 3 or TP(b)	a. Balcony, verandah, carport, covered way or loading bay, which has at least one longer side wholly or permanently open	6 m	6 m	20 m
	b. Detached swimming pool			
	c. Conservatory, garage or outbuilding, with a maximum floor area of 40 m²			
	d. Circulation space [4] (except a protected stairway)	6 m [5]	6 m [5]	20 m [5]
	e. Room [4]			

Notes:

na Not applicable

(1) See also the guidance to AD B2.

(2) The classification of external roof surfaces is explained in Appendix A of AD B.

(3) No limit in the case of any space described in 2a. b. and c.

(4) Single-skin roof-light only, in the case of non-thermoplastic material.

(5) The roof-light should also meet the provisions of Diagram 42 of AD B4.

Polycarbonate and PVC roof-lights which achieve a Class 1 rating by test, see paragraph 14.6 of AD B4, may be regarded as having an AA designation.

None of the above designations are suitable for protected stairways – see paragraph 6.10 of AD B2.

Products may have upper and lower surfaces with different properties if they have double skins or are laminates of different materials.

Internal fire spread (structure)

A number of factors need to be considered in order to reduce the effects of fire spread throughout the structure of a building as follows: **Regs Sch. 1 B3**

- The building must be so designed and constructed that its stability will be maintained for a reasonable period during a fire. **B3(1)**
- Walls which are common to two or more buildings must be designed and constructed so that they resist the spread of fire between those buildings. Semi-detached and terraced houses are treated as separate buildings for the purposes of this requirement. **B3(2)**
- The building must be sub-divided by fire-resisting construction, depending on its size and intended use, where this is necessary to inhibit the spread of **B3(3)**

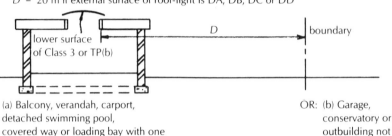

D = no limit if external surface of roof-light is TP(a)
D = 6 m if external surface of roof-light is AD, BD, CA, CB, CC, CD or TP(b)
D = 20 m if external surface of roof-light is DA, DB, DC or DD

lower surface of Class 3 or TP(b)

D boundary

(a) Balcony, verandah, carport,
detached swimming pool,
covered way or loading bay with one
longer side permanently open

OR: (b) Garage,
conservatory or
outbuilding not
exceeding 40 m²
floor area

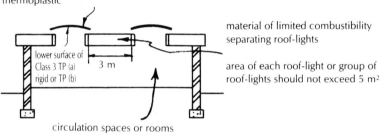

TP(a) or TP(b) thermoplastic material or roof-light of single-skin material if non-thermoplastic

lower surface of Class 3 TP (a) rigid or TP (b)

3 m

material of limited combustibility separating roof-lights

area of each roof-light or group of roof-lights should not exceed 5 m²

circulation spaces or rooms

distance D as above except that for roof-lights of TP (a) material, D = 6 m

Note: see also Fig. 7.22 and Table 18 above for more details of plastics roof-lights

Fig. 7.23 Plastics roof-lights.

fire. (This requirement does not apply to material alterations to prisons provided under section 33 of the Prisons Act 1952.)

B£(4)
● Fire and smoke may spread unseen through concealed spaces in the structure and fabric of a building. The building must be designed and constructed so that this fire and smoke spread is inhibited.

Fire resistance and structural stability

If the structural elements of a building can be satisfactorily protected against the effects of fire for a reasonable period, it will be possible for the occupants to be evacuated safely and also the spread of fire throughout the building will be kept to a minimum. The risk to firefighters (who may have to search for or rescue people who are trapped) will be reduced and there will be less risk to **AD B3** people in the vicinity of the building from falling debris or as a result of an **7.1** impact with an adjacent building from the collapsing structure.

One way to measure the standard of protection to be provided is by reference to the fire resistance of the elements under construction. A number of

AD B, Appendix A

Table **A3** **Limitations on fire-protecting suspended ceilings**

Height of building or separated part (m)	Type of floor	Provision for fire-resistance of floor (minutes)	Description of suspended ceiling
Less than 20	Not compartment	60 or less	Type A, B, C or D
	Compartment	Less than 60	
		60	Type B, C, or D
20 or more	Any	60 or less	Type C or D
No limit	Any	More than 60	Type D

Notes:

Ceiling type and description:

A.   Surface of ceiling exposed to the cavity should be Class 0 or Class 1.

B.   Surface of ceiling exposed to the cavity should be Class 0.

C.   Surface of ceiling exposed to the cavity should be Class 0. Ceiling should not contain easily openable access panels.

D.   Ceiling should be of a material of limited combustibility and not contain easily openable access panels. Any insulation above the ceiling should be of a material of limited combustibility.

Any access panels provided in fire-protecting suspended ceilings of type C or D should be secured in position by releasing devices or screw fixings, and they should be shown to have been tested in the ceiling assembly in which they are incorporated.

factors which have a bearing on fire resistance are considered in Appendix A of AD B including:

AD B
Appendix A
A3, A4

- Fire severity – estimated from the purpose group (and therefore, the use) of the building. This assumes that the contents (which constitute the fire load) are the same for buildings of similar usage and that the contents of some building types will be more hazardous than others.
- Height of the top floor above ground – affects ease of escape, firefighting and the consequences of a large-scale collapse.
- Building occupancy – influences the speed of evacuation.
- The presence of basements – lack of venting may increase heat build-up and the duration of a fire, and hinder firefighting.
- The number of floors – escape from single-storey buildings is easier and a structural failure is unlikely to happen before evacuation has taken place.

It can be seen from the foregoing that an assessment of the standard of fire resistance in a building is a complicated matter. It is further complicated by the fact that there can be little control exercised over a building's future fire load unless a material change of use occurs. If a fire engineering approach is adopted to the assessment of fire resistance, then future changes in use should also be borne in mind.

AD B
Appendix A
A5

The method of assessment of fire resistance contained in AD B is based on the performance of an element of structure, door or other part of a building by reference to standard tests contained in BS 476: Parts 20–24: 1987 (or to BS 476: Part 8: 1972, for items tested prior to 1 January 1988).

The tests relate to the ability of the element:

- to resist a fire without collapse (loadbearing capacity),
- to resist fire penetration (integrity),
- to resist excessive heat penetration so that fire is not spread by radiation or conduction (insulation).

Clearly the criteria of resistance to fire and heat penetration are applicable only to fire-separating elements, such as walls and floors. The criterion of resistance to collapse is applicable to all load-bearing elements, such as columns and beams, in addition to floors and load-bearing walls; however, it does not apply to external curtain walling or other claddings which transmit only self-weight and wind loads.

Table A1 to Appendix A of AD B (see below) shows the method of exposure required for the various elements of structure and other forms of construction, together with the BS 476 requirements which should be satisfied in terms of load-bearing capacity, integrity and insulation. For some items the table indicates the actual period of fire resistance recommended under each heading, but for others Table A2 of Appendix A gives the detailed recommendations in respect of fire resistance periods. The performance standards for doors are contained in Table B1 of AD B which is reproduced in the section on fire doors below.

In addition to the elements of structure defined on p. 7.5 and illustrated in Fig. 7.3 above, there are requirements for some other elements of the building to be of fire-resisting construction. Included in this category are some doors, pipe casings and cavity barriers. These are considered later under the actual element references.

Minimum period of fire resistance

AD B
Appendix A
section 7
7.2 to 7.6

In order to establish the minimum period of fire resistance for the elements of structure of a building, it is necessary, first, to determine the building's use or purpose group. The fire resistance period will then depend on the height of the top storey of the building above ground or the depth of the lowest basement storey below ground.

It will be seen that the fire resistance recommendations for basements are generally more onerous than for ground floors in the same building. This reflects the greater difficulty experienced in dealing with a basement fire. However, it is sometimes the case that due to the slope of the ground, at least one side of a basement is accessible at ground level. This gives opportunities for smoke venting and firefighting and in these circumstances it may be reasonable to adopt the less onerous fire resistance provisions of the upper elements of the construction for the elements of structure in the basement.

The minimum periods of fire resistance recommended for the elements of

AD B, Appendix A

Table **A1**   **Specific provisions of test for fire resistance of elements of structure etc.**

Part of building	Minimum provisions when tested to the relevant part of BS 476[1] (minutes)			Method of exposure
	Load-bearing capacity [2]	**Integrity**	**Insulation**	
1. Structural frame, beam or column	See Table A2	Not applicable	Not applicable	Exposed faces
2. Load-bearing wall (which is not also a wall described in any of the following items)	See Table A2	Not applicable	Not applicable	Each side separately
3. Floors: a. in upper storey of 2 storey dwellinghouse (but not over garage)	30	15	15	From underside [3]
b. any other floor including compartment floors	See Table A2	See Table A2	See Table A2	
4. Roofs: a. any part forming an escape route	30	30	30	From underside [3]
b. any roof that performs the function of a floor	See Table A2	See Table A2	See Table A2	
5. External walls: a. any part less than 1 m from any point on the relevant boundary	See Table A2	See Table A2	See Table A2	Each side separately
b. any part 1 m or more from the relevant boundary	See Table A2	See Table A2	15 [4]	From inside
6. Compartment walls separating occupancies (see 8.10)	60 or see Table A2 (whichever is less)	60 or see Table A2 (whichever is less)	60 or see Table A2 (whichever is less)	Each side separately
7. Compartment wall other than 6.	See Table A2	See Table A2	See Table A2	Each side separately

Table **A1**   **Continued**

Part of building	Minimum provisions when tested to the relevant part of BS 476[1] (minutes)			Method of exposure
	Load-bearing capacity[2]	Integrity	Insulation	
8. Protected shafts, excluding any firefighting shaft: a. any glazing described in Approved Document B3 Diagram 26	Not applicable	30	No provision[5]	Each
b. any other part between the shaft and a protected lobby/corridor described in Diagram 26	30	30	30	side
c. any part not described in (a) or (b) above	See Table A2	See Table A2	See Table A2	separately
9. Enclosure (which does not form part of a compartment wall or a protected shaft) to: a. a protected stairway	30	30	30[6]	Each
b. a lift shaft	30	30	30	side
c. a service shaft	30	30	30	separately
10. Firefighting shafts: a. construction separating firefighting shaft from rest of building	120	120	120	From side remote from shaft
	60	60	60	From shaft side
b. construction separating firefighting stairway, firefighting lift shaft and firefighting lobby	60	60	60	Each side separately
11. Enclosure (which is not a compartment wall or described in item 8) to: a. a protected lobby	30	30	30[6]	Each side separately
b. a protected corridor	30	30	30[6]	
12. Sub-division of a corridor	30	30	30[6]	Each side separately
13. Wall separating an attached or integral garage from a dwellinghouse	30	30	30[6]	From garage side

Table **A1** Continued

Part of building	Minimum provisions when tested to the relevant part of BS 476[1] (minutes)			Method of exposure
	Load-bearing capacity[2]	Integrity	Insulation	
14. Enclosure in a flat or maisonette to a protected entrance hall, or to a protected landing	30	30	30[6]	Each side separately
15. Fire-resisting construction: a. enclosing communal areas in sheltered housing	30	30	30	Each side separately
b. in dwellings not described elsewhere	30	30	30[6]	
16. Cavity barrier	Not applicable	30	15	Each side separately
17. Ceiling described in Section 9 Diagram 29 or Diagram 31	Not applicable	30	30	From underside
18. Duct described in paragraph 9.14(e)	Not applicable	30	No provision	From outside
19. Casing around a drainage system described in Section 10 Diagram 34	Not applicable	30	No provision	From outside
20. Flue walls described in Section 10 Diagram 35	Not applicable	See Table A2	See Table A2	From outside
21. Fire doors		See Table B1		

Notes:

1. Part 21 for load-bearing elements, Part 22 for non-load-bearing elements, Part 23 for fire-protecting suspended ceilings, and Part 24 for ventilation ducts. BS 476: Part 8 results are acceptable for items tested or assessed before 1 January 1988.

2. Applies to load-bearing elements only.

3. A suspended ceiling should only be relied on to contribute to the fire resistance of the floor if the ceiling meets the appropriate provisions given in Table A3.

4. Thirty minutes for any part adjacent to an external escape route (but no provision for glazed elements in respect of insulation).

5. Except for any limitations on glazed elements given in Table A4.

6. See Table A4 for permitted extent of uninsulated glazed elements.

structure in the basements, ground or upper storeys of a building are given in Table A2 from Appendix A of AD B which is reproduced below.

The following points should also be taken into account when using Table A2:

- Any element of structure should have fire resistance at least equal to the fire resistance of any element which it carries, supports or to which it gives stability. This principle may be varied where:

  (a) the supporting structure is in the open air and would be unlikely to be affected by a fire in the building, or
  (b) the supporting and supported structures are in different compartments (the separating element between the compartments would have to have the higher standard of fire resistance, see next item).

- If an element of structure forms part of more than one building or compartment, and is thus subject to two or more different fire resistances, it is the greater of these which applies.

- A structural frame, beam, column or load-bearing wall of a *single-storey building* (or which is part of the ground storey of a building that consists of a ground storey and one or more basement storeys) is generally not required to have fire resistance. (This reflects the view that, given satisfactory means of escape, and the restricted use of combustible materials as wall and ceiling linings, fire resistance in the elements of structure in the ground storey will contribute little to the safety of the occupants.) However, the above concession will only apply if the element of structure:

  (a) is part of or supports an external wall and is sufficiently far from its relevant boundary to be regarded as a totally unprotected area;
  (b) is not part of and does not support a compartment wall or a wall which is common to two or more buildings;
  (c) is not a wall between a house and an attached or integral garage; or
  (d) does not support a gallery.

- Single-storey buildings should comply with the fire resistance periods under the heading 'not more than 5'. Where they have basements, these should, of course, comply with the recommendations for basement storeys depending on their depth below ground level.

- Further fire resistance provisions relating to the following elements may be found in the relevant sections below:

AD B3
section 8

  (a) Compartment walls, external walls and the wall between a dwelling-house and a domestic garage.
  (b) Walls which enclose a firefighting shaft or protect a means of escape.
  (c) Compartment floors.

Meeting the performance recommendations

AD B
Appendix A
A2

Reference has been made above to the BS 476 tests, where the fire resistance period of a material, product, structure or system may be assessed. It should be realised that the aim of the standard fire tests is to measure or assess the response of the sample to one or more aspects of fire behaviour under standardised conditions. The tests cannot normally measure fire hazard and represent only one aspect of the total fire safety package.

AD B, Appendix A

Table A2 Minimum periods of fire resistance

Purpose group of building	Minimum periods (minutes) for elements of structure in:					
	Basement storey[$] including floor over		Ground or upper storey			
	Depth (m) of a lowest basement		Height (m) of top floor above ground, in building or separated part of building			
	more than 10	not more than 10	not more than 5	not more than 20	not more than 30	more than 30
1. Residential (domestic):						
a. flats and maisonettes	90	60	30*	60**†	90**	120**
b. and c. dwellinghouses	Not relevant	30*	30*	60††	Not relevant	Not relevant
2. Residential:						
a. institutional~	90	60	30*	60	90	120#
b. other residential	90	60	30*	60	90	120#
3. Office:						
– not sprinklered	90	60	30*	60	90	Not permitted
– sprinklered[(2)]	60	60	30*	30*	60	120#
4. Shop and commercial:						
– not sprinklered	90	60	60	60	90	Not permitted
– sprinklered[(2)]	60	60	30*	60	60	120#
5. Assembly and recreation:						
– not sprinklered	90	60	60	60	90	Not permitted
– sprinklered[(2)]	60	60	30*	60	60	120#
6. Industrial:						
– not sprinklered	120	90	60	90	120	Not permitted
– sprinklered[(2)]	90	60	30*	60	90	120#
7. Storage & other non-residential:						
a. any building or part not described elsewhere:						
– not sprinklered	120	90	60	90	120	Not permitted
– sprinklered[(2)]	90	60	30*	60	90	120#
b. car park for light vehicles:						
i. open-sided park[(3)]	Not applicable	Not applicable	15*+	15*+	15*+	60
ii. any other park	90	60	30*	60	90	120#

Modifications referred to in Table A2:

$ The floor over a basement (or if there is more than one basement, the floor over the topmost basement) should meet the provisions for the ground and upper storeys if that period is higher.

* Increased to a minimum of 60 minutes for compartment walls separating buildings.

** Reduced to 30 minutes for any floor within a maisonette, but not if the floor contributes to the support of the building.

~ Multi-storey hospitals designed in accordance with the NHS Firecode documents should have a minimum 60 minutes standard.

Reduced to 90 minutes for elements not forming part of the structural frame.

+ Increased to 30 minutes for elements protecting the means of escape.

✧ Refer to p. 7.113 regarding the acceptability of 30 minutes in flat conversions.

†† 30 minutes in the case of three storey dwelling houses increased to 60 minutes minimum for compartment walls separating buildings.

Notes:

1. Refer to Table A1 for the specific provisions of test.

2. 'Sprinklered' means that the building is fitted throughout with an automatic sprinkler system meeting the relevant recommendations of BS 5306: Part 2; i.e. the relevant occupancy rating together with the additional requirements for life safety.

3. The car park should comply with the relevant provisions in the guidance on requirement B3, section 11 (see p. 7.143).

In a real fire the conditions will not be standard and there is always the possibility of premature failure of a particular component due to faulty design or workmanship. Therefore, the periods stated in Table A2 should be used for guidance and not taken as 'cast in tablets of stone', there being little correlation between the period derived from the standard test and the performance in a real fire. They do however, enable different systems to be compared under similar circumstances and are useful in this sense.

Therefore, in order that a material, product or structure may be used in a building it should:

AD B
Appendix A
A1

- Be assessed against appropriate standards and have been shown by test to be capable of meeting the relevant performance standard (suitably qualified fire safety engineers, laboratories accredited by the National Measurement Accreditation Service (NAMAS), the BRE and/or bodies such as the British Board of Agrément (BBA) might be expected to have the necessary expertise), or
- Comply with an appropriate specification given in relevant tables of notional performance included in AD B. (Over the years since the 1976 Regulations, there has been a tendency to reduce the practical guidance given on forms of construction which will satisfy the fire resistance requirements. The 1992 edition of AD B contains no such examples, the reader is merely referred to the publication listed in the next point.) or
- For fire-resisting elements, conform with an appropriate specification from Part II of the BRE Report *Guidelines for the construction of fire-resisting structural elements* (BRE, 1988).

In order to provide some assistance to readers, Table 7.1 gives notional periods of fire resistance for some common floor and wall constructions. It is based on a selection of constructions from Table A3 of DOE's 1985 Approved Document B2/3/4, which in turn is based on an earlier edition of the BRE Report. The latest BRE Report mentioned above also contains much information on fire protection to structural frameworks of beams and columns. Additionally, there are available various mineral based insulating boards which are capable of providing differing degrees of fire protection depending on their thickness and method of fixing. Reference should be made to individual manufacturers or their trade associations for further details.

For the first two items in Table 7.1 it should be noted that the upper floor of a two-storey dwelling is regarded as a special case. Such a floor, when tested for fire resistance from the underside is required only to provide

(a) stability for 30 minutes; and,
(b) integrity for 15 minutes; and,
(c) insulation for 15 minutes.

This is termed 'modified 30 minutes' fire resistance (see Table A1, Appendix A, item 3, 'Floors').

Compartmentation

AD B3
section 8

In order to prevent the rapid spread of fire within buildings (which could trap the occupants) and to restrict the size of any fires which do occur, AD B3

Table 7.1 Notional periods of fire resistance of some common constructions.

These constructions are a selection from Table A3 of DOE's 1985 Approved Document B2/3/4.

A large number of constructions other than those shown are capable of providing the fire resistance looked for. For example, various mineral based insulating boards can be used. Because their performance varies and is dependent on their thickness, it is not possible to give specific thicknesses in this table. However, manufacturers will normally be able to say what thickness would be needed to achieve the particular performance.

Floors: timber joist

Modified 30 minutes (stability 30 minutes) (integrity 15 minutes) (insulation 15 minutes)	1		any structurally suitable flooring: floor joists at least 37 mm wide ceiling: (a) 12.5 mm plasterboard[a] with joints taped and filled and backed by timber, or (b) 9.5 mm plasterboard[a] with 10 mm lightweight gypsum plaster finish
	2		at least 15 mm t&g boarding or sheets of plywood or wood chipboard, floor joists at least 37 mm wide ceiling: (a) 12.5 mm plasterboard[a] with joints taped and filled, or (b) 9.5 mm plasterboard[a] with at least 5 mm neat gypsum plaster finish
30 minutes	3		at least 15 mm t&g boarding or sheets of plywood or wood chipboard, floor joists at least 37 mm wide ceiling: 12.5 mm plasterboard[a] with at least 5 mm neat gypsum plaster finish
	4		at least 21 mm t&g boarding or sheets of plywood or wood chipboard, floor joists at least 37 mm wide ceiling: 12.5 mm plasterboard[a] with joints taped and filled

Table 7.1 *continued*

60 minutes	5		at least 15 mm t&g plywood or wood chipboard, floor joists at least 50 mm wide ceiling: not less than 30 mm plasterboard[a] with joints staggered and exposed joints taped and filled

Floors: concrete

60 minutes	6		reinforced concrete floor not less than 95 mm thick, with not less than 20 mm cover on the lowest reinforcement

Walls: internal

30 minutes load-bearing	7		framing members at least 44 mm wide[b] and spaced at not more than 600 mm apart, with lining (both sides) of 12.5 mm plasterboard[a] with all joints taped and filled
	8		100 mm reinforced concrete wall[c] with minimum cover to reinforcement of 25 mm
60 minutes load-bearing	9		framing members at least 44 mm wide[b] and spaced at not more than 600 mm apart, with lining (both sides) at least 25 mm plasterboard[a] in 2 layers with joints staggered and exposed joints taped and filled
	10		solid masonry wall (with or without plaster finish) at least 90 mm thick (75 mm if non-load-bearing) *Note:* for masonry cavity walls, the fire resistance may be taken as that for a single wall of the same construction, whichever leaf is exposed to fire
	11		120 mm reinforced concrete wall[c] with at least 25 mm cover to the reinforcement

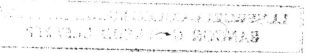

Table 7.1 *continued*

Walls: external

Modified 30 minutes (stability 30 minutes) (integrity 30 minutes) (insulation 15 minutes) load-bearing wall 1 m or more from relevant boundary	**12**		any external weathering system with at least 8 mm plywood sheathing, framing members at least 37 mm wide and spaced not more than 600 mm apart internal lining: 12.5 mm plasterboard[a] with at least 10 mm lightweight gypsum plaster finish
30 minutes load-bearing wall less than 1 m from the relevant boundary	**13**		100 mm brickwork or blockwork external face (with, or without, a plywood backing); framing members at least 37 mm wide and spaced not more than 600 mm apart internal lining: 12.5 mm plasterboard[a] with at least 10 mm lightweight gypsum plaster finish
60 minutes load-bearing wall less than 1 m from the relevant boundary	**14**		solid masonry wall (with or without plaster finish) at least 90 mm thick (75 mm if non-load-bearing) *Note:* for masonry cavity walls, the fire resistance may be taken as that for a single wall of the same construction, whichever leaf is exposed to fire

Notes

[a] Whatever the lining material, it is important to use a method of fixing that the manufacturer says would be needed to achieve the particular performance. For example, if the lining is plasterboard, the fixings should be at 150 mm centres as follows (where two layers are being used each should be fixed separately):
9.5 mm thickness, use 30 mm galvanised nails
12.5 mm thickness, use 40 mm galvanised nails
19 mm–25 mm thickness, use 60 mm galvanised nails.

[b] Thinner framing members, such as 37 mm, may be suitable depending on the loading conditions.

[c] A thinner wall may be suitable depending on the density of the concrete and the amount of reinforcement. (See *Guidelines for the construction of fire-resisting structural elements* (BRE, 1988).)

contains provisions for sub-dividing a building into compartments separated from one another by fire-resisting walls and floors.

AD B3 8.1

The extent of the sub-divisions will depend on the same factors which were considered for fire resistance, namely:

- The severity of the fire.
- The height of the top floor above ground.
- The building occupancy.

AD B3 8.2

- The presence of basements.
- The number of floors.

AD B3
8.3

These items are considered more fully on p. 7.77 above. Additionally, the provision of a sprinkler system can affect the growth rate of a fire and may suppress it altogether.

The sub-division is achieved by means of compartment walls and compartment floors and since these come under the definition of elements of structure they should be fire resisting (but not, of course, the lowest floor in the building.)

AD B3
8.15

Most multi-storey buildings are required to be compartmented because of the increased risk to life should a fire occur. However, this risk is obviously less in single-storey buildings where compartmentation is required only for single-storey hospitals with a floor area limit of 3000 m^2.

Generally, two main principles are adopted when deciding how to compartment a building:

AD B3
8.18

(a) For non-residential multi-storey buildings, the compartment sizes should be restricted to the maximum dimensions shown in Table 12 of AD B3, which is reproduced below, and the following points should be noted:

- Limits are given by reference to maximum floor areas for differing top storey heights for assembly and recreation, shop and commercial, and industrial buildings.
- For storage and other non-residential buildings, the limits are according to maximum compartment volumes for differing top storey heights.
- The installation of sprinklers allows the compartment size limits to be doubled.
- All floors which are more than 30 m above ground or more than 10 m below ground should be constructed as compartment floors, as should the ground floor over a basement.

AD B3
8.4

(b) For all buildings, the following walls and floors which separate buildings in different ownerships, uses or occupancies should be constructed as compartment walls and floors:

AD B3
8.8, 8.11,
8.20

- Walls which are common to two or more buildings (including walls between semi-detached and terraced houses). These walls should run the full height of the building in a continuous vertical plane, the adjoining buildings being separated only by walls, not floors.

AD B3
8.9

- Walls and floors dividing parts of buildings used for different purposes. This does not apply where one use is ancillary to another. (See p. 7.11 above.)

AD B3
8.10

- Walls and floors dividing buildings into different occupancies.

AD B3
8.4, 8.21

- Walls dividing buildings into separated parts. These walls should also run the full height of the building in a continuous vertical plane, in a similar manner to common walls above. (See p. 7.8.)

AD B3
8.6

- Walls and floors bounding a protected shaft. (Protected shafts are considered in more detail below.)

Finally, the following walls and floors in particular purpose groups should be constructed as compartment walls and floors:

AD B3, Section 8

Table **12**   **Maximum dimensions of buildings or compartment (multi-storey non-residential buildings)**

Purpose group of building (or part)	Height of floor of top storey above ground level (m)	Floor area of any one storey in the building or compartment (m²)
Office	No limit	No limit
Assembly & Recreation Shop & Commercial:		
not sprinklered	No limit	2 000
sprinklered [1]	No limit	4 000
Industrial [3]:		
not sprinklered	Not more than 20	7 000
	More than 20	2 000 [2]
sprinklered [1]	Not more than 20	14 000
	More than 20	4 000 [2]

	Height of floor of top storey above ground level (m)	Maximum compartment volume (m³)
Storage [3] & Other non-residential:		
a. car park for light vehicles	No limit	No limit
b. any other building or part:		
not sprinklered	Not more than 20	20 000
	More than 20	4 000 [2]
sprinklered [1]	Not more than 20	40 000
	More than 20	8 000 [2]

Notes:

1.   'Sprinklered' means that the building is fitted throughout with an automatic sprinkler system meeting the relevant recommendations of BS 5306: Part 2, i.e. the relevant occupancy rating together with the additional requirements for life safety.

2.   This reduced area limit applies only to storeys that are more than 20 m above ground level.

3.   There may be additional limitations on floor area and/or sprinkler provisions in certain industrial and storage uses under other legislation, for example, in respect of storage of LPG and certain chemicals.

It should also be noted that although multi-storey hospitals are not included in this table, there is a maximum compartment floor area limit of 2000 m² placed on them by AD B3.

- Any floor or wall separating an attached or integral domestic garage from a dwellinghouse should be constructed as shown in Fig. 7.24.   **AD B3 8.12**
- Any floor in an institutional or other residential building.   **AD B3 8.14, 8.17**
- Any floor in flats or maisonettes, except an internal floor in an individual dwelling.   **AD B3 8.13**

Section

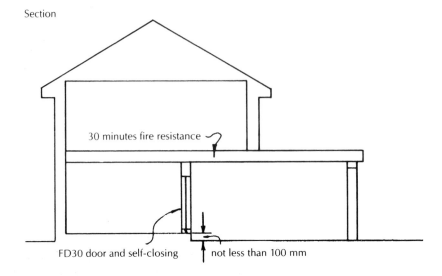

30 minutes fire resistance

FD30 door and self-closing   not less than 100 mm

Plan

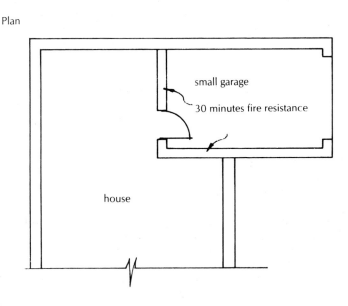

small garage

30 minutes fire resistance

house

Fig. 7.24 Attached small garages.

- Any wall separating a flat or maisonette from any other part of the same building.
- Any wall enclosing a refuse storage chamber in flats or maisonettes.

AD B3
8.18
- Any wall or floor in a shopping complex referred to in Section 5 of BS 5588: Part 10 as needing to be constructed to the compartmentation standard.

AD B3
8.16
- Any wall between compartments in health-care premises used for progressive horizontal evacuation (see p. 7.64, above).

Compartment walls and floors – construction details

Since the purpose of compartment walls and floors is to form a complete barrier to the passage of fire between the compartments which they separate, it follows that they should have the appropriate standards of fire resistance indicated in Tables A1 and A2 above.

AD B3 8.19

It also follows that any points of weakness in compartment walls and floors should be adequately protected. These points of weakness occur:

(a) at junctions with other compartment walls, external walls and roofs; and
(b) at openings for doors, pipes and ducts of various kinds, refuse chutes and protected shafts.

Junction details

Where a compartment wall and roof meet, the wall may be carried at least 375 mm above the roof covering surface, measured at right angles to the roof slope, or the wall may be taken up to meet the underside of the roof covering or deck, and the junction fire-stopped. Acceptable design solutions are illustrated in Fig. 7.25 for buildings in different purpose groups.

AD B3 8.22 to 8.26

Generally, the covering in a 1.5 m wide zone on either side of the wall should be designated AA, AB or AC (see below for designations) and it should be carried on a substrate or deck consisting of a material of limited combustibility.

Some exceptions are permitted in the case of certain roof constructions in buildings not more than 15 m high. In the roof shown at (b) in Fig. 7.25, the following combustible materials may be carried over the top of the compartment wall:

- Roof boarding serving as a base for roof covering; *or*
- Woodwool slabs; *or*
- Timber slating or tiling battens.

In each case the materials should be fully bedded in mortar or similar material. Any roof support members which pass through the wall may need to have fire protection on their undersides for at least 1.5 m on either side of the wall to delay distortion at the junction.

Where a compartment wall or floor meets another compartment wall or an external wall, the junction should maintain the fire resistance of the compartmentation. This will normally mean that the various structures should be bonded together or the junction fire-stopped (see Fig. 7.25).

Openings in compartment walls and floors

Generally, the only openings permitted in compartment walls and floors are one or more of the following:

AD B3 8.29

- An opening fitted with a door which has the appropriate fire resistance given in Table B1 of Appendix B to AD B and is fitted in accordance with Appendix B of AD B (see Fire doors below).

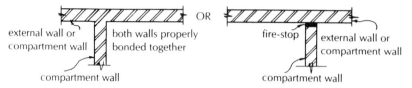

external wall or compartment wall — both walls properly bonded together

compartment wall

OR

fire-stop

external wall or compartment wall

compartment wall

Junction of external and compartment walls

EITHER

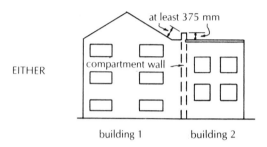

at least 375 mm

compartment wall

building 1   building 2

(a) Building or compartment of any use or height

roof covering: AA, AB or AC on a substrate or deck of material of limited combustibility

OR

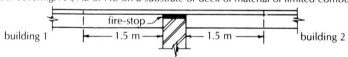

fire-stop

building 1   1.5 m   1.5 m   building 2

(b) Dwellinghouse, office, assembly and recreation, or residential (not institutional) nor more than 15 m high

roof covering: AA, AB or AC

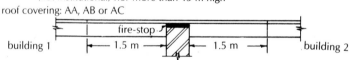

fire-stop

building 1   1.5 m   1.5 m   building 2

see below for materials which may be carried over or through compartment wall

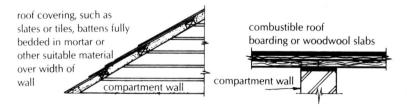

roof covering, such as slates or tiles, battens fully bedded in mortar or other suitable material over width of wall

compartment wall

combustible roof boarding or woodwool slabs

compartment wall

Combustible materials permitted over wall

Fig. 7.25 Junction details – compartment walls.

- An opening for a protected shaft (see Protected shafts on p. 7.97).
- An opening for a refuse chute of non-combustible construction.
- An opening for a pipe, ventilation duct, chimney, appliance, ventilation duct or duct encasing one or more flue pipes, provided it complies with the relevant parts of Section 10 of AD B (see Pipes, ventilation ducts and flues on p. 7.108).

In the case of compartment walls which are common to two or more buildings or which separate different occupancies in the same building, it would not be sensible or necessary to allow all of the above openings to exist. Therefore such walls should be imperforate except for:

AD B3
8.28

- An opening for a door which is needed as a means of escape in case of fire. The door should have the same fire resistance as the wall and should be fixed in accordance with the provisions of Appendix B to AD B (see below).
- An opening for a pipe complying with the provisions of Section 10 of AD B (see p. 7.108).

Fire doors

All fire doors should be fitted with an automatic self-closing device unless they are to cupboards or service ducts which are normally kept locked shut. Rising butt hinges are not considered as automatic self-closing devices unless the door is:

AD B
Appendix B
B2

- To or within a dwelling.
- In a cavity barrier.
- Between a dwellinghouse and a garage.

AD B
Appendix E

As a general rule, no device should be provided to hold a door open. However, in some cases a self-closing device may be considered a hindrance to normal use. In such cases a fire-resisting door may be held open by:

AD B
Appendix B
B3

- A fusible link device, provided that the door is not fitted in an opening used as a means of escape. (This does not apply where two doors are provided in the opening as mentioned below.)
- A door closure delay device.
- An automatic release mechanism, provided that the door can also be readily closed manually.

In this context an automatic release mechanism is one which automatically closes a door in the event of each or any one of:

AD B
Appendix E

(a) smoke detection by appropriate apparatus;
(b) manual operation by a suitably located switch;
(c) failure of the electricity supply to the device, smoke detector, or switch;
(d) operation of a fire alarm system, if fitted.

Automatic release mechanisms are not permitted on any door to:

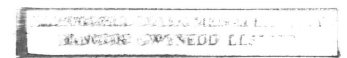

AD B
Appendix B
B3

(a) The only escape stair in a building or part of a building.
(b) A firefighting stair.
(c) An escape stair in any building in the Residential purpose groups (PGs 1(a), 1(b), 1(c), 2(a) and 2(b)).

AD B
Appendix B
B1

All fire-resisting doors should have the appropriate fire resistance described in Table B1 of Appendix B (see below). In the table the fire doors are given a rating in terms of their performance under test to BS 476: Part 22. This rating relates to the ability of the door to maintain its integrity for a specified period in minutes, e.g. FD30. Doors should be tested from each side separately; however, since lift doors are only at risk from one side in the event of a fire, it is only necessary to test these from the landing side. The rating for some doors has the suffix S added where restricted smoke leakage is needed at ambient temperatures. The leakage rate should not exceed 3m/m/hour (head and jambs only) when tested at 25 Pa to BS 476: Section 31.1, unless pressurisation techniques complying with BS 5588: Part 4 are used.

AD B
Appendix B
B5

No fire door should be hung on hinges made of a material with a melting point less than 800°C, unless the hinges can be shown to be satisfactory when tested as part of a fire door assembly.

AD B
Appendix B
B4

Although each fire door should have the appropriate period of fire resistance defined in Table B1, it is permissible for two fire doors to be fitted in an opening if each door is capable of closing the opening and the required level of fire resistance can be achieved by the two doors together. However, if these two doors are fitted in an opening provided for a means of escape, both doors should be self-closing. One of them may be held open by a fusible link and be fitted with an automatic self-closing device if the other is easily openable by hand and has at least 30 minutes fire resistance.

AD B
Appendix B
B7

Doors which are to be kept closed or locked when not in use, or which are held open by an automatic release mechanism, should be marked with the appropriate fire safety sign in accordance with BS 5499 *Fire safety signs, notices and graphic symbols*, Part 1: 1990 *Specification for fire safety signs*. The signs should be on both sides of the fire doors, except for cupboards and service ducts where it is only necessary to mark the doors on the outside. This recommendation does not apply to:

AD B
Appendix B
B8

● Fire doors within dwellinghouses.
● Fire doors to and within flats or maisonettes.
● Fire doors in other residential buildings (PG 2(b)).
● Lift entrance doors.

AD B
Appendix B
B6

Ironmongery used on fire doors can significantly affect their performance in a fire. Reference should be made to the *Code of practice for hardware essential to the optimum performance of fire-resisting timber doorsets* published by the Association of Builders' Hardware Manufacturers, 1983, where guidance may be obtained on the selection of suitable hardware.

Glazing to fire doors

AD B
Appendix B
B9

It is often desirable to provide glazed vision panels in fire doors. Where the glazing can satisfy the relevant insulation criteria from Table A1 of Appendix A, there are no limitations on its use in fire doors. Where this is not the case, the uninsulated glazing should comply with the recommendations of Table A4 of Appendix A which is reproduced below. Table A4 also contains details of

AD B, Appendix B

Table **B1** **Provisions for fire doors**

Position of door	Minimum fire resistance of door in terms of integrity (minutes)[1]
1. In a compartment wall separating buildings	As for the wall in which door is fitted, but a minimum of 60
2. In a compartment wall:	
a. if it separates a flat or maisonette from a space in common use,	FD 30S
b. enclosing a protected shaft forming a stairway situated wholly or partly above the adjoining ground in a building used for Flats, Other Residential, Assembly & Recreation, or Office purposes,	FD 30S
c. enclosing a protected shaft forming a stairway not described in (b) above,	Half the period of fire resistance of the wall in which it is fitted but 30 minimum and with suffix S[2]
d. not described in (a), (b) or (c) above	As for the wall it is fitted in, but add S if the door is used for progressive horizontal evacuation under guidance to B1
3. In a compartment floor	As for the floor in which it is fitted
4. Forming part of the enclosure of:	
a. a protected stairway (except where described in item 9),	FD 30S
b. lift shaft, or	FD 30
c. service shaft, which does not form a protected shaft in 2(c) above	FD 30
5. Forming part of the enclosures of:	
a. a protected lobby approach (or protected corridor) to a stairway	FD 30S
b. any other protected corridor	FD 20S
6. Affording access to an external escape route	FD 30
7. Sub-dividing:	
a. corridors connecting alternative exits,	FD 20S
b. dead-end portions of corridors from the remainder of the corridor	FD 20S
8. Any door:	
a. within a cavity barrier,	FD 30
b. between a dwellinghouse and a garage,	FD 30
c. forming part of the enclosure to a communal area in sheltered housing	FD 30S
9. Any door:	
a. forming part of the enclosures to a protected stairway in a single family dwellinghouse	FD 20
b. forming part of the enclosure to a protected entrance hall or protected landing in a flat or maisonette	FD 20
c. within any other fire-resisting construction in a dwelling not described elsewhere in this table	FD 20

Notes:
1. To BS 476: Part 22 (or BS 476: Part 8 for items tested or assessed prior to 1 Jan. 1988).

2. See note on p. 7.94 regarding this suffix.

AD B, Appendix A

Table **A4**   **Limitations on the use of uninsulated glazed elements on escape routes.**
(These limitations do not apply to glazed elements which satisfy the relevant insulation criterion, see Table A1)

Position of glazed element	Maximum total glazed area in parts of a building with access to:			
	a single stairway		more than one stairway	
	walls	door leaf	walls	door leaf
1. Single family dwellinghouses within the enclosures of a protected stairway or within fire-resisting separation shown in Fig. 7.12	Fixed fanlights only	Unlimited	Fixed fanlights only	Unlimited
2. Within the enclosures of a protected entrance hall or protected landing of a flat or maisonette	Fixed fanlights only	Unlimited above 1.1 m from floor	Fixed fanlights only	Unlimited above 1.1 m from floor
3. Between residential/sleeping accommodation and a common escape route (corridor, lobby or stair)	Nil	Nil	Nil	Nil
4. Between a protected stairway[1] and: i. the accommodation; or ii. a corridor which is not a protected corridor, other than in item 3 above	Nil	25% of door area	Unlimited above 1.1 m [2]	50% of door area
5. Between: i. a protected stairway and a protected lobby or protected corridor; or ii. accommodation and a protected lobby, other than in item 3 above	Unlimited above 1.1 m from floor	Unlimited above 0.1 m from floor	Unlimited above 0.1 m from floor	Unlimited above 0.1 m from floor
6. Between the accommodation and a protected corridor forming a dead end, other than in item 3 above	Unlimited above 1.1 m from floor	Unlimited above 0.1 m from floor	Unlimited above 0.1 m from floor	Unlimited above 0.1 m from floor
7. Between accommodation and any other corridor; or sub-dividing corridors, other than in item 3 above.	Not applicable	Not applicable	Unlimited above 0.1 m from floor	Unlimited above 0.1 m from floor

Notes:
1. If the protected stairway is also a protected shaft (see paragraph 8.30 of AD B3) or a firefighting stair (see Section 17 of AD B5) there may be further restrictions on the uses of glazed elements.

2. Measured vertically from the landing floor level or the stair pitch line.

the use of uninsulated glazing in protected stairways, lobbies and corridors. This is described more fully under Protected shafts below.

Protected shafts

Protected shafts are needed when it is necessary to pass persons, things or air between compartments. Therefore, they should only be used to accommodate:

AD B3
8.31

- Stairs, lifts and escalators.
- Pipes, ducts or chutes.
- Sanitary accommodation and/or washrooms.

They should form a complete barrier between the different compartments which they connect and, except for glazed screens which should meet the recommendations referred to below, should have fire resistance as specified in Table A1 of Appendix A (see above).

AD B3
8.32

A protected shaft containing a stairway is often approached by way of a corridor or lobby. It is sometimes desirable to glaze the wall between the shaft and the corridor or lobby in order to allow light and visibility in both directions.

AD B3
8.33

This glazing is permitted provided that it has at least 30 minutes fire resistance in terms of integrity and the following conditions are met:

- The stair enclosure is not required to have more than 60 minutes fire resistance.
- The corridor or lobby has at least 30 minutes fire separation from the rest of the floor (including doors).
- The protected shaft is not a firefighting shaft.

These recommendations are illustrated in Fig. 7.26. Where these provisions cannot be met, the guidance shown in Table A4 of Appendix A relating to the limits on areas of uninsulated glazing will apply.

AD B3
8.34

There should be no-oil pipe or ventilating duct within any protected shaft which contains any stairway and/or lift (although pipes which convey oil for hydraulic lift mechanisms and ducts used in pressurisation systems aimed at keeping stairs smoke-free are permitted). Where a protected shaft contains a pipe carrying natural gas, the pipe should be of screwed steel or of all welded steel construction in accordance with the Gas Safety Regulations S.I. 1972 No. 1178, and Gas Safety (Installation & Use) Regulations 1984 as amended 1990. The shaft should be adequately ventilated direct to external air by ventilation openings at high and low level in the shaft.

AD B3
8.35, 8.36

Ideally, protected shafts should be imperforate except for certain openings mentioned below. The number of openings permitted will depend, to a great extent, on the function of the wall surrounding the protected shaft.

Generally, external walls to protected shafts do not need to be fire-resisting and hence there are no restrictions on the number of openings in such walls. This would not be so if the external wall was part of a firefighting shaft; in this case, reference should be made to BS 5588: Part 5: 1991 *Code of practice for firefighting stairs and lifts.*

AD B3
8.37

Where part of the enclosure to a protected shaft consists of a wall which is common to two or more buildings, the only openings permitted are those referred to on p. 7.93 above.

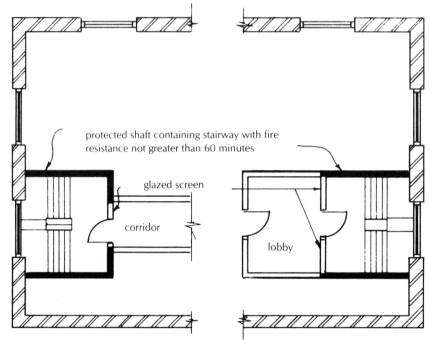

protected shaft containing stairway with fire
resistance not greater than 60 minutes

glazed screen

corridor

lobby

glazed screen, walls to corridor or lobby and doors to have at least 30 minutes fire
resistance

Fig. 7.26 Glazed screen separating protected shaft from lobby or corridor.

Any other walls which make up the enclosure should have no openings
other than those referred to below:

- A protected shaft containing one or more lifts may have openings to allow
  lift cables to pass to the lift motor room. If this is at the bottom of the shaft,
  the openings should be kept as small as possible.
- Where a protected shaft contains or is itself a ventilating duct, any inlets to,
  outlets from and openings for the duct should comply with the guidance in
  Section 10 of AD B3 (see Pipes, ventilation ducts and flues on p. 7.108).
- It is permissible to form an opening for pipes (other than those specifically
  forbidden above), provided that the pipe complies with Section 10 of AD
  B3.
- Any opening, other than those detailed above, should be fitted with a fire
  door complying with Table B1 of Appendix B of AD B.

Figure 7.27 illustrates the principles of protected shafts.

Concealed spaces

AD B3
9.1
Many buildings constructed today contain large hidden void spaces within
floors, walls and roofs. This is particularly true of system-built housing, schools
and other local authority buildings such as old people's homes.

These buildings may also contain combustible wall panels, frames and

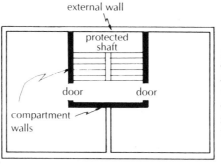

protected shaft
bounded on three
sides by compartment
walls and fourth
side by external
wall

protected shaft
containing stairway
and/or lift should
not contain oil pipe
or ventilating duct;
gas pipes to comply
with Gas Safety Regulations

Plan

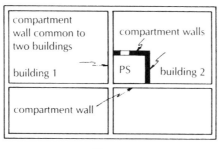

protected shaft bounded
on four sides by
compartment walls, used
as services duct

Plan

Fig. 7.27 Protected shafts.

insulation, thereby increasing the risk of unseen smoke and flame spread through these concealed spaces.

Therefore, despite compartmentation and the use of fire-resistant construction, many buildings have been destroyed as a result of fire spreading through cavities formed by, or in, constructional elements, and by-passing compartment walls/floors, etc.

Section 9 of AD B contains provisions designed to reduce the chance of hidden fire spread by making sure that cavities are:

- Interrupted if there is a chance that the cavity could form a route around a barrier to fire (such as a compartment wall or floor),
- Sub-divided if they are very large.

This interruption or sub-division of concealed spaces is achieved by using *cavity barriers*. These are defined in Appendix E of AD B as any form of construction which is intended to close a cavity (concealed space) and prevent the penetration of smoke or flame, or is fitted inside a cavity (concealed space) in order to restrict the movement of smoke or flame within the cavity. Therefore, providing it meets the requirements for cavity barriers, a form of construction designed for some other use (such as a compartment wall) may be acceptable as a cavity barrier.

Interruption of cavities

AD B3
9.4

Where an element which is required to form a barrier to fire abuts another element containing a cavity, there is a risk that smoke and flame could by-pass the fire barrier via the cavity. Cavity barriers should, therefore, be provided to interrupt the cavity at the point of contact between the fire barrier and the element containing the cavity (see Fig. 7.28).

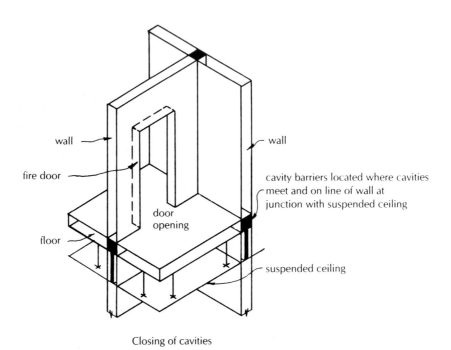

wall

wall

fire door

cavity barriers located where cavities meet and on line of wall at junction with suspended ceiling

door opening

floor

suspended ceiling

Closing of cavities

fire-resisting element

cavity barriers located in same plane as fire-resisting element (prevents fire by-passing element)

Includes: wall, floor, roof or other structure

element with cavity

Excludes: wall required to have fire resistance only because it is load-bearing

Interrupting cavities

Fig. 7.28 Cavity barriers.

The degree to which cavity barriers should be provided will depend on the use (or purpose group) of the building concerned and is set out in Table 13 of AD B3. These provisions are summarised below.

Cavities in some elements are excluded from the above provisions regarding interruption, mainly on the grounds that they present little risk to unseen fire spread. These include any cavity:

AD B3 9.11

(a) in an external wall of masonry construction which complies with the guidance shown in Fig. 7.29 below; *and*
(b) in a wall which requires fire resistance only because it is load-bearing (therefore it does not form a barrier to fire).

Cavity sub-division

Any cavity, including a roof space, should generally be sub-divided into separate sections by cavity barriers placed across the cavity at intervals not greater than the distances specified in Table 14 to Section 9 of AD B3, which is reproduced above. These distances are measured along the members bounding the cavity and depend on the cavity location and the class of surface exposed within it (see Fig. 7.30).

AD B3 9.3

Section 9 of AD B3 permits certain variations to the dimensions given in Table 14 as follows:

- If a room under a ceiling cavity exceeds the dimensions given in Table 14, then cavity barriers need only be placed on the line of the enclosing walls or partitions of that room, subject to a maximum cavity barrier spacing of 40 m.

AD B3 9.12

- Cavities over undivided areas which exceed the 40 m limit mentioned above in both directions need not be divided with cavity barriers if the following conditions can be met:

AD B3 9.13

(a) both room and cavity are compartmented from the rest of the building;
(b) an automatic fire detection and alarm system is fitted in the building;
(c) if the cavity is used for ventilating and air-conditioning ductwork this should be in accordance with the recommendations of BS 5588: Part 9;
(d) the ceiling surface exposed in the cavity is Class 0 and any supports or fixings for the ceiling are non-combustible;
(e) any pipe insulation system should have a Class 1 flame spread rating;
(f) electrical wiring in the void should be laid in metal trays or metal conduit;
(g) any other materials in the cavity should be of limited combustibility.

Additionally, the following low-risk cavities are excluded from the provisions of Table 14 (as well as those specified above for cavity interruption):

AD B3 9.11

(a) under a floor next to the ground or oversite concrete, provided that either:

(i) there is no access to the cavity for persons; *or*
(ii) the height of the cavity is not more than 1 m.

This exclusion does not apply if it is possible for combustible material to accumulate in the cavity through openings in the floor (such as happened at the Bradford City Football Club fire). In this case the cavity should be accessible for cleaning and should contain cavity barriers as described in Table 14.

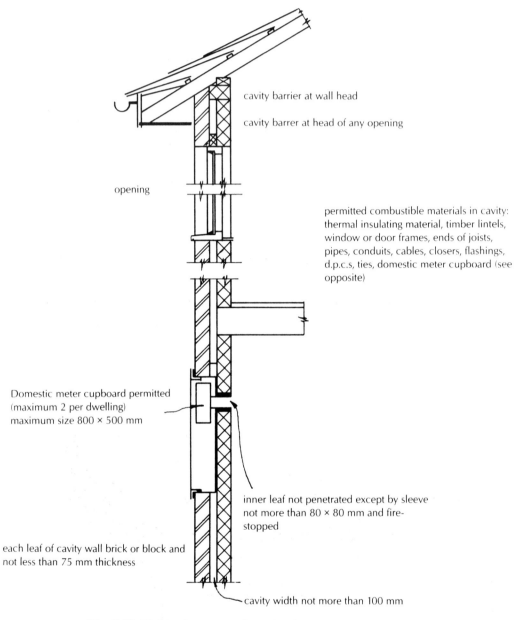

cavity barrier at wall head

cavity barrer at head of any opening

opening

permitted combustible materials in cavity:
thermal insulating material, timber lintels,
window or door frames, ends of joists,
pipes, conduits, cables, closers, flashings,
d.p.c.s, ties, domestic meter cupboard (see
opposite)

Domestic meter cupboard permitted
(maximum 2 per dwelling)
maximum size 800 × 500 mm

inner leaf not penetrated except by sleeve
not more than 80 × 80 mm and fire-
stopped

each leaf of cavity wall brick or block and
not less than 75 mm thickness

cavity width not more than 100 mm

Fig. 7.29 Wall cavity exempt from Section 9 of AD B3, Table 13.

(b) between double-skinned corrugated or profiled insulated roof sheeting
consisting of materials of limited combustibility, provided that the sheets
are separated by insulating material having a surface of Class 0 or Class 1,
and that insulating material is in contact with both sheets in line with the
corrugations.

AD B3, Section 9

Table **A14**   **Maximum dimensions of cavities in non-domestic buildings (purpose groups 2–7)**

Location of cavity	Class of surface exposed in cavity (excluding surface of any pipe, cable or conduit, or insulation to any pipe)	Maximum dimension in any direction (m)
Between a roof and a ceiling	Any	20
Any other cavity	Class 0 or Class 1	20
	Class other than Class 0 or 1	10

Note

Exceptions to these provisions are given in paragraphs 9.11–9.13 of AD B3 and are summarised on pp. 7.101 and 7.102.

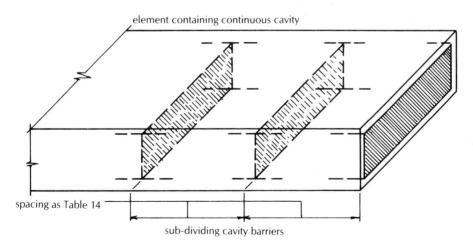

This requirement does not apply to:

- certain cavity walls in masonry construction (see Fig. 7.29)
- floors next to ground or oversite concrete if no access or not more than 1 m high space and no possibility of rubbish accumulating below floor
- cavity between non-combustible roof sheeting under certain conditions

Fig. 7.30 Sub-division of cavities.

(c) within a floor or roof, or enclosed by a roof, provided the lower side of that cavity is enclosed by a ceiling which:

  (i) extends throughout the whole building or compartment;
  (ii) is not designed to be demountable;
  (iii) has at least 30 minutes fire resistance;
  (iv) is imperforate except for any openings permitted in a cavity barrier (see below);
  (v) has an upper surface facing the cavity of at least Class 1 surface spread of flame;
  (vi) has a lower surface of Class 0.

Cavities above such fire-resisting ceilings are subject to an overall limit of 30 m in extent (see Fig. 7.31).

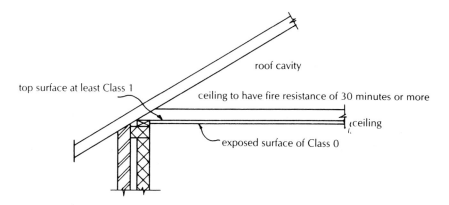

(i)   ceiling should extend throughout building or compartment
(ii) ceiling should be imperforate except for any allowable openings (see text)
(ii) ceiling should not be demountable
(iv) maximum cavity dimension – 30 m in any direction

Fig. 7.31 Fire-resisting ceilings.

Provision of cavity barriers

AD B3
Table 13 Bearing in mind the exclusions already mentioned above, cavity barriers should be provided as indicated below.

(a) In all buildings:

  • At the junction between a compartment wall which separates buildings and an external cavity wall. The top of the external wall in this case should also be closed with cavity barriers.

(b) In dwellinghouses of three or more storeys:

  • Above the enclosure to a protected stairway unless a fire-resisting ceiling is provided as shown in Figs. 7.31 and 7.32.

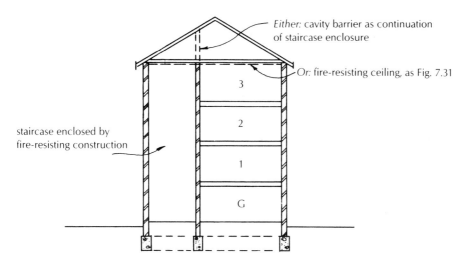

Fig. 7.32 Provision of cavity barriers in dwellinghouses of three or more storeys.

(c) In flats and maisonettes:

- At the junction between every compartment wall and compartment floor and an external cavity wall.
- At the junction between a cavity wall and every compartment wall, compartment floor or other wall or door assembly which forms a fire-resisting barrier.
- Above fire-resisting construction in a protected escape route if this is not carried up to the full storey height (or to the underside of the roof, if it is in the top storey).

(d) In office, shop and commercial, assembly and recreation, industrial, storage and other non-residential buildings:

- All as recommended for flats and maisonettes in (c) above.
- Above corridor enclosures where the corridor (which is not a protected corridor) is sub-divided to prevent smoke or fire affecting two escape routes simultaneously. The cavity barriers are not needed if the corridor enclosure is carried up to the full storey height (or to the underside of the roof, if it is in the top storey) or if the cavity is enclosed on the lower side by a fire-resisting ceiling complying with Fig. 7.31. (See also Fig. 7.33.)
- To sub-divide extensive cavities in floors, roofs, walls, etc., so that the distance between cavity barriers does not exceed the dimensions given in Table 14 (see above).

(e) In other residential and institutional buildings:

- All as recommended for office, etc., buildings in (d) above.
- Above any bedroom partitions if these are not carried up to the full storey height (or to the underside of the roof, if they are in the top storey).

To prevent both storey exits from becoming blocked by smoke:

EITHER

(a) sub-divide storey with fire-resisting construction

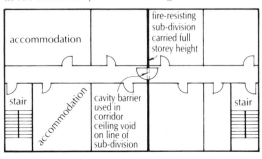

Plan

OR

(b) provide cavity barriers above corridor enclosure if
enclosures are not carried up to full storey height

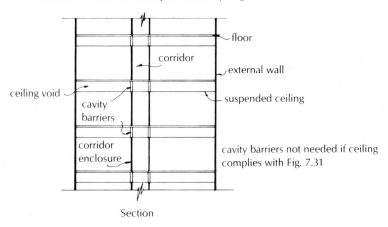

Section

Fig. 7.33 Alternative forms of corridor enclosure.

Rain screen cladding

AD B3
9.1, 9.11
Rain screen cladding is a form of construction often used in refurbishment
works to improve the weather resistance and thermal performance of the
external walls of buildings. It has been used regularly to upgrade the performance of blocks of high-rise flats (often built in the 1960s), where an external
weatherproof cladding is lined internally with combustible insulation and this
is fixed to the outer surface of the external walls. There have been occasions
when fire has occurred within the void space containing the insulation and fire
has spread upwards within the cladding, eventually affecting the interior of the
building.

Table 13 of AD B3 recommends that cavity barriers be provided within the
void behind the external face of rain screen cladding at every floor level, and

on the line of compartment walls abutting the external wall, where the building has any floor which is more than 20 m above ground level. This applies to flats, maisonettes, other residential buildings and institutional buildings. The spacing recommendations of Table 14 do not apply to such rain screen cladding, or to the over-cladding of an external masonry (or concrete) external wall, or to an over-clad concrete roof where the cavity does not contain combustible insulation and the cavity barriers are positioned as described above.

Construction of cavity barriers

A cavity barrier may be formed by construction provided for another purpose provided that it meets the recommendations for cavity barriers. However, where compartment walls are provided these should be taken up the full storey height to a compartment floor or roof and not completed by cavity barriers above them. This is because the fire resistance standards for compartment walls are higher than those for cavity barriers, and compartment walls should therefore be continued through the cavity to maintain the fire resistance standard.

AD B3 9.5, 9.7

Table A1 of Appendix A recommends that all cavity barriers should have a minimum standard of fire resistance of 30 minutes with regard to integrity and 15 minutes with regard to insulation. The only exception to this is in the case of a cavity barrier in a stud wall or partition which is permitted to be formed of:

AD B3 9.6

- Steel at least 0.5 mm thick.
- Timber at least 38 mm thick.
- Polythene sleeved mineral wool, or mineral wool slabs, in either case under compression when installed in the cavity.
- Calcium silicate, cement based or gypsum based boards at least 12.5 m thick.

Cavity barriers should be tightly fitted against rigid construction and mechanically fixed in position where possible. Where they abut against slates, tiles, corrugated sheeting and similar non-rigid construction, the junctions should be fire-stopped as described below.

AD B3 9.8

Cavity barriers should also be fixed in such a way that their performance is unlikely to be affected by:

(a) building movements due to shrinkage, subsidence, thermal change, or the movement of the external envelope due to wind; *or*
(b) failure of their fixings, or any construction or material which they abut, due to fire; *or*
(c) collapse in a fire of any services which penetrate them.

AD B3 9.9

Openings in cavity barriers

Cavity barriers should be imperforate except for one or more of the following openings:

AD B3 9.14

- For a pipe which complies with Section 10 of AD B3.
- For a cable, or for a conduit containing one or more cables.

- If fitted with a suitably mounted fire shutter.
- For a duct (unless it is fire-resisting), fitted with an automatic fire shutter where it passes through the barrier.
- If fitted with a door which complies with Appendix B of AD B, and having at least 30 minutes fire resistance.

Pipes, ventilation ducts and flues

AD B3
10.1 to 10.3
It is impossible to construct a building without passing some pipes or ducts through the walls and floors, and such penetration of the elements of structure is a potential source of flame and smoke spread. Section 10 of AD B3 therefore attempts to control the specifications of such pipes and ducts and of their associated enclosing structures. The measures in Section 10 are primarily designed to delay the passage of fire. They may also have the added benefit of retarding smoke spread, but the integrity test specified in Appendix A of AD B does not cover criteria for the passage of smoke as such.

Pipes

AD B
Appendix E
For the purposes of Section 10, the term 'pipe' includes a ventilating pipe for an above-ground drainage system, but does not include any flue pipe or other form of ventilating pipe.

As is usual, the expression 'pipe' here may be read as 'pipeline', and should be taken to include all pipe fittings and accessories.

Requirements

AD B3
10.5, 10.6,
10.7
Where a pipe as defined passes through an opening in:

  (i) a compartment wall or compartment floor (unless the pipe is wholly enclosed within a protected shaft); *or*

(ii) a cavity barrier;

then either a proprietary sealing system should be used which will maintain the fire resistance of the floor, wall or cavity barrier (and has been shown by test to do so) or, the nominal internal diameter of the pipe should not exceed the relevant dimension listed in Table 15 of Section 10 of AD B3 (reproduced above), the opening should be as small as practicable and fire-stopped around the pipe.

Where a pipe of specification (b) of Table 15 penetrates a structure, it is permissible to pass it through or connect it to a pipe or sleeve of specification (a) of Table 15 provided the pipe or sleeve of specification (a) extends on both sides of the structure for a minimum distance of 1m. The sleeve should be in contact with the pipe. (See Fig. 7.34.)

The following above-ground drainage system pipes complying with specification (b) of Table 15 may be passed through openings in a wall which separates houses, or through openings in a compartment wall or compartment floor in other buildings:

AD B3
10.8, 10.9

  (a) a stack pipe of not more than 160 mm nominal internal diameter, provided it is contained within an enclosure in each storey; *and*

(b) a branch pipe of not more than 110 mm nominal internal diameter, provided it discharges into a stack pipe which is contained in an enclosure,

AD B3, Section 10

Table **15**   **Maximum nominal internal diameter of pipes passing through a compartment wall/floor**

Situation	Pipe material and maximum nominal internal diameter (mm)		
	Specification (a)	Specification (b)	Specification (c)
	Non-combustible material[1]	Lead, aluminium, aluminium alloy, PVC[2] fibre-cement	Any other material
1. Structure (but not a wall separating buildings) enclosing a protected shaft which is not a stairway or a lift shaft	160	110	40
2. Wall separating dwellinghouses, or compartment wall or compartment floor between flats	160	160 (stack pipe)[3] 110 (branch pipe)[3]	40
3. Any other situation	160	40	40

Notes:

1. A non-combustible material (such as cast iron or steel) which if exposed to a temperature of 800°C, will not soften or fracture to the extent that flame or hot gas will pass through the wall of the pipe.

2. PVC pipes complying with BS 4514: 1983 and PVC pipes complying with BS 5255: 1989.

3. These diameters are only in relation to pipes forming part of an above-ground drainage system and enclosed as shown in Fig. 7.34. In other cases the maximum diameters against situation 3 apply.

the enclosure being partly formed by the wall penetrated by the branch pipe.

The enclosure referred to in both (a) and (b) immediately above should comply with the following requirements:

(a) In any storey, the enclosure should extend from floor to ceiling or from floor to floor if the ceiling is suspended.
(b) Each side of the enclosure should be formed by a compartment wall, external wall or casing.
(c) The internal surface of the enclosure should meet the requirements of Class 0, except for any supporting members.
(d) No access panel to the enclosure should be fitted in any bedroom or circulation space.
(e) The enclosure should not be used for any purpose except to accommodate drainage or water supply pipes.

The 'casing' referred to in Section 10, Diagram 34 and (b) immediately above, should provide at least 30 minutes fire resistance, including any access panel, and it should not be formed of sheet metal. The only openings permitted in a casing are openings for the passage of a pipe, or openings fitted

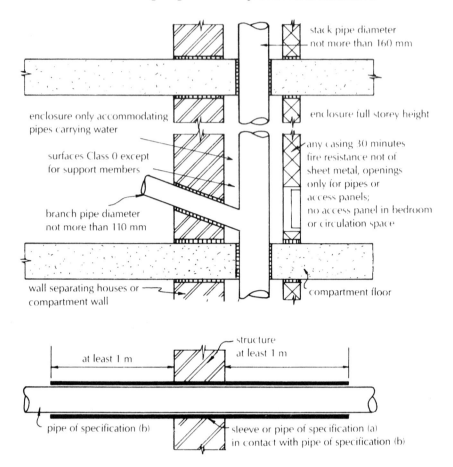

stack pipe diameter not more than 160 mm

enclosure only accommodating pipes carrying water

enclosure full storey height

surfaces Class 0 except for support members

any casing 30 minutes fire resistance not of sheet metal, openings only for pipes or access panels; no access panel in bedroom or circulation space

branch pipe diameter not more than 110 mm

wall separating houses or compartment wall

compartment floor

at least 1 m

structure at least 1 m

pipe of specification (b)

sleeve or pipe of specification (a) in contact with pipe of specification (b)

openings in structure to be as small as possible and fire-stopped around pipes – see Table 15 of AD B3 for pipe material specifications

Fig. 7.34 Penetration of elements of structure by pipes.

with an access panel. The pipe opening, whether it be in the structure or the casing, should be as small as is practicable and fire-stopped around the pipe (see Fig. 7.34). A casing to the drainage or water supply pipes should always be provided if a wall which separates houses is penetrated by a branch pipe in the top storey.

Ventilating ducts

AD B3
10.10

Ventilating ducts, normally forming part of an air-conditioning system, convey air to various parts of a building. It is, therefore, inevitable that they will need to pass through compartment walls and floors at some stage and it is important that the integrity of these fire separating elements is maintained. This can be achieved by following the guidance in Part 9 of BS 5588 where alternative ways of protecting compartmentation are described when air handling ducts pass from one compartment to another.

Flues

Where any flue, appliance ventilation duct or duct containing one or more flues: **AD B3** 10.11

(a) passes through a compartment wall or floor, *or*
(b) is built into a compartment wall,

the flue or duct walls should be separated from the compartment wall or floor by non-combustible construction of fire resistance equal to at least half that required for the compartment wall or floor (see Fig. 7.35).

For the purposes of the above, an appliance ventilation duct is a duct provided to convey combustion air to a gas appliance.

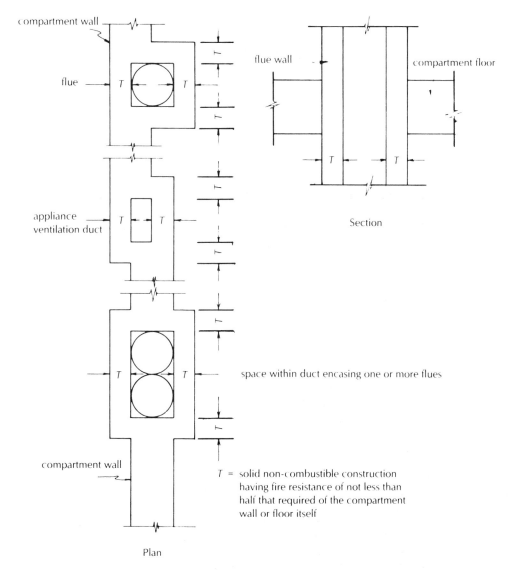

Fig. 7.35 Flues, etc., contained in compartment walls and floors.

Fire stops

AD B3
10.12

A fire stop is defined in Appendix E of AD B as a seal provided to close an imperfection of fit or design tolerance between elements or components, to restrict the passage of fire and smoke.

Therefore, fire stops should be provided:

(a) at junctions or joints between elements which are required to act as a barrier to fire; *and*
(b) where pipes, ducts, conduits or cables pass through openings in cavity barriers (see p. 7.108 above), or elements which serve as a barrier to fire.

In case (b) above, the openings should be kept as small and as few in number as possible, and the fire-stopping should not restrict the thermal movement of pipes or ducts.

AD B3
10.13

Fire-stopping materials should be reinforced with or supported by materials of limited combustibility to prevent displacement if:

(a) the unsupported span exceeds 100 mm; *or*
(b) in any other case, non-rigid materials have been used (unless these have been shown by test to be satisfactory).

AD B3
10.14

Suitable fire-stopping materials include:

- Cement mortar.
- Gypsum based plaster.
- Cement or gypsum based vermiculite/perlite mixes.
- Glassfibre, crushed rock, blast furnace slag or ceramic based products (with or without resin binders).
- Intumescent mastics.
- Any proprietary fire-stopping or sealing systems (including those designed for penetration by services) capable of maintaining the fire resistance of the element concerned. (Test results would be necessary to prove acceptibility.)

These materials should be used in appropriate situations i.e. they may not all be suitable in every situation.

Variations to the provisions of Approved Document B3

As has been mentioned in the introduction to this chapter (see p. 7.2 above), some difficulty may be encountered when trying to apply the provisions of AD B to existing buildings. Accordingly, AD B3 contains a number of specific recommendations related to raised storage areas, floors in domestic loft conversions and to the conversion of buildings to flats, where the 'normal' provisions are somewhat reduced.

Varying the provisions – raised storage areas

AD B3
7.8

The recommendations regarding the fire resistance of elements of structure apply to raised storage areas. These are usually erected in single-storey industrial buildings and may be supported by racking.

Often, they consist of automated storage systems where it is not necessary for people to go onto the raised storage tiers. In these cases it is not necessary to provide fire resistance to the storage structure since there is no risk to the occupants of the building.

AD B3 7.9

Where it is essential that people go onto the storage tiers in the normal course of their work, it may be possible to reduce the level of fire resistance or even allow unprotected steelwork if the following precautions are taken:

AD B3 7.10, 7.11

- The structure should be used for storage purposes only and should contain only one tier.
- The number of people using the floor at any time should be limited.
- Members of the public should not be admitted.
- At least one of the stairs which serves the floor should discharge within 4.5 m of an exit from the building.
- If the floor is more than 10m in length and width an automatic fire detection and alarm system should be provided to give early warning.
- The layout and construction should be such as to warn persons using the floor of any outbreak of fire below (such as providing a perforated floor, or leaving a gap between the edge of the platform and the walls of the building housing it).

Varying the provisions – dwellinghouses

Under the recommendations for means of escape in case of fire (AD B1) certain special provisions apply where it is proposed to construct one or two rooms in the roof space of a two-storey dwelling thereby creating a three-storey dwelling (see p. 7.29 above).

AD B3 7.7

Floors in an existing two-storey dwelling may only be capable of achieving a modified 30 minutes fire resistance (see p. 7.84 above). However, AD B3 requires that floors in a dwelling of three or more storeys should have a full 30 minutes fire resistance.

It is considered reasonable to relax the requirement for the *existing* floor (thereby allowing the modified 30 minutes standard) provided the following provisions are complied with by way of compensation:

- Only one storey is being added with a floor area not exceeding 50 m.
- No more than two rooms are provided in the new storey.
- The existing floor should only separate rooms (not circulation spaces).
- The means of escape provisions in AD B1 should be complied with.

It is sometimes the case that a floor is only capable of achieving a modified 30 minutes standard of fire resistance because it is constructed with plain edged boarding on the upper surface. The addition of a 3.2 mm thickness of standard hardboard nailed to the floor boards can usually upgrade the floor to the full 30 minutes standard (see Fig. 7.36).

Other methods of upgrading existing timber floors can be found in *BRE Digest 208*.

Varying the provisions – conversion to flats

If it is proposed to convert a building into flats, Approved Document B of Schedule 1 of the 1991 Regulations will apply due to the material change of use.

AD B3 7.12, 7.13

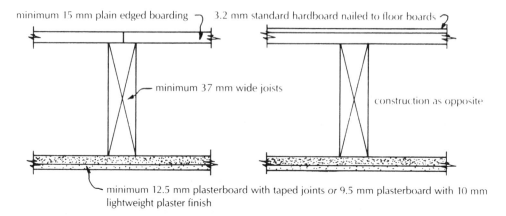

minimum 15 mm plain edged boarding   3.2 mm standard hardboard nailed to floor boards

minimum 37 mm wide joists

construction as opposite

minimum 12.5 mm plasterboard with taped joints *or* 9.5 mm plasterboard with 10 mm lightweight plaster finish

Modified 30 minutes floor   Full 30 minutes floor

Fig. 7.36 Upgrading of existing floors in dwellings.

It is often the case that the existing floors are of timber construction and have insufficient fire resistance for the proposed change of use.

The provision of an adequate, fully protected means of escape which complies with the recommendations of Section 2 of AD B1 will allow a 30 minute standard of fire resistance in the elements of structure in a building of not more than three storeys.

The full standard of fire resistance given in Table A2 of Appendix A would normally be required if the converted building contained four or more storeys.

External fire spread

Regs Sch. 1
B4

B$(1)

B4(2)

The external walls of a building are required to resist the spread of fire over their surfaces and from one building to another. In assessing the adequacy of resistance to fire spread, regard must be given to the height, use and position of the building.

The roof of a building must also offer adequate resistance to the spread of fire across its surface and from one building to another, having regard to the use and position of the building.

External walls

External walls serve to restrict the outward spread of fire to a building beyond the property boundary and also help resist fire from outside the building. This is achieved by ensuring that the walls have adequate fire resistance and external surfaces with restricted fire spread and low rates of heat release. Fire spread between buildings usually occurs by radiation through openings in the external walls (termed 'unprotected areas'). The risk of fire spread and its consequences are related to:

AD B4
0.49

- The severity of the fire.
- The fire resistance offered by the facing external walls including the number and disposition of the unprotected areas.

- The distance between the buildings.
- The risk presented to people in the opposite building.

In general, the severity of a fire will be related to the amount of combustible material contained in the building per unit of floor area (termed the 'fire load density'). Certain types of buildings, such as shops, industrial buildings and warehouses, may contain large quantities of combustible materials and are usually required to be sited further from their boundaries than other types of buildings.

External walls – general constructional recommendations

External walls are elements of structure and therefore they should have the relevant period of fire resistance specified in Appendix A of AD B. However, the provisions for space separation mentioned below allow increasingly large areas of the external walls of a building to be unprotected as the distance to the relevant boundary increases. A point will eventually be reached where the whole of a wall may be unprotected. In such a case only the load-bearing parts of the wall would need fire resistance. Similarly, where a wall is 1 m or more from the relevant boundary it only needs to resist fire from the inside and the insulation criteria of fire resistance is, in most cases, not applied. **AD B4 12.3**

AD B4 12.1

The combustibility of the external envelope of a building is also controlled in certain circumstances. The limiting factors are the height of the building, its use and its distance from the relevant boundary. Table 7.2 sets out the recommendations for the external surfaces of walls and it is generally the case that buildings which are less than 1 m from the relevant boundary should have external surfaces of Class 0. For buildings which are 20 m or more in height, there are restrictions on the external surface materials irrespective of the distance to the boundary. **AD B4 12.2, 12.5**

After the disastrous fire at the Summerland Leisure complex on the Isle of Man, special recommendations were introduced to prevent other assembly and recreation buildings from suffering a similar fate. The Summerland centre was constructed largely of plastics materials which extended to ground level. A fire was deliberately started adjacent to the building which, because of its rapid surface spread of flame characteristics, quickly became engulfed in flames. Therefore, any assembly and recreation building which has more than one storey (galleries counted, but not basements) should have only those external surfaces indicated in Fig. 7.37 below.

The provisions described above for the combustibility of external wall surfaces may, of course, be affected by the recommendations for space separation and the limits on unprotected areas contained in Section 13 of AD B and described below.

Mention has already been made of the risks involved with the use of combustible insulation in rain screen cladding of buildings (see p. 7.106). In such a system the surface of the outer cladding which faces the cavity should comply with the provisions of Table 7.2 and Fig. 7.37. Furthermore, any insulation used in the external walls of a building over 20m in height should be composed of materials of limited combustibility, although this restriction does not apply to insulation in masonry cavity walls which comply with Fig. 7.29 above. (Reference should also be made to the BRE Report *Fire performance of external thermal insulation for walls of multi-storey buildings*, BR 135, 1988.) **AD B4 12.6, 12.7**

Table 7.2 Limitations on external wall surfaces (all buildings).

Maximum height of building (m)	Distance of cladding from any point on the relevant boundary[a]	
	Less than 1 m	*1 m or more*
Up to 20	Class 0	No provision
Over 20	Class 0	Any surface less than 20 m above the ground — Timber at least 9 mm thick; or any material with an index of performance (*I*) not more than 20
		Any surface 20 m or more above the ground — Class 0

Notes

For meaning of class 0 and index of performance (*I*) see p. 7.66 above.

[a] The relevant boundary might be a notional boundary.

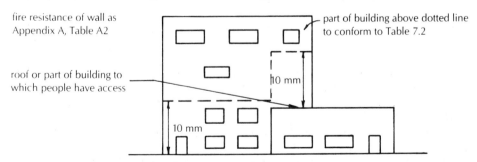

Assembly and recreation buildings of more than one storey

fire resistance of wall as Appendix A, Table A2

part of building above dotted line to conform to Table 7.2

roof or part of building to which people have access

10 mm

10 mm

Part of building below dotted lines:
surfaces
(a) Class 0, if less than 1 m to boundary
(b) index of performance (*I*) not more than 20, or timber cladding at least 9 mm thick, if more than 1 m from the boundary

Fig. 7.37 External walls – special provisions for assembly and recreation buildings.

External walls and steel portal frames

Steel portal frames are commonly used in single-storey industrial and commercial buildings. Structurally, the portal frame acts as a single member. Therefore, where the column sections are built into the external walls, collapse of the roof sections may result in destruction of the walls.

If the building is so situated that the external walls cannot be totally unprotected, the provisions of AD B may recommend that both rafter and column sections be fire protected. This would result in an uneconomic building which would defeat the object of using a portal frame.

Investigations have been carried out into the behaviour of steel portal frames in fire. Provided that the connection between the frame and its foundation can be made sufficiently rigid to transfer the over-turning moment caused by collapse, in a fire, of the rafters, purlins and some of the roof cladding, it may be possible to remove the fire protection to the rafters and purlins while still allowing the external wall to perform its structural function.

AD B4 12.4

Additional measures may be necessary in certain circumstances to ensure the stability of the external walls. Full details of the design method may be found in the publication *Fire and Steel Construction: The Behaviour of Steel Portal Frames in Boundary Conditions*, 1990 (2nd edition) which is available from the Steel Construction Institute, Silwood Park, Ascot, Berks, SL5 7QN. The publication also contains guidance on many aspects of portal frames including multi-storey types.

Normally, reinforced concrete portal frames can support external walls without specific measures at the base to resist overturning.

Space separation – permitted limits of unprotected areas

Unprotected areas in the external walls of a building are those areas which have less fire resistance than that recommended by Table A2 of Appendix A of AD B. Areas such as doors, windows, ventilators or combustible cladding are permitted in the external walls but their extent is limited depending on the use of the building and its distance from the relevant boundary. In order that a reasonable standard of space separation may be specified for buildings, the following basic assumptions are made in AD B, Section 13:

AD B4 13.7

AD B4 13.1

- A fire in a compartmented building will be restricted to that compartment and will not spread to adjoining compartments.
- The intensity of the fire is related to the purpose group of the building and it can be moderated by a sprinkler system.
- Residential, and assembly and recreation purpose groups represent a greater risk to life than other uses.
- Where buildings are on the same site, the spread of fire between them represents a low risk to life and can be discounted. This does not apply to buildings in the residential, and assembly and recreation purpose groups.
- There is a building on the far side of the boundary situated an equal distance away with an identical elevation to the building in question.
- The amount of thermal radiation that passes through an external wall which has fire resistance may be discounted.

AD B4 13.2

It follows from the above that reduced separation distances (or increased amounts of unprotected areas) may be obtained by dividing a building into compartments.

Boundaries

It is clear from AD B4 that the separation distances referred to are those to the relevant boundaries of the site of the building in question. (Relevant boundary is defined on p.7.7 above and is illustrated in Fig. 7.6.) Where the site boundary adjoins an area which is unlikely to be developed, such as a street, canal, railway or river, then the relevant boundary is usually taken as the centreline of that area.

AD B4
13.4, 13.5
13.6
Where buildings share the same site, the separation distance between them is usually discounted. However, if either or both of the buildings are in the residential, or assembly and recreation purpose groups, then a notional boundary is assumed to exist between them such that they both comply with the space separation recommendations of AD B4. (Notional boundary is defined on p. 7.5 and is illustrated in Fig. 7.5.)

Unprotected areas that can be discounted

Certain openings, etc., in walls have little effect on fire protection. Accordingly, AD B4 provides that four areas may be discounted when calculating the permitted limits of unprotected areas in the external walls of a building:

AD B4
13.10
- Any unprotected area of not more than $0.1 \, m^2$ which is at least 1.5 m away from any other unprotected area in the same side of the building or compartment, except an area of external wall forming part of a protected shaft.
- One or more unprotected areas, with a total area of not more than $1 \, m^2$, which is at least 4 m away from any other unprotected area in the same side of the building or compartment, except a small area of not more than $0.1 \, m^2$ as described above.

AD B4
13.8
- Any unprotected area in an external wall forming part of a protected shaft. (But see Fig. 7.21 above for further provisions affecting stairways.)

AD B4
13.12
- An unprotected area in the side of an uncompartmented building, if the area is at least 30 m above the ground adjoining the building.

AD B4
13.9
Where part of an external wall is regarded as an unprotected area merely because of combustible cladding more than 1mm thick, the unprotected area presented by that cladding is to be calculated as only half the actual cladding area (see Fig. 7.38).

AD B4
13.13
Therefore any wall which is situated within 1 m of the relevant boundary should contain only those unprotected areas listed above, and illustrated in Fig. 7.38. The rest of the wall will need to meet the fire resistance requirements contained in Table A2 of Appendix A of AD B.

Unprotected areas – methods of calculation

AD B4
13.14
Where a wall is situated 1 m or more from the relevant boundary, the permitted limit of unprotected areas may be determined by either of two

small openings to be discounted

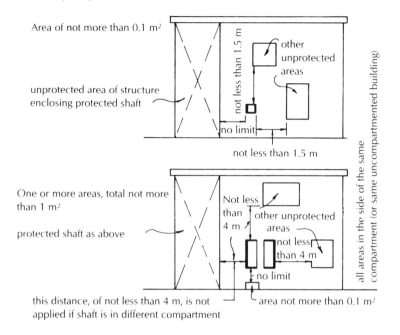

Area of not more than 0.1 m²

unprotected area of structure
enclosing protected shaft

One or more areas, total not more
than 1 m²

protected shaft as above

this distance, of not less than 4 m, is not
applied if shaft is in different compartment

no limit on distance between areas if compartment floor or wall in between

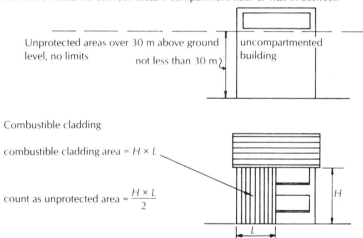

Unprotected areas over 30 m above ground
level, no limits

Combustible cladding

combustible cladding area = *H* × *L*

count as unprotected area = $\dfrac{H \times L}{2}$

Fig. 7.38 Calculation of unprotected area limit, small openings and
combustible cladding.

methods described in full in AD B4. The Approved Document also permits   **AD B4**
other methods, described in a BRE Report *External fire spread: Building*   **13.15**
separation and boundary distances, BRE, 1991. Part 1 of this report covers the
'Enclosing Rectangle' and 'Aggregate Notional Area' methods which were
originally contained in the 1985 edition of AD B2/3/4 and which are also
described below. An applicant may use whichever of these methods gives the

AD B4
13.16

most favourable result for his own building. Again, the rest of the wall should meet the fire resistance recommendations of Table A2 of Appendix A.

The basis of the two methods described in AD B4 is contained in the BRE Report mentioned above. The building should be separated from its boundary by at least half the distance at which the total thermal radiation intensity received from all unprotected areas in the wall would be 12.6 kW/m^2 in still air, assuming that the radiation intensity at each unprotected area is:

- 84 kW/m^2 for buildings in the residential, office or assembly and recreation purpose groups, and
- 168 kW/m^2 for buildings in the shop and commercial, industrial, storage or other non-residential purpose groups.

This clearly illustrates the different fire load densities assumed for the two groups of buildings.

AD B4
13.17

Where a sprinkler system complying with BS 5306: Part 2 is installed throughout a building, it is reasonable to assume that the extent and intensity of a fire will be reduced. In these circumstances the permitted boundary distances may be halved subject to a minimum distance of 1 m.

AD B, Section 4

Diagram **41**   **Permitted unprotected areas in small residential buildings**

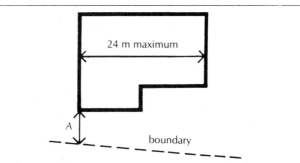

Minimum distance (*A*) between side of building and relevant boundary (m)	Maximum total area of unprotected areas (m²)
1	5.6
2	12
3	18
4	24
5	30
6	no limit

Method 1 – small residential buildings

Method 1 applies only to dwellinghouses, flats, maisonettes or other residential buildings (not institutional buildings) which:

- are not less than 1m from the relevant boundary,
- are not more than three storeys high (basements not counted),
- have no side which exceeds 24 m in length.

The permitted limit of unprotected area in an external wall of any of these buildings is given in Diagram 41 of AD B4 which is reproduced above. It varies according to the size of the building and the distance of the side from the relevant boundary. Any parts of the side in excess of the maximum unprotected area should have the recommended fire resistance. The small areas referred to above may be discounted.

Method 2 – all buildings or compartments

This method applies to buildings and compartments in any purpose group which:

- are not less than 1m from the relevant boundary,
- are not more than 10m high (except open-sided car parks in purpose group 7(b)).

The permitted limits of unprotected areas given by this method are contained in Table 16 of AD B4 which is reproduced below. It should be noted that actual areas are not given. Column 3 of Table 16 expresses the permitted unprotected areas in percentage terms but fails to state what the percentage relates to. It is assumed that the percentages given refer to the total area of the side of the building in question. Thus, if a shop is at least 2 m from its relevant boundary then it is permitted to have 8% of its external wall area on that side as unprotected. Any other areas (except the small permitted areas) would need the requisite fire resistance.

Additional methods – enclosing rectangles

This method of calculating the permitted limit of unprotected areas is based on the smallest rectangle of a height and width taken from Table 7.3 (which follows and is based on Table J2 of the 1985 edition of AD B2/3/4) which would totally enclose all the relevant unprotected areas in the side of a building or compartment. This is referred to as the *enclosing rectangle* and is usually larger than the actual rectangle that would enclose these areas (see Fig. 7.39).

The unprotected areas are projected at right angles onto a *plane of reference* and Table 7.3 then gives the distance that the relevant boundary must be from the plane of reference according to the *unprotected percentage*, the height and width of the enclosing rectangle and the purpose group of the building. The figures in Table 7.3 relate to shop, industrial or other non-residential buildings, whilst those in brackets relate to residential, office or assembly buildings.

The plane of reference is a vertical plane which touches some part of the outer surface of a building or compartment. It should not pass through any part of the building (except projections such as balconies or copings) and it should not cross the relevant boundary. It can be at any angle to the side of the building and in any position which is most favourable to the building designer,

AD B, Section 4

Table **16**   **Permitted unprotected areas in small buildings or compartments**

Minimum distance between side of building and relevant boundary (m)		Maximum total percentage of unprotected area (%)
Purpose groups		
Residential, Office, Assembly and Recreation	Shop & Commercial Industrial, storage & other Non-residential	
(1)	(2)	(3)
n.a.	1	4
1	2	8
2.5	5	20
5	10	40
7.5	15	60
10	20	80
12.5	25	100

Notes:
n.a. = not applicable

● Intermediate values may be obtained by interpolation.

● For buildings which are fitted throughout with an automatic sprinkler system, meeting the relevant recommendations of BS 5306: Part 2, the values in columns (1) & (2) may be halved, subject to a minimum distance of 1 m being maintained.

● In the case of open-sided car parks in purpose group 7(b), the distances set out in column (1) may be used instead of those in column (2).

although it is usually best if roughly parallel to the relevant boundary. This method can be used to determine the maximum permitted unprotected areas for a given boundary position (Fig. 7.40) *or* how close to the boundary a particular design of building may be (Fig. 7.41).

It is permissible to calculate the enclosing rectangle separately for each compartment in a building. Therefore the provision of compartment walls and floors in a building can effectively reduce the enclosing rectangle thereby decreasing the distance to the boundary without affecting the amount of unprotected areas provided. This technique is demonstrated in Fig. 7.42 which also shows how the enclosing rectangle method is applied in practice.

This method is quick and easy to use in practice but in certain circumstances it may give an uneconomical result with regard to the permitted distance from the boundary and it may unduly restrict the designer's freedom in choice of window areas, etc. It takes no account of the true distance from the boundary of unprotected areas in deeply indented buildings since all unpro-

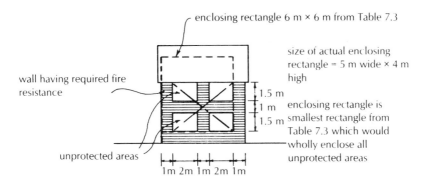

enclosing rectangle 6 m × 6 m from Table 7.3

size of actual enclosing rectangle = 5 m wide × 4 m high

wall having required fire resistance

1.5 m
1 m
1.5 m

enclosing rectangle is smallest rectangle from Table 7.3 which would wholly enclose all unprotected areas

unprotected areas

1m 2m 1m 2m 1m

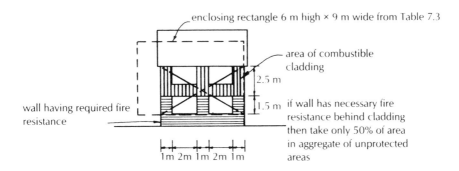

enclosing rectangle 6 m high × 9 m wide from Table 7.3

area of combustible cladding

2.5 m

wall having required fire resistance

1.5 m

if wall has necessary fire resistance behind cladding then take only 50% of area in aggregate of unprotected areas

1m 2m 1m 2m 1m

Note: plane of reference is taken to coincide with surface of wall

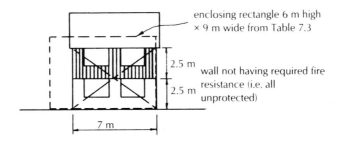

enclosing rectangle 6 m high × 9 m wide from Table 7.3

2.5 m

wall not having required fire resistance (i.e. all unprotected)

2.5 m

7 m

Fig. 7.39 Enclosing rectangles.

tected areas must be projected onto a single plane of reference. It also assumes that the effects of a fire will be equally felt at all points on the boundary from all unprotected areas despite the fact that some windows, for example, may be shielded from certain parts of the boundary by fire resistant walls. For these reasons it may be preferable to use the following method –

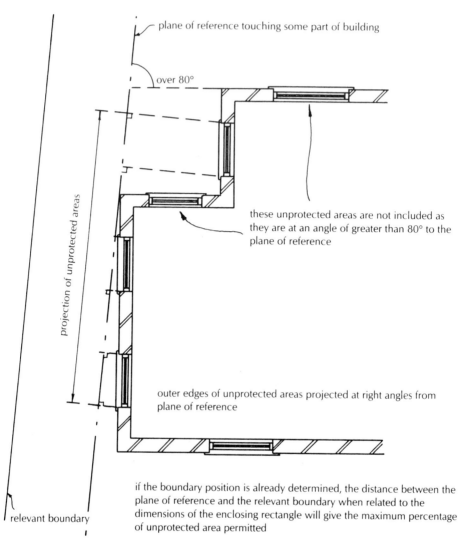

Fig. 7.40 Determination of maximum unprotected areas for given boundary position.

the *aggregate notional areas* technique – as this method is usually more accurate in practice.

Additional methods – aggregate notional areas
The basis of the method of aggregate notional areas is to assess the effect of a building fire at a series of points 3 m apart on the relevant boundary.

A *vertical datum* of unlimited height is set at any position on the relevant boundary (see point P on Fig. 7.43). A *datum line* is drawn from this point to the nearest point on the building or compartment. A base line is then constructed at 90° to the datum line and an arc of 50 m radius is drawn centred on the vertical datum to meet the base line.

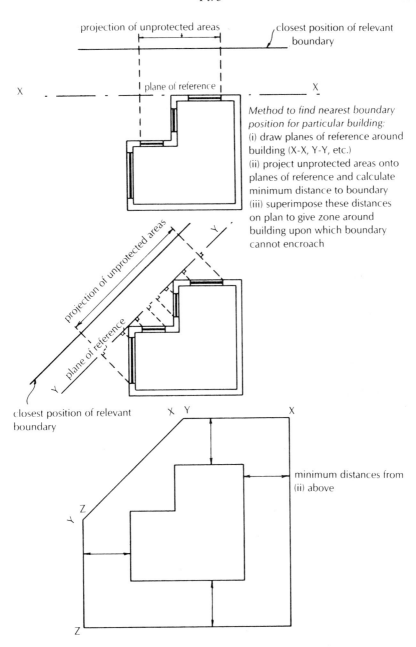

Method to find nearest boundary
position for particular building:
(i) draw planes of reference around
building (X-X, Y-Y, etc.)
(ii) project unprotected areas onto
planes of reference and calculate
minimum distance to boundary
(iii) superimpose these distances
on plan to give zone around
building upon which boundary
cannot encroach

Fig. 7.41 Nearest position to boundary for given building design.

Residential, office or assembly and recreation building

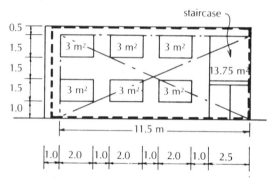

Uncompartmented building
(i) minimum size of rectangle enclosing unprotected areas = 11.5 m wide × 5.5 m high
(ii) from Table 7.3, enclosing rectangle = 12 m wide × 6 m high (take next highest values)
(iii) calculate aggregate of unprotected areas = 18 + 13.75 m² = 31.75 m²
(iv) calculate unprotected percentage (aggregate of unprotected areas as percentage of enclosing rectangle) = $\dfrac{31.75}{12 \times 6}$ × 100 = 44% ∴ use 50% column in Table 7.3

(v) select distance from Table 7.3 (second part of table, fourth column, fourth row, figure in brackets) permitted distance to boundary = 3.5 m

Note:
(a) minimum size of rectangle indicated by diagonal lines
(b) enclosing rectangle indicated by dotted lines
(c) relevant boundary is parallel with wall face, plane of reference coincides with wall face (this need not be the case)

The situation can be improved if the staircase is enclosed in a protecting structure and a compartment floor is provided:

compartment floor provided 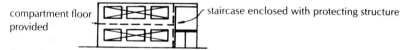 staircase enclosed with protecting structure

(i) each compartment now considered separately, the minimum rectangle being shown by diagonal lines above = 8 m × 1.5 m
(ii) enclosing rectangle from Table 7.3 = 9 m wide × 3 m high
(iii) aggregate of unprotected areas = 9 m²
(iv) unprotected percentage = $\dfrac{9}{9 \times 3}$ × 100 = 33¼% i.e. 40% in Table 7.3
(v) distance from Table 7.3 = 1.5 m

Compartment has therefore reduced the permitted distance to the boundary from 3.5 m to 1.5 m

Fig. 7.42 Enclosing rectangles – effects of compartmentation.

Using this method it is possible to exclude certain unprotected areas that would have to be considered under the enclosing rectangles method (see Fig. 7.43).

For those unprotected areas which remain (that is, those that cannot be excluded), it is necessary to measure the distance of each from the vertical datum. Table J3 from the 1985 edition of AD B2/3/4 is reproduced below as Table 7.4. This contains a series of multiplication factors which are related to the distance from the vertical datum. The table is based on the fact that the amount of heat caused by a fire issuing from an unprotected area will decrease in proportion to its distance from the boundary (it does, in fact, correspond to an inverse square law of the type $y = 1/x^2$).

Therefore, each unprotected area is multiplied by its factor (which depends on its distance from the vertical datum) and these areas are then totalled to give the *aggregate notional area* for that particular vertical datum. The aggregate notional area thus achieved should not exceed:

- $210\,\text{m}^2$ for residential, assembly or office buildings; *or*,
- $90\,\text{m}^2$ for shop, industrial or other non-residential buildings.

In order to confirm that the unprotected areas in the building comply at other points on the boundary, it is necessary to repeat the above calculations at a series of points 3 m apart starting from the original vertical datum. In practice it is usually possible, by observation, to place the first vertical datum at the worst position thereby obviating the need for further calculations.

The series of measurements and calculations mentioned above may be simplified if a number of protractors are made corresponding to different scales (i.e. 1:50, 1:100 and 1:200). A typical example is illustrated in Fig. 7.44.

Canopy structures and space separation

Since Building Regulations apply to the erection of a building, a canopy would need to comply with the provisions concerning space separation (referred to above) unless it falls into one of the exempt classes (see Class VI and Class VII on p. 2.11 above). This might prove unduly onerous in the case of say, petrol pump canopies, since the open sides would be regarded as unprotected areas. Paragraph 13.11 of AD B4 allows space separation recommendations to be disregarded for canopies which are more than 1m from the relevant boundary in view of the high degree of ventilation and heat dissipation achieved by the open-sided construction. However, any plastic roof-lights used in the canopy would be subject to the restrictions contained in Table 18 of AD B4 (see p. 7.75 above).

Roofs

Roofs are not required to provide fire resistance from the inside of a building but should resist fire penetration from outside and the spread of flame over their surfaces. The term roof covering means constructions which may contain one or more layers of material but it does not refer to the roof structure as a whole.

Table 7.3 Permitted unprotected percentages in relation to enclosing rectangles.

Width of enclosing rectangle (m)	Distance from relevant boundary for unprotected percentage not exceeding								
	20%	30%	40%	50%	60%	70%	80%	90%	100%
	Minimum boundary distance (m): figures in brackets are for residential, office or assembly								
Enclosing rectangle 3 m high									
3	1.0 (1.0)	1.5 (1.0)	2.0 (1.0)	2.0 (1.5)	2.5 (1.5)	2.5 (1.5)	2.5 (2.0)	3.0 (2.0)	3.0 (2.0)
6	1.5 (1.0)	2.0 (1.0)	2.5 (1.5)	3.0 (2.0)	3.0 (2.0)	3.5 (2.0)	3.5 (2.5)	4.0 (2.5)	4.0 (3.0)
9	1.5 (1.0)	2.5 (1.0)	3.0 (1.5)	3.5 (2.0)	4.0 (2.5)	4.0 (2.5)	4.5 (3.0)	5.0 (3.0)	5.0 (3.5)
12	2.0 (1.0)	2.5 (1.5)	3.0 (2.0)	3.5 (2.0)	4.0 (2.5)	4.5 (3.0)	5.0 (3.0)	5.5 (3.5)	5.5 (3.5)
15	2.0 (1.0)	2.5 (1.5)	3.5 (2.0)	4.0 (2.5)	4.5 (2.5)	5.0 (3.0)	5.5 (3.5)	6.0 (3.5)	6.0 (4.0)
18	2.0 (1.0)	2.5 (1.5)	3.5 (2.0)	4.0 (2.5)	5.0 (2.5)	5.0 (3.0)	6.0 (3.5)	6.5 (4.0)	6.5 (4.0)
21	2.0 (1.0)	3.0 (1.5)	3.5 (2.0)	4.5 (2.5)	5.0 (3.0)	5.5 (3.0)	6.0 (3.5)	6.5 (4.0)	7.0 (4.5)
24	2.0 (1.0)	3.0 (1.5)	3.5 (2.0)	4.5 (2.5)	5.0 (3.0)	5.5 (3.5)	6.0 (3.5)	7.0 (4.0)	7.5 (4.5)
27	2.0 (1.0)	3.0 (1.5)	4.0 (2.0)	4.5 (2.5)	5.5 (3.0)	6.0 (3.5)	6.5 (4.0)	7.0 (4.0)	7.5 (4.5)
30	2.0 (1.0)	3.0 (1.5)	4.0 (2.0)	4.5 (2.5)	5.5 (3.0)	6.0 (3.5)	6.5 (4.0)	7.5 (4.0)	8.0 (4.5)
40	2.0 (1.0)	3.0 (1.5)	4.0 (2.0)	5.0 (2.5)	5.5 (3.0)	6.5 (3.5)	7.0 (4.0)	8.0 (4.0)	8.5 (5.0)
50	2.0 (1.0)	3.0 (1.5)	4.0 (2.0)	5.0 (2.5)	6.0 (3.0)	6.5 (3.5)	7.5 (4.0)	8.0 (4.0)	9.0 (5.0)
60	2.0 (1.0)	3.0 (1.5)	4.0 (2.0)	5.0 (2.5)	6.0 (3.0)	7.0 (3.5)	7.5 (4.0)	8.5 (4.0)	9.5 (5.0)
80	2.0 (1.0)	3.0 (1.5)	4.0 (2.0)	5.0 (2.5)	6.0 (3.0)	7.0 (3.5)	8.0 (4.0)	9.0 (4.0)	9.5 (5.0)
no limit	2.0 (1.0)	3.0 (1.5)	4.0 (2.0)	5.0 (2.5)	6.0 (3.0)	7.0 (3.5)	8.0 (4.0)	9.0 (4.0)	10.0 (5.0)

Enclosing rectangle 6 m high

3	1.5 (1.0)	2.0 (1.0)	2.5 (1.5)	3.0 (2.0)	3.0 (2.0)	3.5 (2.0)	3.5 (2.5)	4.0 (2.5)	4.0 (3.0)
6	2.0 (1.0)	3.0 (1.5)	3.5 (2.0)	4.0 (2.5)	4.5 (3.0)	5.0 (3.0)	5.5 (3.5)	5.5 (4.0)	6.0 (4.0)
9	2.5 (1.0)	3.5 (2.0)	4.5 (2.5)	5.0 (3.0)	5.5 (3.5)	6.0 (4.0)	6.0 (4.5)	7.0 (4.5)	7.0 (5.0)
12	3.0 (1.5)	4.0 (2.5)	5.0 (3.0)	5.5 (3.5)	6.5 (4.0)	7.0 (4.5)	7.5 (5.0)	8.0 (5.0)	8.5 (5.5)
15	3.0 (1.5)	4.5 (2.5)	5.5 (3.0)	6.0 (4.0)	7.0 (4.5)	7.5 (5.0)	8.0 (5.5)	9.0 (5.5)	9.0 (6.0)
18	3.5 (1.5)	4.5 (2.5)	5.5 (3.5)	6.5 (4.0)	7.5 (4.5)	8.0 (5.0)	9.0 (5.5)	9.5 (6.0)	10.0 (6.5)
21	3.5 (1.5)	5.0 (2.5)	6.0 (3.5)	7.0 (4.0)	8.0 (5.0)	9.0 (5.5)	9.5 (6.0)	10.0 (6.5)	10.5 (7.0)
24	3.5 (1.5)	5.0 (2.5)	6.0 (3.5)	7.0 (4.5)	8.5 (5.0)	9.5 (5.5)	10.0 (6.0)	10.5 (7.0)	11.0 (7.0)
27	3.5 (1.5)	5.0 (2.5)	6.5 (3.5)	7.5 (4.5)	8.5 (5.0)	9.5 (6.0)	10.5 (6.5)	11.0 (7.0)	12.0 (7.5)
30	3.5 (1.5)	5.0 (2.5)	6.5 (3.5)	8.0 (4.5)	9.0 (5.0)	10.0 (6.0)	11.0 (6.5)	12.0 (7.0)	12.5 (8.0)
40	3.5 (1.5)	5.5 (2.5)	7.0 (3.5)	8.5 (4.5)	10.0 (5.5)	11.0 (6.5)	12.0 (7.0)	13.0 (8.0)	14.0 (8.5)
50	3.5 (1.5)	5.5 (2.5)	7.5 (3.5)	9.0 (4.5)	10.5 (5.5)	11.5 (6.5)	13.0 (7.5)	14.0 (8.0)	15.0 (9.0)
60	3.5 (1.5)	5.5 (2.5)	7.5 (3.5)	9.5 (5.0)	11.0 (5.5)	12.0 (6.5)	13.5 (7.5)	15.0 (8.5)	16.0 (9.5)
80	3.5 (1.5)	6.0 (2.5)	7.5 (3.5)	9.5 (5.0)	11.5 (6.0)	13.0 (7.0)	14.5 (7.5)	16.0 (8.5)	17.5 (9.5)
100	3.5 (1.5)	6.0 (2.5)	8.0 (3.5)	10.0 (5.0)	12.0 (6.0)	13.5 (7.0)	15.0 (8.0)	16.5 (8.5)	18.0 (10.0)
120	3.5 (1.5)	6.0 (2.5)	8.0 (3.5)	10.0 (5.0)	12.0 (6.0)	14.0 (7.0)	15.5 (8.0)	17.0 (8.5)	19.0 (10.0)
no limit	3.5 (1.5)	6.0 (2.5)	8.0 (3.5)	10.0 (5.0)	12.0 (6.0)	14.0 (7.0)	16.0 (8.0)	18.0 (8.5)	19.0 (10.0)

Table 7.3 *(contd)* Permitted unprotected percentages in relation to enclosing rectangles.

Width of enclosing rectangle (m)	Distance from relevant boundary for unprotected percentage not exceeding								
	20%	30%	40%	50%	60%	70%	80%	90%	100%
	Minimum boundary distance (m); figures in brackets are for residential, office or assembly								
Enclosing rectangle 9 m high									
3	1.5 (1.0)	2.5 (1.0)	3.0 (1.5)	3.5 (2.0)	4.0 (2.5)	4.0 (2.5)	4.5 (3.0)	5.0 (3.0)	5.0 (3.5)
6	2.5 (1.0)	3.5 (2.0)	4.5 (2.5)	5.0 (3.0)	5.5 (3.5)	6.0 (4.0)	6.5 (4.5)	7.0 (4.5)	7.0 (5.0)
9	3.5 (1.5)	4.5 (2.5)	5.5 (3.5)	6.0 (4.0)	6.5 (4.5)	7.5 (5.0)	8.0 (5.5)	8.5 (5.5)	9.0 (6.0)
12	3.5 (1.5)	5.0 (3.0)	6.0 (3.5)	7.0 (4.5)	7.5 (5.0)	8.5 (5.5)	9.0 (6.0)	9.5 (6.5)	10.5 (7.0)
15	4.0 (2.0)	5.5 (3.0)	6.5 (4.0)	7.5 (5.0)	8.5 (5.5)	9.5 (6.0)	10.0 (6.5)	11.0 (7.0)	11.5 (7.5)
18	4.5 (2.0)	6.0 (3.5)	7.0 (4.5)	8.5 (5.0)	9.5 (6.0)	10.0 (6.5)	11.0 (7.0)	12.0 (8.0)	12.5 (8.5)
21	4.5 (2.0)	6.5 (3.5)	7.5 (4.5)	9.0 (5.5)	10.0 (6.5)	11.0 (7.0)	12.0 (7.5)	13.0 (8.5)	13.5 (9.0)
24	5.0 (2.0)	6.5 (3.5)	8.0 (5.0)	9.5 (5.5)	11.0 (6.5)	12.0 (7.5)	13.0 (8.0)	13.5 (9.0)	14.5 (9.5)
27	5.0 (2.0)	7.0 (3.5)	8.5 (5.0)	10.0 (6.0)	11.5 (7.0)	12.5 (7.5)	13.5 (8.5)	14.5 (9.5)	15.0 (10.0)
30	5.0 (2.0)	7.0 (3.5)	9.0 (5.0)	10.5 (6.0)	12.0 (7.0)	13.0 (8.0)	14.0 (9.0)	15.0 (9.5)	16.0 (10.5)
40	5.5 (2.0)	7.5 (3.5)	9.5 (5.5)	11.5 (6.5)	13.0 (7.5)	14.5 (8.5)	15.5 (9.5)	17.0 (10.5)	17.5 (11.5)
50	5.5 (2.0)	8.0 (4.0)	10.0 (5.5)	12.5 (6.5)	14.0 (8.0)	15.5 (9.0)	17.0 (10.0)	18.5 (11.5)	19.5 (12.5)
60	5.5 (2.0)	8.0 (4.0)	11.0 (5.5)	13.0 (7.0)	15.0 (8.0)	16.5 (9.5)	18.0 (11.0)	19.5 (11.5)	21.0 (13.0)
80	5.5 (2.0)	8.5 (4.0)	11.5 (5.5)	13.5 (7.0)	16.0 (8.5)	17.5 (10.0)	19.5 (11.5)	21.5 (12.5)	23.0 (13.5)
100	5.5 (2.0)	8.5 (4.0)	11.5 (5.5)	14.5 (7.0)	16.5 (8.5)	18.5 (10.0)	21.0 (11.5)	22.5 (12.5)	24.5 (14.5)
120	5.5 (2.0)	8.5 (4.0)	11.5 (5.5)	14.5 (7.0)	17.0 (8.5)	19.5 (10.0)	21.5 (11.5)	23.5 (12.5)	26.0 (14.5)
no limit	5.5 (2.0)	8.5 (4.0)	11.5 (5.5)	15.0 (7.0)	17.5 (8.5)	20.0 (10.5)	22.5 (12.0)	24.5 (12.5)	27.0 (15.0)

Enclosing rectangle 12 m high

3	2.0 (1.0)	2.5 (1.5)	3.0 (2.0)	3.5 (2.0)	4.0 (2.5)	4.5 (3.0)	5.0 (3.0)	5.5 (3.5)	5.5 (3.5)
6	3.0 (1.5)	4.0 (2.5)	5.0 (3.0)	5.5 (3.5)	6.5 (4.0)	7.0 (4.5)	7.5 (5.0)	8.0 (5.0)	8.5 (5.5)
9	3.5 (1.5)	5.0 (3.0)	6.0 (3.5)	7.0 (4.5)	7.5 (5.0)	8.5 (5.5)	9.0 (6.0)	9.5 (6.5)	10.5 (7.0)
12	4.5 (1.5)	6.0 (3.5)	7.0 (4.5)	8.0 (5.0)	9.0 (6.0)	9.5 (6.5)	11.0 (7.0)	11.5 (7.5)	12.0 (8.0)
15	5.0 (2.0)	6.5 (3.5)	8.0 (5.0)	9.0 (5.5)	10.0 (6.5)	11.0 (7.0)	12.0 (8.0)	13.0 (8.5)	13.5 (9.0)
18	5.0 (2.5)	7.0 (4.0)	8.5 (5.0)	10.0 (6.0)	11.0 (7.0)	12.0 (7.5)	13.0 (8.5)	14.0 (9.0)	14.5 (10.0)
21	5.5 (2.5)	7.5 (4.0)	9.0 (5.5)	10.5 (6.5)	12.0 (7.5)	13.0 (8.5)	14.0 (9.0)	15.0 (10.0)	16.0 (10.5)
24	6.0 (2.5)	8.0 (4.5)	9.5 (6.0)	11.5 (7.0)	12.5 (8.0)	14.0 (8.5)	15.0 (9.5)	16.0 (10.5)	16.5 (11.5)
27	6.0 (2.5)	8.0 (4.5)	10.5 (6.0)	12.0 (7.0)	13.5 (8.0)	14.5 (9.0)	16.0 (10.5)	17.0 (11.0)	17.5 (12.0)
30	6.5 (2.5)	8.5 (4.5)	10.5 (6.5)	12.5 (7.5)	14.0 (8.5)	15.0 (9.5)	16.5 (10.5)	17.5 (11.5)	18.5 (12.5)
40	6.5 (2.5)	9.5 (5.0)	12.0 (6.5)	14.0 (8.0)	15.5 (9.5)	17.5 (10.5)	18.5 (12.0)	20.0 (13.0)	21.0 (14.0)
50	7.0 (2.5)	10.0 (5.0)	13.0 (7.0)	15.0 (8.5)	17.0 (10.0)	19.0 (11.0)	20.5 (13.0)	23.0 (14.0)	23.0 (15.0)
60	7.0 (2.5)	10.5 (5.0)	13.5 (7.0)	16.0 (9.0)	18.0 (10.5)	20.0 (12.0)	21.5 (13.5)	23.5 (14.5)	25.0 (16.0)
80	7.0 (2.5)	11.0 (5.0)	14.5 (7.0)	17.0 (9.0)	19.5 (11.0)	21.5 (13.0)	23.5 (14.5)	26.0 (16.0)	27.5 (17.0)
100	7.5 (2.5)	11.5 (5.0)	15.0 (7.5)	18.0 (9.5)	21.0 (11.5)	23.0 (13.5)	25.5 (15.0)	28.0 (16.5)	30.0 (18.0)
120	7.5 (2.5)	11.5 (5.0)	15.0 (7.5)	18.5 (9.5)	22.0 (11.5)	24.0 (13.5)	27.0 (15.0)	29.5 (17.0)	31.5 (18.5)
no limit	7.5 (2.5)	12.0 (5.0)	15.5 (7.5)	19.0 (9.5)	22.5 (12.0)	22.5 (14.0)	28.0 (15.5)	30.5 (17.0)	34.0 (19.0)

Table 7.3 *(contd)* Permitted unprotected percentages in relation to enclosing rectangles.

Width of enclosing rectangle (m)	Distance from relevant boundary for unprotected percentage not exceeding								
	20%	30%	40%	50%	60%	70%	80%	90%	100%
	Minimum boundary distance (m); figures in brackets are for residential, office or assembly								

Enclosing rectangle 15 m high

Width	20%	30%	40%	50%	60%	70%	80%	90%	100%
3	2.0 (1.0)	2.5 (1.5)	3.5 (2.0)	4.0 (2.5)	4.5 (2.5)	5.0 (3.0)	5.5 (3.5)	6.0 (3.5)	6.0 (4.0)
6	3.0 (1.5)	4.5 (2.5)	5.5 (3.0)	6.0 (4.0)	7.0 (4.5)	7.5 (5.0)	8.0 (5.5)	9.0 (5.5)	9.0 (6.0)
9	4.0 (2.0)	5.5 (3.0)	6.5 (4.0)	7.5 (5.0)	8.5 (5.5)	9.5 (6.0)	10.0 (6.5)	11.0 (7.0)	11.5 (7.5)
12	5.0 (2.0)	6.5 (3.5)	8.0 (5.0)	9.0 (5.5)	10.0 (6.5)	11.0 (7.0)	12.0 (8.0)	13.0 (8.5)	13.5 (9.0)
15	5.5 (2.0)	7.0 (4.0)	9.0 (5.5)	10.0 (6.5)	11.5 (7.0)	12.5 (8.0)	13.5 (9.0)	14.5 (9.5)	15.0 (10.0)
18	6.0 (2.5)	8.0 (4.5)	9.5 (6.0)	11.0 (7.0)	12.5 (8.0)	13.5 (8.5)	14.5 (9.5)	15.5 (10.5)	16.5 (11.0)
21	6.5 (2.5)	8.5 (5.0)	10.5 (6.5)	12.0 (7.5)	13.5 (8.5)	14.5 (9.5)	16.0 (10.5)	16.5 (11.0)	17.5 (12.0)
24	6.5 (3.0)	9.0 (5.0)	11.0 (6.5)	13.0 (8.0)	14.5 (9.0)	15.5 (10.0)	17.0 (11.0)	18.0 (12.0)	19.0 (13.0)
27	7.0 (3.0)	9.5 (5.5)	11.5 (7.0)	13.5 (8.5)	15.0 (9.5)	16.5 (10.5)	18.0 (11.5)	19.0 (12.5)	20.0 (13.5)
30	7.5 (3.0)	10.0 (5.5)	12.0 (7.5)	14.0 (8.5)	16.0 (10.0)	17.0 (11.0)	18.5 (12.0)	20.0 (13.5)	21.0 (14.0)
40	8.0 (3.0)	11.0 (6.0)	13.5 (8.0)	16.0 (9.5)	18.0 (11.0)	19.5 (12.5)	21.0 (13.5)	22.5 (15.0)	23.5 (16.0)
50	8.5 (3.5)	12.0 (6.0)	15.0 (8.5)	17.5 (10.0)	19.5 (12.0)	21.5 (13.5)	23.0 (15.0)	25.0 (16.5)	26.0 (17.5)
60	8.5 (3.5)	12.5 (6.5)	15.5 (8.5)	18.0 (10.5)	21.0 (12.5)	23.5 (14.0)	25.0 (15.5)	27.0 (17.0)	28.0 (18.0)
80	9.0 (3.5)	13.5 (6.5)	17.0 (9.0)	20.0 (11.0)	23.0 (13.5)	25.5 (15.0)	28.0 (17.0)	30.0 (18.5)	31.5 (20.0)
100	9.0 (3.5)	14.0 (6.5)	18.0 (9.0)	21.5 (11.5)	24.5 (14.0)	27.5 (16.0)	30.0 (18.0)	32.5 (19.5)	34.5 (21.5)
120	9.0 (3.5)	14.0 (6.5)	18.5 (9.0)	22.5 (11.5)	25.5 (14.0)	28.5 (16.5)	31.5 (18.5)	34.5 (20.5)	37.0 (22.5)
no limit	9.0 (3.5)	14.5 (6.5)	19.0 (9.0)	23.0 (12.0)	27.0 (14.5)	30.0 (17.0)	34.0 (19.0)	36.0 (21.0)	39.0 (23.0)

Enclosing rectangle 18 m high

3	2.0 (1.0)	2.5 (1.5)	3.5 (2.0)	4.0 (2.5)	5.0 (2.5)	5.0 (3.0)	6.0 (3.5)	6.5 (4.0)	6.5 (4.0)
6	3.5 (1.5)	4.5 (2.5)	5.5 (3.5)	6.5 (4.0)	7.5 (4.5)	8.0 (5.0)	9.0 (5.5)	9.5 (6.0)	10.0 (6.5)
9	4.5 (2.0)	6.0 (3.5)	7.0 (4.5)	8.5 (5.0)	9.5 (6.0)	10.0 (6.5)	11.0 (7.0)	12.0 (8.0)	12.5 (8.5)
12	5.0 (2.5)	7.0 (4.0)	8.5 (5.0)	10.0 (6.0)	11.0 (7.0)	12.0 (7.5)	13.0 (8.5)	14.0 (9.0)	14.5 (10.0)
15	6.0 (2.5)	8.0 (4.5)	9.5 (6.0)	11.0 (7.0)	12.5 (8.0)	13.5 (8.5)	14.5 (9.5)	15.5 (10.5)	16.5 (11.0)
18	6.5 (2.5)	8.5 (5.0)	11.0 (6.5)	12.0 (7.5)	13.5 (8.5)	14.5 (9.5)	16.0 (11.0)	17.0 (11.5)	18.0 (13.0)
21	7.0 (3.0)	9.5 (5.5)	11.5 (7.0)	13.0 (8.0)	14.5 (9.5)	16.0 (10.5)	17.0 (11.5)	18.0 (12.5)	19.5 (13.0)
24	7.5 (3.0)	10.0 (5.5)	12.0 (7.5)	14.0 (8.5)	15.5 (10.0)	16.5 (11.0)	18.5 (12.0)	19.5 (13.0)	20.5 (14.0)
27	8.0 (3.5)	10.5 (6.0)	12.5 (8.0)	14.5 (9.0)	16.5 (10.5)	17.5 (11.5)	19.5 (12.5)	20.5 (13.5)	21.5 (14.5)
30	8.0 (3.5)	11.0 (6.5)	13.5 (8.0)	15.5 (9.5)	17.0 (11.0)	18.5 (12.0)	20.5 (13.5)	21.5 (14.5)	22.5 (15.5)
40	9.0 (4.0)	12.0 (7.0)	15.0 (9.0)	17.5 (11.0)	19.5 (12.0)	21.5 (13.5)	23.5 (15.0)	25.0 (16.5)	26.0 (17.5)
50	9.5 (4.0)	13.0 (7.0)	16.5 (9.5)	19.0 (11.5)	21.5 (13.0)	23.5 (15.0)	26.0 (16.5)	27.5 (18.0)	29.0 (19.0)
60	10.0 (4.0)	14.0 (7.5)	17.5 (10.0)	20.5 (12.0)	23.0 (14.0)	26.0 (16.0)	27.5 (17.5)	29.5 (19.5)	31.0 (20.5)
80	10.0 (4.0)	15.0 (7.5)	19.0 (10.0)	22.5 (13.0)	26.0 (15.0)	28.5 (17.0)	31.0 (19.0)	33.5 (21.0)	35.0 (22.5)
100	10.0 (4.0)	16.0 (7.5)	20.5 (10.0)	24.0 (13.5)	28.0 (16.0)	31.0 (18.0)	33.5 (20.5)	36.0 (22.5)	38.5 (24.0)
120	10.0 (4.0)	16.5 (7.5)	21.0 (10.0)	25.5 (14.0)	29.5 (16.5)	32.5 (19.0)	35.5 (21.0)	39.0 (23.5)	41.5 (25.5)
no limit	10.0 (4.0)	17.0 (8.0)	22.0 (10.0)	26.5 (14.0)	30.5 (17.0)	34.0 (19.5)	37.0 (22.0)	41.0 (24.0)	43.5 (26.5)

Table 7.3 *(contd)* Permitted unprotected percentages in relation to enclosing rectangles.

Minimum boundary distance (m): figures in brackets are for residential, office or assembly

Enclosing rectangle 21 m high

Width of enclosing rectangle (m)	Distance from relevant boundary for unprotected percentage not exceeding								
	20%	*30%*	*40%*	*50%*	*60%*	*70%*	*80%*	*90%*	*100%*
3	2.0 *(1.0)*	3.0 *(1.5)*	3.5 *(2.0)*	4.5 *(2.5)*	5.0 *(3.0)*	5.5 *(3.0)*	6.0 *(3.5)*	6.5 *(4.0)*	7.0 *(4.5)*
6	3.5 *(1.5)*	5.0 *(2.5)*	6.0 *(3.5)*	7.0 *(4.0)*	8.0 *(5.0)*	9.0 *(5.5)*	9.5 *(6.0)*	10.0 *(6.5)*	10.5 *(7.0)*
9	4.5 *(2.0)*	6.5 *(3.5)*	7.5 *(4.5)*	9.0 *(5.5)*	10.0 *(6.5)*	11.0 *(7.0)*	12.0 *(7.5)*	13.0 *(8.5)*	13.5 *(9.0)*
12	5.5 *(2.5)*	7.5 *(4.0)*	9.0 *(5.5)*	10.5 *(6.5)*	12.0 *(7.5)*	13.0 *(8.5)*	14.0 *(9.0)*	15.0 *(10.0)*	16.0 *(10.5)*
15	6.5 *(2.5)*	8.5 *(5.0)*	10.5 *(6.5)*	12.0 *(7.5)*	13.5 *(8.5)*	14.5 *(9.5)*	16.0 *(10.5)*	16.5 *(11.0)*	17.5 *(12.0)*
18	7.0 *(3.0)*	9.5 *(5.5)*	11.5 *(7.0)*	13.0 *(8.0)*	14.5 *(9.5)*	16.0 *(10.5)*	17.0 *(11.5)*	18.0 *(12.5)*	19.5 *(13.0)*
21	7.5 *(3.0)*	10.0 *(6.0)*	12.5 *(7.5)*	14.0 *(9.0)*	15.5 *(10.0)*	17.0 *(11.0)*	18.5 *(12.5)*	20.0 *(13.5)*	21.0 *(14.0)*
24	8.0 *(3.5)*	10.5 *(6.0)*	13.0 *(8.0)*	15.0 *(9.5)*	16.5 *(10.5)*	18.0 *(12.0)*	20.0 *(13.0)*	21.0 *(14.0)*	22.0 *(15.0)*
27	8.5 *(3.5)*	11.5 *(6.5)*	14.0 *(8.5)*	16.0 *(10.0)*	18.0 *(11.5)*	19.0 *(13.0)*	21.0 *(14.0)*	22.5 *(15.0)*	23.5 *(16.0)*
30	9.0 *(4.0)*	12.0 *(7.0)*	14.5 *(9.0)*	16.5 *(10.5)*	18.5 *(12.0)*	20.5 *(13.0)*	22.0 *(14.5)*	23.5 *(16.0)*	25.0 *(16.5)*
40	10.0 *(4.5)*	13.5 *(7.5)*	16.5 *(10.0)*	19.0 *(12.0)*	21.5 *(13.5)*	23.0 *(15.0)*	25.5 *(16.5)*	27.0 *(18.0)*	28.5 *(19.0)*
50	11.0 *(4.5)*	14.5 *(8.0)*	18.0 *(11.0)*	21.0 *(13.0)*	23.5 *(14.5)*	25.5 *(16.5)*	28.0 *(18.0)*	30.0 *(20.0)*	31.5 *(21.0)*
60	11.5 *(4.5)*	15.5 *(8.5)*	19.5 *(11.5)*	22.5 *(13.5)*	25.5 *(15.5)*	28.0 *(17.5)*	30.5 *(19.5)*	32.5 *(21.0)*	33.5 *(22.5)*
80	12.0 *(4.5)*	17.0 *(8.5)*	21.0 *(12.0)*	25.0 *(14.5)*	28.5 *(17.0)*	31.5 *(19.0)*	34.0 *(21.0)*	36.5 *(23.5)*	38.5 *(25.0)*
100	12.0 *(4.5)*	18.0 *(9.0)*	22.5 *(12.0)*	27.0 *(15.5)*	31.0 *(18.0)*	34.5 *(20.5)*	37.0 *(22.5)*	40.0 *(25.0)*	42.0 *(27.0)*
120	12.0 *(4.5)*	18.5 *(9.0)*	23.5 *(12.0)*	28.5 *(16.0)*	32.5 *(18.5)*	36.5 *(21.5)*	39.5 *(23.5)*	43.0 *(26.5)*	45.5 *(28.5)*
no limit	12.0 *(4.5)*	19.0 *(9.0)*	25.0 *(12.0)*	29.5 *(16.0)*	34.5 *(19.0)*	38.0 *(22.0)*	41.5 *(25.0)*	45.5 *(26.5)*	48.0 *(29.0)*

Enclosing rectangle 24 m high

3	2.0 (1.0)	3.0 (1.5)	3.5 (2.0)	4.5 (2.5)	5.0 (3.0)	5.5 (3.5)	6.0 (3.5)	7.0 (4.0)	7.5 (4.5)
6	3.5 (1.5)	5.0 (2.5)	6.0 (3.5)	7.0 (4.5)	8.5 (5.0)	9.5 (5.5)	10.0 (6.0)	10.5 (7.0)	11.0 (7.0)
9	5.0 (2.0)	6.5 (3.5)	8.0 (5.0)	9.5 (5.5)	11.0 (6.5)	12.0 (7.5)	13.0 (8.0)	13.5 (9.0)	14.5 (9.5)
12	6.0 (2.5)	8.0 (4.5)	9.5 (6.0)	11.5 (7.0)	12.5 (8.0)	14.0 (8.5)	15.0 (9.5)	16.0 (10.5)	16.5 (11.5)
15	6.5 (3.0)	9.0 (5.0)	11.0 (6.5)	13.0 (8.0)	14.5 (9.0)	15.5 (10.0)	17.0 (11.0)	18.0 (12.0)	19.0 (13.0)
18	7.5 (3.0)	10.0 (5.5)	12.0 (7.5)	14.0 (8.5)	15.5 (10.0)	16.5 (11.0)	18.5 (12.0)	19.5 (13.0)	20.5 (14.0)
21	8.0 (3.5)	10.5 (6.0)	13.0 (8.0)	15.0 (9.5)	16.5 (10.5)	18.0 (12.0)	20.0 (13.0)	21.0 (14.0)	22.0 (15.0)
24	8.5 (3.5)	11.5 (6.5)	14.0 (8.5)	16.0 (10.0)	18.0 (11.5)	19.5 (12.5)	21.0 (14.0)	22.5 (15.0)	24.0 (16.0)
27	9.0 (4.0)	12.5 (7.0)	15.0 (9.0)	17.0 (11.0)	19.0 (12.5)	20.5 (13.5)	22.5 (15.0)	24.0 (16.0)	25.5 (17.0)
30	9.5 (4.0)	13.0 (7.5)	15.5 (9.5)	18.0 (11.5)	20.0 (13.0)	21.5 (14.0)	23.5 (15.5)	25.0 (17.0)	26.5 (18.0)
40	11.0 (4.5)	14.5 (8.5)	18.0 (11.0)	20.5 (13.0)	23.0 (14.5)	25.0 (16.0)	27.5 (18.0)	29.0 (19.0)	30.5 (20.5)
50	12.0 (5.0)	16.0 (9.0)	19.5 (12.0)	22.5 (14.0)	25.5 (16.0)	27.5 (17.5)	30.0 (19.5)	32.0 (21.0)	33.5 (22.5)
60	12.5 (5.0)	17.0 (9.5)	21.0 (12.5)	24.5 (15.0)	27.5 (17.0)	30.0 (19.0)	32.5 (21.0)	35.0 (23.0)	36.5 (24.5)
80	13.5 (5.0)	18.5 (10.0)	23.5 (13.5)	27.5 (16.5)	31.0 (18.5)	34.5 (21.0)	37.0 (23.5)	39.5 (25.5)	41.5 (27.5)
100	13.5 (5.0)	20.0 (10.0)	25.0 (13.5)	29.5 (17.0)	33.5 (20.0)	37.0 (20.0)	40.0 (25.0)	43.0 (27.5)	45.5 (29.5)
120	13.5 (5.5)	20.5 (10.0)	26.5 (13.5)	31.0 (17.5)	36.0 (20.5)	39.5 (23.5)	43.0 (26.5)	46.5 (29.0)	49.0 (31.0)
no limit	13.5 (5.5)	21.0 (10.0)	27.5 (13.5)	32.5 (18.0)	37.5 (21.0)	42.0 (24.0)	45.5 (27.5)	49.5 (30.0)	52.0 (32.5)

Table 7.3 *(contd)* Permitted unprotected percentages in relation to enclosing rectangles.

Width of enclosing rectangle (m)	Distance from relevant boundary for unprotected percentage not exceeding								
	20%	30%	40%	50%	60%	70%	80%	90%	100%
	Minimum boundary distance (m); figures in brackets are for residential, office or assembly								
Enclosing rectangle 27 m high									
3	2.0 (1.0)	3.0 (1.5)	4.0 (2.0)	4.5 (2.5)	5.5 (3.0)	6.0 (3.5)	6.5 (4.0)	7.0 (4.0)	7.5 (4.5)
6	3.5 (1.5)	5.0 (2.5)	6.5 (3.5)	7.5 (4.5)	8.5 (5.0)	9.5 (6.0)	10.5 (6.5)	11.0 (7.0)	12.0 (7.5)
9	5.0 (2.0)	7.0 (3.5)	8.5 (5.0)	10.0 (6.0)	11.5 (7.0)	12.5 (7.5)	13.5 (8.5)	14.5 (9.5)	15.0 (10.0)
12	6.0 (2.5)	8.0 (4.5)	10.5 (6.0)	12.0 (7.0)	13.5 (8.0)	14.5 (9.0)	16.0 (10.5)	17.0 (11.0)	17.5 (12.0)
15	7.0 (3.0)	9.5 (5.5)	11.5 (7.0)	13.5 (8.5)	15.0 (9.5)	16.5 (10.5)	18.0 (11.5)	19.0 (12.5)	20.0 (13.5)
18	8.0 (3.5)	10.5 (6.0)	12.5 (8.0)	14.5 (9.0)	16.5 (10.5)	17.5 (11.5)	19.5 (12.5)	20.5 (13.5)	21.5 (14.5)
21	8.5 (3.5)	11.5 (6.5)	14.0 (8.5)	16.0 (10.0)	18.0 (11.5)	19.0 (13.0)	21.0 (14.0)	22.5 (15.0)	23.5 (16.0)
24	9.0 (3.5)	12.5 (7.0)	15.0 (9.0)	17.0 (11.0)	19.0 (12.5)	20.5 (13.5)	22.5 (15.0)	24.0 (16.0)	25.5 (17.0)
27	10.0 (4.0)	13.0 (7.5)	16.0 (10.0)	18.0 (11.5)	20.0 (13.0)	22.0 (14.0)	24.0 (16.0)	25.5 (17.0)	27.0 (18.0)
30	10.0 (4.0)	13.5 (8.0)	17.0 (10.0)	19.0 (12.0)	21.0 (13.5)	23.0 (15.0)	25.0 (17.0)	26.5 (18.0)	28.0 (19.0)
40	11.5 (5.0)	15.5 (9.0)	19.0 (11.5)	22.0 (14.0)	24.5 (15.5)	26.5 (17.5)	29.0 (19.0)	30.5 (20.5)	32.5 (22.0)
50	12.5 (5.5)	17.0 (9.5)	21.0 (12.5)	24.0 (15.0)	27.0 (17.0)	29.5 (19.0)	32.0 (21.0)	34.5 (22.5)	36.0 (24.0)
60	13.5 (5.5)	18.5 (10.5)	22.5 (13.5)	26.5 (16.0)	29.5 (18.5)	32.0 (20.5)	35.0 (22.5)	37.0 (24.5)	39.0 (26.5)
80	14.5 (6.0)	20.5 (11.0)	25.0 (14.5)	29.5 (17.5)	33.0 (20.5)	36.5 (22.5)	39.5 (25.0)	42.0 (27.5)	44.0 (29.5)
100	15.5 (6.0)	21.5 (11.0)	27.0 (15.5)	32.0 (19.0)	36.5 (21.5)	40.5 (24.5)	43.0 (27.0)	46.5 (30.0)	48.5 (32.0)
120	15.5 (6.0)	22.5 (11.5)	28.5 (15.5)	34.0 (19.5)	39.0 (22.5)	43.0 (26.0)	46.5 (28.5)	50.5 (32.0)	53.0 (34.0)
no limit	15.5 (6.0)	23.5 (11.5)	29.5 (15.5)	35.0 (20.0)	40.5 (23.5)	44.5 (27.0)	48.5 (29.5)	52.0 (33.0)	55.5 (35.0)

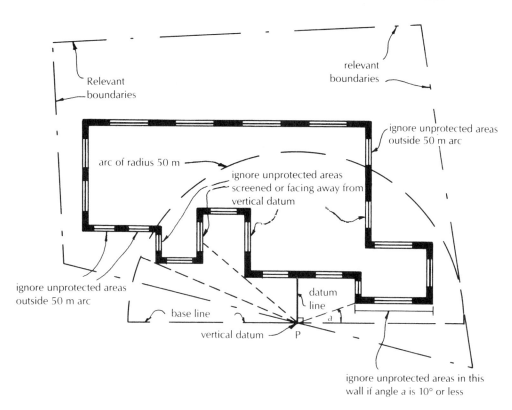

Exclude unprotected areas which:
- are outside 50 m arc
- are screened or face away from the vertical datum
- make an angle of 10° or less with a line drawn from the vertical datum to the unprotected area
- are shown in Fig. 7.38

Fig. 7.43 Aggregate notional areas.

In addition to the recommendations for roof coverings contained in AD B4, reference should also be made to AD B1 (Roofs as part of a means of escape), AD B2 (Internal surfaces of roof-lights) and AD B3 (Roofs used as part of a floor and roofs passing over the top of compartment walls).

The type of construction permitted for a roof depends on the purpose group and size of the building and its distance from the boundary.

Types of construction are specified by the two-letter designations from BS 476: Part 3: 1958 *External fire exposure roof test*.

The first letter refers to flame penetration:

A – Not penetrated within one hour.
B – Penetrated in not less than half an hour.
C – Penetrated in less than half an hour.
D – Penetrated in preliminary flame test.

Table 7.4 Multiplication factors for aggregate notional area.

Distance of unprotected area from vertical datum (m)		
Not less than	*Less than*	*Multiplication factor*
1.0	1.2	80.0
1.2	1.8	40.0
1.8	2.7	20.0
2.7	4.3	10.0
4.3	6.0	4.0
6.0	8.5	2.0
8.5	12.0	1.0
12.0	18.5	0.5
18.5	27.5	0.25
27.5	50.0	0.1
50.0	no limit	0.0

The second letter refers to the surface spread of flame test:

A – No spread of flame.
B – Not more than 21 inches (533.4 mm) spread.
C – More than 21 inches (533.4 mm) spread.
D – Those continuing to burn for five minutes after withdrawal of the test flame, or with a spread of more than 15 inches (381 mm) across the region of burning in the preliminary test.

Example
Roof surface classified AA. This means no fire penetration within one hour and no spread of flame.

Table A5 to Appendix A of AD B (reproduced below) gives a series of roof constructions together with their two-letter notional designations. In the example shown above, a roof constructed in accordance with Part 1 of Table A5 would satisfy the AA rating if it was of natural slates, fibre reinforced cement slates, clay tiles or concrete tiles and it was supported as shown in the table.

Table 17 of AD B4 is reproduced below and gives the notional two-letter designations for roofs in different buildings according to the distance of the roof from the relevant (or notional) boundary. Once the two-letter designation has been established a form of construction may be chosen from Table A5. Where it has been decided to use a different form of roof construction, the manufacturer's details should be consulted to confirm that the necessary designation will be achieved. It should be noted that there are no restrictions on the use of roof coverings which are designated AA, AB or AC. Also, the boundary formed by the wall separating two semi-detached houses may be disregarded for the purposes of roof designations.

Where plastic roof-lights form part of a roof structure they should comply with the provisions of AD B2, paragraph 6.10 and Table 18 of AD B4 (see p.

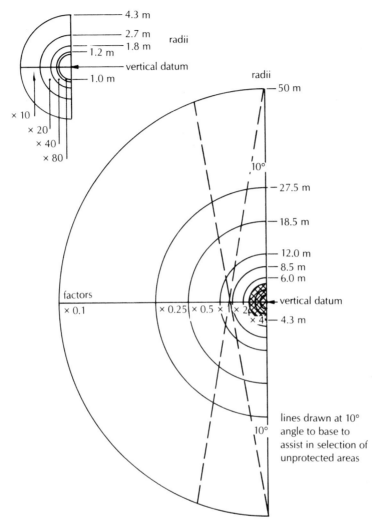

radii

4.3 m
2.7 m   radii
1.8 m
1.2 m
vertical datum
1.0 m

× 10
× 20
× 40
× 80

50 m

10°

27.5 m

18.5 m

12.0 m
8.5 m
6.0 m

factors   vertical datum

× 0.1   × 0.25 × 0.5 ×   × 2
                              × 4 4.3 m

lines drawn at 10°
angle to base to
assist in selection of
unprotected areas

10°

Note: the figure shown above is an enlargement to show radii and factors applicable near to the vertical datum (covering the portion shown hatched)

Fig. 7.44 Aggregate notional areas protractor.

7.75 above) with regard to their separation, area and disposition. The following roof-light materials may be regarded as having an AA designation:

- Rigid thermoplastic sheet made from polycarbonate or unplasticised PVC, which achieves a Class 1 surface spread of flame rating when tested to BS476: Part 7: 1971 or 1987.
- Unwired glass at least 4 mm thick.

AD B4

Table 17   Limitations on roof coverings*

Designation of covering of roof or part of roof	Minimum distance from any point on relevant boundary			
	Less than 6 m	At least 6 m	At least 12 m	At least 20 m
AA, AB, or AC	●	●	●	●
BA, BB, or BC	○	●	●	●
CA, CB, or CC	○	● (1)	● (2)	●
AD, BD, or CD	○	● (1)	● (2)	● (2)
DA, DB, DC, or DD	○	○	○	● (1)
Thatch or wood shingles, if performance under BS 476: pt 3: 1958 cannot be established	○	● (1)	● (2)	● (2)

Notes:
Separation distance considerations do not apply to roofs of a pair of semi-detached houses.

*   See below for limitations on glass, and on plastics roof-lights.

●   Acceptable

○   Not acceptable

(1)   Not acceptable on any of the following buildings:
    a.   Houses in terraces of three or more houses,
    b.   Industrial, Storage or Other non-residential purpose group buildings of any size,
    c.   Any other buildings with a cubic capacity of more than 1500 m³.

And only acceptable on other buildings if the part of the roof is no more than 3 m² in area and is at least 1.5 m from any similar part, with the roof between the parts covered with a material of limited combustibility.

(2)   Not acceptable on any of the buildings listed under a. b. or c. above.

Special provisions relating to shopping complexes and buildings used as car parks

AD B3 section11 11.1
Section 11 of AD B3 describes additional considerations which apply to the design and construction of buildings used as car parks, and to shopping complexes. Although these provisions are nominally placed in AD B3 (Internal fire spread – structure), most parts of AD B do have a bearing on these structures. Accordingly, the recommendations are dealt with in this separate section.

Car parks

AD B3 11.2
A considerable amount of research has been carried out into the behaviour of fire in buildings used as parking for cars and light vans, with the following results:

AD B4, Appendix A

Table A5   Notional designations of roof coverings

Part i: Pitched roofs covered with slates or tiles

Covering material	Supporting structure	Designation
1. Natural slates 2. Fibre reinforced cement slates 3. Clay tiles 4. Concrete tiles	1. Timber rafters with or without underfelt, sarking, boarding, woodwool slabs, compressed straw slabs, plywood, wood chipboard, or fibre insulating board	AA
5. Bitumen felt strip slates Type 2E, with Type 2B underlayer bitumen felt	3. Timber rafters and boarding, plywood, woodwool slabs, wood chipboard, or fibre insulating board	BB
6. Strip slates of bitumen felt, class 1 or 2	2. Timber rafters and boarding, plywood, woodwool slabs, compressed straw slabs, wood chipboard, or fibre insulating board	CC

Note:
Any reference in this table to bitumen felt of a specified type is a reference to bitumen felt as so designated in BS 747: 1977.

Part ii: Pitched roofs covered with self-supporting sheet

Roof covering material	Construction	Supporting structure	Designation
1. Profiled sheet of galvanised steel, aluminium, fibre reinforced cement, or pre-painted (coil coated) steel or aluminium with a pvc or pvf2 coating	1. Single skin without underlay, or with underlay of plasterboard, fibre insulating board, woodwool slab	Structure of timber, steel or concrete	AA
2. Profiled sheet of galvanised steel, aluminium, fibre reinforced cement, or pre-painted (coil coated) steel or aluminium with a pvc or pvf2 coating	2. Double skin without interlayer, or with interlayer of resin bonded glass fibre, mineral wool slab, polystyrene, or polyurethane	Structure of timber, steel or concrete	AA

Part iii: Flat roofs covered with bitumen felt

A flat roof comprising bitumen felt should (irrespective of the felt specification) be deemed to be of designation AA if the felt is laid on a deck constructed of any of the materials prescribed in part iv, and has a surface finish of:

a.   bitumen-bedded stone chippings covering the whole surface to a depth of at least 12.5 mm

b.   bitumen-bedded tiles of a non-combustible material

c.   sand and cement screed, or

d.   macadam

Table A5   continued

Part iv: Pitched roofs covered with bitumen felt

Number of layers	Type of upper layer	Type of underlayer	Deck of 6 mm plywood, 12.5 mm wood chipboard, 16 mm (finished) T&G or 19 mm (finished) plain edged timber boarding	Deck of compressed straw slab	Deck or screeded woodwool slab	Profiled fibre reinforced cement or steel deck (single or double skin) with or without fibre insulating board overlay	Profiled aluminium deck (single or double skin) with or without fibre insulating board overlay	Concrete or clay pot slab (in situ or pre-cast)
	Type 1E	Type 1B minimum mass 13 kg/10 m^2	CC	AC	AC	AC	AC	AB
2 or 3 layers built up in accordance with CP 144: Part 3: 1970	Type 2E	Type 1B minimum mass 13 kg/10 m^2	BB	AB	AB	AB	AB	AB
	Type 2E	Type 2B	AB	AB	AB	AB	AB	AB
	Type 3E	Type 3B or 3G	BC	AC	AB	AB	AB	AB

Note:
Any reference in this table to bitumen felt of a specified type is a reference to bitumen felt as so designated in BS 747: 1977.

Part v: Pitched or flat roofs covered with fully supported material

Covering material	Supporting structure	Designation
1. Aluminium sheet 2. Copper sheet 3. Zinc sheet 4. Lead sheet 5. Mastic asphalt 6. Vitreous enamelled steel 7. Lead/tin alloy coated steel sheet 8. Zinc/aluminium alloy coated steel sheet 9. Pre-painted (coil coated) steel sheet including liquid-applied pvc coatings	1. Timber joists and: tongued and grooved boarding, or plain edged boarding	AA*
	2. Steel or timber joists with deck of: woodwool slabs, compressed straw slab, wool chipboard, fibre insulating board, or 9.5 mm plywood	AA
	3. Concrete or clay pot slab (in situ or pre-cast) or non-combustible deck of steel, aluminium, or fibre cement (with or without insulation)	AA

Note:
* Lead sheet supported by timber joists and plain edged boarding may give a BA designation.

- The fire load is not particularly high and is well defined.
- The likelihood of fire spread between one vehicle and another is remote if the car park is well ventilated.
- Similarly, there is a low risk of fire spread from one storey to another in a well ventilated building.

The best natural ventilation is achieved in open-sided car parks. Where this cannot be attained, heat and smoke will not be as readily dissipated and fewer concessions will apply.

Whatever standard of ventilation is achieved, certain provisions are common to all car parks as follows:

(a) The relevant provisions for means of escape in case of fire in AD B1 will apply.   **AD B3**
    11.3
(b) The recommendations of AD B5 regarding access and facilities for the fire service will apply.
(c) All materials used in the construction of the car park building should be non-combustible except for:

    (i) any floor or roof surface finish,
    (ii) any fire door,
    (iii) any attendant's kiosk not greater than $15\,m^2$ in area.

(d) Surface finishes in buildings, compartments or separated parts which are within the structure enclosing the car park do not need to be constructed from non-combustible materials but they should comply with the relevant provisions of AD B2 and AD B4.

Open-sided car parks

To be regarded as open-sided, the car park should comply with the following provisions: **AD B3**
11.4

- It should comply with the recommendations in (a) to (d) immediately above.
- It should contain no basement storeys.
- Natural ventilation should be provided to each storey at each car parking level by permanent openings having an aggregate area of at least 5% of the floor area at that level, and at least half of the ventilation area should be in opposing walls.
- Where the building containing the car park is also used for other purposes, the part which contains the car park should be a separated part (for definition of 'separated part' see p. 7.8 above).

If the above provisions can be met, the car park may be regarded as a small building or compartment for the purposes of space separation in Table 16 of AD B4 above (see p. 7.122) and column (1) of that table may be used. Effectively, this halves the required distance to the relevant boundary (or doubles the permitted limit of unprotected areas). Additionally, the fire resistance recommendations of section 7b(i) of Table A2 of Appendix A (see p. 7.83) will apply, reducing the fire resistance period to only 15 minutes in many cases.   **AD B3**
11.5

Car parks not regarded as open-sided

If the ventilation recommendations mentioned above cannot be achieved, the car park cannot be regarded as open-sided and the fire resistance recommendations of section 7b(ii) of Table A2 of Appendix A (see p. 7.83) will apply without any concessions. Additionally, all car parks require some ventilation and this may be provided by natural or mechanical means as follows:

AD B3
11.6

- Natural ventilation; provide either:

    (a) permanent openings to each storey at each car parking level with an aggregate area of at least 2.5% of the floor area at that level. At least half of the ventilation should be in opposing walls, or
    (b) smoke vents at ceiling level with an aggregate area of permanent opening of at least 2.5% of the floor area, arranged to give a through draught.

AD B3
11.7

- Mechanical ventilation systems for basements and enclosed car parks should be:

    (a) independent of any other ventilating system.
    (b) designed to operate at six air changes per hour for normal extraction of petrol vapour and ten air changes per hour in a fire situation.
    (c) designed to run in two parts, each capable of extracting half of the amount which would be extracted at the rates set out in (b) above and designed so that each part may operate singly or simultaneously.
    (d) provided with an independent power supply for each part of the system, capable of operating in the event of failure of the main supply.
    (e) provided with extract points arranged with half the points at high level and half the points at low level.
    (f) provided with fans rated at 300°C for a minimum of 60 minutes.
    (g) provided with ductwork and fixings constructed with materials having a melting point of at least 800°C.

Shopping complexes

AD B3
11.8

Individual shops contained in single separate buildings should generally be capable of conforming to the recommendations of AD B. However, where a shop unit forms part of a covered shopping complex, certain difficulties may arise. Such complexes often include covered malls providing access to a number of shops and shared servicing areas. Clearly, it is not practical to compartment a shop from a mall serving it and provisions dealing with maximum compartment sizes may be difficult to meet. Certain other problems may arise concerning fire resistance, walls separating shop units, surfaces of walls and ceilings, and distances to boundaries.

AD B3
11.9

In order to achieve a satisfactory standard of fire safety certain alternative arrangements and compensatory features to those set out in AD B may be appropriate. Reference should be made to sections 5 and 6 of BS 5588: Part 10: 1991 *Code of practice for enclosed shopping complexes* and the relevant recommendations of those sections should be followed.

Access and facilities for the fire service

Introduction

Part B5 of Schedule 1 to the 1991 Regulations contains totally new require-ments which deal with access and facilities for the fire service. It reflects guidance produced by the Home Office Fire Department for the Fire Service which has been in use for many years, and replaces goodwill recommendations with statutory requirements. Many local authorities have, through the medium of private Acts of Parliament, applied means of access regulations in their own districts or boroughs for a considerable number of years. These regulations have tended to vary from authority to authority so the new requirements will bring consistency to this important area of control.

Interpretation

The following terms occur throughout AD B5:

ACCESS LEVEL – the level at which the fire service gain access to a building. This may not always be at ground level, e.g. in a podium design the access may be above ground level.

AD B5
15.1

FIREFIGHTING LIFT – a lift which is designed to have additional fire protection in which the controls may be overriden by the fire service so that they may control it directly for use in fighting fires (see Fig. 7.45).

AD B
Appendix E

FIREFIGHTING LOBBY – a protected lobby usually situated between a firefighting stair and the accommodation of a building which may also give access to any associated firefighting lift (see Fig. 7.45).

AD B
Appendix E

FIREFIGHTING SHAFT – a fire protected enclosure which contains a firefighting stair and firefighting lobbies. It may also contain a firefighting lift, if this is included in the building, together with its machine room. (See Fig. 7.45.)

AD B
Appendix E

FIREFIGHTING STAIR – a fire protected stair which is separated from the accommodation of the building by a firefighting lobby (see Fig. 7.45).

AD B
Appendix E

PERIMETER (of building) – the maximum aggregate plan perimeter. This is found by vertical projection of the building onto a horizontal plane and is illustrated in Fig. 7.46.

AD B
Appendix E

Access and facilities for the fire service – the statutory requirements

Buildings must be designed and constructed so as to provide facilities to assist firefighters in the protection of life. There must also be provision made within the site of the building to enable fire appliances to gain access to the building.

Regs Sch. 1
B5(1)

B5(2)

These requirements are interesting in that they apply provisions to the site of the building and not just to the building itself. This approach may be compared to that for access for disabled people in Chapter 17 where site

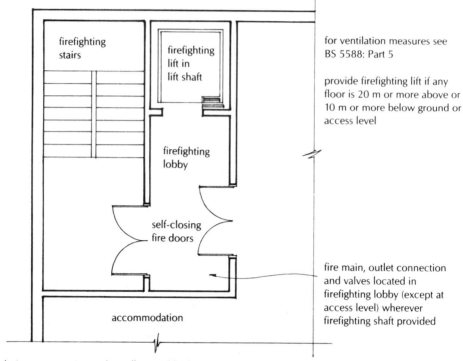

firefighting stairs

firefighting lift in lift shaft

firefighting lobby

self-closing fire doors

accommodation

for ventilation measures see BS 5588: Part 5

provide firefighting lift if any floor is 20 m or more above or 10 m or more below ground or access level

fire main, outlet connection and valves located in firefighting lobby (except at access level) wherever firefighting shaft provided

for design, construction and installation of firefighting shafts see BS 5588: Part 5: 1991

Fig. 7.45 Firefighting shafts and components.

recommendations are also made. It should be noted that there is no definition of 'site' contained in the regulations or approved documents.

AD B5
0.54
The main factor that determines the facilities which are needed to assist the fire service is the size of the building, since it is the philosophy of firefighting in the United Kingdom that this be carried out inside the building if it is to be effective. This philosophy ensures that the water used for firefighting actually reaches the fire and means that effective search and rescue can be carried out as close as possible to the source of the fire since it is at this point that trapped people will be in most peril. Therefore, in order to meet these statutory requirements it is necessary to provide:

(a) in most buildings:

AD B5
0.53
- sufficient means of vehicular access across the site of the building to enable fire appliances to be brought near to the building for effective use,
- sufficient means of access for firefighting personnel into and within the building so that they may effect rescue and fight fire, and

(b) in large buildings and/or buildings with basements:

- sufficient fire mains and other facilities, such as firefighting shafts, to assist firefighters, and
- adequate means of venting heat and smoke from basement fires.

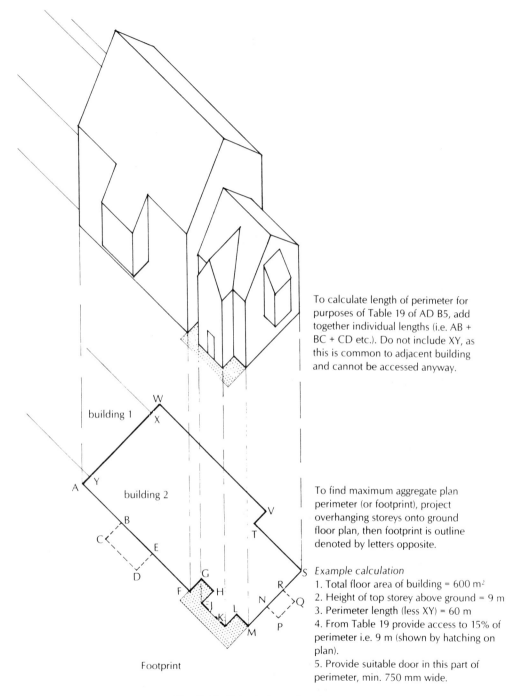

To calculate length of perimeter for purposes of Table 19 of AD B5, add together individual lengths (i.e. AB + BC + CD etc.). Do not include XY, as this is common to adjacent building and cannot be accessed anyway.

To find maximum aggregate plan perimeter (or footprint), project overhanging storeys onto ground floor plan, then footprint is outline denoted by letters opposite.

Example calculation
1. Total floor area of building = 600 m²
2. Height of top storey above ground = 9 m
3. Perimeter length (less XY) = 60 m
4. From Table 19 provide access to 15% of perimeter i.e. 9 m (shown by hatching on plan).
5. Provide suitable door in this part of perimeter, min. 750 mm wide.

Fig. 7.46 Calculation of perimeter.

It should be noted that these arrangements for access and firefighting facilities are required in order to secure reasonable standards of health and safety for people (including firefighters) in or about buildings and for the purposes of protecting life by assisting the fire service.

Access facilities for fire appliances

AD B5
16.1

Vehicle access to the exterior of a building is needed:

- to enable high-reach appliances (i.e. turntable ladders and hydraulic platforms) to be used; *and*
- to enable pumping appliances to supply water and equipment for rescue activities and firefighting.

Clearly, the requirements for access to buildings increase with building size and height. In large buildings it may be necessary to provide firefighting shafts and fire mains. Fire mains are provided in buildings to enable firefighters to connect their hoses to a convenient water supply at the floor level of the fire. Therefore, where these are fitted, it will be necessary for pumping appliances to gain access to the perimeter of the building at points near to the mains. This is especially so in the case of dry mains since these will need to be connected by hose to the pumping appliance. The provision of fire mains and firefighting shafts is described more fully below.

Buildings not fitted with fire mains

AD B5
16.3

Fire mains need only be provided where there is a necessity to provide a firefighting shaft (see p. 7.151 below). Therefore, in buildings not fitted with fire mains, access for fire service vehicles should be provided in accordance with Table 19 of AD B5 which is reproduced below. It should be noted that Table 19 does not apply to buildings with fire mains.

AD B5
16.2

Buildings with a total aggregate floor area of up to 2000 m^2 and a top storey less than 9 m above ground level are referred to as 'small buildings' in AD B5 (see the first row of Table 19). There should be vehicle access to 15% of, or to within 45 m of any point on, the maximum aggregate plan perimeter (or footprint, see Fig. 7.46) of such buildings, whichever figure is the less onerous. It is likely that there will be extreme difficulty in interpreting this recommendation since the statement 'to within 45 m of any point on' can mean either, that at least some point on the perimeter should be within 45 m of the pumping appliance, or that all points on the perimeter should be within 45 m of the pumping appliance. If the first alternative is assumed, then this will always be less onerous than the 15% recommendation, and if the second alternative is assumed, then this will always be more onerous than the 15% recommendation!

AD B5
16.4

In Table 19, the key figure to remember is 9 m above ground level for the top storey of the building. Buildings above this height not fitted with fire mains will need access for high-reach appliances as well as pumping appliances. There should be a suitable door giving access to the interior of the building, at least 750 mm wide, situated in any elevations which are required to be accessed by virtue of Table 19. A typical example of this is shown in Fig. 7.46.

AD B

Table **19** **Fire service vehicle access to buildings not fitted with fire mains**

Total floor [1] area of building (m²)	Height of floor of top storey above ground [2] (m)	Provide vehicle access to:	Type of appliance
Up to 2000	Up to 9 Over 9	See paragraph 16.2 15% of perimeter [3]	Pump High-reach
2000–8000	Up to 9 Over 9	15% of perimeter 50% of perimeter [3]	Pump High-reach
8000–16 000	Up to 9 Over 9	50% of perimeter [3] 50% of perimeter [3]	Pump High-reach
16 000–24 000	Up to 9 Over 9	75% of perimeter [3] 75% of perimeter [3]	Pump High-reach
Over 24 000	Up to 9 Over 9	100% of perimeter [3] 100% of perimeter [3]	Pump High-reach

Notes:

1. The total floor area is the aggregate of all the floors in the building.

2. In the case of purpose group 7(a) (storage) buildings, height should be measured to mean roof level, see methods of measurement in Appendix C of AD B.

3. Perimeter is described in Fig. 7.46.

Buildings fitted with fire mains

As mentioned above, buildings provided with firefighting shafts should also have fire mains.

    Where dry fire mains are fitted:

AD B5
16.5

- Access should be provided for a pumping appliance to within 18 m of each fire main inlet connection point.
- The inlet should be visible from the appliance.

AD B5
16.6

Where wet fire mains are fitted:

- Access should be provided for a pumping appliance to within 18 m of a suitable entrance giving access to the main.
- The entrance should be visible from the appliance.
- The inlet for the emergency replenishment of the suction tank for the main should be visible from the appliance.

AD B5
16.7

Access routes and hardstandings

AD B5
16.8 In order to provide access for fire service vehicles across the site of a building, it is necessary to design a road or other route which is wide enough and has sufficient load-carrying capacity (including manhole covers, etc.) to take the necessary vehicles. Unfortunately, fire appliances are not standardised and so it is a wise precaution to check with the local fire service in a particular area in order to establish their weight and size requirements.

Some design guidance is given in Table 20 of AD B5 (reproduced below), where typical vehicle access route specifications are shown. It should be noted that the typical minimum carrying capacity for a high-reach appliance is 17 tonnes. A roadbase designed to take 12.5 tonnes should be satisfactory for high-reach vehicles since the use would be infrequent and the weight of the vehicle is distributed over a number of axles. Structures such as bridges should

AD B5
16.9 still be designed to take the full 17 tonnes, however. Provision of access to the required elevations may not be enough on its own, since it is possible that overhead obstructions such as cables and branches might interfere with the setting of ladders or the swing of high-reach appliances. Therefore, a zone should be established, in accordance with Fig. 7.47, which should be kept clear of overhead obstructions.

AD B5
16.10 Any dead-end access route which is more than 20 m long should be provided with a turning point, such as a hammerhead or turning circle, designed in accordance with dimensions given in Table 20 of AD B5 (see below).

AD B5
16.2 In all the above considerations for site access it should be borne in mind that requirements cannot be made under the Building Regulations for work to be done outside the site of the works shown on the submitted plans, building notice or initial notice.

AD B5

Table **20   Typical vehicle access route specification**

Appliance type	Minimum width of road between kerbs (m)	Minimum width of gateways (m)	Minimum turning circle between kerbs (m)	Minimum turning circle between walls (m)	Minimum clearance height (m)	Minimum carrying capacity (tonnes)
Pump	3.7	3.1	16.8	19.2	3.7	12.5
High-reach	3.7	3.1	26.0	29.0	4.0	17.0

Access for firefighting personnel to buildings

AD B5
17.1 As has been mentioned above, it is important that fire service personnel are able to gain access to buildings in order to reach the seat of the fire and to carry out effective search and rescue. In low-rise buildings without deep basements, this may be achieved by using the normal means of escape facilities within the building and by ensuring that firefighting appliances can get sufficiently close to facilitate ladder access to upper storeys.

In high buildings and/or buildings with deep basements, there will be a need for additional facilities contained within a protected firefighting shaft, such as firefighting lifts, firefighting stairs and firefighting lobbies. These facilities are

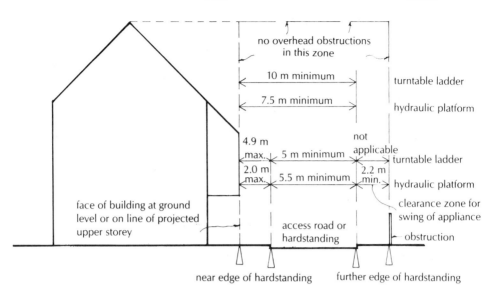

Fig. 7.47 Overhead obstructions – access dimensions for high-reach fire appliances.

necessary in order to avoid delay in tackling the fire and to provide a safe working base from which effective action may be taken.

Firefighting shafts

Firefighting shafts should be provided to serve the storeys indicated in the following list:

(a) Buildings which contain two or more basement storeys each exceeding 900 m^2 in area.   AD B5 17.4

(b) Buildings with any storey of 600 m^2 or more in area situated more than 7.5 m above ground or fire service access level. This does not apply to open-sided car parks.   AD B5 17.3

(c) Buildings with any floor more than 20 m above ground or fire service access level.   AD B5 17.2

(d) Buildings with any floor more than 10 m below ground or fire service access level.   AD B5 17.2

(e) Shopping complexes, in accordance with the recommendations of Section 3 of BS 5588: Part 10: 1991 *Code of practice for enclosed shopping complexes*.   AD B5 17.6

Firefighting shafts should conform to the following recommendations:

- Those provided in (c) and (d) above should also contain firefighting lifts.
- Where provided to serve a basement under (a) or (d) above, there is no need to serve the upper floors also unless they qualify in their own right because of the size or height of the building.   AD B5 17.5

- Similarly, where a shaft is provided to serve upper floors in (b) or (c) above, it need not also serve a basement which is not large or deep enough to qualify on its own.
- Where they are provided, firefighting shafts and lifts should serve all intermediate floors between the lowest and the highest in the building.

Standard of provision of firefighting shafts

AD B5
17.7

The number of firefighting shafts which needs to be provided in a building may be obtained from Table 7.5 below. It can be seen that if the building is fitted with a sprinkler system meeting the relevant recommendations of BS 5306: Part 2: 1990 *Specification for sprinkler systems*, then it is possible to reduce the number of firefighting shafts which is provided.

AD B5
17.8

Firefighting shafts should be located so that every part of each storey is within 60 m of the entrance to a firefighting lobby measured along a route which is suitable for laying fire hoses. This figure is reduced to 40 m where the internal layout of the building is not known at the design stage, the 40 m being measured in a straight line from every point in the storey to the entrance of the firefighting lobby. These distance recommendations do not apply to accommodation situated at fire service access level.

Table 7.5 Provision of firefighting shafts in buildings.

Area of largest qualifying floor $(m^2)^a$	Number of firefighting shafts to be provided
A. With sprinklers fitted in the building (except basements)	
Under 900	1
900 to 2000	2
Over 2000	2 (plus 1 shaft for every extra 1500 m² or part thereof)
B. Without sprinklers and in any qualifying basement	
Up to 900	1
Over 900	2 (plus 1 shaft for every extra 900 m² or part thereof)

Notes
^{a} This is the largest floor area which is situated:
   (a) over 20 m above ground level, or
   (b) over 7.5 m above ground level and 600 m² or more in area, or
   (c) in any qualifying basement (see p. 7.151 above).

Layout and construction of firefighting shafts

Firefighting shafts should be designed and constructed to encompass the following recommendations:

- Access to the accommodation in a building from a firefighting lift or stair should be through a firefighting lobby. AD B5 17.9
- Every firefighting shaft should be equipped with a fire main. AD B5 17.10
- All fire mains should have outlet connections and valves situated in firefighting lobbies except at fire service access level.
- Firefighting shafts should comply with the following parts of BS 5588: Part 5: 1991 *Code of practice for firefighting stairs and lifts:* AD B5 17.11

  (a) section 2: *Planning and construction,*
  (b) section 3: *Firefighting lift installation,*
  (c) section 4: *Electrical services.*

The various components of a firefighting shaft are illustrated in Fig. 7.45 above.

Provision of fire mains

Fire mains are provided to enable the fire service to fight fires inside the building. They are equipped with valves which permit direct connection of fire hoses. This assists firefighters by making it unnecessary to take hoses up stairways from the pumping appliance at ground or access level, thus saving time and avoiding blockage of the escape route. AD B5 15.1

Fire mains which serve floors above ground or access level are commonly known as rising mains and those which serve floors below ground or access level (such as basements) are usually referred to as falling mains.

Where it is necessary to provide firefighting shafts in a building (see p. 7.151 above), each shaft should contain a fire main. AD B5 15.2, 15.4

There are two types of fire main:

- Wet mains (often called 'wet risers') are usually kept full of water by header tanks and pumps in the building. Since there is the danger that the water supply may run out in a serious fire, there should be a facility to replenish the wet main from the pumping appliance in an emergency. Wet risers should be provided in any building which has a floor situated more than 60 m above ground or access level. They may, of course, serve lower floors if so desired. AD B5 15.3
- Dry mains ('dry risers') are normally kept empty and are charged with water from a fire service pumping appliance in the event of a fire. Where provided, they may serve any floor which is less than 60 m above ground or access level.

The outlets from fire mains at each floor level in the building should be situated in firefighting lobbies giving access to the accommodation from a firefighting shaft. AD B5 15.5

Further guidance on the design and construction of fire mains may be obtained from sections 2 and 3 of BS 5306 *Fire extinguishing installations and* AD B5 15.6

equipment on premises, Part 1: 1976 (1988) *Hydrant systems, hose reels and foam inlets.*

Smoke and heat venting of basements

AD B5
18.1

A basement fire differs from a fire in another part of a building in that it is difficult for heat and smoke to be adequately vented. Products of combustion from basement fires tend to escape via stairways making it difficult for fire service personnel to gain access to the fire. Consequently, visibility will be reduced and temperatures will tend to be higher in a basement fire making search, rescue and firefighting more difficult.

AD B5
18.2

If smoke outlets (or smoke vents) are installed they can provide a route for heat and smoke, enabling it to escape direct to outside air from the basement and permitting the ingress of cooler air. Two typical designs for smoke outlet shafts are shown in Fig. 7.48.

Standard of provision for smoke outlets

AD B5
18.4, 18.5

Basement storeys which exceed $200\,m^2$ in area or are more than 3 m below ground level should be provided with smoke outlets connected directly to outside air. Some basements are excepted from this rule as follows:

- Any basement in a single family dwellinghouse (PG 1(b) or 1(c)).
- Any strong room.

AD B5
18.3

If possible each basement room or space should have one or more smoke outlets. In some basements the plan may be too deep, or there may be insufficient external wall areas to permit this. An acceptable solution might be to vent perimeter spaces directly and to allow internal spaces to be vented indirectly by the fire service by means of connecting doors. This solution is not acceptable if the basement is compartmented since each compartment should have direct access to venting without the use of intervening doors between compartments.

Means of venting

Smoke venting may be by natural or mechanical means.

Natural smoke venting may be achieved by providing smoke outlets which conform to the following recommendations:

AD B5
18.7

- The total clear cross-sectional area of all the smoke outlets should be at least 2.5% of the floor area of the storey they serve.

AD B5
18.8

- Places of special fire risk should be provided with separate outlets.

AD B5
18.11

- Outlets should be positioned so that they do not compromise escape routes from the building.

AD B5
18.10

- Outlets which terminate in readily accessible positions may be covered by pavement lights, stallboards or panels which can be broken out or opened. They should be suitably marked to indicate their position.

AD B5
18.9

- Outlets which terminate in less accessible positions should be kept unob-

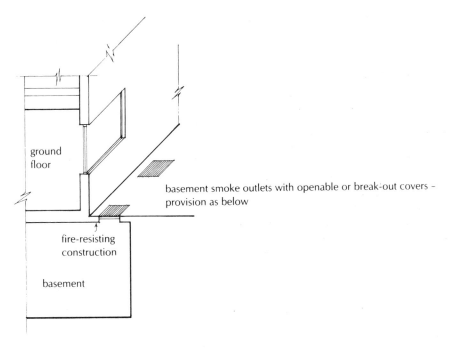

basement smoke outlets with openable or break-out covers –
provision as below

fire-resisting
construction

outlets evenly distributed around perimeter and not placed where
they would jeopardise escape routes

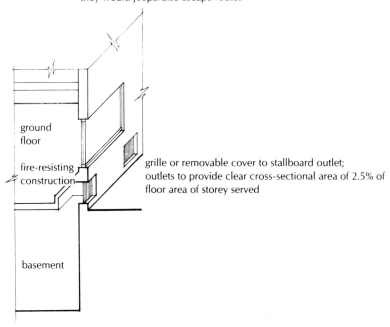

grille or removable cover to stallboard outlet;
outlets to provide clear cross-sectional area of 2.5% of
floor area of storey served

Fig. 7.48 Construction of smoke outlets.

structed and should only be covered by a louvre or grille which is non-combustible.

**AD B5
18.6**

● Smoke outlets should be sited at high level (i.e. in the ceiling or wall of the space they serve) and should be distributed evenly around the perimeter of the building so as to discharge into open air outside the building.

**AD B5
18.12**

Mechanical smoke extraction may be used as an alternative to natural venting if a sprinkler system conforming to BS 5306: Part 2 is installed in the basement. Unless needed for other reasons, it is not necessary to install sprinklers on the other storeys in the building merely because they are provided to allow mechanical smoke extraction in the basement to be used.

**AD B5
18.13**

Any mechanical smoke extraction system should:

● Achieve at least ten air changes per hour.
● Be capable of handling gas temperatures of 400°C for at least one hour.
● Come into operation automatically on activation of the sprinkler system, or be activated by an automatic fire detection system conforming to BS 5839: Part 1: 1988 (at least L3 standard).

Construction of smoke vent outlet ducts and shafts

**AD B5
18.14**

Outlet ducts and shafts for smoke and heat venting should be enclosed in non-combustible fire-resisting construction. This applies equally to any bulkheads over the ducts or shafts as indicated in Fig. 7.48.

**AD B5
18.15**

Natural smoke outlets from different compartments in the same basement storey or from different basement storeys should also be separated from each other by non-combustible fire-resisting construction.

Ventilation of basement car parks

**AD B5
18.16**

The provisions contained in Section 11 of AD B (see p. 7.144 above) regarding the ventilation of basement car parks, may be regarded as satisfying the requirements contained in AD B5 for the smoke venting of any basement which is used as a car park.

Chapter 8

Materials, workmanship, site preparation and moisture exclusion

Materials and workmanship

Introduction

Regulation 7 of the 1991 Regulations is concerned with the fitness and use of the materials necessary for carrying out building work. It is supported by its own Approved Document entitled, rather aptly, *Approved Document to support Regulation 7* (AD Regulation 7). Apart from dealing generally with the standards of materials and workmanship needed for building work, AD Regulation 7 is also concerned with:

- The use of materials which are susceptible to changes in their properties,
- Resistance to moisture and deleterious substances in the subsoil, *and*
- Short-lived materials.

The use of materials which are unsuitable for permanent buildings is covered by section 19 of the Building Act 1984. Local authorities are enabled to reject plans for the construction of buildings of short-lived or otherwise unsuitable materials, or to impose a limit on their period of use. The Secretary of State may, by Building Regulations, prescribe materials which are considered unfit for particular purposes. Tables 1 and 2 of the 1976 Regulations listed materials which were considered unfit for the weather-resisting part of any external wall or roof. Neither the 1991 regulations nor AD Regulation 7 prescribe any materials as unfit for particular purposes as yet, however the AD does lay down some general criteria against which materials may be judged (see p. 8.9). Bearing this in mind, it is unlikely that section 19 can be used by local authorities to proscribe certain materials. Since it is now possible to use materials and methods of workmanship which comply with European Standards or Technical Approvals, all references to Agrément Certificates in the 1992 edition of AD Regulation 7 have been replaced in the 1999 edition by references to national or European certificates issued by a European Technical Approvals issuing body.

Interpretation

A large number of terms and abbreviations appear in AD Regulation 7 as follows:

AD Reg. 7,
Appendix

BS – British Standard, issued by the British Standards Institution (BSI). To achieve British Standard status a draft document is prepared by relevant experts and is submitted for public consultation. Comments received are considered and consensus is reached before the proposed document is issued. More information on British Standards may be obtained from the BSI at 389 Chiswick High Road, London W4 4AL (Internet: www.bsi.org.uk).

BUILDING CONTROL BODY – includes both local authorities and approved inspectors.

CE MARKING – materials bearing the CE mark (see Fig. 8.1) are presumed to comply with the minimum legal requirements as set out in the Construction Products Regulations 1991 (see Chapter 5). This is described more fully on page 8.6.

Fig. 8.1 The CE mark.

CEN – Comité Européen de Normalisation. This is the body recognised by the European Commission (see below) to prepare harmonised standards to support the Construction Products Directive (CPD) (see p. 8.4). The committee comprises representatives of the standards bodies of participating members of the EU and EFTA (European Free Trade Association).

CPD – Construction Products Directive (see p. 8.4).

EEA (EUROPEAN ECONOMIC AREA) – those states which signed the Agreement at Oporto on 2 May 1992 plus the Protocol adjusting that Agreement signed in Brussels on 17 March 1993. The states are Austria, Belgium, Denmark, Finland, France, Germany, Greece, Iceland, Ireland, Italy, Luxembourg, Liechtenstein, Netherlands, Norway, Portugal, Spain, Sweden, and United Kingdom.

EOTA – European Organisation for Technical Approvals. This is the umbrella organisation for bodies which issue European Technical Approvals for individual products. Whilst EOTA operates over the same area as CEN, it complements their work by producing guidelines for innovative and other products for which standards do not exist.

EN – these letters indicate that a European standard has been implemented in a particular Member State. Thus, in the United Kingdom the designation will be BS EN, followed by the relevant standard number. A British Standard shown in this way will be identical to the standards of other Member States, but will also include additional guidance regarding its use and its relationship with other standards in the same group. An EN does not have a separate existence as a formally published document.

EUROPEAN COMMISSION – based in Brussels, this is the executive organisation of the EU. The Commission ensures that Community rules are implemented and observed and it alone has power to propose legislation based on the Treaties. It also executes decisions taken by the Council of Ministers.

EUROPEAN TECHNICAL APPROVAL – a technical assessment which specifies that a construction product is fit for its intended use. It is issued for the purposes of the Construction Products Directive (CPD) and the issuing body must be authorised by a Member State and be notified to the European Commission under section 10 of the CPD. Details of the approval issuing bodies are published in the 'C' series of the Official Journal of the European Communities. For the United Kingdom, in addition to the British Board of Agrément the current listing also includes WIMLAS Ltd, St Peter's House, 6–8 High Street, Iver, Buckinghamshire, SL0 9NG. A current list can be found on the Building Regulations pages of the DETR website at: http://www.detr.gov.uk

EU – the 15 countries of the European Union, *viz*. Austria, Belgium, Denmark, Finland, France, Germany, Greece, Ireland, Italy, Luxembourg, Netherlands, Portugal, Spain, Sweden, and United Kingdom.

ISO – International Organisation for Standardisation. This is the worldwide standards organisation and it is likely that some ISO standards may be adapted for use with the CPD. Such standards are identified by 'ISO' and a number and in the UK they may appear as BS ISO or, if they are adopted as European standards, as BS EN ISO. Unlike ENs, ISOs are published separately.

TECHNICAL SPECIFICATION – a standard or European Technical Approval Guide. A document against which compliance can be shown (e.g. for standards) and against which an assessment is made in order to deliver the European Technical Approval.

UKAS – United Kingdom Accreditation Service. An accreditation body for quality assurance and management schemes. Further details may be obtained from UKAS, 21–47 High Street, Feltham, Middlesex, TW3 4UN.

The influence of European standards

In order to understand the changes which are coming about with the advent of the Single European Market, the following brief summary sets out the context of these changes and their influence on the building control system in England and Wales.

A main goal of the European Community is to allow free movement among the Member States of goods, services, people and capital. Free movement of goods may be hampered by physical, technical or fiscal barriers.

Significant technical barriers arise from the use of different technical requirements or regulations in Member States. This results in the necessity to produce slightly different versions of the same product to satisfy different markets and different methods of test for suitability must be used, resulting in undue expense and waste of resources by manufacturers and suppliers.

To overcome these difficulties, early directives were issued for which it was necessary to resolve technical issues with unanimous agreement by all Member States. This turned out to be a slow and cumbersome process.

The New Approach

The Single European Act in 1986 declared an agreement to establish the Single European Market by the end of 1992. Progress to the single market has been aided by a radical change in the approach to the writing of directives and European standards – the so-called New Approach. This recognises that EU legislation should only apply to areas already subject to existing national laws or regulations and also allows qualified majority voting into the decision-making process.

The New Approach Directives express requirements in broad terms, called the Essential Requirements. Member States presume conformity with these requirements where a product satisfies a harmonised European technical specification or, as an interim measure, a national standard accepted by the Commission. The advent of European standards will prevent Member States from using their own standards to protect their own markets.

A number of new approach product directives have been adopted which are relevant to the construction industry, the most significant being the Construction Products Directive 89/106 EEC.

The Construction Products Directive (CPD)

The full title of this directive reflects its objectives:

> 'Council Directive of 21 December 1988 on the approximation of laws, regulations and the administrative provisions of the Member States relating to construction products.'

The CPD was implemented in the UK on 27 December 1991 by the Construction Products Regulations 1991 (see also DOE circular 13/91).

Construction products are defined as:

Those produced for incorporation in a permanent manner in construction works, in so far as the Essential Requirements (ERs) relate to them.

There are six Essential Requirements as follows:

(1) Mechanical resistance and stability.
(2) Safety in case of fire.

(3) Hygiene, health and the environment.
(4) Safety in use.
(5) Protection against noise.
(6) Energy economy and heat retention.

The link between the Essential Requirements and the product on the market is made in the Interpretative Documents (IDs).

Interpretative Documents

There is one ID for each Essential Requirement. These interpret the ERs more fully and indicate:

- appropriate product characteristics,
- appropriate topics for harmonised technical specifications,
- the need for different levels or classes of performance to allow for different regulation requirements in different Member States.

The UK is represented by experts from the DETR and BRE on the technical committees and drafting panels for the IDs. The IDs are intended mainly for standards' writers and enforcement authorities and will be of less use to manufacturers and suppliers.

Technical specifications

Sometimes referred to as harmonised technical specifications, these are deemed to comply with the Essential Requirements and so products meeting the specifications will immediately demonstrate their fitness to be placed on the market.
Three types exist:

(1) *Harmonised Standards* – ideally the best route for demonstrating compliance for a product. These standards are developed mainly by CEN on the basis of standardisation requests (called mandates) from the Commission. Only those parts of standards which relate to the ERs are mandated and these are the parts which support fixing of the CE mark to products.
(2) *European Technical Approvals* – these have replaced the Agrément certificate form of product assessment and are carried out by bodies designated by individual Member States which are then notified by that Member State to the European Commission. In the UK the British Board of Agrément and WIMLAS Ltd have been designated as the bodies authorised to issue approvals. All such bodies are members of the European Organisation for Technical Approvals (EOTA), and they operate under a common set of rules.
(3) *National Specifications* – although a possibility, the recognition of national specifications at community level is likely to be rare. The procedures have to be initiated by the Commission and their use may be expected only in situations where a barrier to trade has been demonstrated and the production of a Technical Specification is some way off.

The CE mark and attestation of conformity

AD Reg. 7,
sec. 1; 1.2 d

The purpose of attestation of conformity with technical specifications is to assure purchasers and regulators that products placed on the market comply with the ERs. Such products may carry the CE mark.

Each Member State is required to maintain a register of designated or notified bodies which identifies:

- Test laboratories – for testing samples;
- Inspection bodies – for inspecting factories and processes;
- Certification authorities – to interpret results;

so as to allow the manufacturer to affix the CE mark to his product. The marking may be placed on the product itself, a label, the packaging or on the accompanying commercial information. It will be accompanied by a reference to the technical specification to which it conforms and, where appropriate, by an indication of its characteristics. In the UK the register of designated bodies is maintained by the DETR.

It is important to appreciate that the CE mark is not a quality mark. It signifies only that the product satisfies the requirements. The CE mark is therefore primarily intended for enforcement officers and only states that the product may legally be placed on the market.

Interim procedures

AD Reg. 7,
sec. 1; 1.2 a

It will be apparent that in order to harmonise the standards of all the countries in the European Union it is necessary for there to be a transitional period during which existing standards will have to co-exist with new standards.

With specific reference to British Standards, nearly all those which are related to construction products will be revised to become the British 'transposition' of the new European Standards (ENs) which are currently being drafted. In the past it has been the practice to adopt the old British Standard number when transposing an EN which is based on substantially the same material. Current BSI numbering policy is to adopt the CEN numbering, prefaced with BS.

Although British Standards are normally withdrawn when their equivalent European Standards are published the following circumstances may require a deferred withdrawal of the British Standard:

- When it is necessary for the BS to remain available for work which has already commenced.
- When a BS is called up in an Approved Document which has yet to be revised.

In practice it may be necessary for some BSs to remain valid for a number of years, fully maintained alongside the new transposed standards of European origin. Therefore, it will be necessary for controllers, designers, etc., to check the applicability of the standard in each context. Where an old standard is retained it is reasonable to presume that it will satisfy the requirements of regulation 7, and where a new standard is introduced it must be checked for applicability during the transitional period and if found suitable, compliance may reasonably also be presumed.

Standards of European origin have clauses specifically identified which relate to the 'harmonised' requirements relevant to the Building Regulations (i.e. those dealing with health and safety matters) and 'non-harmonised' requirements which contain additional material relating to trading requirements of the construction industry. The non-harmonised requirements are not the concern of regulation 7 and are not covered by the AD Regulation 7 recommendations.

Interestingly, it is possible for a product to be tested and certified as complying with a British Standard by an approved body in another Member State of the European Community under the procedure covered by article 16 of the CPD. In this case it should normally be accepted by the building control body as complying with that standard. If there are reasons to doubt its acceptance then the burden of proof is on the controlling body which is obliged to notify the Trading Standards Officer so that the UK Government can notify the Commission.

With regard to CE marking the UK Construction Product Regulations state only that products must be 'fit for intended use', without reference to how this is demonstrated. The Regulations also apply only to products supplied after 27 December 1991. Therefore, any product which was legally usable before that date can continue to be sold. **AD Reg. 7, sec. 1; 1.2 d**

Where a manufacturer identifies an existing barrier to trade with another Member State, a procedure exists under article 16 of the CPD to overcome this. This involves the testing of products against the requirements of the importing country by a notified body in the exporting country. This procedure will not lead to a CE mark.

Not all materials will necessarily be CE marked under the CPD and, from a practical viewpoint, it will not be possible for all products to be CE marked until all the relevant technical specifications are available. However, it should be noted that for some products CE marking is compulsory under other Directives (such as gas boilers).

Materials and workmanship generally

Building work must be carried out:

- With proper and adequate materials which are: **Reg. 7**

    (1) appropriate for the circumstances in which they are used,
    (2) adequately mixed and prepared, *and*
    (3) applied, used or fixed so as adequately to perform the functions for which they are designed, *and*

- In a workmanlike manner.

Guidance on the choice and use of materials, and on ways of establishing the adequacy of workmanship, is given in AD Regulation 7. It should be noted however, that materials and workmanship are controlled only to the extent of: **AD Reg. 7, Performance 0.3**

- Securing reasonable standards of health and safety for persons in or about buildings for Parts A to K and N of Schedule 1;

- Conserving fuel and power in Part L, *and*
- Providing access and facilities for disabled people in Part M.

Therefore, although it may be desirable for reasons of consumer satisfaction or protection to require higher standards, this cannot be required under building regulations.

AD Reg. 7,
Performance
0.1

In order to achieve a satisfactory standard of performance **materials** should be:

- Suitable in nature and quality in relation to the purposes for which, and the conditions in which, they are used.

Additionally, **workmanship** should be such that, where relevant, materials are:

- Adequately mixed and prepared (for example, in concrete mixes, the correct proportions must be used, there should be an appropriate water/cement ratio, mixing should be thorough, etc.), *and*
- Applied, used or fixed so as to adequately perform their intended functions (for example, for reinforced concrete this would give control over the actual placing of the concrete, positioning of reinforcement, curing, etc.).

The definition of materials is quite broad and covers products, components, fittings, naturally occurring materials (such as timber, stone and thatch), items of equipment, and materials used in the backfilling of excavations in connection with building work. It should be noted that Building Regulations do not seek to control the use of materials after completion of the building work.

AD Reg. 7,
Performance
0.4

AD Reg. 7,
Performance
0.2

In order to reduce the environmental impact of building work careful thought should be given to the choice of materials, and where appropriate, recycled or recyclable materials should be considered. Obviously, the use of such materials must not have an adverse effect on the health and safety standards of the building work.

Fitness of materials

AD Reg. 7,
sec. 1; 1.1,
1.2

A number of ways of establishing the fitness of materials are dealt with in AD Regulation 7 and whilst this is mostly by reference to British Standards or certificates issued by European Technical Approvals issuing bodies other materials or products may be suitable in the particular circumstances. The following aids to establishing the fitness of materials are given in the approved document:

- A material may conform to the relevant provisions of an appropriate British Standard (but see the notes under interim procedures on page 8.6).
- A material may conform to the national technical specifications of other Member States which are contracting parties to the European Economic Area. It should be noted that where a person intends to use a product which complies with a national technical specification of another Member State, the onus is on that person to show that the product is equivalent to the relevant British Standard (and it would be necessary to provide a translation).

- A material may be covered by a national or European certificate issued by a European Technical Approvals issuing body. The conditions of use must be in accordance with the terms of the certificate and again, it will be up to the person intending to use the product to demonstrate equivalence and provide a translation.
- A material may bear a CE marking confirming that it conforms with a harmonised European standard or European Technical Approval together with the appropriate attestation procedure. Materials bearing the CE marking must be accepted if they are appropriate for the circumstances in which they are used. AD Regulation 7 qualifies this by saying that a CE marked product can only be rejected by a building control body on the basis that:

  (a) its performance is not in accordance with its technical specification, *or*
  (b) where a particular declared value or class of performance is stated for a product, the resultant value does not meet Building Regulation requirements.

  The burden of proof is on the controlling body which is obliged to notify the Trading Standards Officer so that the UK Government can notify the Commission.

- Independent certification schemes, (e.g. the kitemark scheme operated by the British Standards Institution), may also serve to show that a material is suitable for its purpose. However, some materials which are not so certified may still conform to a relevant standard. In the UK, many certification bodies which approve such schemes are accredited by UKAS.
- A material may be shown to be capable of performing its function by the use of tests, calculations or other means. It is important to ensure that tests are carried out in accordance with recognised criteria. UKAS run an accreditation scheme for testing laboratories, and together with similar schemes run by equivalent certification bodies (including accreditation schemes operated by other Member States of the EU), this ensures that standards of testing are maintained.
- In some cases past experience of a material in use in a building may be relied upon to ensure that it is capable of adequately performing its function.
- Local authorities are entitled under regulation 17 to take and test samples of materials in order to confirm compliance. Approved inspectors are not so entitled, but they may enter into arrangements with their clients whereby samples can be tested as and when deemed appropriate by the approved inspector.

Short-lived materials

Only general guidance is given on the use of short-lived materials. These are materials which may be considered unsuitable due to their rapid deterioration when compared to the life of the building. **AD Reg. 7, sec. 1; 1.3 to 1.6**

The main criteria to be considered are:

- Accessibility for inspection, maintenance and replacement.
- The effects of failure on public health and safety.

Clearly, if a material or component is inaccessible and its failure would create a serious health risk, it is unlikely that the material or component would be suitable. (See also the reference to section 19 of the Building Act 1984 on p. 8.1.)

Materials subject to changes in their properties

AD Reg. 7,
sec. 1; 1.7

Under certain environmental conditions, some materials may undergo a change in their properties which may affect their performance over time. A notable example of this occurred during the 1970s to structures constructed using high alumina cement. The subsequent deterioration of the concrete led to the collapse of a number of long span roof structures and the use of HAC was banned for all work except when the material was used as a heat-resisting material. It is known that a number of other materials, (such as certain stainless steels, structural silicone sealants and intumescent paints) may also be susceptible to changes in their properties under certain environmental conditions. In order to use such materials it will be necessary to estimate their final residual properties (including their structural properties) at the time the materials are incorporated into the work. It should then be shown that these residual properties will be adequate for the building to perform its intended function for its expected life.

Resistance to moisture and soil contaminants

AD Reg. 7,
sec. 1; 1.8,
1.9

Materials which are likely to suffer from the adverse effects of condensation, ground water or rain and snow may be satisfactory if:

(a) The construction of the building is such as to prevent moisture from reaching the materials; *or*
(b) The materials are suitably treated or otherwise protected from moisture.

Similarly, materials which are in contact with the ground will only be satisfactory if they can adequately resist the effects of deleterious substances, such as sulphates (see Site preparation and moisture exclusion on p. 8.12).

Special treatment against house longhorn beetle infestation

The previous section in the 1992 edition of AD Regulation 7 on the house longhorn beetle has been omitted from the 1999 edition. It is intended to incorporate the information into a revised Approved Document A which is currently under development and is expected to be published in 2001. In the interim, the advice given in the 1992 edition is still relevant and is described below.

In specified areas in the south of England all softwood roof timbers, including ceiling joists, should be treated with a suitable preservative against the house longhorn beetle.

The specified areas are as follows:

● The Borough of Bracknell Forest
● The Borough of Elmbridge
● The Borough of Guildford, other than the area of the former borough of Guildford
● The District of Hart other than the area of the former Urban District of Fleet

- The District of Runnymede
- The Borough of Spelthorne
- The Borough of Surrey Heath
- The Borough of Woking
- In the Borough of Rushmoor, the area of the former district of Farnborough
- The Borough of Waverley, other than the parishes of Godalming and Haslemere
- In the Royal Borough of Windsor and Maidenhead, the Parishes of Old Windsor, Sunningdale and Sunninghill.

No specific forms of treatment are recommended, however most reputable timber treatment companies (and members of such organisations as the British Wood Preserving and Dampproofing Association) have been providing treatment for house longhorn beetle for many years and will be able to recommend suitable treatments.

Adequacy of workmanship

It should be remembered that Building Regulations set different standards of workmanship to those imposed by, for example, a building specification. Building Regulations are not concerned with quality or value for money; they are concerned with public health and safety, the conservation of fuel and power, and access and facilities for disabled people. **AD Reg. 7, sec. 2; 2.1**

Adequacy of workmanship, like that of materials, may be established in a number of ways:

- A British Standard Code of Practice or other equivalent technical specification (e.g. of Member States which are contracting parties to the European Economic Area) may be used. In this context, BS 8000: *Workmanship on Building Sites* may be useful since it gathers together guidance from a number of other BSI Codes and Standards.
- Technical approvals, such as national or European certificates issued by European Technical Approvals issuing bodies, often contain workmanship recommendations. Additionally, it may be possible to use an equivalent technical approval (such as those of a member of EOTA) if this provides an equivalent level of protection and performance. The onus of proof of acceptability rests with the user in this case.
- Workmanship which is covered by a scheme complying with BS EN ISO 9000: *Quality management and quality assurance standards,* will demonstrate an acceptable standard since these schemes relate to processes and products for which there may also be a suitable British or other technical standard. A number of such schemes have been accredited by UKAS. There are also a number of independent schemes for accreditation and registration of installers of materials, products and services and these ensure that work has been carried out to appropriate standards by knowledgeable contractors.
- In some cases past experience of a method of workmanship such as a building in use, may be relied upon to ensure that the method is capable of producing the intended standard of performance.
- Local authorities are empowered under regulation 16 to test drains and sewers in order to establish compliance with Part H of Schedule 1 (see Chapter 13 Drainage and waste disposal). Approved inspectors are not so

empowered, but they may enter into arrangements with their clients whereby drains and sewers can be tested as and when deemed appropriate by the approved inspector.

Site preparation and moisture exclusion

Introduction

Part C of Schedule 1 to the 1991 Regulations is concerned with site preparation and resistance to moisture. In addition to moisture exclusion, paragraph C2 contains provisions controlling sites containing dangerous or offensive substances. This replaces section 29 of the Building Act 1984.

The supporting Approved Document C now contains recommendations relating to the control of radon gas in certain areas of England and Wales, and landfill gases on certain sites near waste disposal tips, etc. The guidance on the control of dampness in buildings in AD C does not extend to that concerning damage from condensation. For this, reference should be made to AD F, Ventilation (see Chapter 11), AD L, Conservation of fuel and power (see Chapter 16), and the BRE publication entitled *Thermal insulation: avoiding risks.*

Certain provisions regarding the damp-proofing and weather resistance of floors and walls do not apply to buildings used solely for storage of plant or machinery in which the only persons habitually employed are storemen, etc. Other similar types of buildings where the air is so moisture-laden that any increase would not adversely affect the health of the occupants are also excluded. These buildings are referred to as 'excepted buildings' throughout this chapter.

Preparation of site

Regs Sch. 1 C1

The ground to be covered by the building is required to be reasonably free from vegetable matter.

AD C1/3

Decaying vegetable matter could be a danger to health and it could also cause a building to become unstable if it occurred under foundations. AD C1/3, therefore, recommends that the site should be cleared of all turf and vegetable matter at least to a depth to prevent future growth. This does not apply to excepted buildings.

AD C1/3 sec. 1 1.2

Below-ground services (such as foul or surface water drainage) should be designed to resist the effects of tree roots. This can be achieved by making services sufficiently robust or flexible and with joints that cannot be penetrated by roots.

AD C1/3 sec. 1 1.3

Dangerous and offensive substances

Regs Sch. 1 C2

Precautions must be taken to prevent any substances found on or in the ground from causing a danger to health and safety. This is, of course, the ground covered by the building and includes the area covered by the foundations.

AD C Introduction 0.4

There is a special definition of *contaminant* for the purposes of AD C – any material (including faecal or animal matter) and any substance which is or could become toxic, corrosive, explosive, flammable or radioactive and therefore likely to be a danger to health or safety. This material must be in or on the ground to be covered by the building.

Contaminants can be liquids or solids arising out of a previous use of land, or they may be gases. In recent years problems have arisen from the emission of landfill gas from waste disposal sites. Additionally, in certain parts of the country, contamination by the naturally-occurring radioactive gas radon and its products of decay has lead to concern over the long-term health of the occupants of affected buildings.

AD C2
sec. 2
2.1

Where a site is being redeveloped, knowledge of its previous use, from planning or other local records, may indicate a possible source of contamination. Table 1 to AD C2 (reproduced below) lists a number of site uses that are likely to contain contaminants.

Where the presence of contaminants has not been identified at an early stage, a later site survey may indicate possible contamination. Table 2 to AD C2 (reproduced below) shows the signs to be looked for, indicates which materials may be responsible and suggests relevant actions that may be taken. The Environmental Health Officer should always be notified if contamination is suspected. He may then agree on the remedial measures necessary to make the site safe.

AD C2
sec. 2
2.2 & 2.3

Some contaminants present such hazardous conditions that only complete removal of the offending substances can provide a complete remedy, while in other cases the risks may be reduced to acceptable levels by remedial measures. This may necessitate the removal of large quantities of material and such works or remedial actions should only be undertaken with the benefit of expert advice.

AD C2
sec. 2
2.5

For guidance on site investigation, reference should be made to BS Draft for Development DD 175: 1988 *Code of practice for the identification of potentially contaminated land and its investigation*. Further information on sites in general may be obtained from BS 5930: 1981 *Code of practice for site investigations*.

AD C2
sec. 2
2.6

AD C2

Table 1   Sites likely to contain contaminants

Asbestos works
Chemical works
Gas works, coal carbonisation plants and ancillary byproduct works
Industries making or using wood preservatives
Landfill and other waste disposal sites
Metal mines, smelters, foundries, steel works and metal finishing works
Munitions production and testing sites
Nuclear installations
Oil storage and distribution sites
Paper and printing works
Railway land, especially the larger sidings and depots
Scrap yards
Sewage works, sewage farms and sludge disposal sites
Tanneries

AD C2

Table 2   Possible contaminants and actions

Signs of possible contaminants	Possible contaminant	Relevant action
Vegetation (absence, poor or unnatural growth)	Metals Metal compounds*	None
	Organic compounds Gases	Removal[1]
Surface materials (unusual colours and contours may indicate wastes and residues)	Metals Metal compounds*	None
	Oily and tarry wastes	Removal, filling or sealing
	Asbestos (loose) Other mineral fibres	Filling[2] or sealing[3] None
	Organic compounds including phenols	Removal or filling
	Combustible material including coal and coke dust	Removal or filling
	Refuse and waste	Total removal or see guidance
Fumes and odours (may indicate organic chemicals at very low concentrations)	Flammable explosive and asphyxiating gases including methane and carbon dioxide	Removal
	Corrosive liquids	Removal, filling or sealing
	Faecal animal and vegetable matter (biologically active)	Removal or filling
Drums and containers (whether full or empty)	Various	Removal with all contaminated ground

Notes:

Liquid and gaseous contaminants are mobile and the ground covered by the building can be affected by such contaminants from elsewhere. Some guidance on landfill gas and radon is given in AD C; other liquids and gases should be referred to a specialist.

* Special cement may be needed with sulphates.

Actions assume that ground will be covered with at least 100 mm in-situ concrete.

[1] Removal – the contaminant and any contaminated ground removed to a depth of 1 m below lowest floor (or less if local authority agrees) to place named by local authority.

[2] Filling – Area of building covered to a depth of 1 m (or less if local authority agrees) with suitable material. Filling material and ground floor design considered together. Combustible materials adequately compacted to avoid combustion.

[3] Sealing – Imperforate barrier between contaminant and building sealed at joints, edges and service entries. Polythene may not always be suitable if contaminant is tarry waste or organic solvent.

AD C2
sec. 2
2.4

Radon gas contamination

Radon is a naturally-occurring, colourless and odourless gas which is radio-active. It is formed in small quantities by the radioactive decay of uranium and radium, and thus travels through cracks and fissures in the subsoil until it reaches the atmosphere or enters spaces under or in buildings.

AD C2
sec. 2
2.7

It is recognised that radon gas occurs in all buildings; however the concentration may vary from below 20 Bq/m^3 (the national average for houses in the UK) to more than 100 times this value. The National Radiological Protection Board (NRPB) has recommended an action level of 200 Bq/m^3 for houses. The lifetime risk of contracting lung or other related cancers at the action level is about 3%. Geographical distribution of houses at or above the action level is very uneven with about two-thirds of the total being in Devon and Cornwall.

The DETR is reviewing the areas where preventative measures should be taken as information becomes available from the NRPB. This information has been placed in the BRE guidance document *Radon: guidance on protective measures for new dwellings* (BRE Report BR 211, obtainable from Building Research Establishment, Garston, Watford, WD2 7JR) which will be updated as necessary. The report identifies certain Districts and Boroughs in Devon, Cornwall, Somerset, Northamptonshire and Derbyshire (listed in Table 8.1 below) where full radon precautions are necessary. This involves both primary measures (radon-proof barrier) and secondary measures (radon sump and extract pipe, or ventilated subfloor void). In Table 8.2 are listed those parishes

BR 211

Table 8.1 Areas where full radon precautions are required for new dwellings.

Districts and Boroughs	Parishes and Towns		
Cornwall			
Caradon	Boconnoc	Liskeard	St Keyne
	Broadoak	Looe	St Martin
	Callington	Menheniot	St Mellion
	Calstock	Morval	St Neot
	Dobwalls and Trewidland	Pelynt	St Pinnock
	Duloe	Pillaton	St Veep
	Landrake with St Erney	Quethiock	St Winnow
	Lanreath	St Cleer	Sheviock
	Lansallos	St Dominick	South Hill
	Lanteglos	St Germans	Warleggan
	Linkinhorne	St Ive	
Carrick	All		
Kerrier	Breage	Gweek	Portreath
	Budock	Helston	Redruth
	Camborne	Illogan	St Anthony in Meneage
	Carharrack	Lanner	St Day
	Carn Brea	Mabe	St Gluvias
	Constantine	Manaccan	St Martin in Meneage
	Crowan	Mawgan in Meneage	Stithians
	Cury	Mawnan	Sithney
	Germoe	Porthleven	Wendron
	Gunwalloe		
Somerset			
Mendip	Cranmore		
	Doulting		
	Evercreech		
West Somerset	Skilgate		
	Upton		

Table 8.1 (*continued*)

Northamptonshire			
Kettering	Broughton	Grafton Underwood	Pytchley
	Burton Latimer	Kettering	Thorpe Malsor
	Cranford	Loddington	Warkton
	Cransley	Orton	Weekley
Wellingborough	Finedon	Isham	Orlingbury
	Great Harrowden	Little Harrowden	Sywell
	Hardwick	Mears Ashby	
Daventry	Boughton	Harlestone	Overstone
	Brixworth	Holcot	Pitsford
	Chapel Brampton	Lamport	Scaldwell
	Church Brampton	Moulton	Spratton
	Hannington	Old	Walgrave
Northampton		All	

Derbyshire			
Derbyshire Dales	Abney and Abney Grange	Great Hucklow	Offerton
	Aldwark	Great Longstone	Over Haddon
	Ashford in the Water	Grindlow	Parwich
	Bakewell	Harthill	Pilsley
	Ballidon	Hartington Middle Quarter	Rowland
	Birchover	Hartington Nether Quarter	Sheldon
	Blackwell in the Peak	Hartington Tow Quarter	Stanton
	Bradwell	Hassop	Stoney Middleton
	Brushfield	Hazelbadge	Taddington
	Calver	Highlow	Thorpe
	Chelmorton	Lea Hall	Tideswell
	Eaton and Alsop	Little Hucklow	Tissington
	Edensor	Little Longstone	Wardlow
	Elton	Litton	Wheston
	Eyam	Middleton and Smerrill	Winster
	Flagg	Monyash	Youlgreave
	Foolow	Nether Haddon	
	Gratton	Newton Grange	
High Peak	Aston	Green Fairfield	Peak Forest
	Brough and Shatton	Hartington Upper Quarter	Thornhill
	Buxton	Hope	Wormhill
	Castleton	King Sterndale	

Table 8.2 Areas where secondary radon precautions are required for new dwellings.

Districts and Boroughs	Parishes and Towns		
Cornwall			
Caradon	Antony	Maker with Rame	Saltash
	Botusfleming	Millbrook	Torpoint
	Landulph	St John	
Kerrier	Grade Ruan	Mullion	
	Landewednack	St Keverne	
North Cornwall	Boyton	St Ervan	St Merryn
	Jacobstow	St Eval	St Minver Lowlands
	North Petherwin	St Gennys	Warbstow
	North Tamerton	St Issey	Week St Mary
	Otterham	St Juliot	Whitstone
	Padstow		

Table 8.2 (*continued*)

Devon

Mid Devon			
	Bickleigh	Cruwys Morchard	Shobrooke
	Bow	Down St Mary	Stockleigh English
	Brushford	Eggesford	Stockleigh Pomeroy
	Cadbury	Hittisleigh	Stoodleigh
	Cadeleigh	Hockworthy	Templeton
	Chawleigh	Huntsham	Thelbridge
	Cheriton Bishop	Kennerleigh	Thorverton
	Cheriton Fitzpaine	Lapford	Tiverton
	Clannaborough	Loxbeare	Upton Hellions
	Clayhanger	Morchard Bishop	Washfield
	Coldridge	Newton St Cyres	Washford Pyne
	Colebrooke	Nymet Rowland	Wembworthy
	Copplestone	Poughill	Woolfardisworthy
	Crediton	Puddington	Zeal Monachorum
	Crediton Hamlets	Sandford	

North Devon			
	Ashford	Filleigh	North Molton
	Atherington	Fremington	Parracombe
	Barnstaple	Georgeham	Pilton West
	Berrynarbor	Georgenympton	Queensnympton
	Bishops Nympton	Goodleigh	Rackenford
	Bishops Tawton	Heanton Punchardon	Romansleigh
	Brauton	Ilfracombe	Rose Ash
	Brayford	Kingsnympton	Satterleigh and Warkleigh
	Burrington	Knowstone	Shirwell
	Challacombe	Landkey	South Molton
	Chittlehamholt	Mariansleigh	Stoke Rivers
	Chittlehampton	Martinhoe	Swimbridge
	Chulmleigh	Marwood	Tawstock
	Combe Martin	Meshaw	Trentishoe
	East Anstey	Molland	Twitchen
	East and West Buckland	Mortehoe	West Anstey
	East Worlington	Newton Tracey	Witheridge

South Hams			
	Ashprington	Dartington	Salcombe
	Berry Pomeroy	East Portlemouth	South Pool
	Bickleigh	Frogmore and Sherford	Staverton
	Brixton	Harberton	Stoke Gabriel
	Charleton	Littlehempston	Stokenham
	Chivelstone	Malborough	Totnes
	Cornworthy	Marldon	Wembury

Teignbridge			
	Abbotskerswell	Holcombe Burnell	Stokinteignhead
	Ashton	Ide	Tedburn St Mary
	Bishopsteignton	Ideford	Teigngrace
	Broadhempston	Ipplepen	Teignmouth
	Chudleigh	Kingkerswell	Torbryan
	Coffinswell	Kingsteignton	Trusham
	Doddiscombsleigh	Newton Abbot	Whitestone
	Dunchideock	Ogwell	Woodland
	Haccombe-with-Combe	Shaldon	

Torridge			
	Abbots Bickington	Great Torrington	Pancrasweek
	Alverdiscott	Halwill	Parkham
	Alwington	Hartland	Peters Marland
	Ashreigney	High Bickington	Petrockstowe
	Beaford	Hollacombe	Pyworthy
	Black Torrington	Holsworthy	Roborough
	Bradford	Holsworthy Hamlets	St Giles in the Wood
	Bradworthy	Huish	St Giles in the Heath
	Bridgerule	Huntshaw	Shebbear
	Broadwoodwidger	Landcross	Sheepwash
	Buckland Brewer	Langtree	Sutcombe
	Buckland Filleigh	Littleham	Tetcott
	Bulkworthy	Little Torrington	Thornbury
	Clawton	Luffincott	Virginstowe
	Clovelly	Merton	Weare Gifford
	Cookbury	Milton Damerel	Welcombe
	Dolton	Monkleigh	Winkleigh
	Dowland	Newton St Petrock	Woolfardisworthy
	East and West Putford	Northcott	Yarnscombe
	Frithelstock		

Table 8.2 (*continued*)

West Devon	Beaworthy	Germansweek	Jacobstowe
	Bondleigh	Hatherleigh	Meeth
	Bratton Clovelly	Highampton	Monkokehampton
	Broadwoodkelly	Iddesleigh	Northlew
	Exbourne	Inwardleigh	North Tawton
Torbay	Paignton		
	Torquay		
Plymouth		All	

Somerset

South Somerset	Alford	Huish Episcopi	North Perrott
	Aller	Ilchester	Pitcombe
	Ansford	Isle Abbots	Pitney
	Ash	Isle Brewers	Puckington
	Babcary	Iton	Queen Camel
	Barrington	Keinton Mandeville	Rimpton
	Barton St David	Kingbory Episcopi	Seavington St Mary
	Bruton	Kingsdon	Seavington St Michael
	Castle Cary	Kingstone	Shepton Beauchamp
	Charton Mackerell	King Weston	Shepton Montigue
	Chilton Cantelo	Langport	Somerton
	Compton Dundon	Limington	South Barrow
	Corton Denham	Long Load	South Petherton
	Crewkerne	Long Sutton	South Cadbury
	Curry Mallett	Lopen	Sparkford
	Curry Rivel	Marston Magna	Stocklinch
	Dinnington	Martock	Tintinhull
	Drayton	Merriott	Wayford
	Fivehead	Misterton	West Camel
	Hambridge & Westport	Muchelney	West Crewkerne
	High Ham	North Barrow	White Lackington
	Hinton St George	North Cadbury	Yeovilton
West Somerset	Brompton Ralph	Dulverton	Luxborough
	Brompton Regis	Exford	Oare
	Brushford	Exmoor	Treborough
	Clatworthy	Exton	Winsford
	Cutcombe	Huish Champflower	Withypoole
Taunton Deane	Ashbrittle	Chipstable	Stock St Gregory
	Bathealton	North Curry	Wiveliscombe
	Burrowbridge		
Mendip	Ashwick	Ditcheat	Priddy
	Batcombe	Downhead	Pylle
	Binegar	Emborough	Shepton Mallet
	Butleigh	Holcombe	Stoke St Michael
	Chewton Mendip	Lamyat	Ston Easton
	Chilcompton	Leigh on Mendip	Stratton on the Fosse
	Coleford	Litton	Street
	Croscombe	Milton Clevedon	Walton
Sedgemoor	Lyng	Middlezoy	Othery

Northamptonshire

Daventry	Althorp	Draughton	Naseby
	Arthingworth	East Haddon	Newnham
	Badby	Everdon	Norton
	Brington	Farthingstone	Preston Capes
	Brockhall	Fawsley	Ravensthorpe
	Byfield	Flore	Staverton
	Canons Ashby	Great Oxendon	Stowe Nine Churches
	Catesby	Guilsborough	Thornby
	Charwelton	Haselbech	Watford
	Clipston	Hellidon	Weedon Bec
	Cold Ashby	Holdenby	Welton
	Cottesbrooke	Hollowell	West Haddon
	Creaton	Kelmarsh	Whilton
	Daventry	Long Buckby	Winwick
	Dodford	Maidwell	Woodford Cum Membris

Table 8.2 (*continued*)

South Northamptonshire	Abthorpe	Eydon	Potterspury
	Adstone	Farthinghoe	Quinton
	Ashton	Gayton	Radstone
	Aston le Walls	Greatworth	Roade
	Aynho	Grafton Regis	Rothersthorpe
	Blakesley	Greens Norton	Shutlanger
	Blisworth	Hackleton	Silverstone
	Boddington	Harpole	Slapton
	Brackley	Hartwell	Stoke Bruerne
	Braddon	Helmdon	Sulgrave
	Brafield on the Green	Hinton in the Hedges	Syresham
	Bugbrooke	Kings Sutton	Thenford
	Castle Ashby	Kislingbury	Thorpe Mandeville
	Chacombe	Litchborough	Tiffield
	Chipping Warden	Little Houghton	Towcester
	Cogenhoe and Whiston	Maidford	Upper Heyford
	Cold Higham	Marston St Lawrence	Wappenham
	Courteenhall	Middleton Cheney	Warkworth
	Croughton	Milton Malsor	Weston and Weedon
	Culworth	Moreton Pinkney	Whitfield
	Denton	Nether Heyford	Whittlebury
	Easton Neston	Newbottle	Woodend
	Edgcote	Pattishall	Yardley Gobion
	Evenley	Paulerspury	Yardley Hastings
East Northamptonshire	Apethorpe	Higham Ferrers	Ringstead
	Blatherwycke	Irthlingborough	Rushden
	Chelveston Cum Caldecott	Islip	Stanwick
	Collyweston	Kings Cliffe	Thrapston
	Denford	Laxton	Twywell
	Duddington-with-Fineshade	Little Addington	Wakerley
	Easton on the Hill	Lowick	Warmington
	Fotheringhay	Nassington	Woodford
	Great Addington	Newton Bromswold	Woodnewton
	Hargrave	Raunds	Yarwell
Wellingborough	Bozeat	Great Doddington	Wellingborough
	Earls Barton	Grendon	Wilby
	Easton Maudit	Irchester	Wollaston
	Ecton	Strixton	
Corby	Corby	East Carlton	
	Cottingham	Middleton	
Kettering	Braybrooke	Harrington	Rothwell
	Desborough	Newton	Rushton
	Geddington		
Northampton	Billing	Great Houghton	Wootton
	Collingtree	Hardingstone	Upton

Derbyshire

Derbyshire Dales	Atlow	Darley Dale	Kniveton
	Baslow and Bubnell	Edensor	Mappleton
	Beeley	Fenny Bentley	Matlock Bath
	Bonsall	Froggatt	Matlock Town
	Bradbourne	Grindleford	Northwood and Tinkersley
	Brassington	Hathersage	Offcote and Underwood
	Callow	Hognaston	Outseats
	Carsington	Hopton	Rowsley
	Chatsworth	Ible	South Darley
	Cromford	Ivonbrook Grange	Tansley
	Curbar	Kirk Ireton	Wirksworth
High Peak	Bamford	Derwent	New Mills
	Chapel en le Frith	Edale	Whaley Bridge
	Chinley, Buxworth and Brownside	Hayfield	
North-East Derbyshire	Calow	Killamarsh	Unstone
	Eckington	Sutton Cum Duckmanton	

Table 8.2 (*continued*)

Bolsover	Ault Hucknall	Glapwell	Scarcliffe
	Barlborough	Old Bolsover	Shirebrook
	Clowne	Pleasley	Whitwell
	Elmton		
Chesterfield	Staveley		
Amber Valley	Ashlyhay		
	Dethick Lea and Holloway		

where there should be provision made in all new dwellings for future secondary measures.

As more information becomes available from NRPB it is likely that further areas will be covered by the need for radon precautions. The local building control authority should be contacted for confirmation.

Primary protection against radon

Primary protection may be provided by an airtight, and therefore radon-proof, barrier across the whole of the building including the floor and walls. This could consist of:

- Polyethylene (polythene) sheet membrane of at least 300 micrometre (1200 gauge) thickness.
- Flexible sheet roofing materials.
- Prefabricated welded barriers.
- Liquid coatings.
- Self-adhesive bituminous-coated sheet products.
- Asphalt tanking.

It is important to have adequately sealed joints and the membrane must not be damaged during construction. Where possible, penetration of the membrane by service entries should be avoided. With careful design it may be possible for the barrier to serve the dual purpose of damp-proofing and radon protection. Some typical details are shown in Fig. 8.2.

Secondary protection against radon

In practical terms, a totally radon-proof barrier may be difficult to achieve. Therefore, in high-risk areas it is necessary to provide additional secondary protection. This might consist of:

- Natural ventilation of an underfloor space by airbricks or ventilators on at least two sides.
- The addition of an electrically operated fan in place of one of the airbricks to provide enhanced subfloor ventilation.
- A subfloor depressurisation system comprising a sump located beneath the floor slab, joined by pipework to a fan. It may only be necessary to provide the sump and underfloor pipework during construction thus giving the owner the option of connecting a fan at a later stage if necessary.

Examples of these methods are shown in Figs. 8.2 and 8.3.

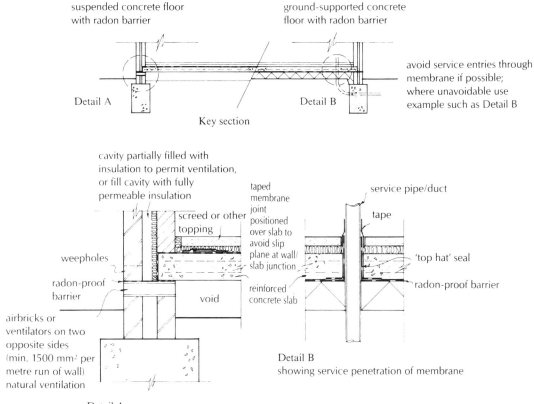

Fig. 8.2 Primary protection against radon.

It should be noted that the above brief notes on BR 211 are intended to give an idea of the content of that document. Designers of buildings in the delimited areas should consult the full report and any other relevant references, such as those listed on p. 8.25 below.

Contamination of landfill gas

Landfill gas is typically made up of 60% methane and 40% carbon dioxide, although small quantities of other gases such as hydrogen, hydrogen sulphide and a wide range of trace organic vapours, may also be present. The gas is produced by the breakdown of organic material by micro-organisms under oxygen-free (anaerobic) conditions. Gases similar to landfill gas can also arise naturally from coal strata, river silt, sewage and peat.

AD C2
sec. 2
2.8

Properties of landfill gases

The largest component of landfill gas, methane, is a flammable, asphyxiating gas with a flammable range between 5% and 15% by volume in air. If such a

BR 212

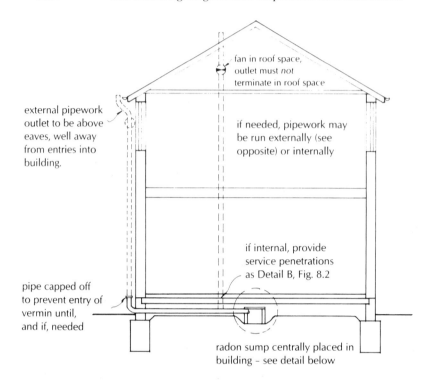

fan in roof space, outlet must *not* terminate in roof space

external pipework outlet to be above eaves, well away from entries into building.

if needed, pipework may be run externally (see opposite) or internally

if internal, provide service penetrations as Detail B, Fig. 8.2

pipe capped off to prevent entry of vermin until, and if, needed

radon sump centrally placed in building – see detail below

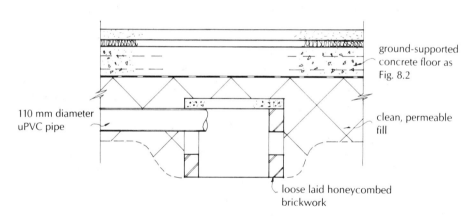

ground-supported concrete floor as Fig. 8.2

110 mm diameter uPVC pipe

clean, permeable fill

loose laid honeycombed brickwork

Fig. 8.3 Secondary protection against radon.

concentration occurs within a building and the gas is ignited it will explode. Methane is lighter than air.

The other major component, carbon dioxide, is a non-flammable, toxic gas which has a long-term exposure limit of 0.5% by volume and a short-term exposure limit of 1.5%. It is heavier than air.

Movement of landfill gases

The proportions of these two main gases and the amount of air mixed with them will largely determine the properties of the landfill gas since they remain mixed and do not separate, although the mixture can remain separate from surrounding air. These landfill gases will migrate from a landfill site as a result of diffusion through the ground and this migration may be increased by rainfall or freezing temperatures as these conditions tend to seal the ground surface. The gases will also follow cracks, cavities, pipelines and tunnels, etc., as these form ideal pathways. Landfill gas emissions can be increased by rapid falls in atmospheric pressure and by a rising water table. Thus, landfill gas may enter buildings and may collect in underfloor voids, drains and soakaways.

Building near landfill sites

AD C2 suggests that further investigations should be made to determine whether protective measures are necessary if:

AD C2
sec. 2
2.9

- the ground to be covered by a building is on, or within 250 m of, a landfill, *or*
- there is reason to suspect that there may be gaseous contamination of the ground, *or*
- the building will be within the likely sphere of influence of a landfill.

Practical guidance on construction methods to prevent the ingress of landfill gas in buildings is not given in ADC2; instead, reference is made to the BRE Report *Construction of new buildings on gas-contaminated land* (BRE Report BR 212, obtainable from Building Research Establishment, Garston, Watford, WD2 7JR).

The report gives examples of ground floor construction and venting details for houses and other similar small buildings. These are suitable where the concentration of methane in the ground is unlikely to exceed 1% by volume.

With regard to carbon dioxide, a concentration greater than 1.5% by volume in the ground indicates that there is a need to consider using the construction details described in the report to prevent the ingress of gas. Where this is as high as 5% then the floor constructions are required.

AD C2
sec. 2
2.10

Figure 8.4 gives typical examples of some of the constructional details contained in the report. It should be noted that the use of continuous mechanical ventilation for the removal of landfill gases in dwellings is not recommended since maintenance of the system cannot be guaranteed and a failure might result in a sharp increase in indoor methane concentration with the possibility of an explosion occurring.

In buildings other than dwellings, expert advice should be sought. This might include:

AD C2
sec. 2
2.11

- A complete investigation into the source and nature of any hazardous gases present.
- The potential for future gas generation from the landfill site.
- An assessment of the present and future risk posed by the gas, including the need for extended monitoring.
- The design of protective measures incorporated into the overall building design, including arrangements for maintenance and monitoring.

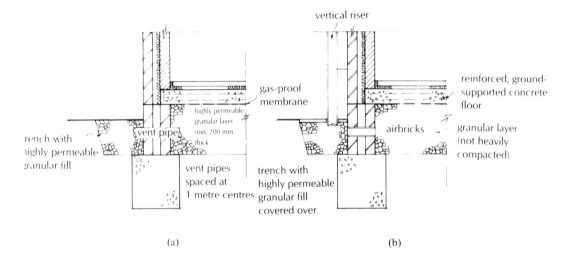

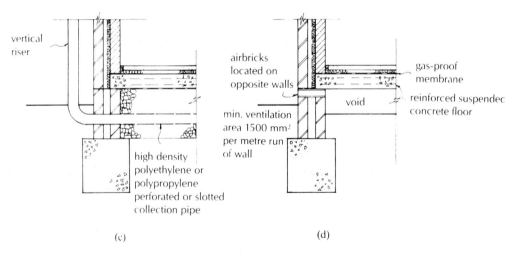

Fig. 8.4 Landfill gas protection details. (a) External trench ventilation.
(b) Sealed trench with vertical riser. (c) Perforated pipe ventilation.
(d) Suspended floor – airbrick ventilation.

The BRE Report mentioned above also includes practical details for buildings other than dwellings.

It should be noted that the above brief notes on BR 212 are intended to give an idea of the content of that document. Designers of buildings which are likely to be affected by landfill gas should consult the full report and any of the references in the following list, which is reproduced from AD C2, as this gives further information regarding documents concerned with site investigations and the development of contaminated land.

Reference list to sources on site investigations

(a) Department of Environment (Her Majesty's Inspectorate of Pollution). *The Control of Landfill Gas*, Waste Management Paper No. 27 (1989), HMSO.
(b) British Standards Institution Draft for Development: *Code of Practice for the Identification of Potentially Contaminated Land and its Investigation* DD 175: 1988.
(c) Crowhurst, D. *Measurement of gas emissions from contaminated land.* BRE Report 1987. HMSO, ISBN 0 85125 246 X.
(d) ICRCL 17/78 *Notes on the development and after-use of landfill sites*, 8th edition, December 1990. Interdepartmental Committee on the Redevelopment of Contaminated Land.
(e) ICRCL 59/83 *Guidance on the assessment and redevelopment of contaminated land*, 2nd edition, July 1987. Interdepartmental Committee on the Redevelopment of Contaminated Land.
(f) Institute of Wastes Management. *Monitoring of Landfill Gas.* September 1989.
(g) BS 5930: 1981 *Code of practice for site investigations.*
(h) BRE. *Radon: guidance on protective measures for new dwellings.* BRE Report. Garston, BRE, 1991. ISBN 0 85125 511 6.
(i) BRE. *Construction of new buildings on gas contaminated land.* BRE Report. Garston, BRE, 1991. ISBN 0 85125 513 2.

(d) and (e) are obtainable from: Department of Environment Publication Sales Unit, Building 1, Victoria Road, South Ruislip, Middlesex HA4 0NZ.

Subsoil drainage

Subsoil drainage must be provided *if* it is necessary to avoid:

(a) the passage of moisture from the ground to the inside of the building; or,
(b) damage to the fabric of the building.

Regs Sch. 1
C3

There arc no provisions in AD C1/3 concerning the flooding of sites. If the site of a building is subject to flooding it merely assumes that suitable steps are being taken.

AD C1/3
sec. 1
1.4

Subsoil water may cause problems in the following cases:

- Where there is a high water table (i.e. within 0.25 m of the lowest floor in the building).
- Where surface water may enter or damage the building fabric.
- Where an active subsoil drain is severed during excavations.

Where problems are anticipated it will usually be necessary either to drain the site of the building or to design and construct it to resist moisture penetration.

Severed subsoil drains should be intercepted and continued in such a way that moisture is not directed into the building. Figure 8.5 illustrates a number of possible solutions.

AD C1/3
sec. 1
1.5 to 1.7

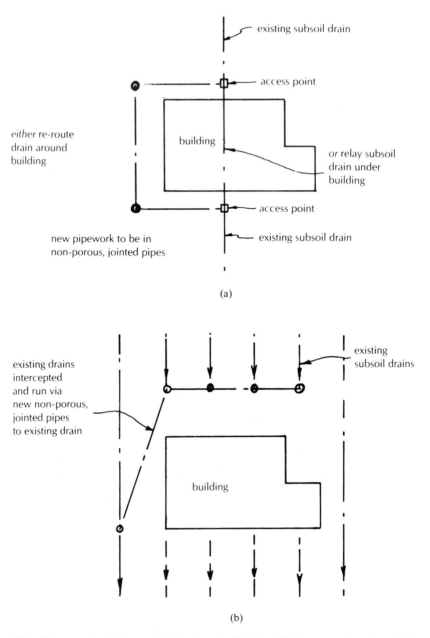

(a)

(b)

Fig. 8.5 Subsoil drainage. (a) Single subsoil drain. (b) Interception of multiple subsoil drains.

Resistance to weather and ground moisture

The floors, walls and roof of a building are required to resist the passage of moisture to the inside of the building.

Reg Sch. 1
C4

Protection of floors next to the ground

The term *floor* means the lower surface of any space in a building and includes any surface finish which is laid as part of the permanent construction. This would, presumably, exclude carpets, lino, tiles, etc., but would include screeds and granolithic finishes.

AD C
Introduction
0.4

A ground floor should be constructed so that:

- The passage of moisture to the upper surface of the floor is resisted. (This does not apply to excepted buildings.)
- It will not be adversely affected by moisture from the ground.
- It will not transmit moisture to another part of the building that might suffer damage (see Fig. 8.6).

AD C
Performance

AD C4
sec. 3
3.2

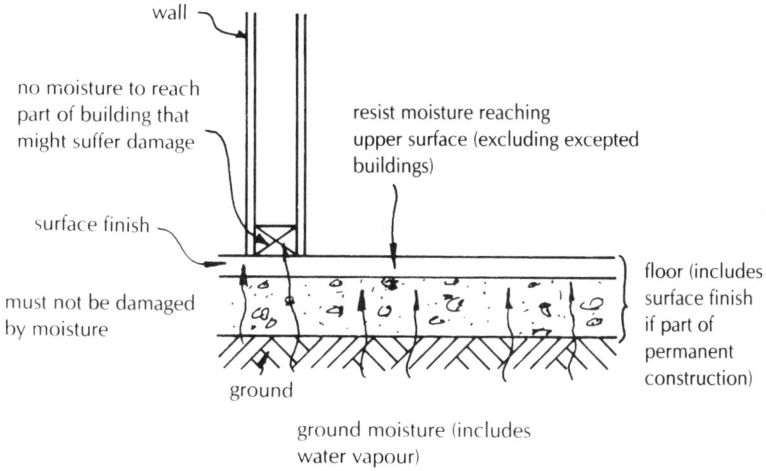

Fig. 8.6 Floors – functional requirements.

The term *moisture* is taken to include water vapour as well as liquid water. Damage caused by moisture is only significant if it would cause a material or structure to deteriorate to such a point that it would present an imminent danger to public health and safety or it would permanently reduce the performance of an insulating material.

AD C
Introduction
0.3, 0.4

Floors supported directly by the ground

The requirements mentioned above can be met, for ground supported floors, by covering the ground with dense concrete incorporating a damp-proof

AD C4
sec. 3
3.2

membrane, laid on a hardcore bed. If required, insulation may also be incorporated in the floor construction.

This form of construction is illustrated in Fig. 8.7 below. However, the following points should also be considered.

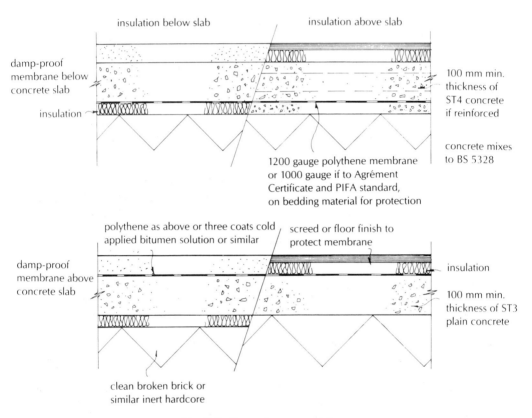

Fig. 8.7 Ground supported floor.

- Hardcore laid under the floor next to the ground should not contain water-soluble sulphates or deleterious matter in such quantities as might cause damage to the floor.

    Broken brick or stone are the best hardcore materials. Clinker is dangerous unless it can be shown that the actual material proposed is free from sulphates, etc., and colliery shales should likewise be avoided. In any event, the builder might well be liable for breach of an implied common law warranty of fitness of materials (see *Hancock* v. *B. W. Brazier (Anerley) Ltd* [1966] 2 All ER 901), where builders were held liable for subsequent damage caused by the use of hardcore containing sulphates.
- A damp-proof membrane (DPM) may be provided above or below the concrete floor slab and should be laid continuous with the damp-proof courses in walls, piers, etc.

    If laid below the concrete, the DPM should be at least equivalent to 300 μm (1200 gauge) polyethylene (e.g. polythene). It should have sealed joints and should be supported by a layer of material that will not cause damage to the polythene. Polythene of 250 μm (1000 gauge) is also satis-factory if it is to the PIFA standard or is used in accordance with an appropriate BBA certificate.

If a polyethylene sheet membrane is laid above the concrete, there is no need to provide the bedding material. It is also possible to use a three-coat layer of cold applied bitumen emulsion in this position. These materials should be protected by a suitable floor finish or screed. Surface protection does not need to be provided where the membrane consists of pitchmastic or similar material which also serves as a floor finish.

- The minimum thickness for the concrete slab is 100 mm, although the structural design for the slab may require it to be thicker. Unreinforced concrete should be composed of 50 kg cement to maximum 0.11 m^3 fine aggregate to maximum 0.16 m^3 coarse aggregate or BS 5328 mix ST3. Reinforced concrete should be composed of 50 kg cement to maximum 0.08 m^3 fine aggregate to maximum 0.13 m^3 coarse aggregate or BS 5328 mix ST4.
- AD C4 gives no guidance on the position of the floor relative to outside ground level. Since this type of floor is unsuitable if subjected to water pressure, it is reasonable to assume that the top surface of the slab should not be below outside ground level unless special precautions are taken.

AD C4
sec. 3
3.4 to 3.6

If it is proposed to lay a timber floor finish directly on the concrete slab, it is permissible to bed the timber in a material that would also serve as a damp-proof membrane.

No guidance is given regarding suitable DPM materials. However 12.5 mm of asphalt or pitchmastic will usually be satisfactory for most timber finishes and it may be possible to lay wood blocks in a suitable adhesive DPM. If a timber floor finish is fixed to wooden fillets embedded in the concrete, the fillets should be treated with a suitable preservative unless they are above the DPM (see BS 1282: 1975 *Guide to the choice, use and application of wood preservatives* for suitable preservative treatments).

AD C4
sec. 3
3.7

Clause 11 of CP 102: 1973 *Code of practice for protection of buildings against water from the ground* may be used as an alternative to the above. Where ground water pressure is evident, recommendations may be found in BS 8102: 1990 *Code of practice for protection of structures against water from the ground*.

AD C4
sec. 3
3.8

Suspended timber floors

The performance requirements mentioned above may be met for suspended timber ground floors by:

- Covering the ground with suitable material to resist moisture and deter plant growth.
- Providing a ventilated space between the top surface of the ground covering and the timber.
- Isolating timber from moisture-carrying materials by means of damp-proof courses.

AD C4
sec. 3
3.9

A suitable form of construction is shown in Fig. 8.8 and is summarised below.

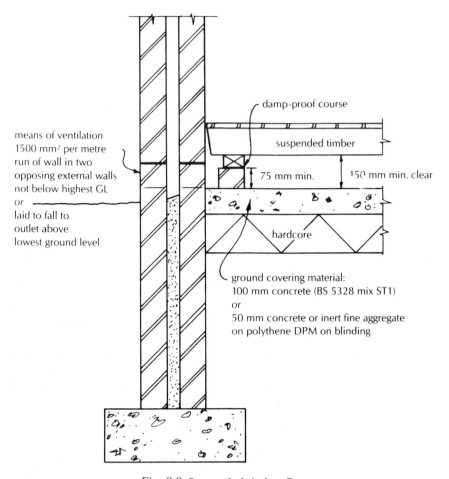

means of ventilation
1500 mm² per metre
run of wall in two
opposing external walls
not below highest GL

or

laid to fall to
outlet above
lowest ground level

damp-proof course

suspended timber

75 mm min.

150 mm min. clear

hardcore

ground covering material:
100 mm concrete (BS 5328 mix ST1)
or
50 mm concrete or inert fine aggregate
on polythene DPM on blinding

Fig. 8.8 Suspended timber floor.

- The ground surface should be covered with at least 100 mm of concrete composed of 50 kg cement to maximum 0.13 m³ fine aggregate to maximum 0.18 m³ coarse aggregate or BS 5328 mix ST1, if unreinforced. It should be laid on clean broken brick or similar inert hardcore not containing harmful quantities of water-soluble sulphates or other materials which might damage the concrete. (The Building Research Establishment suggests that over 0.5% of water-soluble sulphates would be a harmful quantity.)
  Alternatively, the ground surface may be covered with at least 50 mm of concrete, as described above, or inert fine aggregate, laid on a polythene DPM as described for ground supported floors above. The joints should be sealed and the membrane should be laid on a protective bed.
- The ground covering material should be laid so that *either* its top surface is not below the highest level of the ground adjoining the building *or* it falls to an outlet above the lowest level of the adjoining ground.
- There should be a space above the top of the concrete of at least 75 mm to any wall-plate and 150 mm to any suspended timber. There should be ventilation openings in two opposing external walls allowing free ventila-

tion to all parts of the floor. An actual ventilation area equivalent to 1500 mm^2 per metre run of wall should be provided and any ducts needed to convey ventilating air should be at least 100 mm in diameter.

- Damp-proof courses of sheet materials, slates or engineering bricks bedded in cement mortar should be provided between timber members and supporting structures to prevent transmission of moisture from the ground.

AD C4
sec. 3
3.10

Again, the recommendations of Clause 11 of CP 102: 1973 may be used instead of the above, especially if the floor has a highly vapour-resistant finish.

AD C4
sec. 3
3.11

Suspended concrete ground floors

Moisture should be prevented from reaching the upper surface of the floor and the reinforcement should be protected against moisture if the construction is to be considered satisfactory.

AD C4
sec. 3
3.12

Suspended concrete ground floors may be of pre-cast construction with or without infilling slabs or they may be cast in-situ. A damp-proof membrane should be provided if the ground below the floor has been excavated so that it is lower than outside ground level and it is not effectively drained.

The space between the underside of the floor and the ground should be ventilated where there is a risk that an accumulation of gas could cause an explosion. The space should be at least 150 mm in depth (measured from the ground surface to the underside of the floor or insulation, if provided) and the ventilation recommendations should be as for suspended timber floors (see bottom of previous page). These recommendations are summarised in Fig. 8.9.

AD C4
sec. 3
3.13, 3.14

Protection of walls against moisture from the ground

The term *wall* means vertical construction, which includes piers, columns and parapets and may include chimneys if they are attached to the building. Windows, doors and other openings are not included.

Walls should be constructed so that:

- the passage of moisture from the ground to the inside of the building is resisted (this does not apply to excepted buildings),
- they will not be adversely affected by moisture from the ground,
- they will not transmit moisture from the ground to another part of the building that might be damaged.

AD C4
sec. 4
4.1, 4.2

The requirements mentioned above can be met for internal and external walls by providing a damp-proof course of suitable materials in the required position.

Figure 8.10 illustrates the main provisions, which are summarised below:

AD C4
sec. 4
4.3

- The damp-proof course may be of any material that will prevent moisture movement. This would include bituminous sheet materials, engineering bricks or slates laid in cement mortar, polythene or pitch polymer materials.
- The damp-proof course and any damp-proof membrane in the floor should be continuous.
- Unless an external wall is suitably protected by another part of the building,

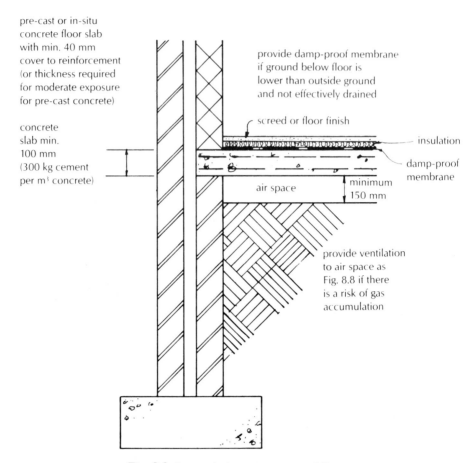

pre-cast or in-situ
concrete floor slab
with min. 40 mm
cover to reinforcement
(or thickness required
for moderate exposure
for pre-cast concrete)

concrete
slab min.
100 mm
(300 kg cement
per m³ concrete)

provide damp-proof membrane
if ground below floor is
lower than outside ground
and not effectively drained

screed or floor finish

insulation

damp-proof
membrane

air space

minimum
150 mm

provide ventilation
to air space as
Fig. 8.8 if there
is a risk of gas
accumulation

Fig. 8.9 Suspended concrete ground floors.

the damp-proof course should be at least 150 mm above outside ground
level.

- Where a damp-proof course is inserted in an external cavity wall, the cavity
should extend at least 150 mm below the lowest level of the damp-proof
course. However, where a cavity wall is built directly off a raft foundation,
ground beam or similar supporting structure, it is impractical to continue
the cavity down 150 mm. The supporting structure should therefore be
regarded as bridging the cavity, and protection be provided by a flashing or
damp-proof course as required (see Fig. 8.10).

AD C4
sec. 4
4.4

Alternatively, the provisions for protection of walls against moisture from
the ground may be met by following the relevant recommendations of Clauses
4 and 5 of BS 8215: 1991 *Code of practice for design and installation of damp-
proof courses in masonry construction.* BS 8102: 1990 *Code of practice for
protection of structures against moisture from the ground* may be followed
especially in the case of walls (including basement walls) subjected to ground
water pressure.

AD C4
sec. 4
4.5

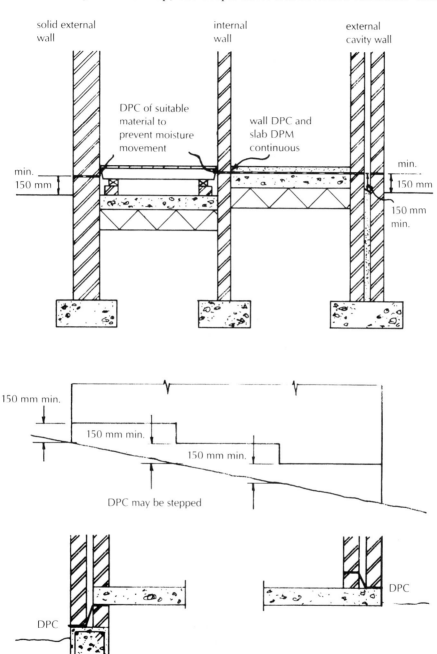

Fig. 8.10 Protection of walls against moisture from the ground.

Weather resistance of external walls

In addition to resisting ground moisture, external walls should:

- resist the passage of rain or snow to the inside of the building (this does not apply to excepted buildings),
- not be damaged by rain or snow,

AD C4
sec. 4
4.2

- not transmit moisture due to rain or snow to another part of the building that might be damaged.

There are a number of forms of wall construction which will satisfy the above requirements:

- A solid wall of sufficient thickness holds moisture during bad weather until it can be released in the next dry spell.
- An impervious cladding prevents moisture from penetrating the outside face of the wall.
- The outside leaf of a cavity wall holds moisture in a similar manner to a solid wall, the cavity preventing any penetration to the inside leaf.

AD C4
sec. 4
4.6

These principles are illustrated in Fig. 8.11.

Solid external walls

AD C4
sec. 4
4.7

The construction of a solid external wall will depend on the severity of exposure to wind-driven rain. This may be assessed for a building in a given area by using BSI Draft for Development DD93: 1984: *Method for assessing exposure to wind-driven rain.* Reference may also be made to BS 5628 *Code of practice for use of masonry*, Part 3: 1985 *Materials and components, design and workmanship* and the publication by the BRE entitled *Thermal insulation – avoiding the risks.* (See also Chapter 16, p.16.6.)

In conditions of *very* severe exposure it may be necessary to use an external cladding. However, in conditions of severe exposure a solid wall may be constructed as shown in Fig. 8.12. The following points should also be considered:

- The brickwork or blockwork should be rendered or given an equivalent form of protection.
- Rendering should have a textured finish and be at least 20mm thick in two coats. This permits easier evaporation of moisture from the wall.
- The bricks or blocks and mortar should be matched for strength to prevent cracking of joints or bricks and joints should be raked out to a depth of at least 10 mm in order to provide a key for the render.
- The render mix should not be too strong or cracking may occur. A mix of 1:1:6 cement:lime:sand is recommended for all walls except those constructed of dense concrete blocks where 1:$\frac{1}{2}$:4 should prove satisfactory.

    Further details of a wide range of render mixes may be obtained from BS 5262: 1976 *Code of practice. External rendered finishes.*
- Where the top of a wall is unprotected by the building structure it should be protected to resist moisture from rain or snow. Unless the protection and joints form a complete barrier to moisture, a damp-proof course should also be provided.

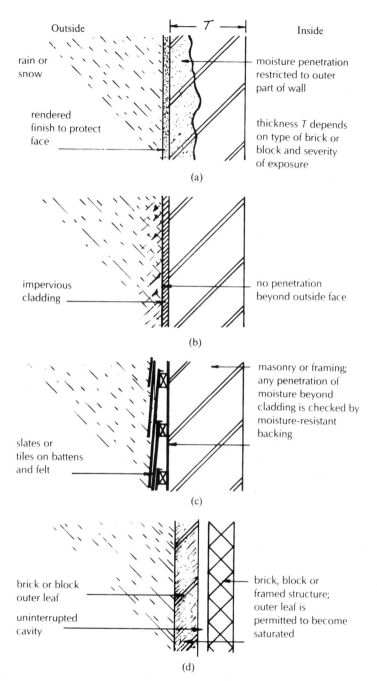

Fig. 8.11 Weather resistance of external walls – principles. (a) Solid external wall. (b) Impervious cladding. (c) Weather-resistant cladding. (d) Cavity wall.

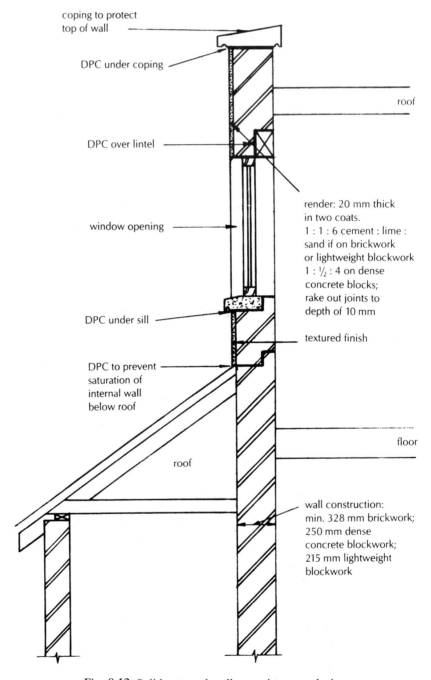

coping to protect
top of wall

DPC under coping

roof

DPC over lintel

render: 20 mm thick
in two coats.
1 : 1 : 6 cement : lime :
sand if on brickwork
or lightweight blockwork
1 : ½ : 4 on dense
concrete blocks;
rake out joints to
depth of 10 mm

window opening

DPC under sill

textured finish

DPC to prevent
saturation of
internal wall
below roof

floor

roof

wall construction:
min. 328 mm brickwork;
250 mm dense
concrete blockwork;
215 mm lightweight
blockwork

Fig. 8.12 Solid external walls – moisture exclusion.

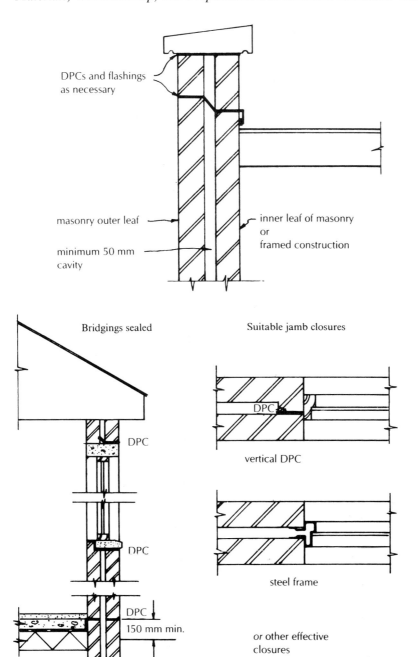

Fig. 8.13 Prevention of dampness in cavity walls.

- Damp-proof courses should be provided to direct moisture towards the outside face of the wall in the positions shown in Fig. 8.12.
- Insulation to solid external walls may be provided on the inside or outside of the wall. Externally placed insulation should be protected unless it is able to offer resistance to moisture ingress so that the wall may remain reasonably dry (and the insulation value may not be reduced). Internal insulation should be separated from the wall construction by a cavity to give a break in the path for moisture. (Some examples of external wall insulation are given in Fig. 8.14 below.)

The performance requirements for solid and cavity external walls can also be met by complying with BS 5628 *Code of practice for use of masonry*, Part 3: 1985 *Materials and components, design and workmanship* or BS 5390: 1976 *Code of practice for stone masonry*.

External cavity walls

In order to meet the performance requirements, an external cavity wall should consist of an internal leaf which is separated from the external leaf by:

(a) a drained air space or,
(b) some other method of preventing moisture from rain or snow reaching the inner leaf.

An external cavity wall may consist of the following:

- An outside leaf of masonry (brick, block, natural or reconstructed stone).
- Minimum 50 mm uninterrupted cavity. Where a cavity is bridged (by a lintel, etc.) a damp-proof course or tray should be inserted in the wall so that the passage of moisture from the outer to the inner leaf is prevented. This is not necessary where the cavity is bridged by a wall tie, or where the bridging occurs, presumably, at the top of a wall and is then protected by the roof. Where an opening is formed in a cavity wall, the jambs should have a suitable vertical damp-proof course or the cavity should be closed so as to prevent the passage of moisture.

- An inside leaf of masonry or framing with suitable lining.

These features are illustrated in Fig. 8.13 opposite.

- Where a cavity is only partially filled with insulation, the remaining cavity should be at least 50mm wide (see Fig. 8.14).

Alternatively, the relevant recommendations of BS 5628 *Code of practice for use of masonry*, Part 3: 1985 *Materials and components, design and workmanship* may be followed. Factors affecting rain penetration of cavity walls are indicated in the Code.

Weather resistance and cavity insulation

Since the installation of cavity insulation effectively bridges the cavity of a cavity wall and could give rise to moisture penetration to the inner leaf, it is

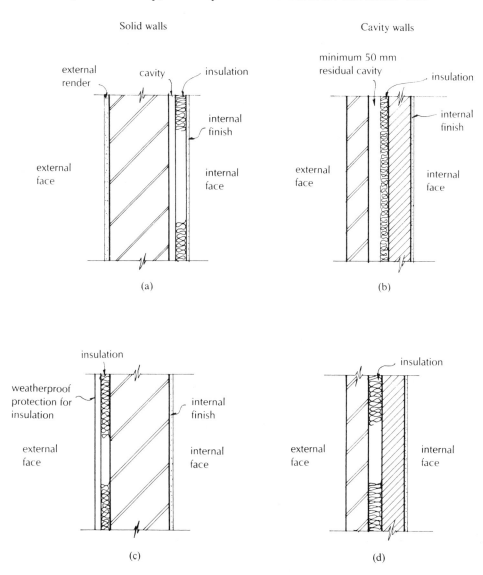

Solid walls

Cavity walls

(a)

(b)

(c)

(d)

Fig. 8.14 Weather resistance and insulation of external walls.
(a) Internal insulation. (b) Partially filled cavity. (c) External insulation.
(d) Fully-filled cavity.

most important that it be carried out correctly and efficiently. AD C4 lists a number of British Standards, Codes of Practice and other documents that cover the various materials that may be incorporated into a cavity wall:

**AD C4
sec. 4
4.14**

- Rigid materials which are built in as the wall is constructed should be the subject of a current British Board of Agrément Certificate or European Technical Approval and the work should be carried out to meet the requirements of that document.

- Urea-formaldehyde foam inserted after the wall has been constructed should comply with BS 5617: 1985 *Specification for urea-formaldehyde (UF) foam systems, etc.* and should be installed in accordance with BS 5618: 1985 *Code of practice for thermal insulation of cavity walls, etc.*
- Other insulating materials inserted after the wall has been constructed should comply with BS 6232 *Thermal insulation of cavity walls by filling with blown man-made mineral fibre*, Parts 1 and 2: 1982, or be the subject of a current British Board of Agrément Certificate, or European Technical Approval. The work should be carried out to meet the requirements of the relevant document by operatives directly employed by the document holder. Alternatively, they may be employed by an installer approved to operate under the document.
- Where materials are inserted into a cavity after the wall has been constructed, the suitability of the wall for filling should be assessed, before installation, in accordance with BS 8208 *Guide to assessment of suitability of external cavity walls for filling with thermal insulants*, Part 1: 1995 *Existing traditional cavity construction*, and the person carrying out the work should operate under a current BSI Certificate of Registration of Assessed Capability or a similar document issued by an equivalent body.

(See Fig. 8.14 for further information regarding the insulation of external walls.)

Claddings for external walls

The principles of external claddings are illustrated in Fig. 8.11(b) and (c) above. Therefore the cladding should be either:

(a) jointless (or have sealed joints) and be impervious to moisture (such as sheets of metal, glass, plastic or bituminous materials); *or*
(b) have overlapping dry joints and consist of impervious or weather-resisting materials (such as natural stone or slate, cement based products, fired clay or wood).

Dry jointed claddings should be backed by a material (such as sarking felt) which will direct any penetrating moisture to the outside surface of the structure.

Moisture-resisting materials consisting of bituminous or plastic products lapped at the joints are permitted but they should be permeable to water vapour unless there is a ventilated space behind the cladding.

Materials that are jointless or have sealed joints should be designed to accommodate structural and thermal movement.

Dry joints between cladding units should be designed either to resist moisture penetration or to direct any moisture entering them to the outside face of the structure. The suitability of dry joints will depend on the design of the joint and cladding and the severity of exposure of the building.

All external claddings should be securely fixed.

AD C4
sec. 5
5.1 to 5.7

Some materials, such as timber claddings, are subject to rapid deterioration unless properly treated. These materials should only be used as the weather-resisting part of a roof or wall if they can meet the conditions specified in AD Regulation 7 described earlier in this chapter. It should be noted that the weather-resisting part of a roof or wall does not include paint or any surface rendering or coating which does not of itself provide all the weather resistance.

Insulation may be incorporated into the roof or wall cladding provided that it is protected from moisture (or is unaffected by it). Possible problems may arise due to interstitial condensation and cold bridges in the construction. Further guidance on this may be found in Ventilation (Chapter 11) and in the BRE publication *Thermal insulation – avoiding risks*.

AD C4 sec. 5 5.8

Weather resistance of roofs

The roof of a building should:

- resist the passage of rain or snow to the inside of the building,
- not be damaged by rain or snow,
- not transmit moisture due to rain or snow to another part of the building that might be damaged.

The requirements for external wall claddings mentioned above apply equally to roof covering materials.

The performance requirements for external wall and roof claddings can also be met if they comply with:

AD C4 sec. 5 5.1 to 5.8

- British Standard Code of Practice 143 *Code of practice sheet roof and wall coverings* (this includes recommendations for aluminium, zinc, galvanised corrugated steel, copper and semi-rigid asbestos bitumen sheet).
- BS 6915: 1988 *Specification for design and construction of fully supported lead sheet roof and wall coverings.*
- BS 5247 *Code of practice for sheet roof and wall coverings*, Part 14: 1975 *Corrugated asbestos-cement.*
- BS 8200: 1985 *Code of practice for design of non-loadbearing external vertical enclosures of buildings.*

AD C4 sec. 5 5.9

The following codes refer to walls only:

- British Standard Code of Practice 297: 1972 *Precast concrete cladding (non-loadbearing).*
- BS 8298: 1989 *Code of practice for design and installation of natural stone cladding and lining.*

The above documents describe the materials to be used and contain design guidance including fixing recommendations.

Chapter 9

Toxic substances

Introduction

In recent years there has been evidence to suggest that fumes from urea-for-maldehyde foam, when used as a cavity wall filling, can have an adverse effect on the health of people occupying the building. This is still a contentious issue, but has nevertheless become a subject for building control.

Cavity insulation

Where insulating material is inserted into a cavity in a cavity wall reasonable precautions must be taken to prevent toxic fumes from penetrating occupied parts of the building.

It should be noted that Approved Document D1 does not require total exclusion of formaldehyde fumes from buildings but merely that these should not increase to an irritant concentration.

The inner leaf of the cavity wall should provide a continuous barrier to the passage of fumes and for this purpose it should be of brick or block construction.

Before work is commenced the wall should be assessed for suitability in accordance with BS 8208 *Guide to assessment of suitability of external walls for filling with thermal insulants* Part 1: 1985 *Existing traditional cavity construction*.

The work should be carried out by a person holding a current BSI Certificate of Registration of Assessed Capability for this particular type of work.

The urea-formaldehyde foam should comply with the requirements of BS 5617: 1985 *Specification for urea-formaldehyde (UF) foam systems etc.* and the installation with BS 5618: 1985 *Code of practice for thermal insulation of cavity walls etc.*

Regs Sch. 1
D1

AD D1

AD D1
1.1
AD D1
1.2(a)

AD D1
1.2(b)

AD D1
1.2(c)

AD D1
1.2(d)
AD D1
1.2(e)

Chapter 10

Sound insulation

Introduction

Sound insulation requirements are covered by Part E of Schedule 1 to the 1991 Regulations. This Part applies only to certain separating walls and floors of dwellings. The purpose of the provisions is to control sound from adjoining parts of buildings. There is no control over sound entering a dwelling through external walls.

The accompanying Approved Document, AD E, gives much more detailed guidance over a wider range of wall and floor constructions than in any previous regulations. The basic requirements for *resistance* to airborne and/or impact sound, however, remain unchanged. This is interesting, since a wide body of opinion is in favour of removing sound insulation from building regulation control. The provision of sound insulation can only be justified if it can be shown that its omission would put at risk the health of the occupants of the building. Since sound levels that would cause actual physical damage are not usually encountered between dwellings, the concern must be for mental rather than physical health.

Therefore, two questions arise:

(i) Are separate requirements for sound insulation in dwellings necessary in building regulations?

Fire resistance and structural stability provisions already require non-combustible, imperforate construction for both separating walls and floors. Since the external walls and roof of a dwelling are not required to be sound resisting a high sound level from traffic, etc. is unavoidable.

(ii) If separate sound insulation requirements are thought to be necessary, how is it possible to judge a reasonable standard of resistance to airborne and/or impact sound?

In the past the standard for resistance to airborne sound for separating walls has been that achieved by a solid brick or block wall with an average mass of $415\,\text{kg/m}^2$ and 12.5 mm of plaster each side. It is interesting to note that this has been reduced to $375\,\text{kg/m}^2$ in AD E. People vary in their response to unwanted sound and the standard mentioned above may prove totally inadequate for many people, especially those who are studying or are unwell.

In fact, the scope of Part E has been extended and now applies to the material change of use of a building to a dwellinghouse or flat (see p. 2.14). Effectively, this means that the requirements of Part E will apply where the material change of use occurs and a dwelling is formed adjacent to or within another building.

Walls

A wall which

**Regs Sch. 1
E1**

(a) separates any dwelling from another dwelling or from another building; *or*,

(b) separates any habitable room or kitchen in a dwelling from any other part of the same building which is not used exclusively with that dwelling;

must be constructed so as to provide resistance to the transmission of airborne sound (see Fig. 10.1).

The Building Regulations are concerned with two types of sound sources in buildings:

**AD E
Introduction
0.4**

- Airborne sources, such as musical instruments, speech or audio-equipment, set up vibrations in the surrounding air which impinge on enclosing walls and floors. These are set into vibration and cause the air to vibrate beside them thereby transmitting the sound to an adjoining space on the other side of the wall or floor. Vibrations may also be transmitted via elements connected to the wall or floor.
- Impact sources such as footsteps or heavy machinery set up vibrations directly in the elements that they strike which then transmit the vibrations to adjacent spaces.

The Building Regulations require walls to provide resistance only to the transmission of *airborne* sound. Floors must reduce *airborne* sound and also, if they are above a dwelling, *impact* sound.

Whatever the source of sound, it may be transmitted through an element in two ways:

**AD E
Introduction
0.6 to 0.8**

- Direct transmission – sound is transmitted directly through the element from one side to the other. This may be reduced by using heavy or stiff materials which are not set into vibration easily or by forming a break in the construction such as in a cavity wall. This relies on the structural isolation of the leaves. However, since the lower stiffness in cavity walls tends to offset the benefit of isolation, they will need to be at least as heavy as solid masonry construction.

**AD E
Introduction
0.11 to 0.13,
0.15**

- Flanking (indirect) transmission – sound is transmitted via flanking elements (i.e. walls, roofs and floors which abut the sound-resisting element). Sound may travel via solid elements or cavities at junction points in the construction. Adequate detailing of junctions is essential in order to reduce flanking transmission and at these points the flanking element either needs sufficient mass to provide resistance to sound waves, or it should be divided

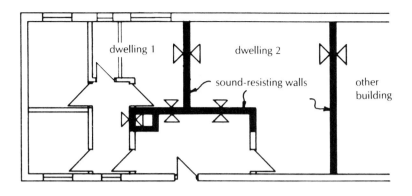

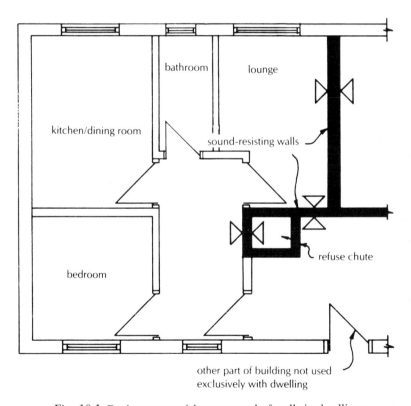

Fig. 10.1 Resistance to airborne sound of walls in dwellings.

up by windows or similar openings into small lengths which do not vibrate freely. The introduction of steps and staggers between buildings can also be of assistance in reducing flanking transmission as can careful consideration of room layouts.

AD E provides three ways of satisfying the regulation requirements for airborne sound insulation of walls:

(i) A series of examples of widely used forms of construction is given which may be adopted by the building designer; *or,*

(ii) A form of construction may be repeated which has been used in a similar building and has been shown by field tests to be acceptable; *or,*

(iii) A part of the construction may be tested in a specified type of acoustic chamber.

Typical wall constructions – new buildings

AD E
sec. 1
1.1, 1.2

Four main types of wall construction are described in section 1 of AD E – solid masonry, cavity masonry, solid masonry core with freestanding panels, and timber frames with absorbent material in the enclosed air space. Each wall construction is specified together with details of how to avoid flanking transmission at junctions between elements.

Solid masonry wall (Type 1) – direct transmission
This type of wall depends mainly on its own weight to resist airborne sound transmission. However, air paths through the wall due to poor workmanship (joints not filled properly and poor bonding at junctions with other elements) can reduce the effectiveness of the wall as a barrier to sound. Table 10.1 gives details of five commonly used solid wall constructions which will give adequate resistance to direct sound transmission.

Solid masonry wall (Type 1) – flanking transmission
It is most important to pay particular attention to detail at junctions between the sound-resisting wall and other elements of the construction. Intermediate floor, ceiling, roof space and external wall junctions all provide weak points where flanking transmission may take place. Figures 10.2 and 10.3 illustrate the precautions that should be taken at these points of junction. The following should also be noted:

AD E
sec. 1
wall Type 1

- Solid lightweight external masonry walls (i.e. less than $120 \, \text{kg/m}^2$) should be divided up by windows or similar openings into small sections which do not vibrate strongly at low frequencies to cause flanking transmission.
- Where lightweight aggregate blocks with a density less than $1200 \, \text{kg/m}^3$ are used to reduce weight in a roof space, these should be treated on one side with cement paint or a plaster skim coat to seal any possible air paths.
- There are no restrictions on the use of masonry as the outside leaf of an external cavity wall.
- No special precautions are necessary at the junction between a sound-resisting wall and a ground floor.
- There are no restrictions on the way in which a partition wall meets a Type 1 sound-resisting wall.
- Floor joists should be carried on joist hangers and not built in unless good workmanship can be ensured. If built in, care should be taken to ensure that there are no air paths through the wall around joist bearings.

AD E
sec. 1
wall Type 2

Cavity masonry wall (Type 2) – direct transmission
This type of wall depends partly on its own weight and partly on the degree of isolation between the leaves in order to resist airborne sound transmission. Cavity walls do not generally behave better than solid walls of similar materials and weight.

Table 10.1 Solid masonry – wall Type 1.

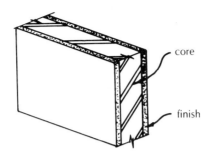

Wall type	Core	Finish	Remarks	Examples
A	Brickwork of min. mass 375 kg/m² (incl. plaster)	Min. 13 mm plaster on each face	Brick bond to include headers	215 mm brick (1610 kg/m³ density) lightweight plaster
B	Concrete blockwork min. mass 415 kg/m² (incl. plaster)	Min. 13 mm plaster on each face	Blocks to extend full thickness of wall	215 mm block (1840 kg/m³ density) lightweight plaster
C	Brickwork of min. mass 375 kg/m² (incl. plasterboard)	Min. 12.5 mm plasterboard each side	Brick bond to include headers	215 mm brick (1610 kg/m³ density)
D	Concrete blockwork min. mass 415 kg/m² (incl. plasterboard)	Min. 12.5 mm plasterboard each side	Blocks to extend full thickness of wall	215 mm block (1840 kg/m³ density)
E	Concrete in-situ or large panels (1500 kg/m³ min. density). Min. mass 415 kg/m² (incl. plaster if used).	Plaster coat optional	Fill panel joints with mortar	190 mm thick wall (2200 kg/m³ density) unplastered

Notes
Bricks should be laid frog up.
All joints should be filled and sealed with mortar.
It is permissible to use wall lining laminates of plasterboard and mineral wool in place of plaster or plasterboard in the above table.

In addition to maintaining properly filled mortar joints with bricks laid frog up, it is also important to space the leaves at least 50 mm apart and to ensure that this cavity does not become blocked. The leaves should be connected with butterfly pattern wall ties spaced as required for structural purposes (see BS 5628 *Code of practice for use of masonry*, Part 3: 1985 *Materials and components, design and workmanship* where the maximum cavity width

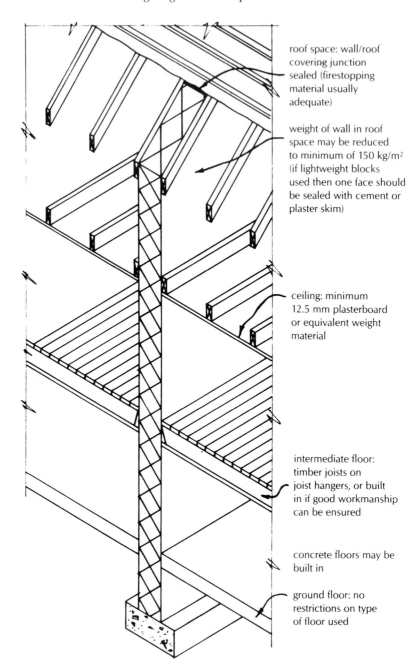

roof space: wall/roof covering junction sealed (firestopping material usually adequate)

weight of wall in roof space may be reduced to minimum of 150 kg/m^2 (if lightweight blocks used then one face should be sealed with cement or plaster skim)

ceiling: minimum 12.5 mm plasterboard or equivalent weight material

intermediate floor: timber joists on joist hangers, or built in if good workmanship can be ensured

concrete floors may be built in

ground floor: no restrictions on type of floor used

Fig. 10.2 Flanking transmission – wall and floor junctions (cavity and solid masonry walls).

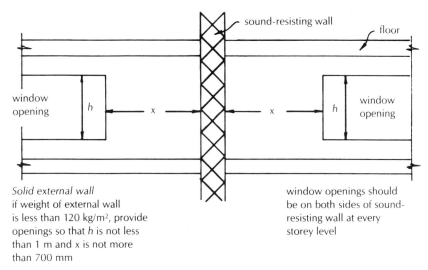

Solid external wall
if weight of external wall
is less than 120 kg/m², provide
openings so that *h* is not less
than 1 m and *x* is not more
than 700 mm

window openings should
be on both sides of sound-
resisting wall at every
storey level

Section A–B

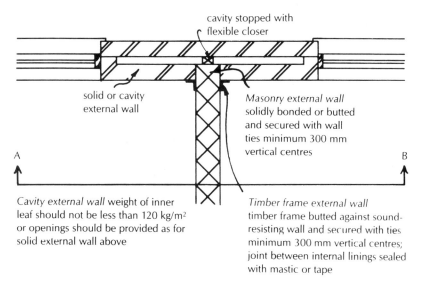

Cavity external wall weight of inner
leaf should not be less than 120 kg/m²
or openings should be provided as for
solid external wall above

Timber frame external wall
timber frame butted against sound-
resisting wall and secured with ties
minimum 300 mm vertical centres;
joint between internal linings sealed
with mastic or tape

Fig. 10.3 Flanking transmission – external wall junction (cavity and masonry external walls).

allowed with butterfly wall ties is 75 mm). Care should be taken to see that cavity insulating materials (other than unbonded particles or loose fibres) inserted in the external cavity wall do not enter the cavity in the sound-resisting wall. (It may be necessary to insert a flexible closer such as mineral wool to prevent this.) The cavity should be maintained up to the underside of the roof.

Table 10.2 gives details of three commonly used cavity wall constructions which will give adequate resistance to direct sound transmission. Additionally,

Table 10.2 Cavity masonry – wall Type 2.

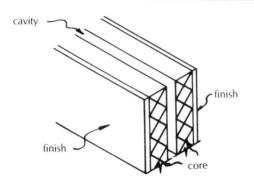

Wall type	Core	Finish	Remarks	Examples
A	Brickwork in two leaves of min. mass 415 kg/m² (incl. plaster)	Min. 13 mm plaster on each face	Min. 50 mm cavity	102 mm brick leaves (1970 kg/m³ density) lightweight plaster
B	Concrete blockwork in two leaves of min. mass 415 kg/m² (incl. plaster)	Min. 13 mm plaster on each face	Min. 50 mm cavity	100 mm block leaves (1990 kg/m³ density) lightweight plaster
C	Two leaves lightweight aggregate blockwork (max. density 1600 kg/m³). Mass incl. finish 300 kg/m².	Min. 12.5 mm plasterboard or 13 mm plaster each side	Min. 75 mm cavity. Use dry lining with lightweight block[a]	100 mm block leaves (1371 kg/m³ density) lightweight plaster
	*Additional constructions to be used where minimum 300 mm step and/or stagger present*			
D	Concrete blockwork in two leaves of min. mass 415 kg/m² (excl. finishes)	Min. 12.5 mm plasterboard each side	Min. 50 mm cavity	100 mm block leaves (1990 kg/m³ density)
E	Two leaves lightweight aggregate blockwork (max. density 1600 kg/m³). Mass incl. finish 250 kg/m².	Min. 12.5 mm plasterboard or 13 mm plaster each side	Min. 75 mm cavity	100 mm block leaves (1105 kg/m³ density) lightweight plaster

Notes
[a] A denser block should not be used with dry lining since the composition of the lightweight aggregate blocks contributes to the performance of this construction

two other constructions are given which may be used where a step and/or stagger of at least 300 mm is present. It is permissible to use wall lining laminates of plasterboard and mineral wool, in place of plaster or plasterboard in Table 10.2.

Cavity masonry wall (Type 2) – flanking transmission
The requirements listed above for solid masonry walls also apply, generally, to cavity masonry walls. The following should also be noted:

AD E
sec. 1
wall Type 2

- Where a concrete intermediate floor construction is used, it should be carried through to the cavity face of the wall leaf.
- A ground bearing concrete slab may be continuous.
- The weight of the inner leaf of an external cavity wall should be at least 120 kg/m^2 unless the sound-resisting wall is of type B (see Table 10.2).
- At the junction between a cavity external wall and a cavity sound-resisting wall the air path should be stopped with a flexible closer, such as mineral wool, to minimise the transmission of sound via the cavities.

(See Figs. 10.2 and 10.3 for details.)

Solid masonry core with freestanding lightweight panels (wall Type 3) – direct transmission
The resistance to airborne sound for this type of wall depends partly on the mass and type of core and partly on the isolation and mass provided by the panels.

AD E
sec. 1
wall Type 3

Again, it is important to ensure that mortar joints in the masonry core are properly filled with bricks laid frog up and that junctions with other elements are solidly bonded. The lightweight panels should be supported at floor and ceiling level only and not tied to the masonry core if adequate isolation is to be achieved. Table 10.3 gives details of four commonly used masonry core constructions and two types of lightweight panel.

Solid masonry core with freestanding lightweight panels (wall Type 3) – flanking transmission
The requirements listed above for solid masonry walls also apply, generally, to the solid masonry core in this form of construction. Additionally, extra precautions need to be taken at the junctions between the lightweight panels and adjacent elements. These precautions are summarised below and illustrated in Fig. 10.4:

AD E
sec. 1
wall Type 3

- In the roof space the gap between the ceiling and the masonry core should be sealed with a timber batten. The panel should also be sealed to the ceiling on the underside with mastic, tape or coving. Where core type D is used, the construction (including the cavity) should be maintained.
- At intermediate floors, timber joists may be carried on joist hangers supported by the masonry core. The spaces between the joists should be sealed with full-depth timber blocking. Again, the gap between ceiling and panel should be sealed with tape, mastic or coving, etc.
- Concrete intermediate floors with a minimum mass of 365 kg/m^2 may be carried through the sound-resisting wall. However, if the core is of type D, the cavity should not be bridged. The panel-to-ceiling junction should be sealed as above.

Table 10.3 Masonry core with lightweight panels – wall Type 3.

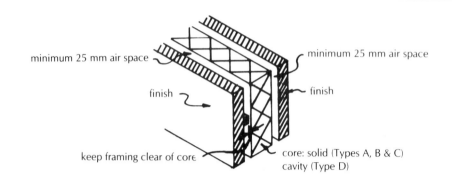

minimum 25 mm air space

minimum 25 mm air space

finish

finish

keep framing clear of core

core: solid (Types A, B & C)
cavity (Type D)

Wall type	Core	Finish	Example
A	Solid brickwork of min. mass 300 kg/m²	Panel type E or F[a]	215 mm brick core, 1290 kg/m³ density, 75 mm coursing
B	Solid concrete blockwork of min. mass 300 kg/m²	Panel type E or F[a]	140 mm block core, 2200 kg/m³ density, 110 mm coursing
C	Lightweight concrete blockwork (max. density 1600 kg/m³). Mass 160 kg/m².	Panel type E or F[a]	200 mm block core, 730 kg/m³ density, 225 mm coursing
D	Cavity blockwork or brickwork of any mass	Panel type E or F[a]	100 mm brick or block leaves, min. 50 mm cavity. Butterfly wall ties.

[a]Panel type	Construction	Remarks
E	Two sheets plasterboard joined by cellular core of min. mass 18 kg/m²	Panel joints taped. Panel spaced min. 25 mm from core. Fix to ceiling and floor only.
F	Two sheets plasterboard with or without framework. Each sheet min. 12.5 mm thick. Overall thickness min. 30 mm if no supporting framework.	Joints between sheets staggered. Sheets fixed min. 25 mm from core. Framing kept clear of core.

- A suspended concrete ground floor should only bear on the core if it has a mass of at least 365 kg/m². Floors of this mass are also permitted to pass through the sound-resisting wall.
- There are no restrictions on the construction of the outer leaf of an external cavity wall.

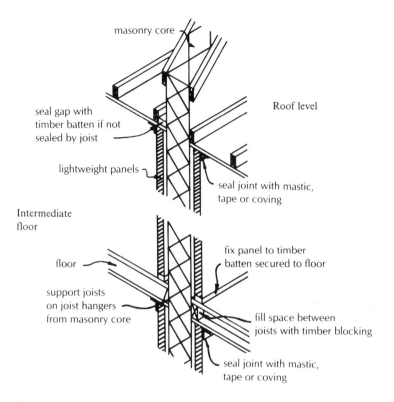

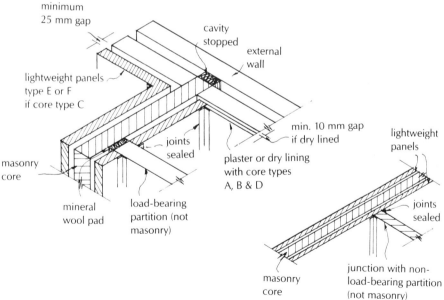

Fig. 10.4 Flanking transmission – wall Type 3.

- There are no restrictions on the construction of the inner leaf of an external cavity wall if it is lined with isolated panels such as types E and F, and it should be so lined if the core of the sound-resisting wall is of type C.
- The inner leaf of the external cavity wall may be finished with plaster or dry lining if the sound-resisting wall has a type A, B or D core. In these cases the inner leaf should have a minimum mass of $120 \, kg/m^2$ and be butt jointed to the sound-resisting wall core with ties at not more than 300 mm vertical centres. The joints in the dry lining should be sealed with tape or caulking and it may be lined with insulation if the 25 mm and 10 mm gaps shown in Fig. 10.4 can be maintained.
- At the junction with a load-bearing partition, the joint should be sealed with mastic or tape. The partition should be fixed to the masonry core through a padding of mineral fibre quilt. It should *not* be of masonry construction.
- A non-load-bearing partition should be tightly butted against the light-weight panel and the joint should be filled with mastic or tape. Again, the partition should not be of masonry construction.

Timber frame (wall Type 4) – direct transmission

**AD E
sec. 1
wall Type 4**

This form of construction relies mainly on the use of two isolated frames together with a degree of sound absorption in the air space in order to mini-mise the transmission of sound.

It is essential to isolate the frames from one another as far as possible. If it is necessary for structural reasons to connect the frames together, this may be done using 14 to 15 gauge metal straps fixed at or just below ceiling level and spaced a minimum horizontal distance of 1.2 m apart.

The detailing of services is important and the following rules should be observed:

- Avoid service penetrations where possible as these may create air paths.
- Electric power points may penetrate the wall cladding if they are backed by a similar thickness of cladding behind the socket box.
- Power points should not be placed back to back.

Fire stops are necessary in this form of construction (see Chapter 7, Fire). These should preferably be flexible but if rigid they should be fixed to only one frame.

Table 10.4 gives details of two basic forms of timber frame construction.

Timber frame (wall Type 4) – flanking transmission

**AD E
sec. 1
wall Type 4**

Many of the principles already discussed for wall Type 2 also apply to timber frame walls. The main requirement is to seal the wall cavity from the rooms on each side at all junctions. Some alternative details are permitted as listed below and illustrated in Fig. 10.5:

- In the roof space both frames may be carried through and the cladding finish reduced to a minimum of 25 mm. Alternatively, the cavity may be closed at ceiling level (but not with a rigid connection), and one frame may be continued through provided it has at least 25 mm of cladding on each side.
- At floor level the cladding may be carried through the floor thickness. Alternatively any detail may be used that will block the air paths between the rooms and the wall cavity. Where the floor joists are at right angles to

Table 10.4 Timber frame – wall Type 4.

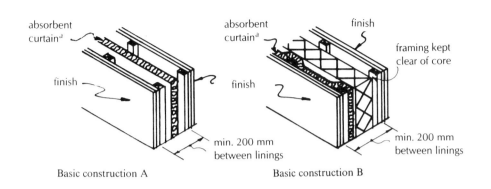

Basic construction A   Basic construction B

Wall specification	Core	Finish	Remarks
Basic construction A	—	Minimum two sheets plasterboard (with or without plywood sheathing) at least 30 mm thickness	Joints between sheets staggered to avoid air paths. Mineral fibre quilt between frames.[a]
Basic construction B	Brickwork or blockwork of any thickness	As above	As above plus core should be connected to only one frame

Notes
[a] Absorbent curtain specification: Unfaced mineral fibre quilt or batts (may be wire reinforced) of minimum density 10 kg/m³.
Minimum thickness: 25 mm if suspended in cavity between frames,
50 mm if fixed to one frame, or 25 mm per quilt if one fixed to each frame.

the sound-resisting wall, the spaces between the joists should be sealed with solid timber blocking to the full floor depth.
- A suspended ground floor slab should have a minimum mass of 365 kg/m².
- At the junction with an external cavity wall, the air path in the cavity should be blocked to minimise transmission of sound via the cavity.

Refuse chutes

Special rules apply to walls separating refuse chutes from dwellings due to the excessive noise that may be generated.

A wall which separates any habitable room or kitchen in a dwelling from any refuse chute in the same building should have an average mass (calculated over any portion of the wall measuring 1 metre square and including the mass of any plaster) of not less than 1320 kg/m².

A wall which separates any part of a dwelling, other than a habitable room, from any refuse chute in the same building, should have an average mass

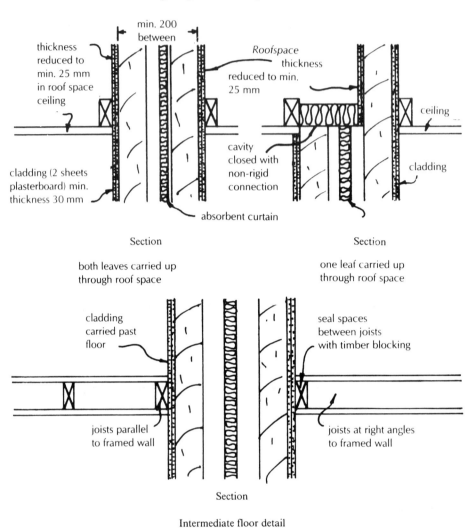

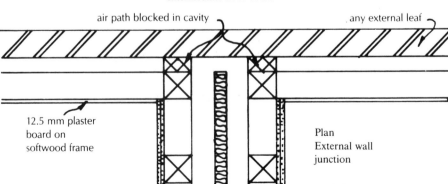

Fig. 10.5 Flanking transmission – wall Type 4.

(calculated over any portion of the wall measuring 1 metre square and including the mass of any plaster) of not less than $220 \, \text{kg/m}^2$.

Thus, if a refuse chute is placed next to a bathroom, hall, etc., the wall between need only be a half brick wall (see Fig. 10.6).

AD E
sec. 1
1.3

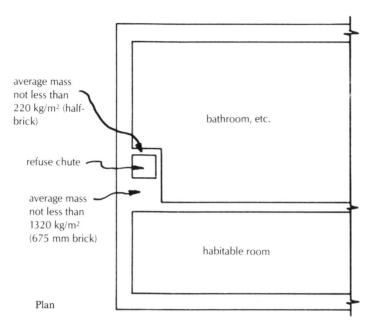

average mass
not less than
220 kg/m² (half-
brick)

bathroom, etc.

refuse chute

average mass
not less than
1320 kg/m²
(675 mm brick)

habitable room

Plan

Fig. 10.6 Refuse chute separation.

Floors

Airborne sound

A floor or stair which

Regs Sch. 1
E2

(a) separates a dwelling from another dwelling; or,
(b) separates a dwelling from another part of the same building which is not used exclusively with that dwelling;

must be so constructed as to provide resistance to the transmission of *airborne* sound.

Impact sound

A floor or stair *above* a dwelling which:

Regs Sch. 1
E3

(a) separates it from another dwelling; *or,*
(b) separates it from another part of the same building which is not used exclusively with that dwelling;

must be constructed so as to provide resistance to the transmission of impact sound. These requirements are illustrated in Fig. 10.7.

AD E provides two ways of satisfying the requirements of the Building Regulations for airborne and impact sound insulation of floors:

(i) A series of examples of widely used forms of construction is given which may be adopted by the building designer; *or,*

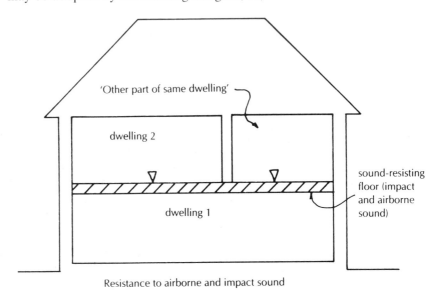

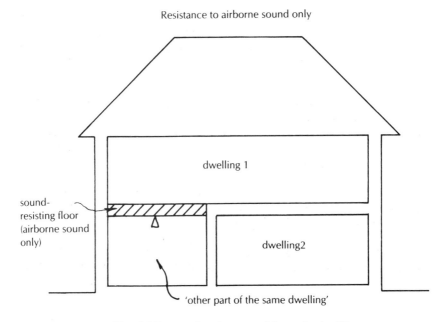

Fig. 10.7 Sound resistance of floors in dwellings.

(ii) A form of construction may be repeated which has been used in a similar building and has been shown by field tests to be acceptable.

The acoustic test chamber method for assessing the adequacy of wall constructions does not apply to floors.

Typical floor constructions

Three main types of floor construction are described in Section 2 of AD E – concrete base with soft covering, concrete base with floating layer, and timber base with floating layer. A selection of examples of each floor construction is given together with details of how to avoid flanking transmission at junctions between elements.

**AD E
sec. 2
2.1**

Concrete based floors (Types 1 and 2) – direct and flanking transmission
The weight of the concrete base (including any bonded screed, ceiling finish or floating layer) is the principal factor in determining resistance to airborne sound for these floor constructions. Impact sound is reduced by the soft covering or floating layer. Where airborne sound insulation alone is required, the soft covering may be omitted from floor Type 1. Table 10.5 gives details of four common floor base constructions which will satisfy the requirements for resistance to direct sound transmission.

**AD E
sec. 2
Floor types
1 and 2**

The soft covering should consist of a resilient material (i.e. one that will return to its original thickness after being compressed) at least 4.5 mm thick. No guidance is given in AD E concerning the actual material to be used; however suitable resilience will be provided if the floor covering has a weighted impact sound improvement (ΔL_W) of at least 17 dB when calculated in accordance with Annex A of BS 5821 *Methods for rating the sound insulation in buildings and of building elements*, Part 2: 1984 *Method for rating the impact sound insulation*. The floating layer should consist of a timber raft or screed on a layer of mineral fibre at least 25 mm thick (with a minimum density of 36 kg/m^3). This may be reduced to 13 mm under a timber raft where the supporting battens have an integral closed cell resilient foam strip. A screeded finish may also rest on:

- 13 mm precompressed expanded polystyrene boarding provided that this is of impact sound duty grade; *or*
- 5 mm extruded (closed cell) polyethylene foam with a density of 30 kg/m^3 to 45 kg/m^3. The material should be laid over a levelling screed to protect it from damage and should have lapped joints.

Figure 10.8 shows typical details for floor Type 2. It should be noted that any of the concrete base specifications shown in Table 10.5 are suitable for use with any of the resilient or floating layers.

Flanking transmission may be avoided by paying attention to detail at the junctions between the sound-resisting floor and external walls, internal walls and sound-resisting separating walls. Pipe penetrations through floors are unavoidable and may create air paths unless properly detailed.

Floor/wall junctions
As has already been mentioned, an external wall which is divided up by

Table 10.5 Concrete base – floor Types 1 and 2.

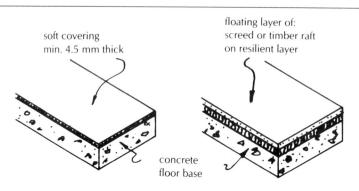

soft covering
min. 4.5 mm thick

floating layer of:
screed or timber raft
on resilient layer

concrete
floor base

Floor Type 1   Floor Type 2

Floor base specification	Floor construction Type 1[a]	Floor construction Type 2[b]	Remarks
A	Solid concrete slab of minimum mass 365 kg/m²	Solid concrete slab of minimum mass 300 kg/m²	Mass of screed or bonded ceiling finish may be included
B	Solid concrete slab with permanent shuttering of min. mass 365 kg/m²	Solid concrete slab with permanent shuttering of min. mass 300 kg/m²	Mass of concrete or metal shuttering, screed or bonded ceiling finish may be included
C	Concrete beams with infilling blocks of minimum mass 365 kg/m²	Concrete beams with infilling blocks of minimum mass 300 kg/m²	Mass of clay or concrete blocks, screed or bonded ceiling finish may be included. Fill joints between beams and blocks.
D	Hollow or solid concrete beams of minimum mass 365 kg/m²	Hollow concrete beams of minimum mass 300 kg/m²	Mass of screed or bonded ceiling finish may be included. Fill joints between beams.

Notes
[a] Full specification includes soft covering.
[b] Full specification includes floating layer.

windows or similar openings will not vibrate freely and will, therefore, be less likely to allow flanking transmission. Accordingly, if the area of the openings is less than 20% of the area of the external wall, the weight of the wall (or the inner leaf if it is a cavity wall) should not be less than 120 kg/m² (including any finishes). For areas in excess of 20% there are no restrictions.

As a general rule the concrete floor base should pass through the wall (this does not apply to the screed which should be stopped off against the wall

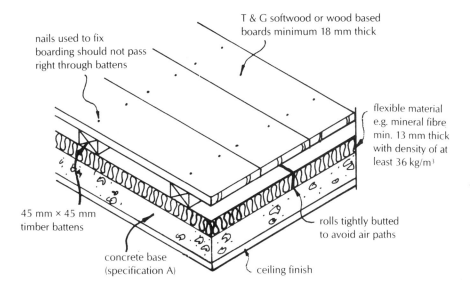

nails used to fix
boarding should not pass
right through battens

T & G softwood or wood based
boards minimum 18 mm thick

flexible material
e.g. mineral fibre
min. 13 mm thick
with density of at
least 36 kg/m³

45 mm × 45 mm
timber battens

concrete base
(specification A)

rolls tightly butted
to avoid air paths

ceiling finish

Floating floor (specification E) with resilient layer (specification G)

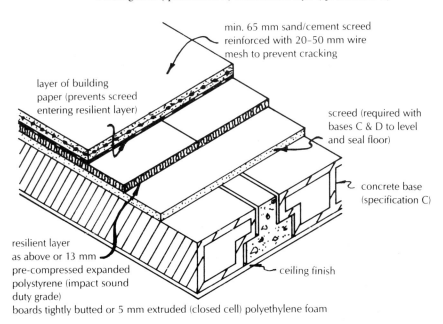

min. 65 mm sand/cement screed
reinforced with 20–50 mm wire
mesh to prevent cracking

layer of building
paper (prevents screed
entering resilient layer)

screed (required with
bases C & D to level
and seal floor)

concrete base
(specification C)

resilient layer
as above or 13 mm
pre-compressed expanded
polystyrene (impact sound
duty grade)
boards tightly butted or 5 mm extruded (closed cell) polyethylene foam

ceiling finish

Floating floor (specification F) with resilient layer (specification H)

Fig. 10.8 Sound insulation – floor Type 2.

surface). If, however, the wall is a sound-resisting or internal solid wall weighing 375 kg/m^2 or more, the floor base *or* the wall may be passed through. In the latter case the floor base should be tied to the wall and the joint grouted.

If the floor base is of concrete beam construction (i.e. specification C or D), the first joint should be at least 300 mm from the outside face of the wall (or the outside face of the inner leaf of a cavity wall).

Special rules apply to floors with floating layers (Type 2) as follows:

- The resilient layer should be carried up against the wall to isolate the floating floor.
- A nominal gap should be left between the floating floor and the skirting (or the resilient layer may be turned under the skirting). Sealing is unnecessary, but where used it should be flexible.

Penetration of floor by pipes
Where a floor separating habitable rooms is penetrated by ducts or pipes these should be in an enclosure both above and below the floor which should be:

- Constructed of material with a mass of at least 15 kg/m^2.
- Lined with 25 mm of unfaced mineral wool (alternatively the pipes or ducts may be wrapped in the same material).

A nominal gap should be left between the floating floor and the enclosure and it should be sealed with neoprene or acrylic caulking.

Pipe penetrations through a floor which separates dwellings will, of course, need to be fire stopped (see Chapter 7, p. 7.112). The fire stopping material should be flexible to prevent rigid contact between the pipe or duct and floor. The Gas Safety Regulations 1972 and the Gas Safety (Installation and Use) Regulations 1984 require ducts containing gas pipes to be ventilated at each floor level. Therefore, gas pipes should be contained in a separate ventilated duct, or they can remain unducted.

These recommendations for floor/wall junctions and pipe penetrations are illustrated in Figs. 10.9 and 10.10 below.

**AD E
sec. 2
floor Type
3** *Timber based floors (Type 3) – direct and flanking transmission*
Since a timber floor radiates sound less efficiently than a concrete floor it may be constructed of lighter materials. The timber floor specifications described in AD E rely on the weights of the elements of the structural floor, pugging and the absorbent blanket to reduce airborne sound and the floating layer to reduce impact sound. They are similar in concept to floor Type 2 but consist entirely of timber.

The three floor specifications described in AD E are similar in form, the main differences being in the positioning of the resilient strip or layer (which reduces impact sound) and in the use of an absorbent blanket or pugging (to reduce airborne sound). Figure 10.11 gives full details of each floor specification.

The pugging may be any of the materials shown in Table 10.6 below; however sand should not be used in kitchens, bathrooms, shower rooms or WC compartments since it may become wet and overload the ceiling.

Floor and wall junctions need careful detailing to avoid flanking transmission. Pipe penetrations can be dealt with as described above for Type 1 and 2 floors, although it will be necessary to seal the joint between the duct casing and the ceiling with, for example, tape or caulking.

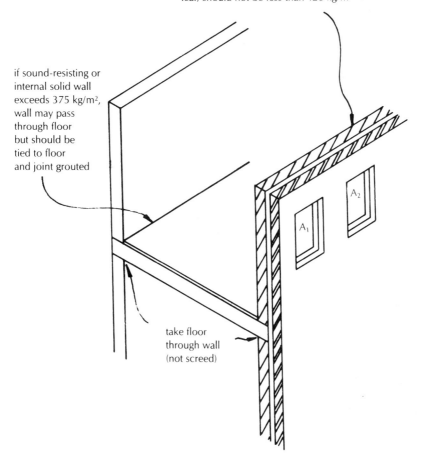

if $A_1 + A_2$ etc. does not exceed 20%
of area of external wall then
weight of external wall (or inner
leaf) should not be less than 120 kg/m²

if sound-resisting or
internal solid wall
exceeds 375 kg/m²,
wall may pass
through floor
but should be
tied to floor
and joint grouted

take floor
through wall
(not screed)

Fig. 10.9 Floor/wall junctions – general rules (Type 1 and 2 floors).

Table 10.6 Pugging materials for floor Type 3.

Material	*Sieve size (mm)*	*Thickness to achieve 80 kg/m² (mm)*
Traditional ash	—	75
Limestone chips	2–10	60
Whin aggregate	2–10	60
Dry sand	—	50

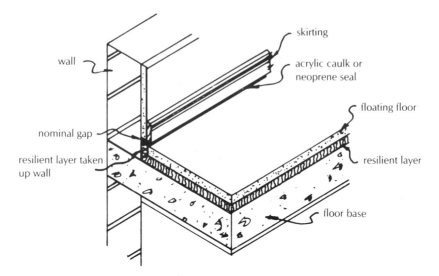

Floor/wall junction floor Type 2

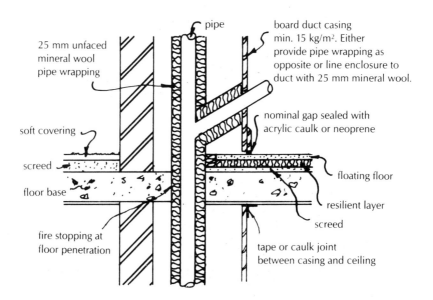

Penetration of floor by pipes

Fig. 10.10 Floor/wall junctions – details (Type 1 and 2 floors).

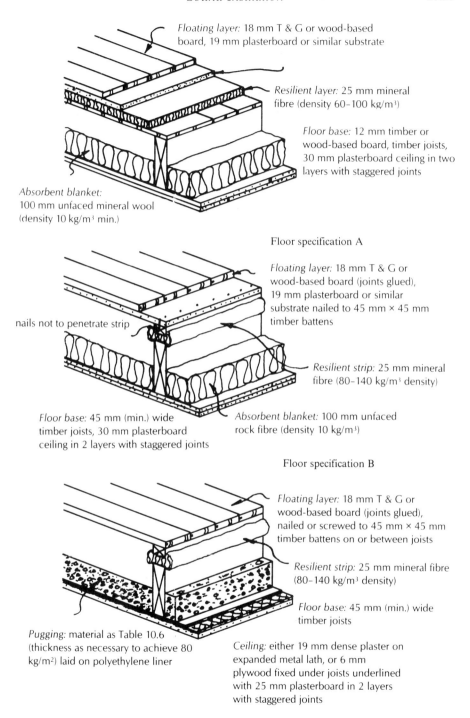

Floating layer: 18 mm T & G or wood-based board, 19 mm plasterboard or similar substrate

Resilient layer: 25 mm mineral fibre (density 60–100 kg/m³)

Floor base: 12 mm timber or wood-based board, timber joists, 30 mm plasterboard ceiling in two layers with staggered joints

Absorbent blanket: 100 mm unfaced mineral wool (density 10 kg/m³ min.)

Floor specification A

Floating layer: 18 mm T & G or wood-based board (joints glued), 19 mm plasterboard or similar substrate nailed to 45 mm × 45 mm timber battens

nails not to penetrate strip

Resilient strip: 25 mm mineral fibre (80–140 kg/m³ density)

Floor base: 45 mm (min.) wide timber joists, 30 mm plasterboard ceiling in 2 layers with staggered joints

Absorbent blanket: 100 mm unfaced rock fibre (density 10 kg/m³)

Floor specification B

Floating layer: 18 mm T & G or wood-based board (joints glued), nailed or screwed to 45 mm × 45 mm timber battens on or between joists

Resilient strip: 25 mm mineral fibre (80–140 kg/m³ density)

Floor base: 45 mm (min.) wide timber joists

Pugging: material as Table 10.6 (thickness as necessary to achieve 80 kg/m²) laid on polyethylene liner

Ceiling: either 19 mm dense plaster on expanded metal lath, or 6 mm plywood fixed under joists underlined with 25 mm plasterboard in 2 layers with staggered joints

Fig. 10.11 Timber base (with floating layer) – Type 3 floor.

The main problem with Type 3 floors is the need to isolate the floating layer from the structural part of the floor (i.e. the floor joist) and the surrounding walls. The gap between the wall and the floating layer should be sealed with a resilient strip glued to the wall. (It should be noted that for floor specification A in Fig. 10.11, the lower density figure for the resilient layer gives the best insulation. Unfortunately, this leads to a softer floor which can be provided with additional support by fixing a timber batten to the wall around the perimeter, with a foam strip along its top edge.) The junction between the ceiling and the wall lining should also be suitably sealed. Figure 10.12 illustrates the main design principles that should be adopted at wall/floor junctions for Type 3 floors. Floor specification A has been used in the diagrams as an example, but any of the three specifications would be satisfactory.

Mass of sound-resisting components

AD E
sec. 1
1.4 to 1.6
& 2.3 to 2.5

Throughout Sections 1 and 2 of AD E, reference is made to the minimum recommended mass for wall and floor constructions (including finishes where applicable). These masses are expressed in kilograms per square metre (kg/m^2) and the densities of the materials used (on which the mass of the wall depends) are expressed in kilograms per cubic metre (kg/m^3). In order that

AD E
Appendix A
A1

these masses and densities may be translated into wall thicknesses, information may be obtained from Agrément Certificates, European Technical Approvals or manufacturers. The Building Control Authority may ask for test evidence to support the claims of a manufacturer if their technical information is used, however.

AD E
Appendix A
A2

As an additional aid, Appendix A of AD E provides a simple approximate method for calculating the mass of a wall. Table A1, which is reproduced below, contains a number of formulae for different coordinating masonry course heights. These formulae are not exact but are accurate enough for the purposes of sound insulation.

The following should also be noted when using Table A1:

- Densities of bricks or blocks should be obtained from current BBA or ETA Certificates, or from manufacturers.
- The density quoted is usually the apparent density i.e. the mass of the brick or block divided by the volume including frog, perforations or voids. This is the density on which the Table A1 formulae are based.
- The mass of any plaster, render or dry lining finish should be included as appropriate.

AD E
Appendix A
A3

- Mortar joints of 10 mm and dry set mortar density of 1800 kg/m^3 are assumed in Table A1. Actual values within 10% of these figures are acceptable.

AD E
Appendix A
A4

No figures are given for concrete floors and screeds. However, Appendix A states that the mass of an in-situ concrete floor or screed may be obtained by multiplying its density by its thickness in metres.

For slabs or composite floor bases the reader is told to divide the total mass of the element by its plan area. Since the purpose of the exercise is to find the mass, this seems particularly unhelpful information.

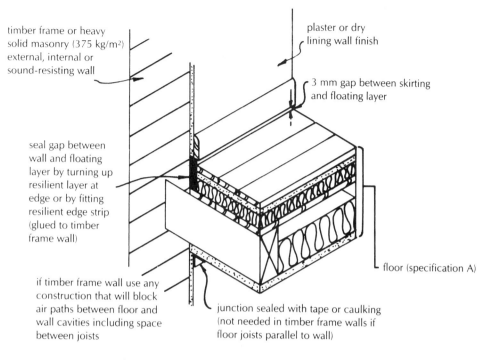

timber frame or heavy solid masonry (375 kg/m²) external, internal or sound-resisting wall

plaster or dry lining wall finish

3 mm gap between skirting and floating layer

seal gap between wall and floating layer by turning up resilient layer at edge or by fitting resilient edge strip (glued to timber frame wall)

floor (specification A)

if timber frame wall use any construction that will block air paths between floor and wall cavities including space between joists

junction sealed with tape or caulking (not needed in timber frame walls if floor joists parallel to wall)

Timber frame or heavy solid masonry wall junction

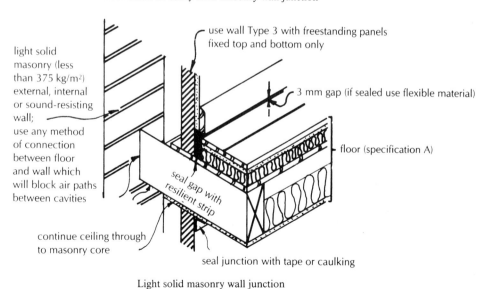

light solid masonry (less than 375 kg/m²) external, internal or sound-resisting wall; use any method of connection between floor and wall which will block air paths between cavities

use wall Type 3 with freestanding panels fixed top and bottom only

3 mm gap (if sealed use flexible material)

floor (specification A)

seal gap with resilient strip

continue ceiling through to masonry core

seal junction with tape or caulking

Light solid masonry wall junction

Fig. 10.12 Wall/floor junctions – Type 3 floor.

AD E, Appendix A

Table A1   Formulae for calculation of wall leaf mass

Coordinating height of masonry course (mm)	Formulae to be used
75	$M = T(0.79D + 380) + NP$
100	$M = T(0.86D + 255) + NP$
150	$M = T(0.92D + 145) + NP$
200	$M = T(0.93D + 125) + NP$

where

$M = $ *mass of 1 m² of leaf in kg/m²*

$T = $ thickness of masonry in metres (i.e. unplastered thickness)

$D = $ density of masonry units in kg/m³ (at 3% moisture content)

$N = $ number of finished faces (if no finish $N = 0$, if finish on one side only $N = 1$, if finish on both sides $N = 2$)

$P = $ mass of 1 m² of wall finish in kg/m² (see below)

Finishes

mass of plaster (assumed thickness 13 mm)

Cement render	29 kg/m²
Gypsum	17 kg/m²
Lightweight	10 kg/m²
Plasterboard	10 kg/m²

Repeated construction

**AD E
sec. 3
3.1**
Sections 1 and 2 of AD E are discussed above and deal with typical wall and floor construction details that may be adopted by a designer in order to satisfy Part E of Schedule 1 to the 1991 Regulations.

Section 3 of AD E provides ways of meeting the regulation requirements by repeating an existing form of construction.

In order to do this, two prerequisites must be met:

(i) the existing wall and floor constructions must have been tested to show that they are capable of achieving the sound insulation values given in Section 3 of AD E; *and,*

(ii) the existing and proposed designs must be sufficiently similar.

It should be noted that in the context of Section 3 of AD E the term 'floor' also includes a stair which has a separating function.

Existing construction – assessment of performance

In order that the performance of an existing wall or floor may be assessed it is necessary to carry out tests in accordance with the method given in BS 2750 *Measurement of sound insulation in buildings and of building elements*, Part 4: 1980 *Field measurements of airborne sound insulation between rooms* and Part 7: 1980 *Field measurements of impact sound insulation of floors.* AD E sec. 3 3.5

These tests allow the Standardised Level Differences D_{nT} for airborne sound insulation and the Standardised Impact Sound Pressure Levels L^1_{nT} for impact sound transmission to be determined.

From these values it is possible to calculate the Weighted Standardised Level Difference $D^1_{nT,w}$ for airborne sound and the Weighted Standardised Impact Sound Pressure Level $L^1_{nT,w}$ for impact sound. These terms are defined in BS 5821: *Methods for rating the sound insulation in buildings and building elements*, Part 1: 1984 *Method for rating the airborne sound insulation in buildings and of interior building elements* and Part 2: 1984 *Method for rating the impact sound insulation.*

Table 10.7 (which is based on Table 2 of AD E) below, shows the values of $D^1_{nT,w}$ and $L^1_{nT,w}$ which should be achieved. For airborne sound insulation the individual values of $D^1_{nT,w}$ should not be less than that given in the individual values column of Table 10.7. For impact sound transmission, the individual values of $L^1_{nT,w}$ should not be more than that given in the individual values column of Table 10.7. The mean values columns in Table 10.7 refer to the arithmetic mean of the individual values obtained from the sets of measurements in the tests. If it is possible to achieve only one set of measurements, the value achieved should be no worse than the mean value. AD E sec. 3 3.6

AD E sec. 3 3.7

The test programme should be carried out in accordance with the following rules: AD E sec. 3 3.5

- The dwellings to be tested should be completed but unfurnished; doors and windows should be closed.

Table 10.7 Sound insulation values – new buildings.

		Mean values	
Type of performance	*Individual values* *(dB)*	*Test in up to 4 pairs of rooms (dB)*	*Test in at least 8 pairs of rooms (dB)*
Airborne sound (minimum values)[a]	49 (walls) 48 (floors)	53 (walls) 52 (floors)	52 (walls) 51 (floors)
Impact sound (maximum values)[b]	65	61	62

Notes
[a] Airborne sound – Weighted Standardised Level Difference ($D^1_{nT,w}$)
[b] Impact sound – Weighted Standardised Sound Pressure Level ($L^1_{nT,w}$)

- Ideally, floors and walls should be tested between eight pairs of rooms. Since most dwellings are not large enough to permit this, the test may be carried out in four pairs of rooms, or as near to four pairs as possible.
- For the measurement of airborne sound transmission, the sound source should be placed in the larger of the two rooms where they are of unequal volume.
- For the measurement of airborne sound insulation between a room and some other space, the sound source should be located in the other space.

Only one set of measurements should be taken between each pair of rooms and the rooms should be as large as possible. Preferably, the pairs of rooms should all be located in dwellings although it is permissible to use pairs consisting of a room and some other space in order to make up the set of four measurements.

In order to satisfy the Building Control Authority, a test report should be provided which describes the performance of the existing construction. The details that this report should contain are set out in Table 1 of AD E, which is reproduced below.

AD E, Section 3

Table 1   Test report details: Test of existing construction

---

1. Organisation conducting test:
   a. name;
   b. address;
   c. NAMAS accreditation number (if appropriate)

2. Name of person in charge of test

3. Date of test

4. Address of building tested

5. Brief details of test:
   a. equipment;
   b. test procedures

6. Description of building:
   a. sketch showing relationship and dimensions of rooms tested;
   b. description of external and separating walls, partitions and floors including details of materials used for their construction and finishes;
   c. estimate of surface mass kg/m^2 of external and separating walls, partitions and floors;
   d. dimensions of any step and stagger between rooms tested;
   e. approximate dimensions of any windows or doors in external walls within 700 mm of the separating wall

7. Results of test, shown in tabular and graphical form:
   a. single number rating;
   b. underlying data from measurements on which the single number rating is based

---

Existing constructions – degree of similarity

The degree of sound insulation provided by a certain form of construction depends not only on the wall or floor specification but also on other factors, such as size and shape of rooms and, for masonry buildings, the positions of the doors and windows.

AD E
sec. 3
3.2

Thus, in order that a satisfactory comparison may be made, the following features of a proposed building should be similar to, but not necessarily identical with, an existing building which has been found by testing to be satisfactory:

AD E
sec. 3
3.3

- Specification of sound-resisting walls and floors provided that there is no reduction in the mass per square metre.
- The construction of walls and floors adjacent to the sound-resisting walls and floors.
- The general arrangement of window and door openings where these are in an external wall with a masonry inner leaf and are adjacent to a sound-resisting wall or floor.
- The general size and shape of the rooms on either side of sound-resisting construction.
- The extent of any step or stagger in a sound-resisting wall. It may be beneficial to provide a step or stagger in a proposed wall if one is not present in the existing building.

AD E specifies certain allowable differences in details that have little effect on the performance of sound-resisting elements. For example, the performance of sound-resisting walls and floors is unlikely to be affected by the construction of a masonry cavity wall provided the inner leaf is of the same general type and its weight is not reduced.

AD E
sec. 3
3.4

In the case of sound-resisting walls, the following differences in construction are unlikely to reduce their performance:

- The material and thickness of the floating layer of a Type 2, or similar, floor.
- A small reduction in the size of a step or stagger.
- The type of timber floor provided, where it is not a sound-resisting floor.

It should be noted that while the test procedure and the values in Table 10.7 are provided to enable an *existing* construction to be assessed before new construction is undertaken, the subsequent failure of the *new* construction to achieve the values in Table 10.7 is not of itself evidence of failure to comply with the requirements of the 1991 Regulations.

AD E
sec. 3
3.8

Test chamber evaluation – new walls

Section 4 of AD E is concerned with the testing of proposed wall constructions in an approved acoustic test chamber.

AD E
sec. 4
4.1

A difficulty encountered with the 1985 edition of AD E was that it made no specific provision to allow innovative new designs to be built. To make good this deficiency the DOE introduced into the 1992 edition of AD E a test for combinations of separating wall and external wall built in a special two-storey test chamber. The external wall is included in the test to ensure that flanking transmission is taken into account. The exact details of the test chamber construction are not contained in AD E; however, they may be obtained from the Building Research Establishment, Garston, Watford, WD2 7JR.

The approved chamber size should be such as to allow a room volume of up to $50\,m^3$. This volume was chosen as being large enough for reliable measurements, yet small enough to be representative of a dwelling. It is possible to use test chambers of other dimensions, as a correction can be applied to make the results comparable between different sized chambers.

The test method

AD E
sec. 4
4.2, 4.3

Once the example wall construction (including the flanking element) has been erected in the test chamber, the degree of insulation against airborne sound is measured between the upper pair of rooms and between the lower pair of rooms in accordance with BS 2750: Part 4: 1980.

Measurement should be made in the one-third octave bands from $100\,Hz$ to $3150\,Hz$, and standardised to a reverberation time of 0.5 second.

In order to pass the test, both pairs of rooms tested should have a modified weighted standardised level difference value (i.e. an insulation value) of at least 55 dB. This figure is a compromise value and when it is compared with the values given in Table 10.7 it can be seen that it is considerably higher than the mean value of 52 dB given there. This apparent discrepancy can be attributed to the fact that test constructions are usually built to an untypically high standard so it is possible that the performance measured in the test will be above average for that type of construction. The BRE has carried out extensive field measurements and these have shown that for most types of construction the best examples are about 4 dB or 5 dB ($D_{nT,w}$) better than the mean for each type. In order to comply with Table 10.7, constructions are required to have a mean insulation value of at least 52 dB ($D_{nT,w}$), so the best examples of a construction that just meets this recommendation will have an insulation value of at least 56 dB. Hence, as a compromise, the figure of 55 dB has been chosen.

AD E
sec. 4
4.4

The modified weighted standardised level difference is derived from the weighted standardised level difference obtained from the test by the addition of a constant which takes account of different room dimensions. The constant K, may be obtained from the formula $K = 10 \log(3/L) + 1$, where L (metres) is the length of the room at right angles to the sound-resisting wall.

Evidence of testing

AD E
sec. 4
4.5

In order to provide evidence to the Building Control Authority that the regulations have been complied with, it may be necessary to obtain a test report from an accredited testing organisation. The test report should contain the information specified in Table 3 of AD E, which is reproduced below. The evidence will only be valid for the specific construction tested; however the following features may be changed without invalidating the results:

- The dimensions of the sound-resisting and flank walls.
- The position of door and/or window openings in the flank wall.
- Internal partitions; they may be positioned against the sound-resisting wall.

AD E, Section 4

Table 3   **Test report details: Test chamber evaluation**

---

1. Organisation operating the test chamber:
    a. name;
    b. address;
    c. NAMAS accreditation number, if appropriate

2. Organisation conducting the acoustic measurements (if different from above):
    a. name;
    b. address;
    c. NAMAS accreditation number, if appropriate

3. Date of test

4. Description of test chamber including method of attaching the test construction

5. Brief details of test:
    a. equipment;
    b. procedure

6. Full details of materials and test construction:
    a. separating wall;
    b. flank wall;
    c. junction between separating wall and flank wall (e.g. bonded or tied);
    d. surface finish;
    e. mass/m^2 of walls;
    f. intermediate floor;
    g. separating wall in loft space;
    h. dimensions;
    j. any other special features

7. Results of test:
    a. for each measurement the standardised level difference ($D_{n,T}$), the weighted standardised level difference, ($D_{nT,w}$) and weighted apparent sound reduction index (R'_w) all according to BS 5821: Part 1: 1984.
    b. the two modified weighted standardised level differences obtained by adding the correction K (see paragraph 4.4 of AD E) to the weighted standardised level difference
    c. statement saying that both values determined in 7b of this section meet the requirements given in paragraph 4.4 of AD E

Conversion to dwellinghouses and flats – improvements to sound insulation

It has already been mentioned (see p.10.2) that the Building Regulations apply sound insulation requirements to the change of use of a building to a dwellinghouse, flat or maisonette, where previously it was used for some other purpose. Sections 5 and 6 of AD E deal with such works of conversion in two ways:

• By reference to a series of constructional forms which may be adopted by the building designer; or,

- By repeating a construction that has been built and tested in a building or a laboratory.

AD E
sec. 5
5.1

Of course, it may be that the existing wall and floor constructions already meet the requirements for sound insulation without the need for works of improvement. This can be demonstrated by:

(a) carrying out a test on the building in accordance with the method specified in Section 6 of AD E (see below); or,

(b) showing that the construction is generally similar to one of the wall or floor types specified in Sections 1 and 2 of AD E.

In the case of (b) above, the existing walls or floors should have a mass which is within 15% of the Section 1 and 2 constructions. Also, flanking transmission may be ignored except for floor penetrations and sealing of joints. This reflects the view that with conversion work, as much of the existing structure as possible must be used in order to make the job economically attractive. This rules out extensive work (such as heavy wall linings) to control flanking transmission.

Conversion work – typical wall and floor constructions

AD E
sec. 5
5.2

If it cannot be demonstrated that the existing construction meets the sound insulation requirements, Section 5 of AD E provides details of one wall treatment, three floor treatments and one stair treatment which may be adopted by the building designer. If a strong case can be made for not using any of the three preferred floor treatments, a further two are given which give lower levels of insulation (but are still an improvement over the existing).

The following general points should also be considered:

AD E
sec. 5
5.3

- The upgrading measures for sound insulation may impose considerably increased loadings on the structure of the building. The structure will need to be assessed to confirm that it can carry these additional loads or whether there is a need for it to be strengthened.

AD E
sec. 5
5.4

- Many old buildings are converted to dwellinghouses or flats and may contain attractive architectural features such as cornices, decorative ceilings or floor finishes. The guidance in AD E should be treated with some degree of flexibility in these circumstances with regard to what is considered reasonable.

Conversion work – upgrading of existing walls

AD E
sec. 5
wall
treatment

The treatment recommended for walls is the familiar independent wall lining with absorbent material in the cavity. In order to resist airborne sound transmission this type of construction depends on the form of the existing construction, the mass and isolation of the independent leaf and some degree of absorption in the cavity. The treatment is illustrated in Fig. 10.13 and may be used on one side only of the existing wall if this:

- is constructed of masonry; *and*
- is plastered on both faces; *and*
- has a thickness of at least 100 mm.

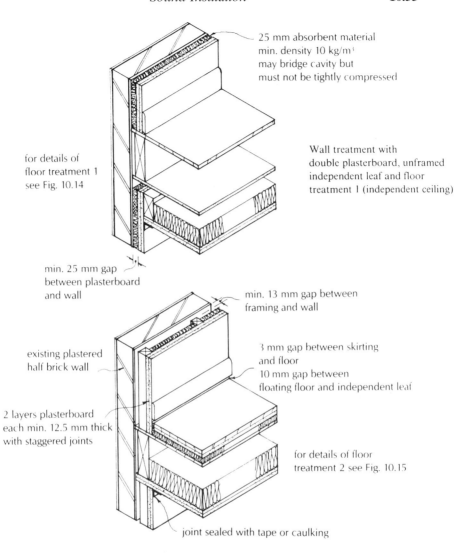

25 mm absorbent material
min. density 10 kg/m³
may bridge cavity but
must not be tightly compressed

for details of
floor treatment 1
see Fig. 10.14

Wall treatment with
double plasterboard, unframed
independent leaf and floor
treatment 1 (independent ceiling)

min. 25 mm gap
between plasterboard
and wall

min. 13 mm gap between
framing and wall

existing plastered
half brick wall

3 mm gap between skirting
and floor
10 mm gap between
floating floor and independent leaf

2 layers plasterboard
each min. 12.5 mm thick
with staggered joints

for details of floor
treatment 2 see Fig. 10.15

joint sealed with tape or caulking

Wall treatment with framed independent leaf and floor treatment 2 (platform floor)

Fig. 10.13 Wall treatment in conversion work.

Where these conditions cannot be met, the wall treatment should be applied to both faces of the existing wall.

Junction details are also important in maintaining the efficacy of the resistance to sound transmission. Details of typical floor/wall junctions are also shown in Fig. 10.13.

Conversion work – upgrading of existing floors

Flanking transmission down the walls often limits the sound insulation that can be achieved with lightweight floors. Therefore, the floor itself should provide better insulation than the flanking construction or the net insulation will be lower than the limit set by the flanking transmission. The three preferred floor treatments recommended in AD E are the independent ceiling, the platform floor and the raft floor.

In many converted buildings the ceiling finish is of lath and plaster. This construction is comparatively heavy and may be retained for sound insulation purposes. However, it should be checked with the recommendations of AD B, Fire (see Chapter 7) to ensure that a satisfactory standard of fire resistance is achieved. If the existing ceiling is not lath and plaster it should be upgraded by applying two or three layers of plasterboard with staggered joints, to give an overall thickness of at least 30 mm.

Floor treatment 1 – independent ceiling with absorbent material

AD E sec. 5 floor treatment 1

In this type of ceiling the resistance to impact and airborne sound depends on:

- The total mass of the existing floor and the new ceiling.
- The isolation of, and absorption of sound within, the independent ceiling.
- The general airtightness of the whole construction.

In essence, this treatment consists of constructing a new insulated ceiling (which is separately supported), below the existing floor construction. It is essential to overhaul the existing floorboards and to seal any gaps with caulking. Alternatively, the floor can be overlaid with hardboard (and this would seem to be the best solution where the floorboards are plain edged). In some buildings the window heads are very close to ceiling level. It is permissible, in these circumstances, to raise the independent ceiling locally, to provide a pelmet recess. Figure 10.14 gives details of the construction of the independent ceiling.

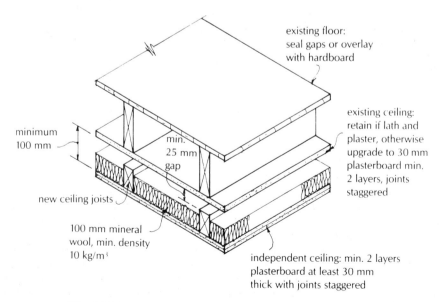

existing floor:
seal gaps or overlay
with hardboard

existing ceiling:
retain if lath and
plaster, otherwise
upgrade to 30 mm
plasterboard min.
2 layers, joints
staggered

minimum
100 mm

min.
25 mm
gap

new ceiling joists

100 mm mineral
wool, min. density
10 kg/m³

independent ceiling: min. 2 layers
plasterboard at least 30 mm
thick with joints staggered

Fig. 10.14 Floor treatment 1 – independent ceiling details.

Floor treatment 2 – platform floor
This treatment consists of placing a resilient layer over the existing floor-boarding and then finishing this with a suitable floating layer to which can be applied the normal floor coverings. The sound insulation which is achieved depends on the total mass of the floor and the effectiveness of the resilient layer which should be of the correct density and able to carry the anticipated load.

AD E
sec. 5
floor
treatment 2

Where the existing floor has to be replaced, advantage should be taken of the opportunity to increase the insulation of the floor by laying 12 mm thick boards and inserting a 100 mm mineral wool absorbent layer between the joists. Alternatively, the mineral wool may be inserted if the ceiling is replaced.

Figure 10.15 shows the typical platform floor construction of floor treatment 2. Junction details with the existing or upgraded walls are illustrated in Fig. 10.13 above.

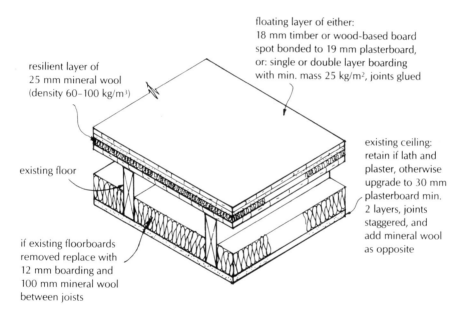

floating layer of either:
18 mm timber or wood-based board
spot bonded to 19 mm plasterboard,
or: single or double layer boarding
with min. mass 25 kg/m², joints glued

resilient layer of
25 mm mineral wool
(density 60–100 kg/m³)

existing ceiling:
retain if lath and
plaster, otherwise
upgrade to 30 mm
plasterboard min.
2 layers, joints
staggered, and
add mineral wool
as opposite

existing floor

if existing floorboards
removed replace with
12 mm boarding and
100 mm mineral wool
between joists

Fig. 10.15 Floor treatment 2 – platform floor details.

Floor treatment 3 – raft floor
The raft floor consists of a floating layer which forms the top surface of the floor separated from the existing floor joists by resilient strips. The space between the joists is infilled with either mineral wool or heavy pugging. Hence, the sound insulation of the floor is dependent on the mass of the total construction, the effectiveness of the resilient strips and the pugging or absorbent material in the floor space.

AD E
sec. 5
floor
treatment 3

With this form of construction it is necessary to remove the existing floor-boarding. At this stage it may be necessary to provide additional lateral strutting between the joists (if this is not present already) since the final construction will be without the benefit of the lateral stability provided by the normal fixing of the floor decking. The existing floor joists should be at least

45 mm wide; however, where heavy pugging is used, it may be necessary to strengthen the floor. (For details of pugging see Table 10.6 above.)

It is particularly important with this type of floor to ensure that the resilient strips have the correct density and can carry the anticipated load without undue compression since the performance of the floor may deteriorate when loaded with furniture if the less dense types of resilient strips are used.

Figure 10.16 shows the typical raft floor construction of floor treatment 3. Junction details with the existing or upgraded walls are illustrated in Fig. 10.13 above.

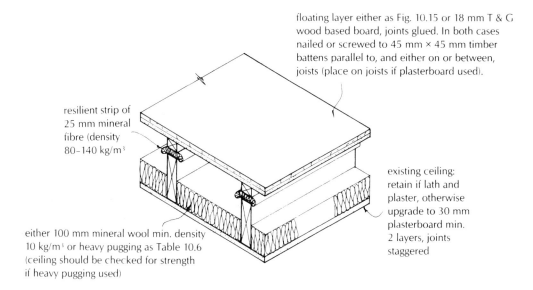

floating layer either as Fig. 10.15 or 18 mm T & G wood based board, joints glued. In both cases nailed or screwed to 45 mm × 45 mm timber battens parallel to, and either on or between, joists (place on joists if plasterboard used).

resilient strip of 25 mm mineral fibre (density 80–140 kg/m³)

existing ceiling: retain if lath and plaster, otherwise upgrade to 30 mm plasterboard min. 2 layers, joints staggered

either 100 mm mineral wool min. density 10 kg/m³ or heavy pugging as Table 10.6 (ceiling should be checked for strength if heavy pugging used)

Fig. 10.16 Floor treatment 3 – raft floor details.

Alternative floor treatments in exceptional circumstances

AD E
sec. 5
5.2

The problem with the solutions in floor treatments 1 to 3 is that either the ceiling level is lowered or the floor level is raised. In some situations, such as in listed buildings, these considerations may make the preferred constructions impractical. In these special cases (and with the Building Control Authority's agreement), either of two lower performance solutions may be used. These are an independent ceiling with the new joists between the original resulting in a lesser reduction in ceiling height, and a simplified floating floor having a thinner floating layer.

Floor treatment 4 – alternative independent ceiling with absorbent material

AD E
sec. 5
floor
treatment 4

With this form of construction it is essential to overhaul the existing floorboards and to seal any gaps with caulking. Alternatively, the floor can be overlaid with hardboard (and this would seem to be the best solution where the floorboards are plain edged). The junction with the walls at the perimeter should be sealed with tape or mastic. Details of the construction are shown in Fig. 10.17 below.

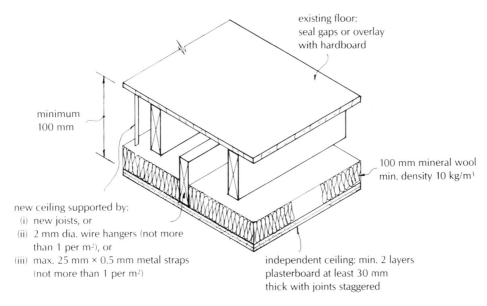

existing floor:
seal gaps or overlay
with hardboard

minimum
100 mm

100 mm mineral wool
min. density 10 kg/m³

new ceiling supported by:
 (i) new joists, or
 (ii) 2 mm dia. wire hangers (not more
      than 1 per m²), or
 (iii) max. 25 mm × 0.5 mm metal straps
      (not more than 1 per m²)

independent ceiling: min. 2 layers
plasterboard at least 30 mm
thick with joints staggered

Fig. 10.17 Floor treatment 4 – alternative independent ceiling details.

Floor treatment 5 – alternative platform floor

This construction is similar to floor treatment 2 but results in a lesser rise in floor level. Where the existing floorboards are removed they should be replaced with boarding at least 12 mm thick. If the existing ceiling is not lath and plaster it should be upgraded in two or three layers of plasterboard to a thickness of at least 30 mm, with staggered joints. The new ceiling should be suspended from suitable resilient hangers or timber cross battens. (See Fig. 10.18 below for constructional details.)

**AD E
sec. 5
floor
treatment 5**

Stair treatment 1 – resilient covering and independent ceiling

Where a timber stair forms a separating function between dwellings it is subject to the same insulation requirements with respect to impact and airborne sound as a floor. (In this case it will also need to satisfy the fire resistance recommendations of AD B described in Chapter 7 above.)

**AD E
sec. 5
stair
treatment 1**

The recommended form of upgrading for timber stairs is illustrated in Fig. 10.19. Impact sound is reduced by the addition of a soft covering, such as a carpet, to the stair treads. The treatment to the underside will depend on whether the stairs are open to the room below or there is an understairs cupboard. In the former case, an independent ceiling, as in floor treatment 1, is constructed under the stairs. Where a cupboard is present, the underside of the stairs within the cupboard may be lined with plasterboard and the remaining space above this filled with mineral wool. The cupboard walls should consist of at least two layers of plasterboard and the door should be small, heavy and well fitted.

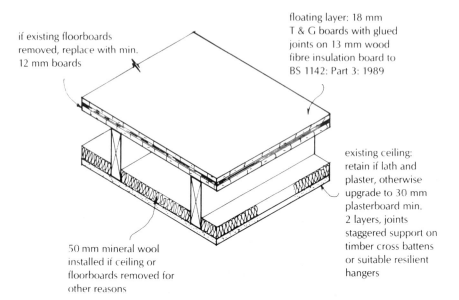

if existing floorboards removed, replace with min. 12 mm boards

floating layer: 18 mm T & G boards with glued joints on 13 mm wood fibre insulation board to BS 1142: Part 3: 1989

existing ceiling: retain if lath and plaster, otherwise upgrade to 30 mm plasterboard min. 2 layers, joints staggered support on timber cross battens or suitable resilient hangers

50 mm mineral wool installed if ceiling or floorboards removed for other reasons

Fig. 10.18 Floor treatment 5 – alternative platform floor details.

Conversion work – penetration of floors by pipes

The recommendations for pipe penetrations of floors in conversion work are generally as for new work (see p. 10.20 above and Fig. 10.9) with the following variations:

- Where floor treatment 1 is used, a 3 mm gap should be left between the pipe enclosure and the floating layer of the floor. This should be sealed with caulking or neoprene.
- If floor treatment 2 is adopted, the enclosure may go down to the floor base but it should be isolated from the floating layer.

Conversion work – laboratory and field tests

If a designer wants to use a novel or innovative method of upgrading the sound insulation in a proposed conversion to dwellings he has the option of building an example, either in the field or in a laboratory, and then having it tested in accordance with the recommendations of Section 6 of AD E. An on-site test can also be carried out if it is felt that an existing construction is capable of satisfying the functional requirements with regard to sound insulation without additional upgrading.

Laboratory testing

**AD E
sec. 6
6.1**

Laboratories should comply with the recommendations of BS 2750 *Measurement of sound insulation in buildings and of building elements*, Part 1: 1980 *Recommendations for laboratories*.

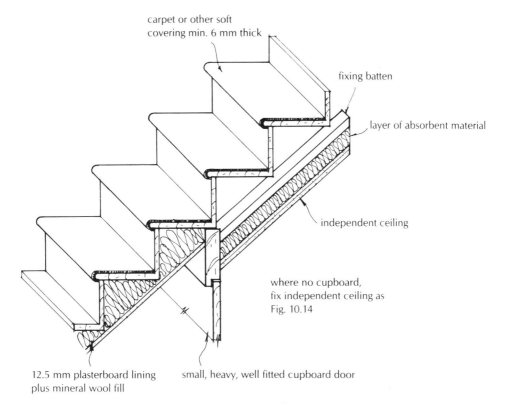

Fig. 10.19 Stair treatment 1.

If it is desired to test some new form of upgrading for a wall or floor, then it will be necessary to assume some kind of 'basic' construction for any existing building so that the results of the tests can be used for comparison purposes. AD E gives the following basic constructions:

AD E sec. 6 6.3

- A typical masonry wall consisting of a half brick leaf, plastered both sides, with a total mass (including the plaster) of not more than 200 kg/m².
- A typical timber floor consisting of 22 mm plain edged boards on joists at 400 mm centres, with a ceiling finish of lath and plaster or plasterboard not more than 30 mm thick.

Clearly, the test results achieved will only be valid for constructions which are similar to those tested. Therefore, if the existing walls and floors in a building are likely to have less resistance to sound than the basic constructions assumed above, the proposed remedial treatments will need to be tested with constructions similar to those actually existing and these would need to be reproduced in the laboratory.

AD E sec. 6 6.4

Laboratory test methods

AD E
sec. 6
6.9 & 6.12

For sound-resisting walls and floors the following procedure should be adopted:

(i) Test basic wall or floor alone.
(ii) Test basic wall or floor with remedial treatment applied.
(iii) Provide both sets of results, for each wall or floor tested, in a test report in accordance with Table 5 of AD E (reproduced below).

AD E, Section 6

Table 5   Test report details: Laboratory tests of remedial treatments

1. Organisation conducting test:
   a. name;
   b. address;
   c. NAMAS accreditation number (if appropriate)

2. Name of person in charge of test

3. Date of test

4. Brief details of test:
   a. equipment;
   b. test procedures

5. Description of the construction tested, including:
   a. description of the base construction, including:
   b. mass/m² of base construction,
   c. description and sketch of the upgraded construction

6. Results of tests shown in tabular and graphical form (including single number rating) for:
   a. the base construction;
   b. the upgraded construction. Single number ratings should be stated for airborne and impact sound as appropriate.

The tests for airborne sound insulation in walls and floors should be conducted in accordance with BS 2750: Part 3: 1980 *Laboratory measurement of airborne sound insulation of building elements*. The insulation value achieved for airborne sound for each combination of wall and/or floor and remedial treatment should be not less than the value given in Table 10.8 below.

AD E
sec. 6
6.10, 6.11

For floors, it is sometimes necessary to replace all the construction apart from the joists. In this case it would only be necessary to test the remedial treatment. Floors should be tested for both impact and airborne sound insulation unless it is only necessary to improve one or the other. If this is the case, then it should be clearly identified in the test report. The tests for insulation against impact sound should be conducted in accordance with BS 2750: Part 6:

AD E
sec. 6
6.12

1980 *Laboratory measurement of impact sound insulation of floors*. The insulation value achieved for impact sound for the floor plus remedial treatment should not be greater than the value given in Table 10.8 below.

Table 10.8 Sound insulation values – conversion works.

Type of performance	Laboratory tests (dB)	Field tests (dB)
Airborne sound (minimum values)[a]	53 (walls) 52 (floors)	49 (walls) 48 (floors)
Impact sound (maximum values)[b]	65	65

Notes:
[a] Airborne sound – weighted standardised level difference for combined wall and/or floor with remedial treatment. (Insulation values should be calculated in accordance with BS 5821: *Methods for rating the sound insulation in buildings and of building elements*, Part 1: 1984 *Method for rating the airborne sound insulation in buildings and of interior building elements.*)
[a] Impact sound – weighted impact sound pressure level for floor with remedial treatment. (Insulation values should be calculated in accordance with BS 5821: Part 2: 1984 *Method for rating the impact sound insulation.*)

AD E, Section 6

Table **4**   **Test report details: Tests of remedial treatments for separating walls, floors and stairs**

1. Organisations conducting test:
   a. name;
   b. address;
   c. NAMAS accreditation number (if appropriate)

2. Name of person in charge of test

3. Date of test
4. Brief details of test:
   a. equipment;
   b. test procedures

5. Description of treatment tested:
   a. sketch showing the relationship and dimensions of rooms tested;
   b. dimensions of any step or stagger between rooms tested;
   c. description of the existing construction (separating and abutting elements);
   d. details of opening (if any) within 700 mm of the separating element;
   e. the mass/m^2 of the existing construction;
   f. description of the materials and methods used to upgrade the existing construction

6. Results of test shown in graphical and tabular form (including single number rating) for:
   a. the existing construction;
   b. the upgraded construction

AD E
sec. 6
6.2
It will be seen from Table 10.8 that the airborne sound insulation values recommended from laboratory conducted tests are higher than those recommended from field tests. This reflects the fact that laboratory tests are carried out with minimal flanking transmission and if the tested construction was erected in a building it might well have less resistance to airborne sound than expected from the test results due to flanking transmission in the building. Accordingly, a better performance is expected of laboratory tests than field tests.

Field test methods

AD E
sec. 6
6.5
For sound-resisting walls and floors the following procedure should be adopted:

(i) Test at least two examples of the basic wall or floor alone.
(ii) Test at least two examples of the basic wall or floor with remedial treatment applied.
(iii) Provide both sets of results for each wall or floor tested, in a test report in accordance with Table 4 of AD E (reproduced below).

AD E
sec. 6
6.5
The tests for airborne sound insulation in walls and floors should be conducted in accordance with BS 2750: Part 4: 1980 *Field measurements of airborne sound insulation between rooms*. The insulation value achieved for airborne sound for each combination of wall and/or floor and remedial treatment should be not less than the value given in Table 10.8 above.

AD E
sec. 6
6.6 to 6.8
For floors, it is sometimes necessary to replace all the construction apart from the joists. In this case it would only be necessary to test at least two examples of the remedial treatment. Floors should be tested for both impact and airborne sound insulation unless it is only necessary to improve one or the other. If this is the case, then it should be clearly identified in the test report. The tests for insulation against impact sound should be conducted in accordance with BS 2750: Part 7: 1980 *Field measurements of impact sound insulation of floors*. The values achieved for impact sound for the floor plus remedial treatment should not be greater than the value given in Table 10.8.

Chapter 11

Ventilation

Introduction

The need to provide adequate ventilation to buildings has long been recognised in building control legislation. Formerly, however, it was restricted to dwellings, and to bathrooms and rooms containing sanitary conveniences in buildings other than dwellings. The building regulations now extend the requirement for adequate means of ventilation to all building types.

It should be noted that if the provisions of Approved Document F are followed, then it would prevent the service of an improvement notice under Section 23(3) of the Health and Safety at Work etc Act 1974. This relates to the requirements for ventilation contained in regulation 6 (1) of the Workplace (Health, Safety and Welfare) Regulations 1992 and is the first time that such a connection between the Building Regulations and the Workplace Regulations has been made.

The 1995 edition of Approved Document F introduced a number of significant changes to the way in which ventilation is treated in domestic buildings including:

- The use of passive stack ventilation or open-flued heating appliances as an alternative to mechanical extract ventilation
- The need to ventilate utility rooms
- Removal of the restrictions on enclosed courtyards, and
- Removal of the need to ventilate common spaces in flats

Means of ventilation

In general, there must be adequate means of ventilation provided for people in buildings. The following are exempted from this rule because providing ventilation in them would not serve to protect the health of the users:

Regs Sch. 1
F1

- Buildings or spaces within buildings where people do not normally go
- Buildings or spaces within buildings used solely for storage, and
- Garages used solely in connection with a single dwelling

The provisions of Approved Document F1 are designed to ensure that suitable air quality is maintained in buildings.

AD F1 –
Performance

Without adequate ventilation, moisture (leading to mould growth) and pollutants (originating inside a building) may accumulate to such levels that they become a hazard to the health of users of the building. For these reasons AD F1 recommends the following methods of ventilation which may be adopted for use in buildings:

- Extract ventilation (either natural or mechanical)

This is used to remove water vapour or pollutants from areas where they are produced in significant quantities and before they become widespread. Clearly, this would apply to kitchens, utility rooms and bathrooms in the case of water vapour. Interestingly, with regard to the extraction of pollutants, AD F1 includes not only rooms containing processes which produce harmful contaminants but also rest rooms where smoking is permitted. This would appear to be yet another victory for the anti-smoking lobby!

- Rapid dilution,

Normally, this would be achieved by providing a door or window which could be thrown open as required. In sanitary accommodation which is not within a bathroom a similar level of rapid dilution may be obtained by mechanical extraction.

- Background ventilation,

The guiding principle here is that a minimum supply of fresh air should be available over a long period of time to disperse residual water vapour as necessary. It is important that the means of ventilation should not compromise security or comfort and should resist rain penetration.

In non-domestic buildings, it is often the case that ventilation is provided by mechanical means or by or air-conditioning systems. These are permissible provided that they achieve the performance listed above, and they are:

(a) designed, installed and commisioned so that their performance will not put at risk the health of people in the building, and
(b) designed to permit necessary maintenance so that all the objectives outlined above may continue to be achieved.

Interpretation

Special definitions apply to AD F1.

- VENTILATION OPENING – includes any permanent or closeable means of ventilation which opens directly to external air as follows:

AD F1
0.1 to 0.9
and Table 1

  - Opening lights in windows,
  - Louvres,
  - Airbricks,
  - Progressively openable ventilators, window trickle ventilators,
  - Doors.

Undoubtedly, the most common way of providing background ventilation is via a trickle ventilator located in or above a window frame and it is also possible to obtain glazing systems containing this facility.

Airbricks, ducted through a wall and finished internally with a 'hit and miss' ventilator are also permissible provided that the main air passages are large enough to minimise resistance to airflow. Therefore, slots should have a minimum dimension of 5 mm and any square or circular holes should be at least 8 mm across (excluding any insect screens or baffles, etc.).

The two methods mentioned above rely on providing means of ventilation which are additional to the windows or external doors of a room. In fact, the windows themselves can be used for background ventilation provided that they are of a suitable type and can be secured in the open position to provide the amount of ventilation recommended in Table 11.1. Vertical sliding sash windows are ideal for this purpose, the background ventilation being provided by opening the top sash and locking it in the required position. Top hung opening casement windows may also be suitable provided that the light can be locked in at least two opening positions. Since these windows are more easily forced they should be restricted to use above ground floor level.

However background ventilation is provided, it should always be located so that it does not cause discomfort from cold draughts. Additionally, it is unlikely to be effective if it is less than 1.75 m above floor level.

Where windows with adjustable locking positions are used to supply background ventilation there is a danger that they would be unusable as a means of escape in case of fire. Approved Document B1 contains details of windows which would need to be used for escape purposes and it is unlikely that these could be used for background ventilation as described above. (See page 7.26 for details of means of escape windows.)

- HABITABLE ROOM – a room used for dwelling purposes which is not solely a kitchen.
- BATHROOM – a room containing a bath or shower with or without sanitary accommodation.
- SANITARY ACCOMMODATION – a room which contains one or more closets or urinals. If sanitary accommodation contains one or more cubicles it is not necessary to provide separate ventilation to each if air is free to circulate throughout the space.
- UTILITY ROOM – a room in which water vapour is likely to be produced in significant quantities because it is designed or intended to be used to contain clothes washing or similar equipment such as a sink, washing machine or tumble drier, etc. It should be noted that ventilation does not need to be provided under Building Regulations if the utility room can be entered solely from outside the building.
- OCCUPIABLE ROOM – includes rooms occupied by people in non-domestic buildings such as offices, workrooms, classrooms, hotel bedrooms etc.

    Excluded from this definition are:

    - Bathrooms, sanitary accommodation and utility rooms, and
    - Rooms or spaces used solely or mainly for circulation, building services plant and storage.

- DOMESTIC BUILDINGS – buildings used for dwelling purposes such as dwelling houses, flats, residential accommodation and student hostels.

Table 11.1 Ventilation recommendations.

1 Room or space	Ventilation recommendations for rooms capable of containing openable windows			Ventilation recommendations for rooms not containing openable windows	*6* Notes
	2 Rapid ventilation	*3* Background ventilation	*4* Extract ventilation (fan rates)	*5* Mechanical extract ventilation (fan rates)	
1 Domestic buildings					
(a) Habitable rooms	Ventilation opening equal to at least $1/20$th room floor area	8000 mm²	* See note column 6	For mechanical ventilation see BS 5720: 1979 and BRE Digest 398	* No recommendation given in AD F1
(b) Kitchens	Opening window (any size)	4000 mm²	30 litres/sec in or adjacent to hob. 60 litres/sec elsewhere	Mechanical extract as column 4, with 15 minute overrun on fan connected to light switch for rooms without natural light	See also text page 11.7 and note at foot of column 4 below for alternatives to mechanical extract
(c) Utility rooms	Opening window (any size)	4000 mm²	30 litres/sec	Mechanical extract as column 4, with 15 minute overrun on fan connected to light switch for rooms without natural light	No ventilation provisions necessary if room entered only from outside
(d) Bathroom	Opening window (any size)	4000 mm²	15 litres/sec	Mechanical extract as column 4, with 15 minute overrun on fan connected to light switch for rooms without natural light	Bathroom may or may not contain WC
(e) Sanitary accommodation if separate from bathroom	$1/20$th room floor area as habitable room in (a) above	4000 mm²	* See note column 6		* See also BS 5720: 1979 and BRE Digest 398
2 Non-domestic buildings					
(f) Occupiable room	$1/20$th room floor area as habitable room in (a) above	4000 mm² for room floor areas up to 10 m². 400 mm²/m² for room floor areas over 10 m²	* See note column 6	For mechanical ventilation allow 8 litres/sec per occupant of fresh air for rooms where no smoking is permitted and 16 litres/sec per occupant for rooms where light smoking is permitted. For rooms designed for heavy smoking, such as rest rooms where smoking is allowed see Table 11.2	* No table recommendation given in AD F1

Table 11.1 Ventilation recommendations (continued).

1 Room or space	Ventilation recommendations for rooms capable of containing openable windows			Ventilation recommendations for rooms not containing openable windows	6 Notes
	2 Rapid ventilation	3 Background ventilation	4 Extract ventilation (fan rates)	5 Mechanical extract ventilation (fan rates)	
(g) Kitchen (For type see note)	Opening window (any size)	4000 mm²	30 litres/sec in or adjacent to hob. 60 litres/sec elsewhere	Mechanical extract as column 4, with 15 minute overrun on fan connected to light switch or occupant detecting sensor for rooms without natural light	Recommendations are for kitchens similar to domestic kitchens i.e. *not* commercial kitchens. See also text page 11.11 for alternatives to mechanical extract
(h) Bathroom	Opening window (any size)	4000 mm² per bath or shower	15 litres/sec per bath or shower	Mechanical extract as column 4, with 15 minute overrun on fan connected to light switch or occupant detecting sensor for rooms without natural light	(Includes shower rooms)
(i) Sanitary accommodation (with or without washing facilities)	$\frac{1}{20}$th room floor area as habitable room in (a) above	4000 mm² per WC	* See note column 6	6 litres/sec per WC mechanical extract or 3 air changes/hour, with 15 minute overrun on fan. Fan may be connected to light switch or occupant detector for rooms without natural light	* See also BS 5720: 1979 and CIBSE Guides A and B
(k) Common spaces	No recommendations for rapid, background or extract ventilation in AD F1. Instead provide either natural ventilation equal to $\frac{1}{50}$th of floor area of common space *or* mechanical ventilation at rate of 1 litre/sec per m² floor area				Applies to spaces where large numbers of people gather, e.g. shopping malls and foyers *not* spaces used principally for circulation
Notes	In domestic buildings the areas given in (a) to (e) above may be varied to give an average of 6000 mm² for the dwelling with at least 4000 mm² per room		Extract ventilation may also be by passive stack ventilation or appropriate open-flued heating appliance – see text page 11.7	In domestic buildings in (b) to (e) and non-domestic buildings in (g) to (j) above, an air inlet should be provided to each room such as a 10 mm gap under the door	

- NON-DOMESTIC – all buildings not contained in the definition of domestic buildings above. To avoid confusion, buildings where people reside only temporarily such as hotels, are regarded as non-domestic buildings.
- PASSIVE STACK VENTILATION (PSV) – a system of ventilation which relies on the natural stack effect (in which warm air rises due to the difference in temperature between the inside and outside of a building) and the effect of wind passing over the roof. The system consists of a series of ceiling outlets connected by ducts to terminals on the roof of a building (see Fig. 11.1).

No positive guidance on PSV is given in AD F1 apart from referring the reader to BRE Information Paper 13/94 (*Passive stack ventilation systems: design and installation*). Additionally, any system with appropriate third party certification (e.g. a British Board of Agrément certificate) would also be acceptable.

It should be noted that where rooms have a double function the individual provisions for rapid, background and extract ventilation shown in Table 11.1 need not be duplicated. Instead, the room will need to be provided with the greater provision for each of the individual functions listed in the Table. Therefore, if a room is a kitchen-diner then the extract ventilation recommendations for the kitchen will need to be included in addition to those for rapid and background ventilation for the dining area.

ridge outlet system with bends   **straight system**

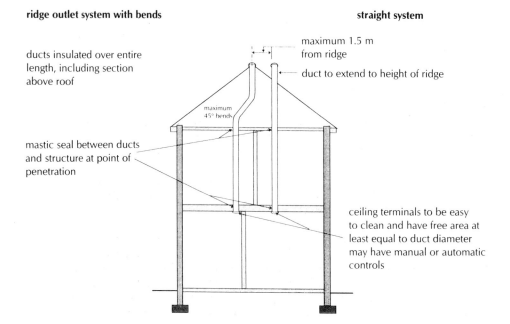

ducts insulated over entire length, including section above roof

maximum 1.5 m from ridge

duct to extend to height of ridge

maximum 45° bends

mastic seal between ducts and structure at point of penetration

ceiling terminals to be easy to clean and have free area at least equal to duct diameter may have manual or automatic controls

Fig. 11.1 Passive stack ventilation – principles.

Ventilation of domestic buildings

Background and rapid ventilation

Kitchens, habitable rooms, utility rooms, bathrooms and sanitary accommodation in domestic buildings should be provided with *background ventilation* and *rapid ventilation*.

AD F1
1.1, 1.2a &
1.2b

Both are achieved by providing a ventilation opening (or openings), some part of which should be at high level (typically 1.75 m above floor level) and with a total area as shown in Table 11.1.

Additionally, background ventilation should be controllable, secure and located to avoid draughts and rain ingress. It might consist of an airbrick, trickle ventilator or suitable opening window as described on page 11.3 above. It should be noted that it is permissible to vary the recommendations of Table 11.1 provided that an average of 6000 mm^2 per room for background ventilation can be achieved with an absolute minimum of 4000 mm^2 in each room.

AD F1,
note to
Table 1

Extract ventilation

In addition to the recommendations for background and rapid ventilation mentioned above, kitchens, utility rooms and bathrooms in domestic buildings should also be provided with *extract ventilation*. Extract ventilation may be achieved using any of the following:

AD F1
1.2c and
Table 1

- mechanical extract ventilation; or
- passive stack ventilation; or
- a suitable open-flued heating appliance.

Mechanical extract ventilation can be operated manually and/or automatically by a controller or sensor. The recommendations are shown in Table 11.1 and fans should be rated at not less than:

- 15 litres/second for a bathroom,
- 30 litres/second for a utility room,
- 30 litres/second for a kitchen where the fan is located within a cooker hood or is not less than 300 mm from the centreline of the hob space, is under humidistat control and is located near the ceiling, or
- 60 litres/second if the fan is located elsewhere in the kitchen.

Passive stack ventilation may be provided by a manual and/or automatic system which uses controllers or sensors to close the system when moisture has been removed (see Fig. 11.1 above).

Open-flued heating appliances can be used as a means of extract ventilation when they are in operation because they take air which they need for combustion from the room or space in which they are installed. When not in operation the appliance should still be capable of providing adequate extract ventilation. Most solid fuel open-flued appliances are acceptable provided that they are used as a primary source of heating, cooking or hot water production. Appliances which burn other fuels may have control dampers which block the air flow when they are not in use or they may have a flue diameter which is insufficient to allow a free flow of air. In these circumstances it is necessary to check that:

(a) the appliance has a flue with a free area which is at least equivalent to a 125 mm diameter duct, and
(b) the appliance has combustion and dilution air inlets which are permanently open when it is not in use so that the ventilation path is unrestricted. These recommendations for rapid, background and extract ventilation are illustrated in Fig. 11.2.

AD F1
1.5

Ventilation of rooms not containing opening windows

In order to make best use of available space in a dwelling, *non-habitable* rooms, such as kitchens, utility rooms, bathrooms and sanitary accommodation are often positioned away from the external walls. This is true especially in flats and since it is not possible to have opening windows in these situations AD F1 permits the use of mechanical and other forms of ventilation as follows:

(a) Mechanical extract ventilation rated as in Table 11.1 operated by a fan, (controlled either automatically using an occupant detector or manually by connection to the light switch) which should continue to run for 15 minutes after the room has been left or the light switched off; *or,*
(b) passive stack ventilation as page 11.6 and Fig. 11.1; *or,*
(c) a suitable open-flued heating appliance.

An air inlet should always be provided to internal rooms ventilated as above and this could be, for example, a 10 mm gap under the door to the room.

AD F1
1.6 and 1.7
An internal *habitable* room may be ventilated through an adjoining room if there is a permanent opening between them with an area equal to $\frac{1}{20}$th of their combined floor areas. Additionally, the ventilation recommendations in Table 11.1 should be based on the combined floor areas of the two rooms.

A habitable room opening onto a conservatory or similar space may be treated as one with the conservatory for the purposes of ventilation. The opening between the room and conservatory (which may contain a door or window, for example) should have an area equal to $\frac{1}{20}$th of the combined floor area of the two rooms (some part of which should be at least 1.75 m above floor level), for rapid ventilation. Background ventilation of 8000 mm should also be provided between the two rooms. Additionally, these provisions for both rapid and background ventilation should be made from the conservatory to outside air (see Fig. 11.2).

Mechanical ventilation

As an alternative to the foregoing methods of ventilation, the requirements of regulation F1 may also be satisfied by:

AD F1
1.9d
(a) following the recommendations of BRE Digest 398 *Continuous mechanical ventilation in dwellings; design installation and operation*. The Digest describes two approaches to the provision of continuous mechanical ventilation which may be applied either to the entire dwelling using a

Kitchen:
Rapid ventilation – opening window (any size); and,
Extract ventilation – 30 l/s via cooker hood otherwise 60 l/s intermittent, or PSV or appropriate open flued heating appliance; and,
Background ventilation – 4000 mm²

Conservatory over habitable room:
Rapid ventilation – provided by openings equal to $\frac{1}{20}$th of combined floor area of conservatory and bedroom; and,
Background ventilation – 8000 mm²
(Note: rapid and background ventilation applies to both outside wall *and* wall between conservatory and bedroom)

Bathroom:
Rapid ventilation – opening window (any size); and,
Extract ventilation – 15 l/s or PSV; and,
Background ventilation – 4000 mm²

Hall:
no requirement for ventilation

Utility Room:
Rapid ventilation (if entered from within dwelling) – opening window (any size); and,
Extract ventilation – 30 l/s or PSV; and,
Background ventilation – 4000 mm²

Habitable rooms:
Rapid ventilation – $\frac{1}{20}$th of floor area some of which is typically 1.75 m above floor; and,
Background ventilation – 8000 mm²

Note: background ventilation may be averaged at 6000 mm² for the dwelling provided that no room has less than 4000 mm²

Sanitary Accommodation (if separate from bathroom):
Rapid ventilation – $\frac{1}{20}$th of floor area some of which is typically 1.75 m above floor or mechanical extract at 6 l/s; and,
Background ventilation – 4000 mm²

Dining room & lounge
counted as one room if area of opening between equals $\frac{1}{20}$th combined floor area

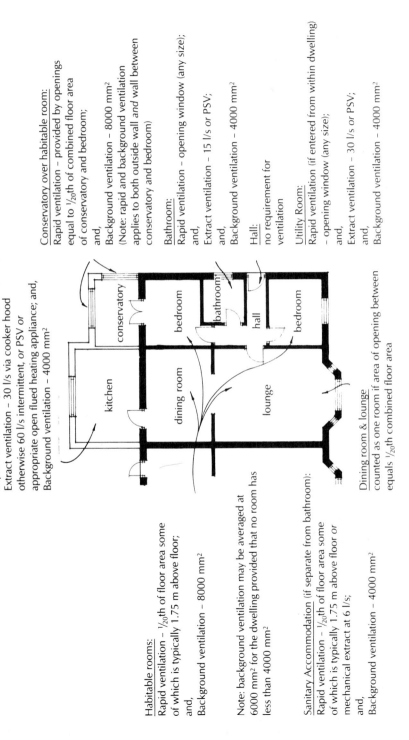

Fig. 11.2 Venilation recommendations for dwellings.

balanced (supply and extract) system or may only apply to the kitchen, utility room, bathroom and sanitary accommodation; *or,*

AD F1
1.9b

(b) following clauses 2.3.2.1, 2.3.3.1, 2.5.2.9, 3.1.1.1, 3.1.1.3 and 3.2.6 of BS 5720: 1979 *Code of practice for mechanical ventilation and air-conditioning in buildings.*

Mechanical extract ventilation and open-flued heating appliances

AD F1
1.8

Mechanical extract ventilation can create dangerous conditions where open-flued appliances are also present due to the spillage of flue gases, whether or not the fans and appliances are in the same room. Where this form of ventilation is provided merely to comply with the recommendations for extract ventilation shown in column 4 of Table 11.1, it is perfectly feasible to use the flue of the combustion appliance for extract ventilation provided it complies with the recommendations shown on page 11.7 above (see *Open-flued heating appliances*).

However, there may still be occasions when it is thought desirable to install open-flued appliances in conjunction with mechanical extract ventilation. In these circumstances it is essential that the appliance is able to operate safely whether or not the fan is running. The risk of danger from the spillage of flue gases will vary according to the type of fuel being burnt. Therefore, whereas mechanical extract ventilation should never be provided in the same room as an open-flued appliance burning *solid* fuel (but for further advice contact the Heating Equipment Testing and Approval Scheme, PO Box 37, Bishop's Cleeve, Gloucestershire, GL52 4TB), it may be possible to use *gas* or *oil* burning appliances in conjunction with mechanical extract ventilation as follows:

(a) *with gas* appliances which are located in a kitchen which is mechanically ventilated, it has been found that an extract rate of not more than 20 litres/sec will be unlikely to cause spillage of flue gases, although it will be necessary to carry out a spillage test in accordance with BS 5440: Part 1, Clause 4.3.2.3. This should be done even though the appliance may be located in a different room to the fan. If this causes spillage then it may be necessary to reduce the extract rate still further until the problem is cured.

(b) advice on the installation of *oil-fired* appliances is contained in Technical Information Note T1/112 which may be obtained from the Oil Firing Technical Association for the Petroleum Industry (OFTEC), Century House, 100 High Street, Banstead Surrey, SM7 2NN.

General information and advice on the subject of the interaction between mechanical extract ventilation and open-flued appliances, including details of the spillage test, may be found in BRE Information Paper 21/92, *Spillage of flue gases from open-flued combustion appliances.*

See also Approved Document J (page 14.2 below) for details of the provision of combustion air to fuel burning appliances.

Ventilation of domestic buildings – alternatives to the AD F1 recommendations

Since AD F1 is not a mandatory document, it is possible to use other advice when providing ventilation in domestic buildings. It is not possible to summarise these other sources of information in this book but the reader may find the following to be of interest:

- BS 5925:1991 *Code of practice for ventilation principles and designing for natural ventilation*, especially clauses 4.4, 4.5, 4.6.1, 4.6.2, 5.1, 6.1, 6.2, 7.2, 7.3, 12 and 13, or, AD F1 1.9a
- BS 5250: 1989 *Code of practice for the control of condensation in buildings*, especially clauses 6, 7, 8, 9.1, 9.8, 9.9.1, 9.9.2, 9.9.3 and Appendix C. AD F1 1.9c

Ventilation of non-domestic buildings – general activities

The ventilation recommendations for general activities in non-domestic buildings in AD F1 follow a similar pattern to those already described for domestic buildings. Provision should be made for rapid, background and extract ventilation and the guidance summarised in Table 11.1 should be read with the following comments: AD F1 2.1a and 2.,2

- An occupiable room in which heavy smoking is to take place (such as a rest room designed for this purpose) should comply with the recommendations shown in Table 11.2. AD F1 2.7e
- The kitchens referred to in the Table are of the domestic type and are not to be construed as commercial kitchens. For further guidance on these see Table 11.2.
- Bathrooms include shower-rooms.
- Sanitary accommodation includes rooms which also contain washing facilities or rooms containing solely washing facilities.
- Extract ventilation can be provided by mechanical means operated manually and/or automatically by a controller or sensor, or by passive stack ventilation for domestic type facilities. The use of open-flued combustion appliances to provide extract ventilation is not mentioned in this part of AD F1 although it is permitted in domestic buildings. Even so, the approved document still recommends that caution be exercised with regard to the use of mechanical extract in a building containing open-flued appliances. AD F1 Table 2
- Background ventilation may be provided by the same means as is described for domestic buildings on page 11.3 above.

Ventilation of communal areas in non-domestic buildings
AD F1 2.5

Many non-domestic buildings have areas where large numbers of people gather, such as foyers in cinemas and theatres, or enclosed shopping malls. Clearly, such spaces need to be ventilated or the air in them will become stale and unhealthy conditions might arise. AD F1 recommends that common spaces should be ventilated either:

- Naturally by means of suitably positioned ventilation opening(s) sufficient to give an opening area equivalent to $\frac{1}{50}$th of the floor area of the common space, or
- Mechanically so that fresh air may be provided at a rate of 1 litre/sec per m^2 of floor area.

Table 11.2 Ventilation recommendations for specialist activities – non-domestic buildings.

1 Use of building or room	2 Approved Document F1-specific recommendations	3 Alternative further guidance documents	4 Notes
(a) School/educational establishment	General areas as Table 11.1. Sanitary accommodation at rate of 6 air changes/hour	See Education (School Premises) Regulations	Fume cupboards complying with Dept of Education Design Note 29 may be needed for areas where noxious fumes generated
(b) Workplaces	—	See Health and Safety Executive Guidance Note EH 22 *Ventilation of the workplace*	
(c) Hospitals	—	See DHSS *Activity Data Base*. For general guidance and standard of provision see individual Dept of Health Building Notes for specific departmental areas	Ventilation needs of different types of accommodation vary with use and may vary throughout year
(d) Building services plant rooms	—	See BS 4434: 1989 *Specification for safety aspects in the design, construction and installation of refrigeration appliances and systems*	Provision may be necessary for emergency ventilation to control dispersal of contaminating gas releases, such as refrigerant leaks. See HSE Guidance Note EH 22 *Ventilation of the workplace,* paragraphs 25 to 27
(e) Rest rooms where smoking allowed	If natural ventilation possible provide both: ● air supply to Table 11.1 for occupiable room, and ● local extraction to remove tobacco smoke. If mechanical ventilation provided allow extract rate of 16 litres/sec per person	—	Workplace (Health and Safety) Regulations 1992 require rest rooms and rest areas to have suitable arrangements to protect non-smokers from discomfort caused by tobacco smoke

Table 11.2 Ventilation recommendations for specialist activities – non-domestic buildings (continued).

1 Use of building or room	*2* Approved Document F1-specific recommendations	*3* Alternative further guidance documents	*4* Notes
(f) Car parks	If *naturally* ventilated provide well distributed permanent ventilation at each level equivalent to $1/20$th floor area at that level with at least 50% in opposing walls. If *mechanically* ventilated provide *either*: ● *both* natural permanent vents not less than $1/40$th of floor area *and* mechanical ventilation of min. 3 air changes/hour, *or* ● 6 air changes per hour for basement car parks and local ventilation at rate of 10 air changes per hour on ramps and exits where cars queue inside building with engines running.	See Association for Petroleum and Explosives Administration publication entitled *Code of practice for ground floor, multi-storey and underground car parks*, or CIBSE Guide B, Section B2.6 and Table B2.7.	Recommendations apply to car parks which are: ● below ground level, or ● enclosed, or ● multi-storey. Instead of provisions in columns 2 and 3 it is also possible to calculate mean predicted pollutant levels and design ventilation system to limit carbon monoxide concentration to: ● not exceeding 50 parts per million average over 8 hour period, and ● not exceeding 100 parts per million for periods not exceeding 15 minutes of peak concentration on ramps and exits.
(g) Commercial Kitchens	—	See Chartered Institution of Building Services Engineers Guide B, Tables B2.3 and B2.11	

It should be noted that the above recommendations do not apply to common spaces used solely or principally for circulation, although AD B1 (means of escape) should also be consulted since it contains certain recommendations regarding ventilation of such spaces (see page 7.39 above).

AD F1
2.6

Ventilation of non-domestic buildings – alternatives to the AD F1 recommendations

As was the case with domestic buildings, it is possible to use certain alternative recommendations to those contained in AD F1 when providing ventilation in non-domestic buildings. The recommendations for rapid, background and extract ventilation shown in Table 11.1, and the guidance given on ventilation to common spaces in non-domestic buildings may also be satisfied by following the advice given in:

- BS 5925: 1991 *Code of practice for ventilation principles and designing for natural ventilation*, clauses 5.1, 5.2, 6.1, 6.2, 7.3, 12 and 13; or
- Chartered Institution of Building Services Engineers (CIBSE) Guide A: *Design data*, section A4 *Air infiltration and natural ventilation*, and CIBSE Guide B: *Installation and equipment data*, section B2 *Ventilation and air-conditioning (requirements)*.

AD F1
2.3

Mechanical ventilation

Many non-domestic buildings contain non-habitable rooms such as, kitchens, bathrooms and sanitary accommodation which are situated away from external walls and are unable to be provided with windows or other ventilation openings. These rooms can be fitted with mechanical extract ventilation at the rates shown in Table 11.1 operated by connection to the light switch or an occupant detector. The fan should have a 15 minute overrun facility and some form of air inlet should be provided to the room, such as a 10 mm gap under the door.

Mechanical ventilation to occupiable rooms should be provided at a rate of:

- 8 litres/sec per occupant for rooms where smoking is not permitted, or
- 16 litres/sec per occupant for rooms designed for light smoking.
- For rooms which are specifically designed for heavy smoking, such as rest rooms where smoking is allowed see Table 11.2.

Mechanical ventilation and air-conditioning plant – design, maintenance and commissioning

AD F1
2.11 to 2.17

Since the air in a building cannot be continuously recycled, at some point in the design of a mechanical ventilation or air-conditioning system it is necessary to introduce fresh air to replace stale air which is being exhausted. Unless care is taken in the siting of the inlets and outlets to the system it is possible that contaminants which are injurious to health may be introduced into the system. Therefore, fresh air inlets should be situated away from areas such as:

- flues,
- exhaust outlets from ventilation systems,
- evaporative cooling towers, and
- areas where vehicles manoevre.

General guidance on how to deal with recirculated air in mechanical ventilation and air-conditioning systems may be found in paragraph 32 of the *Approved Code of Practice and Guidance* L24, issued under the Workplace (Health, Safety and Welfare) Regulations 1992, by the Health and Safety Executive.

Further guidance on the design of mechanical ventilation and air-conditioning systems may be found in BS 5720: 1979 by following clauses 2.3.2, 2.3.3, 2.4.2, 2.4.3, 2.5, 3.2.6, 3.2.8, and 5.5.6 *or* CIBSE Guide B sections B2 and B3.

There is one particular disease which is associated with air-conditioning systems known as legionnaires' disease. This is a form of lung infection caused by the bacteria legionella pneumophilia and is named after an epidemic which affected 182 people attending an American Legion Convention in 1976. In the original outbreak the germ was found to have been transmitted through the cooling and evaporating elements of a large, central air-conditioning system. Since a large number of subsequent outbreaks of the disease have been traced to similar sources it is essential that cooling and heating systems are cleaned regularly and filters are changed often. Further information may be obtained from the guide issued by the Health and Safety Executive, *The control of legionellosis including legionnaires' disease*, paragraphs 71 to 89.

In order to be able to carry out the regular maintenence of cleaning the system and replacing filters it is essential that all parts are available for access and sufficient space is provided, especially in central plant rooms. Normally, special provision for access will be made in the design of the system. Where this is not the case, AD F1 recommends the following minimum dimensions for access passageways and cleaning points in central plant rooms:

- 600 mm wide by 2000 mm high general access passageways for walking between plant, and
- 1100 mm wide by 1400 mm high kneeling spaces for routine cleaning and maintenance of equipment. Additionally, a 690 mm high space should be available for access to low level equipment.

Since these figures are the minimum recommended they do not necessarily include for access doors which may need additional space.

This approved document guidance is very limited and it may prove more useful to consult Building Services Research and Information Association Technical Note TN 10/92: *Space allowances for building services distribution systems – detailed design stage*, Sections A5 and D2. (BSRIA 1992, ISBN 0 86226 350 7).

Mechanical ventilation and air-conditioning systems are complex and need to be commissioned and tested to ensure that they are performing in accordance with their design specifications. Therefore, the local building control authority will need to be satisfied that the installed systems have been commissioned and tested so that they are performing their ventilation functions effectively. This recommendation applies only to a system:

- which is installed in a building to serve a floor area greater than 200 m², and
- in which the other provisions for mechanical ventilation and air-conditioning mentioned above have been followed.

Compliance may be demonstrated to the local authority by presenting them with test reports and commissioning certificates showing that commissioning and testing has been carried out in accordance with the CIBSE commisioning codes.

Ventilation of non-domestic buildings – specialist activities

AD F1
2.7 to 2.10

The recommendations listed in Table 11.1 refer to the ventilation of rooms which are used for activities of a general nature where the production of water vapour or small amounts of tobacco smoke are the main problems, and therefore the activities are not dissimilar to those encountered in domestic premises.

Many non-domestic buildings have rooms or spaces where large quantities of water vapour are produced or where noxious fumes may be generated and the provisions of Table 11.1 may be inadequate in these circumstances. Comprehensive recommendations for the ventilation of these specialist activities is beyond the scope of AD F1 so the reader is referred to a number of additional guidance documents where more detailed guidance may be sought. These are summarised in Table 11.2 and it should be noted that the recommendations for ventilation of car parks shown in the table relate to the provision of air to ensure normal healthy conditions are maintained. Reference should be made also to Approved Document B (see page 7.143 above) for guidance on the design of mechanical ventilation and air-conditioning systems for the purposes of fire safety.

Condensation in roofs

Regs Sch. 1
F2

In buildings, adequate provision must be made to prevent excessive condensation in roofs and roof voids over insulated ceilings.

AD F2

When condensation occurs in roof spaces it can have two main effects:

(a) the thermal performance of the insulant materials may be reduced by the presence of the water; *and,*
(b) the structural performance of the roof may be affected due to increased risk of fungal attack.

Approved Document F2 recommends that, under normal conditions, condensation in roofs and in spaces above insulated ceilings should be limited such that the thermal and structural performance of the roof will not be substantially and permanently reduced.

AD F2 applies only to roofs where the insulation is placed at ceiling level (cold roofs) irrespective of whether the ceiling is flat or pitched. Warm roofs where the insulation is placed above the structural system and roof void do not present the same risks and, therefore, are not covered.

It should be noted that the provisions of AD F2 apply to roofs of any pitch even though a roof which exceeds 70° in pitch is required to be insulated as if it were a wall.

Small roofs over porches or bay windows, etc., may sometimes be excluded from the requirements of regulation F2 if there is no risk to health or safety.

AD F2
0.1 to 0.6

Roofs with a pitch of 15° or more

Pitched roofs should be cross-ventilated by permanent vents at eaves level on the two opposite sides of the roof, the vent areas being equivalent in area to a continuous gap along each side of 10 mm width.

AD F2
1.2

Mono-pitch or lean-to roofs should have ventilation at eaves level as above and also at high level either at the point of junction or through the roof covering at the highest practicable point. The high level ventilation should be equivalent in area to a continuous gap 5 mm wide (see Fig. 11.3).

AD F2
1.4

Roofs with a pitch of less than 15°

In low-pitched roofs the volume of air contained in the void is less and therefore the risk of saturation is greater.

This also applies to roofs with pitch greater than 15° where the ceiling follows the pitch of the roof. High level ventilation should be provided as in 1.4 above.

AD F2
2.1 & 2.5

Cross-ventilation should again be provided at eaves level but the ventilation gap should be increased to 25 mm width.

AD F2
2.2

Where the roof span exceeds 10m or the roof plan is other than a simple rectangle, more ventilation, totalling 0.6% of the roof area, may be needed.

AD F2
2.3

A free airspace of at least 50 mm should be provided between the roof deck and the insulation. This may need to be formed using counter-battens if the joists run at right angles to the flow of air (see Fig. 11.4).

AD F2
2.4

Where it is not possible to provide proper cross-ventilation an alternative form of roof construction should be considered.

AD F2
2.6

It is possible to install vapour checks (called vapour control layers in BS 5250) at ceiling level using polythene or foil-backed plasterboard, etc., to reduce the amount of moisture reaching the roof void. This is not acceptable as an alternative to ventilation unless a complete vapour barrier is installed.

AD F2
2.7

The requirements can also be met for both flat and pitched roofs by following the relevant recommendations of BS 5250: 1989 *Code of practice: the control of condensation in buildings*, Clauses 9.1, 9.2 and 9.4. Further guidance may also be found in the 1994 edition of BRE Report BR 262 *Thermal insulation: avoiding the risks*.

AD F2
1.5 & 2.8

AD F2
0.8

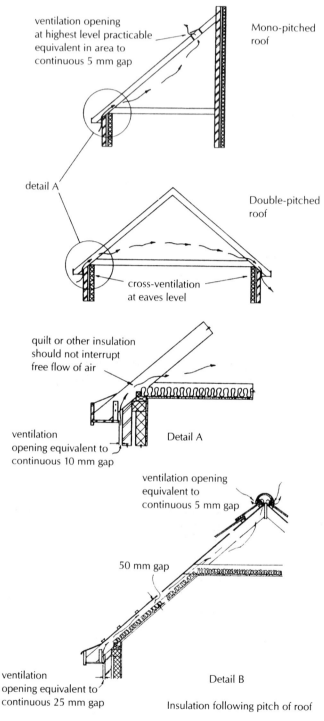

ventilation opening
at highest level practicable
equivalent in area to
continuous 5 mm gap

Mono-pitched
roof

detail A

Double-pitched
roof

cross-ventilation
at eaves level

quilt or other insulation
should not interrupt
free flow of air

ventilation
opening equivalent to
continuous 10 mm gap

Detail A

ventilation opening
equivalent to
continuous 5 mm gap

50 mm gap

ventilation
opening equivalent to
continuous 25 mm gap

Detail B

Insulation following pitch of roof

Fig. 11.3 Roof void ventilation – roofs pitched at 15° or more.

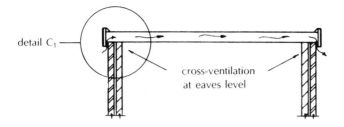

detail C₁

cross-ventilation
at eaves level

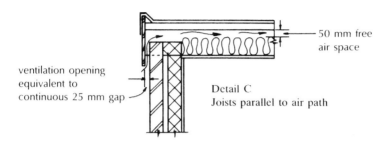

50 mm free
air space

ventilation opening
equivalent to
continuous 25 mm gap

Detail C
Joists parallel to air path

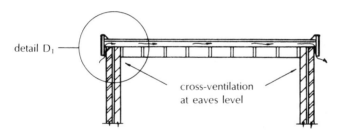

detail D₁

cross-ventilation
at eaves level

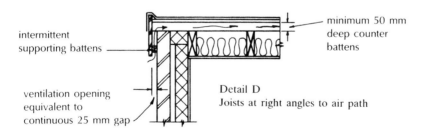

intermittent
supporting battens

minimum 50 mm
deep counter
battens

ventilation opening
equivalent to
continuous 25 mm gap

Detail D
Joists at right angles to air path

Fig. 11.4 Roof void ventilation – roofs pitched at less than 15°.

Chapter 12

Hygiene

Introduction

When first introduced, Part G of Schedule 1 to the Building Regulations 1985 consisted of four requirements grouped under the title 'Hygiene'.

The first of these requirements (G1 – Food storage) required that dwellings be provided with adequate food storage accommodation. Since most people have refrigerators or deep freezers today this regulation has become outdated and food storage is no longer controlled by the building regulations.

Consequently, the former regulation G4 (Sanitary conveniences and washing facilities) has been renumbered G1 in the 1991 Regulations to fill the void left by the now defunct food storage requirements.

The remaining regulations in Part G relate to Bathrooms (G2) and Hot water storage (G3).

Sanitary conveniences and washing facilities

Adequate sanitary conveniences (i.e. closets and urinals) situated in purpose-built accommodation or bathrooms, must be provided in buildings. This requirement replaces section 26 of the Building Act 1984.

Additionally, adequate washbasins with suitable hot and cold water supplies must be provided in rooms containing water-closets or in adjacent rooms or spaces.

These sanitary conveniences and washbasins must be separated from places where food is prepared and must be designed and installed so that they can be cleaned effectively.

It may be noted that section 66 of the 1984 Act enables the local authority to serve a notice on an occupier requiring him to replace any closet provided for his building which is not a water-closet. The notice can only be served where the building has a sufficient water supply and a sewer available. Where a notice requiring closet conversion is served, the local authority must bear half the cost of carrying out the work.

A satisfactory level of performance will be achieved if:

- Sufficient numbers of the appropriate type of sanitary convenience are provided depending on the sex and age of the users of the building.

Regs Sch. 1
G1

1984 Act,
sec. 66

AD G1

- Washbasins with hot and cold water supply are provided either in or adjacent to rooms containing water-closets.

Both sanitary conveniences and washbasins should be sited, designed and installed so as not to be a health risk.

Provision of sanitary conveniences and washbasins

The following definitions apply in AD G1.

SANITARY CONVENIENCE – closets and urinals.

SANITARY ACCOMMODATION – a room containing closets or urinals. Other sanitary fittings may also be present. Sanitary accommodation containing more than one cubicle may be treated as a single room provided there is free air circulation throughout the room.

WATER-CLOSET – is defined by section 126 of the Building Act 1984 as a closet which has a separate fixed receptacle connected to a drainage system and separate provision for flushing from a supply of clean water, either by the operation of mechanism or by automatic action.

AD G1
1.12

AD G1 also permits the use of a chemical or other means of treatment where drains and water supply are not available. It is not clear whether earth-closets would be permitted, but on normal principles of interpretation it is unlikely that they would be.

AD G1
1.1

Houses, flats and maisonettes should have at least one closet and one washbasin. This also applies to houses in multiple occupation (houses where the occupants are not part of a single household), if the facilities are available for the use of all the occupants.

In other types of buildings the scale of provision and the siting of appliances may be the subject of other legislation as follows:

- Workplace (Health, Safety & Welfare) Regulations 1992. Approved Code of Practice & Guidance.

(This document is not referred to in the current edition of ADGI because it was published after that document. However, it repeals those provisions of the Offices, Shops and Railway Premises Act 1963 and the Factories Act 1961 which are referred to in ADGI.)

AD G1
1.4

- The Food Hygiene (General) Regulations 1970
- Part M of Schedule 1 to the 1991 Regulations (Access and facilities for disabled people).

AD G1
1.13

The requirement to provide satisfactory sanitary conveniences can also be met, subject to other legislation, by referring to the relevant clauses of BS 6465 *Sanitary installations*, Part 1: 1984 which contains details of the scale of provision, selection and installation of sanitary appliances.

A room or space containing closets or urinals should be separated by a door from any area in which food is prepared or washing up done. The 1991 edition of AD G1 makes it clear, therefore, that a separate lobby is not required.　AD G1 1.2

Additionally, washbasins should be placed:

(a) in the room containing the closet; *or,*
(b) in the room or space immediately leading to the room containing the closet provided it is not used for food preparation; *or,*
(c) in the case of dwellings, in the room or space adjacent to the room containing the closet.

In this last case it is unclear whether or not the space may be used for the preparation of food, but in all probability it is not.　AD G1 1.3

Closets, urinals and washbasins should have smooth, readily-cleaned, non-absorbent surfaces.　AD G1 1.5

Any flushing apparatus should be capable of cleansing the receptacle effectively. The receptacle should only be connected to a flush pipe or branch discharge pipe.　AD G1 1.6

Any washbasins required by the provisions of regulation G4 should have a supply of hot water from a central source or unit water heater and a piped cold water supply.　AD G1 1.7

Discharge from sanitary conveniences and washbasins

Water-closets should discharge via a trap and branch pipe to a soil stack pipe or foul drain.　AD G1 1.8

In recent years a system of waste disposal has been developed in which the discharge from a waste appliance is fed into a macerator. The liquified contents are then pumped via a small bore pipe to the normal foul drainage system. A closet is permitted to be connected to such a system provided:

(a) a closet discharging directly to a gravity system is also available; *and*　AD G1 1.10
(b) the macerator system is the subject of a current European Technical Approval issued by a member body of the European Organisation for Technical Approvals e.g. the British Board of Agrément. The conditions of use must be in accordance with the terms of the ETA.

Urinals which are fitted with flushing apparatus should have an outlet fitted with an effective grating and trap and should discharge via a branch pipe to a soil stack pipe or foul drain (see Approved Document H1 and Chapter 13 for details of drainage).　AD G1 1.9

Washbasins should discharge via a trap and branch discharge pipe to a soil stack. If on the ground floor, it is permissible to discharge the basin to a gulley or direct to a drain.　AD G1 1.11

Bathrooms

Dwellings are required to be provided with a bathroom containing a fixed bath or shower. Hot and cold water must also be supplied to the bath or shower. This requirement replaces section 27 of the Building Act 1984.　Regs Sch. 1 G2

The foregoing requirements apply to dwellings (i.e. houses, flats and　AD G2 maisonettes) and houses in multiple occupation (houses where the occupants are not part of a single household). In the latter case the facility should be available to all the occupants.

The hot and cold water supplies should be piped to the bath or shower and hot water may come from a central source such as a hot water cylinder or from a unit water heater.

The discharge from the bath or shower should be via a trap and waste pipe to a gulley, soil stack pipe or foul drain direct (see Approved Document H1 and Chapter 13 for details of drainage).

A bath or shower may be connected via a macerator system provided it complies with a current European Technical Approval.

Hot water storage

A hot water storage system incorporating a hot water storage vessel which is not vented to the atmosphere must be installed by a competent person and adequate precautions must be taken to:

(a) prevent the water temperature exceeding 100°C; *and,*
(b) ensure that any hot water discharged from safety devices is conveyed safely to a disposal point where it is visible but will not be a danger to users of the building.

The above requirements do not apply to space heating systems, systems which heat or store water for industrial processes and systems which store 15 litres or less of water.

Approved Document G3 describes the provisions for an unvented hot water storage system. In such a system, the stored hot water is heated in a closed vessel. Without adequate safety devices an uncontrolled heat input would cause the water temperature to rise above the boiling point of water at atmospheric pressure (100°C). At the same time the pressure would increase until the vessel burst. This would result in an almost instantaneous conversion of water to steam with the large increase in volume producing a steam explosion.

Water for domestic use is required at temperatures below 100°C, therefore, an explosion cannot occur if the water is released at these temperatures, however great the pressure, hence the precautions required by regulation G3 to prevent the water temperature exceeding 100°C.

The term 'domestic hot water' is defined in AD G3 as water which has been heated for washing, cooking and cleaning purposes. The term is used irrespective of the type of building in which an unvented hot water storage system is installed.

Figure 12.1 illustrates the three independent levels of protection which should be provided for each source of energy supply to the stored water. These are:

- Thermostatic control (see Part L, Chapter 16).
- Non self-resetting thermal cut-outs to BS 3955: 1986 (electrical controls) or BS 4201: 1979 (for gas burning appliances).
- One or more temperature operated relief valves to BS 6283 *Safety devices for use in hot water systems* Part 2: 1991 or Part 3: 1991.

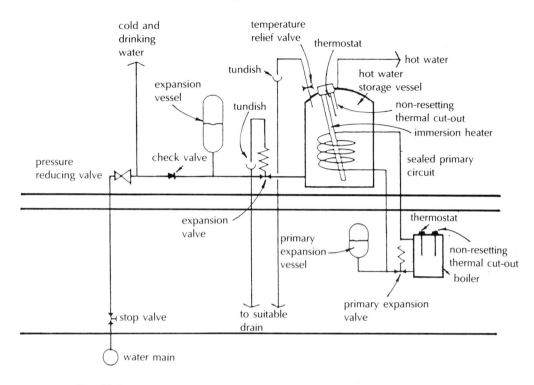

Fig. 12.1 Directly and indirectly heated unvented hot water storage system.

These safety devices are required for both directly and indirectly heated unvented hot water storage systems. The safety devices are designed to work in sequence as the temperature rises. All three means of protection would have to fail for the water temperature to exceed 100°C.

AD G3
Section 3
3.3

AD G3 provides separate recommendations for smaller (usually domestic) systems (not exceeding 500 litres capacity with a heat input below 45 kW) in section 3. Systems which exceed 500 litres capacity or have a heat input in excess of 45 kW are dealt with in Section 4.

AD G3
Section 3

AD G3
Section 4

Section 3 hot water storage systems

Generally, a system covered by Section 3 of AD G3 should be in the form of a unit or package which is:

(a) approved by a member body of the European Organisation for Technical Approvals (EOTA) operating a technical approvals scheme which ensures that the relevant requirements of regulation G3 will be met (e.g. the British Board of Agrément); *or*,

AD G3
Section 3
3.2

(b) approved by a certification body having accreditation from the National Accreditation Council for Certification Bodies (NACCB). This would include testing to the requirements of an appropriate standard to ensure compliance with regulation G3 (e.g. BS 7206: 1990 *Specification for unvented hot water storage units and packages*); *or*,

(c) independently assessed to clearly demonstrate an equivalent level of verification and performance to that in (a) or (b) above.

This means that the system should be factory made and supplied either as a *unit* (fitted with all the safety protection devices mentioned above and incorporating any other operating devices to stop primary flow, prevent backflow, control working pressure, relieve excess pressure and accommodate expansion fitted to the unit by the manufacturer) or as a *package* in which the safety devices are fitted by the manufacturer but the operating devices are supplied in kit form to be fitted by the installer.

This approach ensures that the design and installation of the safety and operating devices are carried out by the manufacturer who is conversant with his own equipment and can control the training and supervision of his staff.

**AD G3
Section 3
3.11**

It should be noted that where a system is subject to the above approvals it is unlikely to need site inspection by the building control authorities.

This may not be the case in other situations.

The recommendations for approval mentioned above ensure that the system is fit for its purpose and that the information regarding installation, maintenance and use of the system is made available to all concerned.

Provision of non self-resetting thermal cut-outs

Storage systems may be heated directly or indirectly. In an unvented, indirectly heated system (see Fig. 12.1) the non self-resetting thermal cut-out should be wired up to a motorised valve or other approved device or should shut off the flow to the primary heater. These devices should be subject to the same approvals as the units or packages.

**AD G3
Section 3
3.6**

Sometimes a unit system may incorporate a boiler. In this case the thermal cut-out may be located on the boiler.

**AD G3
Section 3
3.7**

In many cases an indirect system will also contain an alternative direct method of water heating (such as an immersion heater). This alternative heating source will also need to be fitted with a non self-resetting thermal cut-out. The non self-resetting thermal cut-out should be connected to the direct heat source or the indirect primary flow control device in accordance with the current edition of the Regulations for Electrical Installations of the Institution of Electrical Engineers.

**AD G3
Section
3.10**

Provision of temperature relief valves

Whether the unit or package is directly or indirectly heated the temperature relief valve should be situated directly on the storage vessel in order to prevent the stored water exceeding 100°C.

BS 6283 requires that each valve be marked with a discharge rating (in kW). This rating should never be less than the maximum power input to the vessel which the valve protects. More than one valve may be needed.

Valves should also comply with the following:

- They should not be disconnected except for replacement.
- They should not be relocated in any other position.
- The valve connecting boss should not be used to connect any other devices or fittings.

- They should discharge through a short length of metal pipe (D1) which is of at least the same bore as the valve's nominal outlet size.

  The discharge should either be direct or by way of a manifold which is large enough to take the total discharge of all the pipes connected to it. It should then continue via an air break to a tundish which is located vertically as near as possible to the valve.

AD G3
Section 3
3.5

It may be possible to provide an equivalent degree of safety using other safety devices but these would need to be assessed in a similar manner to units or packages (see above).

AD G3
Section 3
3.4

Installation

The installation of the system should be carried out by a competent person.

  This means a person who holds a current Registered Operative Identity Card for the installation of unvented domestic hot water storage systems issued by:

(a) the Construction Industry Training Board (CITB); *or,*
(b) the Institute of Plumbing; *or,*
(c) the Association of Installers of Unvented Hot Water Systems (Scotland and Northern Ireland); *or,*
(d) designated Registered Operatives who are employed by companies included on the list of Approved Installers published by the BBA up to 31 December 1991; *or,*
(e) an equivalent body.

AD G3
Section 3
3.8

Discharge pipes

Discharge pipe D1 (see above and Fig. 12.2) is usually supplied by the storage system manufacturer. (If not it should be fitted by the installer of the system.)

  In either case the tundish should be:

- vertical;
- located in the same space as the unvented hot water storage system; *and,*
- fitted within 500 mm of the safety device.

  Discharge pipe D2 (see Fig. 12.2) from the tundish should comply with the following:

- it should terminate in a safe place where it cannot present a risk of contact to users of the building;
- it should be laid to a continuous fall;
- the discharge should be visible at either tundish or final outlet, but preferably at both of these locations (Fig. 12.2 shows possible discharge arrangements);
- it should be at least one pipe size larger than the nominal outlet size of the safety device unless its total equivalent hydraulic resistance exceeds that of a straight pipe 9 m long (bends will increase flow resistance, therefore Table 1 from AD G3 is reproduced below and shows how to calculate the minimum size of discharge pipe D2); *and,*

AD G3
Section 3
3.9

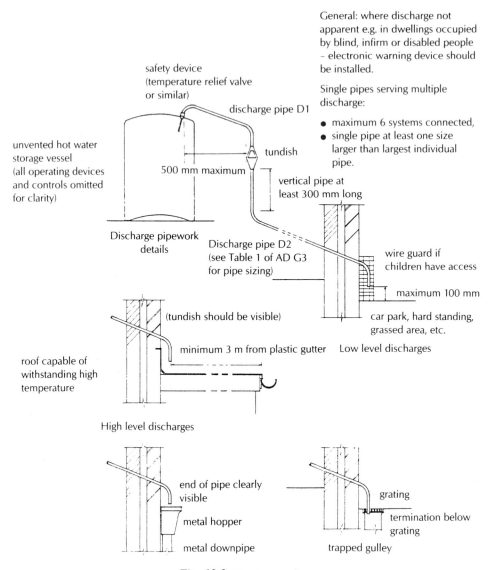

General: where discharge not apparent e.g. in dwellings occupied by blind, infirm or disabled people – electronic warning device should be installed.

Single pipes serving multiple discharge:

- maximum 6 systems connected,
- single pipe at least one size larger than largest individual pipe.

safety device (temperature relief valve or similar)

discharge pipe D1

unvented hot water storage vessel (all operating devices and controls omitted for clarity)

tundish

500 mm maximum

vertical pipe at least 300 mm long

Discharge pipework details

Discharge pipe D2 (see Table 1 of AD G3 for pipe sizing)

wire guard if children have access

maximum 100 mm

car park, hard standing, grassed area, etc.

Low level discharges

(tundish should be visible)

minimum 3 m from plastic gutter

roof capable of withstanding high temperature

High level discharges

end of pipe clearly visible

metal hopper

metal downpipe

grating

termination below grating

trapped gulley

Fig. 12.2 Discharge pipes.

- it should have a vertical section of pipework at least 300 mm long below the tundish before any changes in direction.

Alternative approach

AD G3
Section 3
3.9

Discharge pipes may also be sized in accordance with BS 6700: 1987 *Specification for design, installation, testing and maintenance of services supplying water for domestic use within buildings and their curtilages*, Appendix E, section E2 and table 21.

AD G3, Section 3

Table **1**　Sizing of copper discharge pipe 'D2' for common temperature relief valve outlet sizes

Valve outlet size	Minimum size of discharge pipe D1*	Minimum size of discharge pipe D2* from tundish	Maximum resistance allowed, expressed as a length of straight pipe (i.e. no elbows or bends)	Resistance created by each elbow or bend
G$\frac{1}{2}$	15 mm	22 mm	up to 9 m	0.8 m
		28 mm	up to 18 m	1.0 m
		35 mm	up to 27 m	1.4 m
G$\frac{3}{4}$	22 mm	28 mm	up to 9 m	1.0 m
		35 mm	up to 18 m	1.4 m
		42 mm	up to 27 m	1.7 m
G1	28 mm	35 mm	up to 9 m	1.4 m
		42 mm	up to 18 m	1.7 m
		54 mm	up to 27 m	2.3 m

* see 3.5, 3.9, 3.9(a) and Fig. 12.2

Worked example:
The example below is for a G$\frac{1}{2}$ temperature relief valve with a discharge pipe (D2) having 4 No. elbows and length of 7 m from the tundish to the point of discharge.

From Table 1:
Maximum resistance allowed for a straight length of 22 mm copper discharge pipe (D2) from a G$\frac{1}{2}$ temperature relief valve is: 9.0 m
Subtract the resistance for 4 No. 22 mm elbows at 0.8 m each = 3.2 m

Therefore the maximum permitted length equates to: 5.8 m

5.8 m is less than the actual length of 7 m therefore calculate the next largest size.

Maximum resistance allowed for a straight length of 28 mm pipe (D2) from a G$\frac{1}{2}$ temperature relief valve equates to: 18 m

Subtract the resistance for 4 No. 28 mm elbows at 1.0 m each = 4 m

Therefore the maximum permitted length equates to: 14 m

As the actual length is 7 m, a 28 mm (D2) copper pipe will be satisfactory.

Section 4 hot water storage systems

Systems within the scope of Section 4 exceed 500 litres in capacity or have a power input of more than 45 kW. Generally they will be individual designs for specific projects and therefore, not systems appropriate for EOTA or NACCB certification. Nevertheless, these systems should still conform to the same general safety recommendations as in Section 3 including design by an appropriately qualified engineer and installation by a competent person. *AD G3 Section 4 4.1, 4.2*

　Systems with a storage vessel of more than 500 litres capacity but with a power input of not more than 45kW should have safety devices conforming to BS 6700: 1987 (Section two, Clause 7) or other equivalent practice specifications which recommend a similar operating sequence for the safety devices to prevent the stored water temperature exceeding 100°C. *AD G3 Section 4 4.3*

An unvented hot water storage vessel with a power input which exceeds 45kW should also have an appropriate number of temperature relief valves which:

(a) either comply with BS 6283: Parts 2 or 3 or equivalent giving a combined discharge rating at least equivalent to the power input; *or,*
(b) are equally suitable and marked with the set temperature in °C and a discharge rating marked in kW, measured in accordance with Appendix F of BS 6283: Part 2: 1991 or Appendix G of BS 6283: Part 3: 1991 and certified by a member of EOTA such as BBA or another recognised testing body (e.g. the Associated Offices Technical Committee, AOTC).

AD G3
Section 4
4.4

The temperature relief valves should be factory fitted to the storage vessel and the sensing element located as described in paragraph 3.5 of AD G3.

AD G3
Section 4
4.5, 4.6

The non self-resetting thermal cut-outs should be installed in the system as described in Section 3 and the discharge pipes should also comply with that section.

Chapter 13

Drainage and waste disposal

Introduction

This chapter describes Part H of Schedule 1 to the Building Regulations 1991 and the associated Approved Document H. Together, these documents cover:

- Foul water drainage (H1);
- Cesspools and tanks (H2);
- Rainwater drainage (H3); *and,*
- Solid waste storage (H4).

Formerly, the building regulations dealing with drainage were phrased in functional terms and considerable reliance had to be placed on British Standards and BRE Digests. Much of this information appears to have been included in Approved Document H and, except for very large installations, it would appear possible to design a satisfactory drainage system for a building without reference to other sources of information.

Some sections of the Building Act 1984 are concerned with sanitation and buildings, while others deal with drainage. A number of the relevant provisions of the 1984 Act will be referred to in this chapter.

At this stage three important provisions of the 1984 Act must be noted:

(i) Drainage of new buildings
Section 21 of the 1984 Act makes it unlawful to erect or extend any building unless satisfactory provision is made for the drainage of that building. The local authority *must* reject plans deposited under the regulations if no satisfactory provision for drainage is shown on them.

This provision must be read in light of the decision of the Divisional Court in *Chesterton R.D.C.* v. *Ralph Thompson Ltd*, (1947) KB 300, holding that the local authority are not entitled to reject plans under section 37 on the ground that the sewerage system, into which the drains lead, is unsatisfactory. That is immaterial; the council must consider only the drainage of the particular building.

(ii) Drainage of buildings in combination
On housing estates buildings are invariably drained in combination, and section 22 of the 1984 Act enables the local authority to require the drainage of two or more buildings in combination by means of a private sewer. However, it

should be noted that where plans have already been passed, this power can be exercised only by agreement with the owners.

(iii) Rainwater pipes must not be used to carry soil drainage or to provide ventilation for any system of soil drains

By section 60 of the 1984 Act a pipe for conveying rainwater from a roof may not be used for conveying soil or drainage from a sanitary convenience, or as a ventilating shaft to a foul drain. The practical effect of this provision is that all rainwater pipes must be trapped before entering a foul drain.

With regard to solid waste storage, all dwellings are now required to have satisfactory means of storing solid waste and the provision of sections 23(1) and (2) of the Building Act 1984 which required satisfactory means of access for removal of refuse have been replaced by paragraph H4 of Schedule 1 to the 1991 Building Regulations.

This paragraph of the regulations must be read in light of other legislative provisions in respect of refuse disposal.

In particular, sections 45 to 47 of the Environmental Protection Act 1990 should be referred to (see Chapter 5) since those sections deal with the removal of refuse and allied matters. Thus, under section 45 of the 1990 Act a duty is placed on the local authority to collect all household waste in their area, while sections 46 and 47 make provision for the removal of trade and other refuse. Section 23(3) of the Building Act 1984 requires the local authority's consent to close or obstruct the means of access by which refuse is removed from a house.

Sanitary pipework and drainage

Regs Sch. 1
H1(1)

Any system which carries foul water from appliances in a building to a sewer, cesspool, septic tank or settlement tank is required to be adequate.

FOUL WATER is defined as waste water which comprises or includes:

- Waste from a sanitary convenience or other soil appliance.

Regs Sch. 1
H1(2)

- Water which has been used for cooking or washing.

Further guidance on the meaning of SANITARY CONVENIENCE is given in the guidance to Approved Document 91 where it is defined as a closet or urinal.

AD H1
page 3

FOUL WATER OUTFALL may be a foul or combined sewer, cesspool, septic tank or settlement tank. This term is not specifically defined in AD H1; however the term is inferred from the description of Performance on page 3.

The requirements of Paragraph H1 may be met by any foul water drainage system which:

- Conveys the flow of foul water to a suitable foul water outfall;
- Reduces to a minimum the risk of leakage or blockage;
- Prevents the entry of foul air from the drainage system to the building, under working conditions;
- Is ventilated; *and,*
- Is accessible for clearing blockages.

AD H1 sets out detailed provisions in two sections. Section 1 deals with sanitary pipework (i.e. above ground foul drainage) and section 2 with foul drainage (i.e. below ground foul drainage). There is also an appendix which purports to contain additional guidance for large buildings. It is somewhat surprising to discover, therefore, that it also contains details of special precautions relating to the drains of any building concerning settlement, surcharging, rodent control and ground loads.

AD H1
Appendix

Above ground foul drainage

A number of terms are used throughout AD H1. These are defined below and illustrated in Fig. 13.1. It should be noted that these definitions do *not* appear in the approved document.

DISCHARGE STACK – a ventilated vertical pipe which carries soil and waste water directly to a drain.

VENTILATING STACK – a ventilated vertical pipe which ventilates a drainage system either by connection to a drain or to a discharge stack or branch ventilating pipe.

BRANCH DISCHARGE PIPE (sometimes referred to as a BRANCH PIPE) – the section of pipework which connects an appliance to another branch pipe or a discharge stack if above the ground floor, or to a gully, drain or discharge stack if on the ground floor.

BRANCH VENTILATING PIPE – the section of pipework which allows a branch discharge pipe to be separately ventilated.

STUB STACK – an unventilated discharge stack.

A drainage system, whether above or below ground, should have sufficient *capacity* to carry the anticipated *flow* at any point. The *capacity* of the system, therefore, will depend on the size and gradient of the pipes whereas the *flow* will depend on the type, number and grouping of appliances. Table 13.1 below is based on information from BS 8301 and Table A1 of AD H1, and gives the expected flow rates for a range of appliances.

AD H1
0.1 to 0.3

Since sanitary appliances are seldom used simultaneously the normal size of discharge stack or drain will be able to take the flow from quite a large number of appliances. Table 1 of AD H1 is reproduced below and is derived from BS 5572. It shows the approximate flow rates from dwellings and is based on an appliance grouping per household of 1 WC, 1 bath, 1 or 2 washbasins and 1 sink.

AD H1
0.4

Pipe sizes

Since individual manufacturer's pipe sizes will vary, the sizes quoted in AD H1 are nominal and give a numerical designation in convenient round numbers. Similarly, equivalent pipe sizes for different pipe standards are contained in BS 5572 (Sanitary pipework) and BS 8301 (Building drainage).

AD H1
0.5

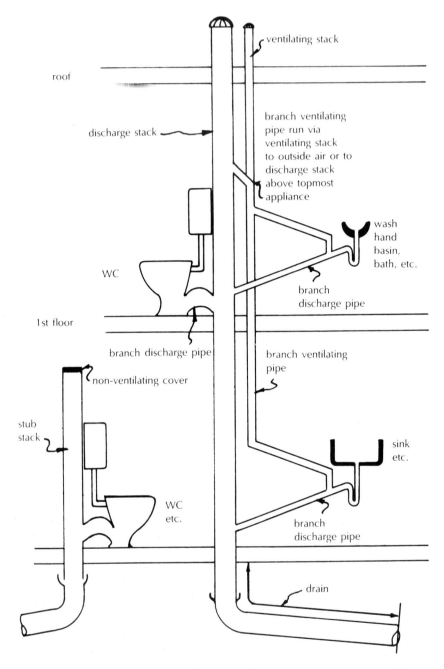

Fig. 13.1 Definitions.

Table 13.1 Appliance flow rates.

Appliance	Flow rate (litres/sec)
WC (9 litre washdown)	2.3
Washbasin	0.6
Sink	0.9
Bath	1.1
Shower	0.1
Washing machine	0.7
Urinal (per person unit)	0.15
Spray tap basin	0.06

AD H1, section 1

Table 1   **Flow rates from dwellings**

Number of dwellings	Flow rate (litres/sec)
1	2.5
5	3.5
10	4.1
15	4.6
20	5.1
25	5.4
30	5.8

Trap water seals

AD H1
sec. 1
1.1

Trap water seals are provided in drainage systems to prevent foul air from the system entering the building. All discharge points into the system should be fitted with traps and these should retain a minimum seal of 25 mm under test and working conditions.

AD H1
sec. 1
1.15

Traditionally the 'one pipe' and 'two pipe' systems of plumbing have required the provision of branch ventilating pipes and ventilating stacks unless special forms of trap are used. The 'single-stack' system of plumbing obviates the need for these ventilating pipes and is illustrated in Fig. 13.2. Table 13.2 below, which is based on Table 2 and Table A2 of AD H1, gives minimum dimensions of pipes and traps where it is proposed to use appliances other than those shown in Fig. 13.2. Additionally, it is permissible to reduce the depth of trap seal to 38mm where sinks, baths or showers are installed on the ground floor and discharge to a gully.

AD H1
sec. 1
1.2 &
Appendix
A2

It should be stressed that the minimum pipe sizes given above relate to branch pipes serving a single appliance. Where a number of appliances are served by a single branch pipe which is unventilated, the diameter of the pipe should be at least the size given in Table 3 to section 1 of AD H1, which is reproduced below.

AD H1
sec. 1
1.12

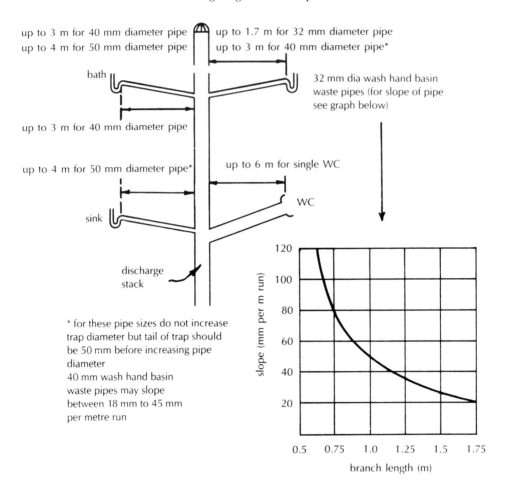

up to 3 m for 40 mm diameter pipe   up to 1.7 m for 32 mm diameter pipe
up to 4 m for 50 mm diameter pipe   up to 3 m for 40 mm diameter pipe*

bath

up to 3 m for 40 mm diameter pipe

32 mm dia wash hand basin
waste pipes (for slope of pipe
see graph below)

up to 4 m for 50 mm diameter pipe*   up to 6 m for single WC

sink   WC

discharge
stack

* for these pipe sizes do not increase
trap diameter but tail of trap should
be 50 mm before increasing pipe
diameter
40 mm wash hand basin
waste pipes may slope
between 18 mm to 45 mm
per metre run

slope (mm per m run)

120

100

80

60

40

20

0.5   0.75   1.0   1.25   1.5   1.75

branch length (m)

Appliance	Minimum diameter of pipe and trap (mm)	Depth of trap seal	Slope (mm/m)
sink	40	75	18−90
bath	40	75	18−90
WC	75 (siphonic only)	50	9
washbasin	32	75	see graph above

Fig. 13.2 Single stack system – design limits.

Table 13.2 Minimum dimensions of branch pipes and traps.

Appliance	Minimum diameter of pipe and trap (mm)	Depth of trap seal (mm)
Bidet	32	75
Shower Food waste disposal unit Urinal bowl Sanitary towel macerator	40	75
Industrial food waste disposal unit	50	75
Urinal (stall, 1 to 7 person positions)	65	50

If it is not possible to comply with the figures given in Table 13.1, Fig. 13.2 or Table 3 above, then the branch discharge pipe should be ventilated in order to prevent loss of trap seals. This is facilitated by means of a *branch ventilating pipe* which is connected to the discharge pipe within 300 mm of the appliance trap. The branch ventilating pipe may be run direct to outside air, where it should finish at least 900 mm above any opening into the building which is nearer than 3 m, or, it may be connected to the discharge stack above the 'spillover' level of the highest appliance served. In this case it should have a continuous incline from the branch discharge pipe to the point of connection with the discharge stack (see Fig. 13.3).

Where a branch ventilating pipe serves only one appliance it should have a minimum diameter of 25 mm. This should be increased to 32 mm diameter if the branch ventilating pipe is longer than 15 m or contains more than five bends.

AD H1
sec. 1
1.18 to 1.20

AD H1, section 1

Table 3   **Common branch discharge pipes (unvented)**

Appliance	Max number to be connected	OR	Max length of branch [m]	Min size of pipe [mm]	Gradient limits (fall per metre)	
					min [mm]	max [mm]
wcs	8		15	100	9 to	90
urinals: bowls	5		*	50	18 to	90
stalls	7		*	65	18 to	90
washbasins	4		4 (no bends)	50	18 to	45

Note
* No limitation as regards venting but should be as short as possible.

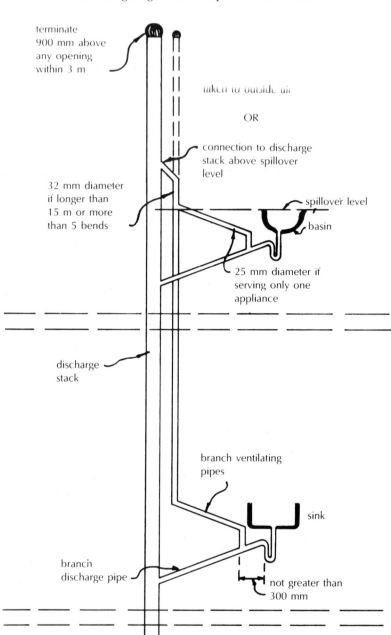

terminate
900 mm above
any opening
within 3 m

taken to outside air

OR

connection to discharge
stack above spillover
level

32 mm diameter
if longer than
15 m or more
than 5 bends

spillover level

basin

25 mm diameter if
serving only one
appliance

discharge
stack

branch ventilating
pipes

sink

branch
discharge pipe

not greater than
300 mm

Fig. 13.3 Branch ventilating pipes.

As appliance traps present an obstacle to the normal flow in a pipe they may be subject to periodic blockages. It is important, therefore, that they be fitted immediately after an appliance and either be removable or be fitted with a cleaning eye. Where a trap forms an integral part of an appliance (such as in a WC pan), the appliance should be removable.

AD H1 sec. 1 1.4

Branch discharge pipes – design recommendations

In addition to size and gradient there are other design recommendations for branch discharge pipes that should be adhered to in order to prevent loss of trap seals. In high buildings especially, back-pressure may build up at the foot of a discharge stack and may cause loss of trap seal in ground floor appliances. Therefore, the following recommendations should be followed:

- For multi-storey buildings up to five storeys high there should be a minimum distance of 750 mm between the point of junction of the lowest branch discharge pipe connection and the invert of the tail of the bend at the foot of the discharge stack. This is reduced to 450 mm for discharge stacks in single dwellings up to three storeys high (see Fig. 13.4).

AD H1 sec. 1 1.8 & Appendix A3

- For appliances above ground floor level the branch pipe should only be run to a discharge stack or to another branch pipe.

AD H1 1.5

- Ground floor appliances may be run to a separate drain, gully or stub stack. (A gully connection should be restricted to pipes carrying waste water

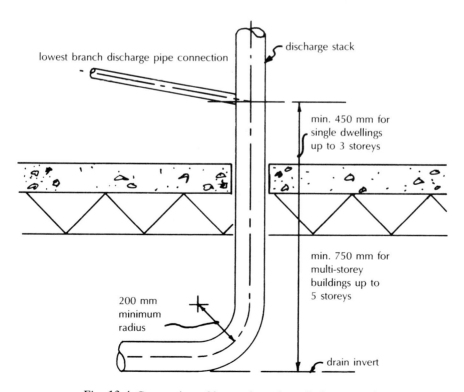

Fig. 13.4 Connection of lowest branch to discharge stack.

only.) They may also be run to a discharge stack in the following circumstances:

(a) In buildings up to five storeys high – without restriction;

AD H1
sec. 1
1.6 & 1.7
Appendix
A4
(b) In buildings with six to twenty storeys – to their own separate discharge stack;

(c) In buildings over twenty storeys – ground and first floor appliances to their own separate discharge stack; (see Fig. 13.5).

AD H1
sec. 1
1.7
Back-pressure and blockages may occur where branches are connected so as to be almost opposite one another. This is most likely to occur where bath and WC branch connections are at or about the same level. Figure 13.6 illustrates ways in which possible cross flows may be avoided.

AD H1
sec. 1
1.9
Additionally, a long vertical drop from a ground floor water closet to a drain may cause self-syphonage of the WC trap. To prevent this the drop should not exceed 1.5 m from crown of trap to invert of drain (see Fig. 13.7).

AD H1
sec. 1
1.11
Similarly, there is a chance of syphonage where a branch discharge pipe connects with a gully. This can be avoided by terminating the branch pipe above the water level but below the gully grating or sealing plate (see Fig. 13.7).

AD H1
sec. 1
1.13
Self-syphonage can also be prevented by ensuring that bends in branch discharge pipes are kept to a minimum. Where bends are unavoidable they should be made with as large a radius as possible. This means that pipes with a diameter of up to 65 mm should have a minimum centreline radius of 75 mm. Junctions on branches should be swept in the direction of flow with a minimum radius of 25 mm or should make an angle of 45° with the discharge stack.

AD H1
sec. 1
1.14
Where a branch diameter is 75 mm or more the sweep radius should be increased to 50 mm (see Fig. 13.6).

AD H1
sec. 1
1.21
Branch discharge pipes should be fully accessible for clearing blockages. Additionally rodding points should be provided so that access may be gained to any part of a branch discharge pipe which cannot be reached by removing a trap or an appliance with an integral trap.

Discharge stacks – design recommendations

The satisfactory performance of a discharge stack will be ensured if it complies with the following rules:

- The foot of the stack should only connect with a drain and should have as large a radius as possible (at least 200 mm at the centreline).
- Ideally, there should be no offsets in the wet part of a stack (i.e. below the highest branch connection).
- If offsets are unavoidable then:

    (a) buildings over three storeys should have a separate ventilation stack connected above and below the offset; *and,*

    (b) buildings up to three storeys should have no branch connection within 750 mm of the offset.

- The stack should be placed inside a building, unless the building has not more than three storeys. This rule is intended to prevent frost damage to discharge stacks and branch pipes.

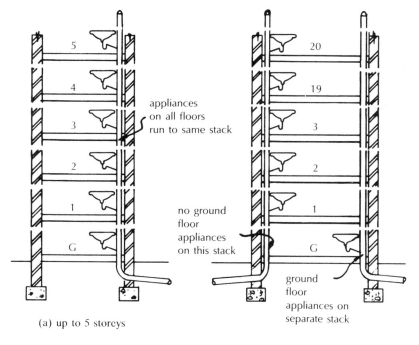

appliances
on all floors
run to same stack

no ground
floor
appliances
on this stack

(a) up to 5 storeys

ground
floor
appliances on
separate stack

(b) 6 to 20 storeys

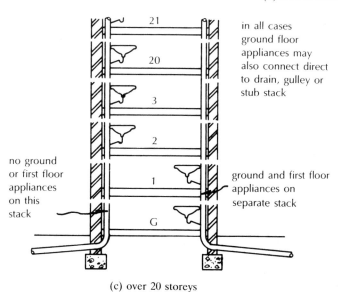

in all cases
ground floor
appliances may
also connect direct
to drain, gulley or
stub stack

no ground
or first floor
appliances
on this
stack

ground and first floor
appliances on
separate stack

(c) over 20 storeys

Fig. 13.5 Provision of discharge stacks to ground floor appliances.

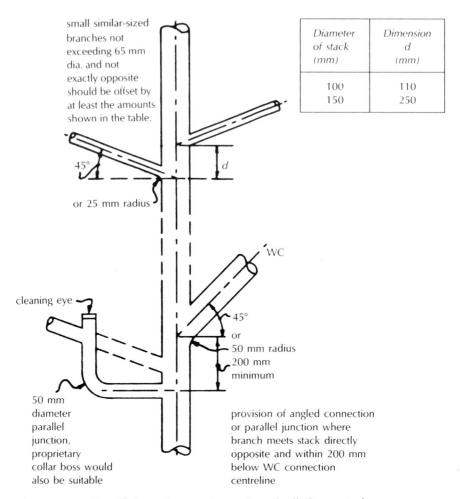

Diameter of stack (mm)	Dimension d (mm)
100	110
150	250

small similar-sized branches not exceeding 65 mm dia. and not exactly opposite should be offset by at least the amounts shown in the table.

45°

or 25 mm radius

d

WC

cleaning eye

45° or 50 mm radius

200 mm minimum

50 mm diameter parallel junction, proprietary collar boss would also be suitable

provision of angled connection or parallel junction where branch meets stack directly opposite and within 200 mm below WC connection centreline

Fig. 13.6 Avoidance of cross flows in discharge stacks.

- The stack should comply with the minimum diameters given in Table 4 to section 1 of AD H1 (see above). Additionally the minimum internal diameter permitted for a discharge stack serving urinals is 50 mm, 75 mm for a siphonic closet and 100 nm for a washdown closet.
- The diameter of a discharge stack should not reduce in the direction of flow.
- Adequate access points for clearing blockages should be provided and all pipes should be reasonably accessible for repairs.

AD H1
sec. 1
1.22 to 1.24
& 1.30

Discharge stacks – ventilation recommendations

In order to prevent the loss of trap seals it is essential that the air pressure in a discharge stack remains reasonably constant. Therefore, the stack should be ventilated to outside air. For this purpose it should be carried up to such a height that its open end will not cause danger to health or a nuisance. AD H1 recommends that the pipe should finish at least 900mm above the top of any

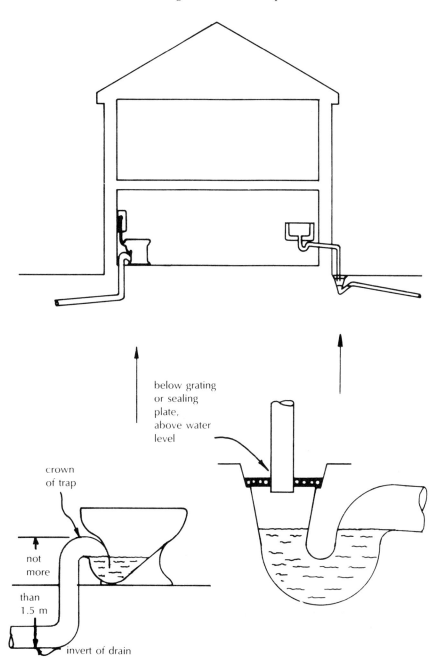

below grating
or sealing
plate,
above water
level

crown
of trap

not
more

than
1.5 m

invert of drain

Fig. 13.7 Ground floor connections for water closets and gullies.

AD H1, section 1

Table **4**   **Maximum capacities for discharge stacks**

Stack size [mm]	Max capacity [litres/sec]
50*	1.2
65*	2.1
75†	3.4
90	5.3
100	7.2

Note
* No wcs.
† Not more than 1 syphonic wc with 75 mm outlet.

opening into the building within 3 m. The open end should be fitted with a durable ventilating cover (see Fig. 13.8).

The dry part of a discharge stack above the topmost branch, which serves only for ventilation, may be reduced in size in one and two storey houses to 75 mm diameter.

It is permissible to terminate a discharge stack inside a building if it is fitted with an air admittance valve. This valve allows air to enter the pipe but does not allow foul air to escape. It should be the subject of a current BBA Certificate and should not adversely affect the operation of the underground

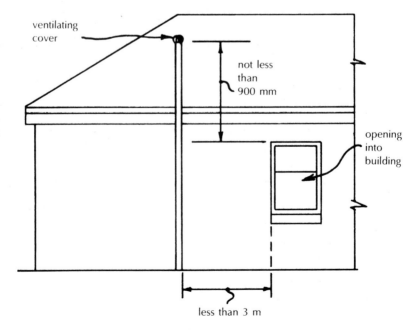

Fig. 13.8 Termination of discharge stacks.

drainage system which normally relies on ventilation from the open stacks of the sanitary pipework.

Some underground drains are subject to surcharging. Where this is the case the discharge stack should be ventilated by a pipe of not less than 50 mm diameter connected at the base of the stack above the expected flood level. This would also apply where a discharge pipe is connected to a drain near an intercepting trap although no minimum dimensions are specified in AD H1.

Stub stacks

There is one exception to the general rule that discharge stacks should be ventilated. This involves the use of an unvented stack (or *stub stack*). A stub stack should connect to a ventilated discharge stack or a drain which is not subject to surcharging and should comply with the dimensions given in Fig. 13.9. It is permissible for more than one ground floor appliance to connect to a stub stack.

AD H1
sec. 1
1.25 to 1.29
& 1.10

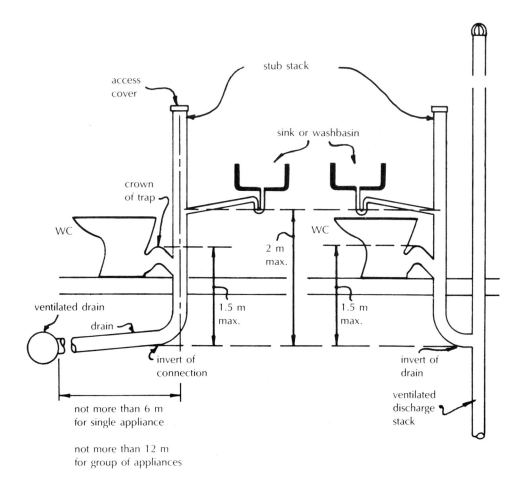

Fig. 13.9 Stub stacks.

Dry ventilating stacks

Where an installation requires a large number of branch ventilating pipes and the distance to a discharge stack is also large it may be necessary to use a dry ventilating stack.

It is normal to connect the lower end of a ventilating stack to a ventilated discharge stack below the lowest branch discharge pipe. It may also be connected directly to a bend as for discharge stacks (see p. 13.17 above).

The upper end of a ventilating stack should either connect back into a ventilated discharge stack above the spillover level of the highest appliance or it should terminate in the outside air as described for discharge stacks above.

AD H1
Appendix
A5 to A8

Ventilating stacks should be at least 32 mm in diameter if serving a building containing dwellings not more than ten storeys high. For all other buildings reference should be made to BS 8301: 1985 *Code of practice for building drainage.*

Materials for above ground drainage systems

Table 5 to section 1 of AD H1, which is reproduced below, gives details of the materials that may be used for pipes, fittings and joints in above ground drainage systems.

To prevent electrolytic corrosion, pipes of different metals should be separated where necessary by non-metallic material. Additionally, pipes should be adequately supported without restricting thermal movement.

AD H1, section 1

Table **5**   **Materials for sanitary pipework**

Material	British Standard
Pipes	.
cast iron	BS 416
copper	BS 864, BS 2871
galvanised steel	BS 3868
uPVC	BS 4514
polypropylene	BS 5254
plastics	BS 5255
ABS	
MUPVC	
polyethylene	
polypropylene	
Traps	
copper	BS 1184
plastics	BS 3943

AD H1
sec. 1
1.31

Note: Some of these materials may not be suitable for conveying trade effluent.

Care should be taken where pipes pass through fire separating elements (see Part B of Schedule 1 to the 1991 Regulations and Approved Document B).

Test for airtightness

In order to ensure that a completed installation is airtight it should be subjected to a pressure test of air or smoke of at least 38 mm water gauge for a maximum of three minutes. A satisfactory installation will maintain a 25 mm water seal in every trap. uPVC pipes should not be smoke tested.

AD H1
sec. 1
1.32

Alternative method of design

The requirements of the 1991 Regulations for above ground drainage can also be met by following clauses 3, 4, and 7 to 12 of BS 5572: 1978 *Code of practice for sanitary pipework.*

AD H1
sec. 1
1.33

Below ground foul drainage

In most modern systems of underground drainage foul water and rainwater are carried separately. Section 2 of AD H1 deals specifically with below ground foul drainage. However, some public sewers are on the combined system taking foul and rainwater in the same pipe. The provisions of AD H1 will apply equally to combined systems although pipe gradients and sizes may have to be adjusted to take the increased flows. Combined systems should never discharge to a cesspool or septic tank.

Where a sewer is above the level of the underground drainage system, sewage pumping equipment will be necessary. Information on these installations can be obtained from BS 8301 *Code of practice for building drainage.*

AD H1
sec. 2
2.1 & 2.2

The performance of a below ground foul drainage system depends on the drainage layout, the pipe cover and bedding, the pipe sizes and gradients, the materials used and the provisions for clearing blockages.

DRAINAGE LAYOUT. The drainage layout should be kept as simple as possible with pipes laid in straight lines and to even gradients. The number of access points provided should be limited to those essential for clearing blockages. If possible, changes of gradient and direction should be combined with access points, inspection chambers or manholes. Junctions between drains or sewers should be made obliquely or in the direction of flow.

A slight curve in a length of otherwise straight pipework is permissible provided the line can still be adequately rodded.

Bends should only be used in or close to inspection chambers and manholes, or at the foot of discharge or ventilating stacks. The radius of any bend should be as large as practicable.

It is important to ventilate an underground foul drainage system with a flow of air. Ventilated discharge pipes may be used for this purpose and should be positioned at the head of each main run and:

- On any branch exceeding 6 m serving a single appliance.
- On any branch exceeding 12 m serving a group of appliances.
- On any drain fitted with an interceptor (especially on a sealed system).

AD H1
sec. 2
2.3 to 2.6

PIPE COVER AND BEDDING. The degree of pipe cover to be provided will usually depend on:

AD H1
sec. 2
2.8 & 2.9

- The invert level of the connections to the drainage system.
- The slope and level of the ground.
- The necessary pipe gradients.
- The necessity for protection to pipes.

AD H1
sec. 2
2.15

In order to protect pipes from damage it is essential that they are bedded and backfilled correctly. The choice of materials for this purpose will depend mainly on the depth, size and strength of the pipes used.

Pipes used for underground drainage may be classed as rigid or flexible. Flexible pipes will be subject to deformation under load and will therefore need more support than rigid pipes so that the deformation may be limited to 5% of the pipe diameter.

Rigid pipes

Table 8 of AD H1 is set out above and contains details of the limits of cover that need to be provided for standard strength rigid pipes in any width of trench. For details of the bedding classes referred to in the table, see Fig. 13.10.

The backfilling materials should comply with the following:

- Granular material should conform to BS 882: 1983 *Specification for aggregates from natural sources for concrete*; Table 4 or BS 8301: 1985 *Code of practice for building drainage*; Appendix D.

AD H1
sec. 2
2.16

- Selected fill should be free from stones larger than 40 mm, lumps of clay over 100 mm, timber, frozen material or vegetable matter. It is possible that ground water may flow in trenches with granular bedding. Provisions may be required to prevent this.

AD H1, section 2

Table **8**   **Limits of cover for standard strength rigid pipes in any width of trench**

Pipe bore	Bedding class	Fields and gardens		Light traffic roads		Heavy traffic roads	
		Min	Max	Min	Max	Min	Max
100	D or N	0.4	4.2	0.7	4.1	0.7	3.7
	F	0.3	5.8	0.5	5.8	0.5	5.5
	B	0.3	7.4	0.4	7.4	0.4	7.2
150	D or N	0.6	2.7	1.1	2.5	—	—
	F	0.6	3.9	0.7	3.8	0.7	3.3
	B	0.6	5.0	0.6	5.0	0.6	4.6

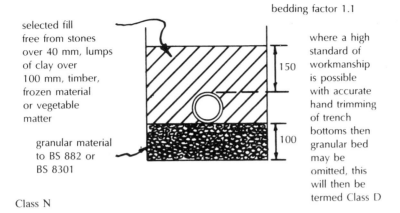

bedding factor 1.1

selected fill
free from stones
over 40 mm, lumps
of clay over
100 mm, timber,
frozen material
or vegetable
matter

150

where a high
standard of
workmanship
is possible
with accurate
hand trimming
of trench
bottoms then
granular bed
may be
omitted, this
will then be
termed Class D

granular material
to BS 882 or
BS 8301

100

Class N

bedding factor 1.5

suitable in all
soil conditions

150

selected fill
as Class N

where socketed
pipes used,
minimum
50 mm
above trench
bottom

100

45°
minimum

granular fill
as Class N

Class F

bedding factor 1.9

suitable in all
soil conditions

150

selected fill as
Class N

granular fill
as Class N

100

granular
fill to half
outside
diameter
of pipe

Class B

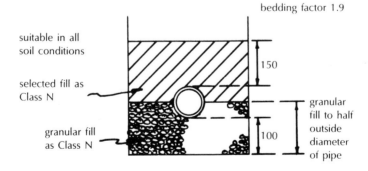

Fig. 13.10 Bedding classes for rigid pipes.

Flexible pipes

Flexible pipes should be provided with a minimum depth of cover of 900 mm under any road. This may be reduced to 600 mm in fields and gardens. The maximum permissible depth of cover is 10 m.

**AD H1
sec. 2
2.17**

Figure 13.11 shows typical bedding and backfilling details for flexible pipes. Where it is necessary to construct a V-shaped trench due to the nature of the subsoil, care should be taken to ensure that the granular bedding material is properly contained.

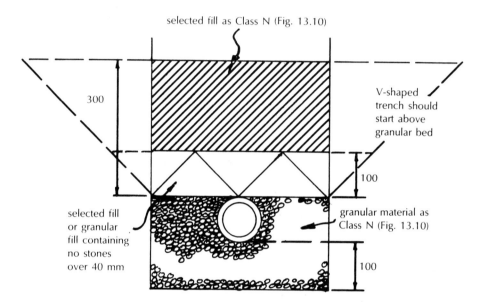

Fig. 13.11 Bedding for flexible pipes.

Special protection to pipes

Where rigid pipes have less cover than is specified above (see Table 8 above), the pipes should be surrounded in concrete to a thickness of at least 100 mm. Expansion joints should also be provided at each socket or sleeve joint face.

Flexible pipes under fields or gardens with less than 600 mm of cover should be bridged by pre-cast concrete paving slabs resting on at least 75 mm of granular fill. The concrete paving slab should be replaced with reinforced concrete surround or bridging for flexible pipes under roads with less than 900 mm of cover (see Fig. 13.12).

**AD H1
Appendix
A15 to A17**

PIPE SIZES AND GRADIENTS. Drains should be laid to falls and should be large enough to carry the expected flow. The rate of flow will depend on the appliances that are connected to the drain (see Table 1, p. 13.5 and Table 13.1, p. 13.5). The capacity will depend on the diameter and gradient of the pipes.

Table 6 to section 2 of AD H1 gives recommended minimum gradients for different sized foul drains and shows the maximum capacities they are capable of carrying. The table is set out above.

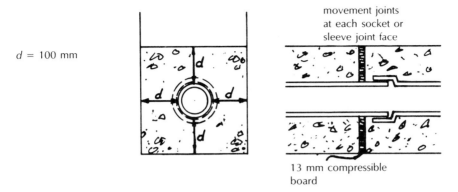

Concrete encasement
for rigid or flexible pipes

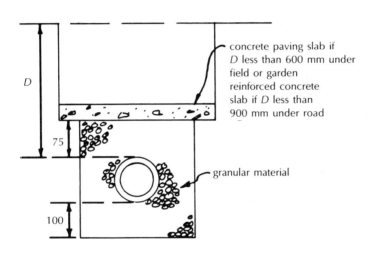

Protection for flexible pipes

Fig. 13.12 Special protection to pipes.

As a further design guide Diagram 7 from AD H1 is reproduced below. This gives discharge capacities for foul drains running at 0.75 proportional depth.

AD H1
sec. 2
2.10 to 2.12

Where foul and rainwater drainage systems are combined, the capacity of the system should be large enough to take the combined peak flow (see Rainwater drainage below).

AD H1
sec. 2
2.13

MATERIALS. Table 7 to section 2 of AD H1, which is reproduced above, gives details of the materials that may be used for pipes, fittings and joints in below ground foul drainage systems.

Joints should remain watertight under working and test conditions and nothing in the joints, pipes or fittings should form an obstruction inside the

AD H1, section 2

Table 6   **Recommended minimum gradients for foul drains**

Peak flow [litres/sec]	Pipe size [mm]	Minimum gradient [1:...]	Maximum capacity [litres/sec]
<1	75	1:40	4.1
	100	1:40	9.2
>1	75	1:80	2.8
	100	1:80*	6.3
	150	1:150†	15.0

Notes
* Minimum of 1 wc.
† Minimum of 5 wcs.

pipeline. To avoid damage by differential settlement pipes should have flexible joints appropriate to the material of the pipes.

AD H1
sec. 2
2.14
To prevent electrolytic corrosion, pipes of different metals should be separated where necessary by non-metallic material.

PROVISIONS FOR CLEARING BLOCKAGES. Every part of a drainage system should be accessible for clearing blockages. The type of access point chosen and its siting and spacing will depend on the layout of the drainage system and the depth and size of the drain runs.

AD H1, section 2

Table 7   **Materials for below ground gravity drainage**

Material	British Standard
Rigid pipes	
asbestos	BS 3656
vitrified clay	BS 65, BSEN 295
concrete	BS 5911
grey iron	BS 437, BS 6087
Flexible pipes	
uPVC	BS 4660
	BS 5481

Note
Some of these materials may not be suitable for conveying trade effluent.

AD H1, section 2

Diagram 7 **Discharge capacities of foul drains running 0.75 proportional depth**

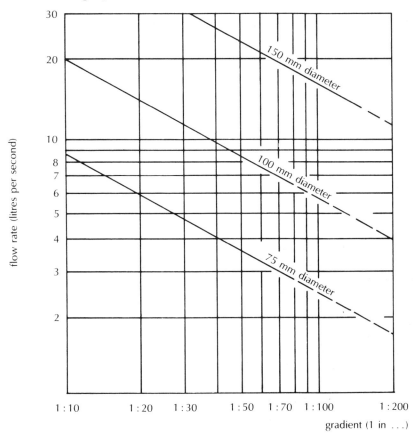

A drainage system designed in accordance with the provisions of AD H1 should be capable of being rodded by normal means (i.e. *not* by mechanical methods).

AD H1
sec. 2
2.18 & 2.19

Access points

Four types of access points are described in AD H1:

- Rodding eyes (or points). These are extensions of the drainage system to ground level where the open end of the pipe is capped with a sealing plate.
- Access fittings. Small chambers situated at the invert level of a pipe and without any real area of open channel.
- Inspection chambers. Chambers having working space at ground level.
- Manholes. Chambers large enough to admit persons to work at drain level.

Some typical access point details are illustrated in Fig. 13.13.

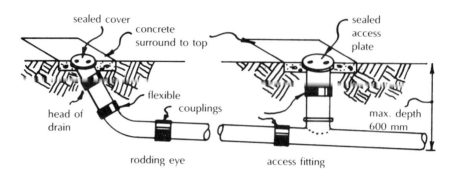

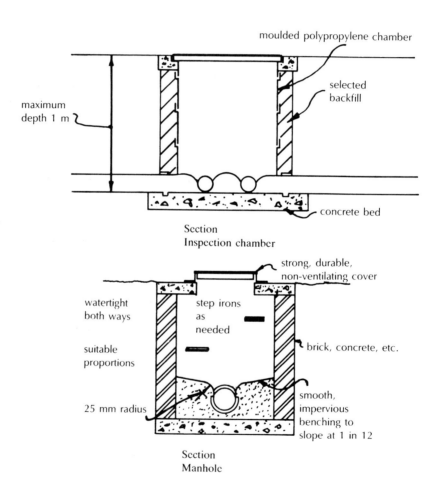

Fig. 13.13 Access points.

Whatever form of access point is used it should be of sufficient size to enable the drain run to be adequately rodded. Table 9 to section 2 of AD H1 sets out the maximum depths and minimum internal dimensions for each type of access point. Where a large number of branches enter an inspection chamber or manhole the sizes given in Table 9 may need to be increased. It is usual to allow 300 mm for each branch connection (thus a 1200 mm long manhole could cater for up to four branch connections on each side). Table 9 is set out below.

AD H1
sec. 2
2.20

AD H1, section 2

Table **9** **Minimum dimensions for access fittings and chambers**

Type	Depth to invert (m)	Internal sizes		Cover sizes	
		Length × width (mm × mm)	Circular (mm)	Length × width (mm × mm)	Circular (mm)
Rodding eye	—	As drain but min 100		—	
Access fitting					
small	0.6 or less	150 × 100	150	150 × 100	150
large		225 × 100	—	225 × 100	—
Inspection chamber	0.6 or less	—	190*	—	190*
	1.0 or less	450 × 450	450	450 × 450	450†
Manhole	1.5 or less	1200 × 750	1050	600 × 600	600
	over 1.5	1200 × 750	1200	600 × 600	600
	over 2.7	1200 × 840	1200	600 × 600	600
Shaft	over 2.7	900 × 840	900	600 × 600	600

Notes
* Drains up to 150 mm.
† For clayware or plastics may be reduced to 430 mm in order to provide support for cover and frame.

Access points – siting and spacing

Access points should be provided:

- At or near the head of any drain run.
- At any change of direction or gradient.
- At a junction, unless each drain run can be rodded separately from another access point.
- At a change of pipe size, unless this occurs at a junction where each drain run can be rodded separately from another access point.
- At regular intervals on long drain runs.

AD H1, section 2

Table **10**   **Maximum spacing of access points in metres**

| From | To | Access Fitting | | Junction | Inspection chamber | Manhole |
		Small	Large			
Start of external drain*		12	12	—	22	45
Rodding eye		22	22	22	45	45
Access fitting small 150 diam						
150 × 100		—	—	12	22	22
large 225 × 100		—	—	22	45	45
Inspection chamber		22	45	22	45	45
Manhole		22	45	45	45	90

Note
* See paragraphs 1.9 and 1.26 of AD H1.

**AD H1
sec. 2
2.21 & 2.22**   The spacing of access points will depend on the type of access used. Table 10 to section 2 of AD H1 gives details of the maximum distances that should be allowed for drains up to 300 mm in diameter and is set out below.

Access points – construction

Generally, access points should:

- Be constructed of suitable and durable materials.
- Exclude subsoil or rainwater.
- Be watertight under working and test conditions.

Table 11 to section 2 of AD H1 is shown below and lists materials which are suitable for the construction of access points.
Inspection chambers and manholes should:

- Have smooth impervious surface benching up to at least the top of the outgoing pipe to all channels and branches. The purpose of benching is to direct the flow into the main channel and to provide a safe foothold. For this reason the benching should fall towards the channel at a slope of 1 in 12 and should be rounded at the channel with a minimum radius of 25 mm (see Fig. 13.13).

AD H1, section 2

Table **11**   **Materials for access points**

Material	British Standard
1 Inspection chambers and manholes	
Clay	
bricks and blocks	BS 3921
vitrified	BS 65
Concrete	
precast	BS 5911
in situ	CP 100
Plastics	BS 7158
2 Rodding eyes and access fittings (excluding frames and covers)	as pipes see Table 7 BBA Certificates

- Be constructed so that branches discharge into the main channel at or above the horizontal diameter where half-round open channels are used. Branches which make an angle of more than 45° with the channel should be formed using a three-quarter section branch bend.
- Have strong, removable, non-ventilating covers of suitable durable material (e.g. cast iron, cast or pressed steel or pre-cast concrete or uPVC).
- Be fitted with step irons, ladders, etc., if over 1.0 m deep.

A manhole or inspection chamber which is situated *within* a building should have an airtight cover that is mechanically fixed (e.g. screwed down with corrosion resistant bolts). This requirement does not apply if the inspection chamber or manhole gives access to part of a drain which itself has inspection fittings and these are provided with watertight covers.

AD H1
sec. 2
2.23 to 2.25

Test for watertightness

After laying and backfilling, gravity below-ground drains and private sewers not exceeding 300 mm in diameter should be pressure tested using air or water. For the air test, a head loss of up to 25 mm at 100 mm water gauge (or 12 mm at 50 mm water gauge) is permitted in a period of five minutes.

Water tested drains using a standpipe which is the same diameter as the drain should be subjected to a pressure of 1.5 m head of water. This should be measured above the invert at the top of the drain run. The section of drain to be tested should be filled up and left to stand for two hours and then topped up. Over the next 30 minutes the leakage should not exceed 0.05 litres per metre run for a 100 mm drain (equivalent to a drop of 6.4 mm per metre) or 0.08 litres per metre run for a 150 mm drain (equivalent to a drop of 4.5 mm per metre).

AD H1
sec. 2
2.26 to 2.28

A drain may be damaged if a head of more than 4 m is applied to the lower end of the run. This may necessitate testing a long drain run in several sections.

Special protection for drains adjacent to or under buildings

Where drains pass under buildings or through foundations and walls there is a risk that settlement of the building may cause pipes to fracture, with consequential blockages and leakage. In the past it was common practice to require pipes (which were rigid jointed) to be encased in concrete. Since the development of flexible pipe systems it has become essential to maintain this flexibility in order that any slight settlement of the building will not cause pipe fracture.

Therefore, drain runs under buildings should be surrounded with at least 100 mm of granular or other flexible filling. On some sites unusual ground conditions may lead to excessive subsidence. To protect drain runs from fracture it may be necessary to have additional flexible joints or use other solutions such as suspended drainage. Shallow drain runs under concrete floor slabs should be concrete encased with the slab where the crown of the pipe is less than 300 mm from the underside of the slab.

Where a drain passes through a wall or foundation the following solutions are possible:

- The wall may be supported on lintels over the pipe. A clearance of 50 mm should be provided round the pipe perimeter and this gap should either be filled with a flexible material such as mineral fibre quilt or it should be masked on both sides of the wall with rigid sheet material to prevent the ingress of fill or vermin.

AD H1
Appendix
A9 & A10

- A length of pipe may be built in to the wall with its joints not more than 150 mm from each face. Rocker pipes not exceeding 600 mm in length should then be connected to each end of the pipe using flexible joints (see Fig. 13.14).

Where a drain or private sewer is laid close to a load-bearing part of a building, precautions should be taken to ensure that the drain or sewer trench does not impair the stability of the building.

AD H1
Appendix
A11

Where any drain or sewer trench is within 1 m of the foundation of a wall, and the bottom of the trench is lower than the wall foundation, the trench should be filled with concrete up to the level of the underside of the foundation.

Where a drain or sewer trench is 1 m or more from a wall foundation, and the trench bottom is lower than the foundation, the trench should be filled with concrete to within a vertical distance below the underside of the foundation of not more than the horizontal distance from the foundation to the trench less 150 mm (see Fig. 13.15).

AD H1
A12

The advice of the local authority should be sought regarding sites where unstable ground is present or there is a risk of drain surcharging or a high water table. They should also be consulted if it is intended to lay pipes on piles or beams or in a common trench.

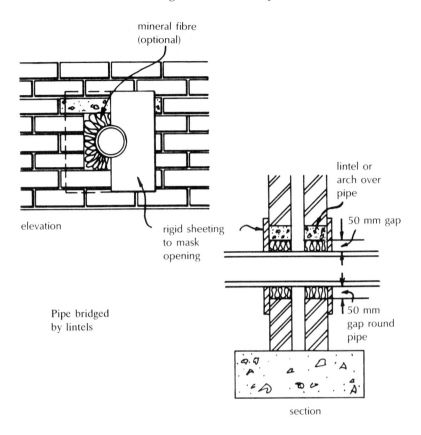

mineral fibre
(optional)

elevation

rigid sheeting
to mask
opening

lintel or
arch over
pipe

50 mm gap

Pipe bridged
by lintels

50 mm
gap round
pipe

section

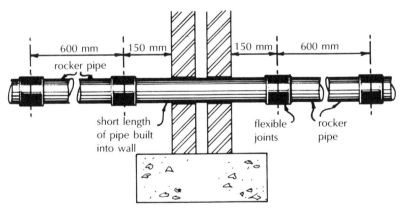

600 mm | 150 mm | 150 mm | 600 mm

rocker pipe

short length
of pipe built
into wall

flexible
joints

rocker
pipe

Pipe built into wall

Fig. 13.14 Drains passing through foundations.

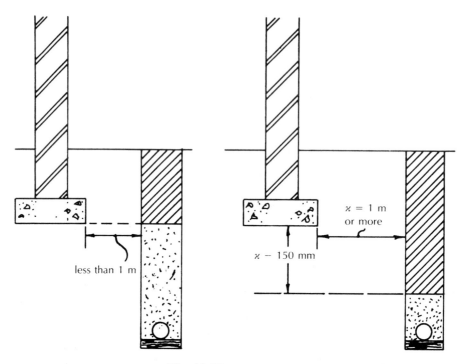

Fig. 13.15 Drain trenches.

Special protection – drain surcharging

Under certain weather conditions drains may be unable to cope with the increased flow and may back up (or become surcharged) creating the need to protect the building from flooding. Some parts of the drainage system may be unaffected by surcharging. These parts should by-pass any protective measures and should discharge into a surcharge free part of the system unless this is unavoidable. Typical protective measures may be obtained from BS 8301. If an anti-flood device is used extra ventilation should be provided to the system to prevent the loss of trap seals.

AD H1
A13

Special protection – rodent control

Generally, rodent infestation (especially by rats) is on the increase. Since rats use drains and sewers as effective communication routes some degree of control may be achieved by providing inspection chambers with screwed access covers on the pipework instead of open channels. Intercepting traps may also be provided as in the past, although they do increase the incidence of blockages unless adequately maintained. The local authority may be able to provide guidance regarding areas where rodent infestation is a problem.

AD H1
A14

Alternative method of design

Additional information on the design and construction of building drainage which meets the requirements of the 1991 Regulations is contained in BS 8301:

1985 *Code of practice for building drainage.* The Code also describes the discharge unit method of calculating pipe sizes.

AD H1
sec. 2
2.29

Cesspools, septic tanks and settlement tanks

Cesspools, septic tanks and settlement tanks should be:

- Ventilated.
- Of adequate capacity.
- Constructed to be impermeable to liquids.

They should also be sited and constructed so that:

- They are accessible for emptying.
- They are not prejudicial to health.
- They will not contaminate any underground water or water supply.

Regs Sch. 1
H2

Capacity

The minimum permitted size for cesspools ($18\,m^3$) is a large capacity tank, and will tend to discourage their use and encourage the use of septic and settlement tanks, which may be much smaller ($2.7\,m^3$).

The septic tank is, of course, the better answer to the problem of sewage disposal for an isolated building. The cesspool was the mediaeval solution.

Minimum capacities are set for cesspools, septic tanks and settlement tanks in order to reduce danger of overflowing and malfunctioning. Septic tanks should only be considered if the subsoil is suitable for disposal of the effluent. Under the Control of Pollution Act 1974 a water authority may require further treatment of effluent discharged from a settlement or septic tank.

AD H2
1.1 & 1.2

Siting and construction

Cesspools and tanks must be periodically desludged. This is usually carried out mechanically using a tanker. Because of the length of piping involved it is necessary that the cesspool or tank be sited within 30 m of a vehicular access. Emptying and cleaning should not involve the contents being taken through a dwelling or place of work, although it is permissible for access to be through an open covered space.

AD H2
1.3

Cesspools and tanks should also be constructed of materials which are impervious to the contents and to ground water. This would include engineering brickwork in 1:3 cement mortar at least 220 mm thick, concrete at least 150 mm thick and glass reinforced concrete.

Prefabricated cesspools and tanks are available made of glass reinforced plastic, polyethylene or steel. These should comply with a BBA Certificate and should be installed strictly in accordance with the manufacturer's instructions. Care should be exercised over the stability of these tanks.

Additionally, cesspools should:

- Be covered and ventilated.
- Have no openings except for the inlet from the drain and the access for emptying. The access should have no dimension smaller than 600 mm where an entry is required, and the inlet should be provided with access for inspection (see Fig. 13.16).

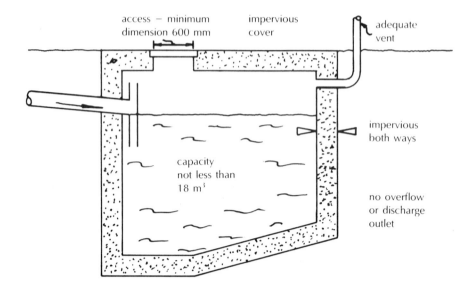

Cesspools

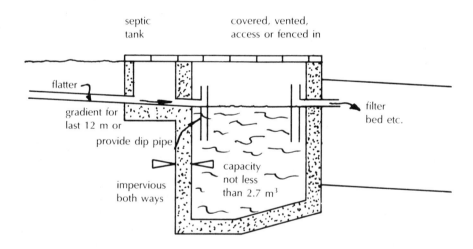

Septic or settlement tanks

Fig. 13.16 Cesspools, septic tanks and settlement tanks.

Septic or settlement tanks should:

- Contain at least two chambers operating in series.
- Be covered with heavy concrete covers or fenced. (If covered they should also be ventilated and provided with access for emptying as for cesspools.)
- Have inlets and outlets provided with access for inspection.

In order to avoid excessive disturbance to the contents of the tank the velocity of flow into the tank should be limited. This may be achieved by laying the last 12 m of the incoming drain at a gradient of 1 in 50 or flatter for all pipes up to 150 mm in diameter. Alternatively, a dip pipe may be provided (see Fig. 13.16) where the tank width does not exceed 1200 mm.

AD H2
1.4 to 1.10

Alternatively, the performance required may also be met by complying with the relevant clauses of BS 6297: 1983 *Code of practice for design and installation of small sewage treatment works and cesspools.*

AD H2
1.11

Rainwater drainage

Any system carrying rainwater from the roof of a building to a sewer, soakaway, watercourse or other suitable outfall is required to be adequate.

Regs Sch. 1
H3

The requirements of Paragraph H3 may be met by any rainwater drainage system which:

AD H3

- Conveys the flow of rainwater to a suitable outfall (surface water or combined sewer, soakaway or watercourse).
- Reduces to a minimum the risk of leakage or blockage.
- Is accessible for clearing blockages.

AD H3
page 20

AD H3 contains no provisions for the drainage of areas such as small roofs and balconies less than 6 m in area unless these areas receive additional flows of water from rainwater pipes or adjacent hard surfaces.

AD H3
0.1

Rainwater or surface water should never be discharged to a cesspool or septic tank.

AD H3
0.5

Gutters and rainwater pipes

A rainwater drainage system should be capable of carrying the anticipated flow at any point in the system. The flow will depend on the area of roofs to be drained and on the intensity of the rainfall. A maximum intensity of 75 mm in any one hour should be assumed in design calculations.

The ultimate capacity of gutters and rainwater pipes depends on their length, shape, size and gradient and on the number, disposition and design of outlets. AD H3 contains design data for half-round gutters up to 150 mm in diameter. They are assumed to be laid level and to have a sharp-edged outlet at one end only. Table 2 to section 1 of AD H3 is reproduced above and gives gutter and outlet sizes for different roof areas for lengths of gutter up to 50 times the water depth. The gutter capacity should be reduced for greater lengths.

AD H3
0.2 & 0.3

AD H3
sec. 1
1.2

The maximum roof areas given in Table 2 are the largest effective areas which should be drained into the gutters given in the table. The effective area of a roof will depend on whether the surface is flat or pitched. Table 1 to section 1 of AD H3 shows how the effective area may be calculated for different roof pitches. The factors given in the table for roof pitches between 30° and 60° appear to be derived by dividing the plan area by the cosine of the angle of pitch, for example, $^1/_{\cos 30°} = 1.1547$, and $^1/_{\cos 45°} = 1.4142$. It would seem reasonable, therefore, to use this relationship for roofs of intermediate pitch. However, no guidance on this is given in the approved document.

AD H3
sec. 1
1.1

AD H3, section 1

Table 2   **Gutter sizes and outlet sizes**

Max effective roof area [m²]	Gutter size [mm dia]	Outlet size [mm dia]	Flow capacity [litres/sec]
6.0	—	—	—
18.0	75	50	0.38
37.0	100	63	0.78
53.0	115	63	1.11
65.0	125	75	1.37
103.0	150	89	2.16

Note
Refers to nominal half round eaves gutters laid level with outlet at one end sharp edged.
Round edged outlets allow smaller downpipe sizes.

Gutters should also be fitted so that any overflow caused by abnormal rainfall will be discharged clear of the building.

Where it is not possible to comply with the conditions assumed in Table 2, further guidance is given in AD H3:

- Where an end outlet is not practicable the gutter should be sized to take the larger of the roof areas draining into it. If two end outlets are provided they may be 100 times the depth of flow apart.
- It may be possible to reduce pipe and gutter sizes if:

AD H3, section 1

Table 1   **Calculation of area drained**

Type of surface	Design area [m²]
1 flat roof	plan area of relevant portion
2 pitched roof at 30° pitched roof at 45° pitched roof at 60°	plan area of portion × 1.15 plan area of portion × 1.40 plan area of portion × 2.00
3 pitched roof over 70° or any wall	elevational area × 0.5

(a) the gutter is laid to fall towards the nearest outlet; or,
(b) a different shaped gutter is used with a larger capacity than the half round gutter; or,
(c) a rounded outlet is used.

In these cases reference should be made to BS 6367: 1983 *Code of practice for drainage of roofs and paved areas.* Rainwater pipes should comply with the following rules:

AD H3 sec. 1 1.3 & 1.4

- Discharge should be to a drain, gully, other gutter or surface which is drained.
- Any discharge into a combined system of drainage should be through a trap (e.g. into a trapped gully).
- Rainwater pipes should not be smaller than the size of the gutter outlet.
- Where more than one gutter serves a rainwater pipe the pipe should have an area at least as large as the combined areas of the gutter outlets.

AD H3 sec. 1 1.5 & 1.6

Materials

Materials used should be adequately strong and durable. Additionally:

- Gutters should have watertight joints.
- Downpipes placed inside a building should be capable of withstanding the test for airtightness described on p. 13.17 above.
- Gutters and rainwater pipes should be adequately supported with no restraint on thermal movement.
- Pipes and gutters of different metals should be separated by non-metallic material to prevent electrolytic corrosion.

AD H3 sec. 1 1.7

If followed, the relevant clauses of BS 6367: 1983 *Code of practice for drainage of roofs and paved areas* will also satisfy the performance requirements for above ground rainwater drainage.

AD H3 sec. 1 1.8

Rainwater drainage below ground

Section 2 of AD H3 deals specifically with drainage systems carrying only rainwater. Combined systems (those carrying both foul and rainwater) are permitted by some drainage authorities where allowance is made for the additional capacity. Where a combined system does not have sufficient capacity, rainwater will need to be taken via a separate system to its own outfall. Pumped systems of surface water drainage may be needed where there is a tendency to surcharging or gravity connections are impracticable. Reference should be made to BS 8301 in these cases.

AD H3 sec. 2 2.1 to 2.3

With the exception of pipe gradients and sizes, the recommendations given above for below ground foul drainage (see pp. 13.17 to 13.31) apply equally to rainwater drainage below ground.

Pipe sizes and gradients

Drains should be laid to falls and should be large enough to carry the expected flow. The rate of flow will depend on the area of the surfaces (including paved or other hard surfaces) being drained. The capacity will depend on the diameter and gradient of the pipes.

AD H3
sec. 2
2.6 & 2.7

The minimum permitted diameter of any rainwater drain is 75 mm. For paved or other hard surfaces a rainfall intensity of 50 mm per hour should be assumed. Diagram 1 to section 2 of AD H3 is reproduced above and gives discharge capacities for rainwater drains running full. As an alternative to section 2 of AD H3 the relevant recommendations of BS 8301: 1985 *Code of practice for building drainage* may be followed.

AD H3
sec. 2
2.12

AD H3, section 2

Diagram 1   Discharge capacities of rainwater drains running full

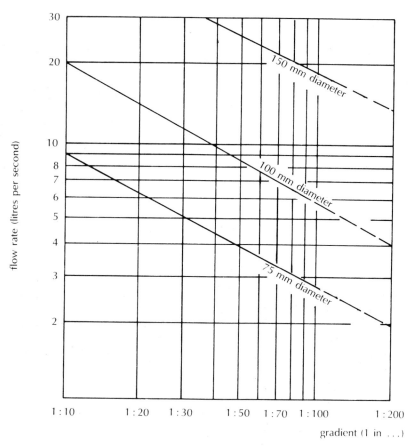

Solid waste storage

Buildings are required to have:

Regs Sch. 1
H4

(a) Adequate means of storing solid waste;
(b) Adequate means of access for the users of the building to a place of storage; and,
(c) Adequate means of access from the place of storage to a street.

The requirements of paragraph H4 may be met by providing solid waste storage facilities which are:

- Large enough, bearing in mind the quantity of refuse generated and the frequency of removal.
- Designed and sited so as not to present a health risk.
- Sited so as to be accessible for filling and emptying.

AD H4
0.1

Therefore the efficacy of the refuse storage system is dependent on its capacity and ease of collection by the relevant authorities.

Domestic buildings

Storage capacity

Assuming weekly collection, dwellinghouses, flats and maisonettes up to four storeys high should have, or have access to, a movable container with a minimum capacity of $0.12\,m^3$ or a communal container with a capacity between $0.75\,m^3$ and $1\,m^3$. These recommended capacities are based on a refuse output of $0.09\,m^3$ per dwelling per week. If weekly collections are not provided by the refuse collection authority then larger capacity containers or more individual containers will need to be provided. Dwellings in buildings over four storeys high should share a container fed by a chute unless this is impracticable. In the latter case suitable management arrangements should be provided for conveying the refuse to the place of storage.

AD H4
1.1 & 1.2

Design and siting

Simple dustbin-type containers should have close fitting lids.

In comparison with Part J of the Building Regulations 1976, AD H4 contains very little information on refuse chutes. This is compensated for by referring the reader to BS 5906: 1980 *Code of practice for storage and on-site treatment of solid waste from buildings*, where full details of refuse chute systems may be obtained.

The provisions of AD H4 require that refuse chutes should be constructed with:

- Smooth, non-absorbent inner surfaces.
- Close fitting access hoppers at each storey containing a dwelling.
- Ventilation at top and bottom.

Containers need not be enclosed, but, if they are, sufficient space should be allowed for filling and emptying. A clear space of 150 mm should be provided between and around containers. The space enclosing communal containers should have a clear headroom of 2 m and be permanently ventilated at top and bottom. Figure 13.17 is based on the recomendations of BS 5906: 1980 and illustrates a typical refuse chute installation.

AD H4
1.3 to 1.5

Refuse containers should comply with the following rules with regard to siting:

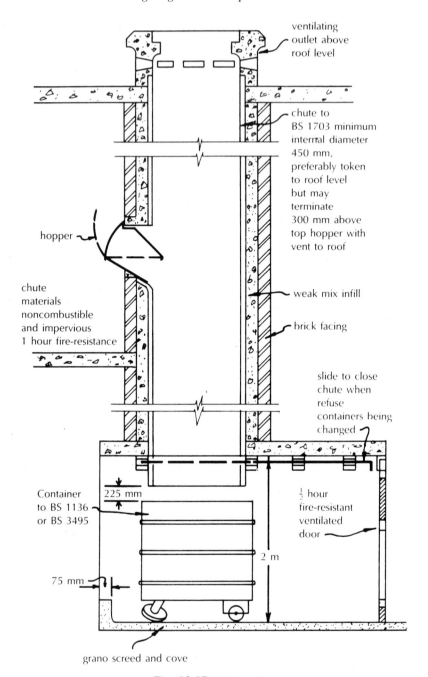

ventilating
outlet above
roof level

chute to
BS 1703 minimum
internal diameter
450 mm,
preferably token
to roof level
but may
terminate
300 mm above
top hopper with
vent to roof

hopper

chute
materials
noncombustible
and impervious
1 hour fire-resistance

weak mix infill

brick facing

slide to close
chute when
refuse
containers being
changed

Container
to BS 1136
or BS 3495

225 mm

½ hour
fire-resistant
ventilated
door

2 m

75 mm

grano screed and cove

Fig. 13.17 Refuse chutes.

- For new buildings it should be possible to collect a container without taking it through a building. (It is permissible to pass through a garage, carport or other covered space.)
- Containers should not be sited more than 25 m from the building which they serve or from any vehicle access.
- Householders should not be required to carry refuse more than 30 m to a container or chute.

AD H4
1.6 & 1.7

Non-domestic buildings

In the development of non-domestic buildings special problems may arise. It is therefore essential to consult the refuse collection authority for their requirements with regard to the following:

- The storage capacity required for the volume and nature of the waste produced. (The collection authority will be able to give guidance as to the size and type of container they will accept and the frequency of collection.)
- Storage method. (This may include details of any proposed on-site treatment and should be related to the future layout of the development and the building density.)
- Location of storage and treatment areas including access for vehicles and operatives.
- Measures to ensure adequate hygiene in storage and treatment areas.
- Measures to prevent fire risks.

AD H4
1.8

Alternative approach

As an alternative to the recommendations listed above for waste disposal, it is permissable to use BS 5906: 1980 especially clauses 3 to 10, 12 to 15 and Appendix A.

AD H4
1.9

Chapter 14

Heat producing appliances

Introduction

Part J of Schedule 1 to the Building Regulations 1991 is concerned with the safe installation of heat producing appliances in buildings.

It is limited to fixed appliances burning solid fuel, oil or gas and to incinerators. (It is assumed that this means incinerators which burn solid fuel, oil or gas since no further guidance is given.) This excludes all electric heating appliances and small portable heaters such as paraffin stoves.

In order that it may function safely a heat producing appliance needs an adequate supply of combustion air and it must be capable of discharging the products of combustion to outside air. This must be achieved without allowing noxious fumes to enter the building and without causing damage by heat or fire to the fabric of the building.

In the accompanying Approved Document J, there has been an attempt at considerable simplification compared with former regulations (Parts L and M of the Building Regulations 1976) which were confusing and repetitive.

The Clean Air Acts 1956–1993

As a direct result of the great London smog of 1952 and after the Report of the Committee on Air Pollution (Cmnd. 9322, November 1954) the Clean Air Act 1956 was passed to give effect to some of the Committee's recommendations. The 1956 Act, amended in 1968, was consolidated in the Clean Air Act 1993. Its main provisions may be summarised briefly, and should be borne in mind in considering the effect of Part J of Schedule 1 to the 1991 Regulations.

The Act makes it an offence to allow the emission of *dark smoke* from a chimney, but certain special defences are allowed, e.g. unavoidable failure of a furnace. DARK SMOKE is defined as smoke which appears to be as dark as, or darker than, shade 2 on the Ringelmann Chart. Regulations made under the Act amplify its provisions in relation to industrial and other buildings, and it should be noted that the prohibition on dark smoke applies to all buildings, railway engines and ships. However, its chief effect is on industrial and commercial premises.

House chimneys rarely emit dark smoke, but local authorities may, by order confirmed by the Secretary of State for the Environment, declare *smoke control areas*. In a smoke control area the emission of smoke from chimneys constitutes an offence, although it is a defence to prove that the emission of smoke was not caused by the use of any fuel other than an authorised fuel. Regulations prescribe the following authorised fuels: anthracite; briquetted fuels carbonised in the process of manufacture; coke; electricity; low temperature carbonisation fuels; low volatile steam coals; and fluidised char binderless briquettes.

The 1956 Act, as amended, provides for the payment of grants by local authorities, and Exchequer contributions, towards the cost of any necessary adaptation or conversion of fireplaces to smokeless forms of heating in private dwellings in smoke control areas.

Heating appliances

Approved Document J is concerned only with heating appliances which produce smoke or gases, and these are divided into the following classes:

(1) Solid fuel and oil-burning appliances with a rated output up to 45 kW (referred to as 'Type 1' appliances in this chapter); *and,*
(2) Gas burning appliances with a rated input up to 60 kW (referred to as 'Type 2' appliances in this chapter).

There are no specific references to incinerators or to the installation of appliances with a higher rating than those given above. Since these will almost invariably be installed under the supervision of a heating engineer it may be considered that sufficient safeguards already exist. However, this does appear to be an area where further guidance could have been given.

Smaller appliances may not receive the attention of a qualified engineer. Accordingly, the requirements for Type 1 and Type 2 appliances are set out in detail in the Approved Document.

Solid fuel and oil-burning appliances may produce smoke and soot and considerable flue temperatures. The requirements for Type 1 appliances are therefore more stringent than for Type 2 (gas) appliances, because the design, manufacture and installation of the latter are more exactly controlled.

Minimum thickness of 100 mm or 200 mm of non-combustible material are required at many points by the Approved Document. This is intended to ensure that if brickwork is used, standard $\frac{1}{2}$ brick or 1 brick thicknesses should be provided in these positions.

Interpretation

Only one definition appears in AD J1/2/3:

AD J1/2/3
sec. 1
1.1

NON-COMBUSTIBLE means capable of being classed as non-combustible when subjected to the non-combustibility test of BS 476, Part 4: 1970 (1984) *Non-combustibility test for materials* (see also Chapter 7, Fire, p. 7.67).

A number of other terms are used in AD J1/2/3 and these are defined below. However, it must be stressed that these definitions do *not* appear in the Approved Document.

FACTORY-MADE INSULATED CHIMNEY means a chimney comprising a flue lining, non-combustible thermal insulation and outer casing.

FLUE means a passage conveying appliance discharge to the external air.

FLUE PIPE means a pipe forming a flue. It does not include a pipe built in as a lining to a chimney.

CHIMNEY includes any part of the structure of a building forming any part of a flue other than a flue pipe.

BALANCED-FLUED or ROOM-SEALED APPLIANCE means a gas appliance which draws its combustion air from a point immediately adjacent to the point where it discharges its combustion products, and is so designed that inlet, outlet and combustion chamber of the installed appliance are isolated from the room or internal space in which the appliance is situated, except for a door for igniting the appliance.

CONSTRUCTIONAL HEARTH means a hearth forming part of the structure of a building. It is usually a concrete slab.

For ease of reference the following terms have been used throughout this chapter. These terms do *not* appear in AD J1/2/3.

TYPE 1 APPLIANCE means a solid fuel or oil-burning appliance of output rating not more than 45 kW.

TYPE 2 APPLIANCE means a gas appliance with a rated input up to 60 kW.

Air supply

Heat producing appliances are required to be provided with an adequate supply of air for combustion of the fuel and for efficient operation of the chimney or flue.

Regs Sch. 1 J1

Air supply to Type 1 appliances

In order to satisfy the requirements of Paragraph J1 the appliance should either be room-sealed or should be situated in a room which has adequate ventilation. The area of ventilation that should be provided is shown in Table 14.1 which is derived from sections 2 and 4 of AD J1/2/3.

Where combustion air is drawn through a permanent air entry opening from an adjacent room or space, outside air should be admitted to that adjacent room or space as required by Table 14.1. (Where an air extract fan is situated in a building containing a heat producing appliance see p. 11.10 above.)

AD J1/2/3 sec. 1 1.2

AD J1/2/3 sec. 1 1.3

Table 14.1 Supply of air for combustion – Type 1 (solid fuel and oil-burning appliances.

Type of appliance	Type of ventilation
1 solid fuel burning open appliance	a permanent air entry opening or openings with a total free area of at least 50% of the appliance throat opening. (See BS 8303: 1986.)
2 other solid fuel appliance	a permanent air entry opening or openings with a total free area of at least 550 mm² per kW of rated output over 5 kW (see also *Note*).
3 oil-burning appliance	a permanent air entry opening with a total free area of at least 550 mm² kW of appliance rated output over 5 kW.

AD J1/2/3
sec. 1, 2.1
and sec. 4
4.1

Note
If the appliance is fitted with a draught stabiliser, then an additional permanent air entry opening (or openings) should be provided with a total free area of at least 300 mm²/kW of appliance rated output.

Air supply to Type 2 appliances

It is possible to install a wide range of gas burning appliances in buildings, some of which discharge the products of combustion into the room in which they are installed (e.g. gas cookers). Other appliances such as boilers and convector heaters may discharge into a conventional open flue or they may be room-sealed.

The performance requirements for air supply will be met if either the appliance is balanced-flued or it is situated in a room which has adequate ventilation.

AD J1/2/3
sec. 2
2.3

Table 14.2 below summarises the requirements for supply of combustion air and shows the areas of ventilation which should be provided.

Discharge of products of combustion

Regs Sch. 1
J2

Heat producing appliances are required to have adequate provision for the discharge of the products of combustion to the outside air.

The requirements of this paragraph may be met by ensuring that flues, flue pipes and chimneys:

- Are of sufficient size.
- Contain only those openings necessary for inspection, cleaning or efficient working of the appliance.
- Are constructed of or lined with suitable materials.
- Are constructed at roof level so as to discharge in a safe manner.

Table 14.2 Air supply to Type 2 (gas burning) appliances.

Type of appliance	Area of permanent ventilation to outside air	Remarks
1 Gas cooker[a]	5000 mm²	For rooms less than 10 m³ in volume
2 Balanced-flued appliance	—	Any appliance situated in a bathroom, shower room or private garage must be balanced-flued[b]
3 Open-flued appliance	450 mm²	per kW of rated input over 7 kW

Notes
[a] Gas cookers should be installed in a room containing an openable window or other opening to outside air. The area of permanent ventilation listed above should also be included if the room volume is less than 10 m³.
[b] Gas Safety (Installation and Use) Regulations 1984.

AD J1/2/3
sec. 3
3.2, 3.3, 3.4

Where an appliance is connected to a chimney built before 1 February 1966 (the date on which the first building regulations became operable) it will not be necessary to comply with certain provisions (see text below) relating to chimneys and flues. If applied, these recommendations would make it virtually impossible to install an appliance without totally rebuilding the chimney.

This provision assumes, of course, that the chimney is performing satisfactorily.

AD J1/2/3
sec. 1
1.7

Discharge of products of combustion – Type 1 appliances

Type 1 appliances are permitted to discharge into balanced or low level flues, chimneys, factory-made insulated chimneys or flue pipes which discharge to external air.

AD J1/2/3
sec. 1
1.4

Oil-burning balanced flued and low level discharge appliances

These appliances should be installed so that:

- The products of combustion may be dispersed externally.
- For a balanced flue, the inlet is situated externally to permit the free intake of air.
- No part of the terminal is within 600 mm of any opening (openable windows, ventilator, etc.) into the building.
- The terminal is protected by a durable guard where it could come into contact with people near the building. This is to prevent danger to the public and damage to the terminal (see Fig. 14.1).
- The entry of any matter which might restrict the flue is prevented.

AD J1/2/3
sec. 4
4.4

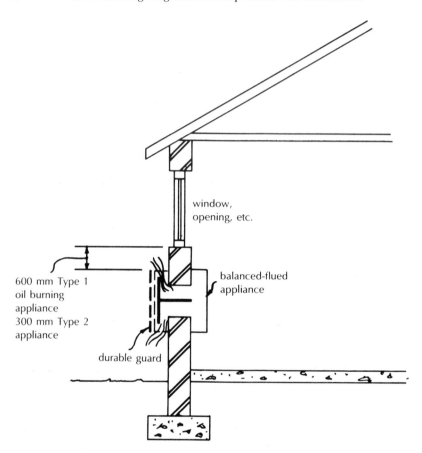

Fig. 14.1 Room sealed appliance.

Flues – general requirements

Whether in chimneys or flue pipes, flues should comply with the following rules:

- SIZE. The size of the flue should be at least that given in Table 2 to AD J1/2/3 (solid fuel) or Table 14.3 (oil) as shown below and never less than the outlet on the appliance.

AD J1/2/3
sec. 2; 2.2
sec. 4; 4.2

- OPENINGS INTO FLUES. Only the following openings into a flue, chimney or flue pipe are permitted:

AD J1/2/3
sec. 1
1.5

(a) An opening for inspection or cleaning, fitted with a non-combustible, rigid, double cased gas tight cover; *and*,

(b) An opening in the same room or internal space as the appliance, fitted with a draught diverter, draught stabiliser or explosion door of non-combustible material.

AD J1/2/3, Section 2

Table 2   Size of flues (solid fuel appliances)

Installation	Minimum flue size
Fireplace recess with an opening up to 500 mm × 550 mm	200 mm diameter or square section of equivalent area
Fireplace recess with an opening in excess of 500 mm × 550 mm	a free area of 15% of the area of the recess opening
Closed appliance up to 20 kW rated output burning bituminous coal	150 mm diameter or square section of equivalent area
Closed appliance up to 20 kW rated output	125 mm diameter or square section of equivalent area
Closed appliance above 20 kW and up to 30 kW rated output	150 mm diameter or square section of equivalent area
Closed appliance above 30 kW and up to 45 kW rated output	175 mm diameter or square section of equivalent area

Note
Should an offset be necessary in a flue run then the flue size should be increased by 25 mm on each dimension (diameter or each side of square flue).

Table 14.3 Size of flues (oil-burning appliances)

Installation	*Minimum size of flue*
(a) flue in a chimney for an appliance with an output rating up to 20 kW	100 mm diameter
(b) flue in a chimney with an output rating between 20 kW and 32 kW	125 mm diameter
(c) flue in a chimney with an output rating between 32 kW and 45 kW	150 mm diameter
(d) flue in flue pipe	at least the size of the outlet from the appliance

Note
For square flues cross-sectional area should be as for equivalent circular flue.

● COMMUNICATIONS WITH FLUES. No flue should communicate with more than one room or internal space in a building. This recommendation does not, however, prohibit the provision of an inspection or cleaning opening allowing access to a flue from a room or space other than the one in which the appliance is installed (see Fig. 14.4 below). A flue may also serve more than one appliance in the same room.

AD J1/2/3
sec. 1
1.6

● OUTLETS OF FLUES. The outlet of any flue in a chimney, or any flue pipe, should be at least:

(a) 1 m above the highest point of contact between the chimney or flue pipe and the roof, for roofs pitched at less than 10°; *and,*
(b) 2.3 m measured horizontally from the roof surface for roofs pitched at 10° or more; *and,*
(c) 1 m above the top of any openable part of a window or skylight, or any ventilator or similar opening which is in a roof or external wall and is not more than 2.3 m horizontally from the top of the chimney or flue pipe; *and,*
(d) 600 mm above the top of any part of an adjoining building which is not more than 2.3 m horizontally from the top of the chimney or flue pipe.

AD J1/2/3
sec. 2
2.3

The outlet from a flue serving an oil-fired pressure jet appliance may terminate anywhere above the roof line.

AD J1/2/3
sec. 4
4.5

Additionally, if the chimney or flue pipe passes through the roof within 2.3 m of the ridge and both slopes are at 10° or more to the horizontal, the top of the chimney or flue pipe may be not less than 600 mm above the ridge (see Fig. 14.2).

● HORIZONTAL FLUE RUNS. Ideally, these should be avoided except in the case of balanced or low level flues for oil-burning appliances and back outlet solid fuel appliances. In the latter case the horizontal section of flue should not exceed 150 mm.

AD J1/2/3
sec. 2; 2.4
sec. 4; 4.3

● BENDS IN FLUES. If possible flues should be vertical. However, if it is necessary to form a bend in a flue it should not be greater than 45° to the vertical for oil-burning appliances and 30° to the vertical for solid fuel appliances (see Fig. 14.3).

AD J1/2/3
sec. 2; 2.5
sec. 4; 4.3

● ACCESS FOR INSPECTION AND CLEANING. This should be provided for any appliance and its chimney and flue pipe. If a flue in a chimney is not directly over an appliance then a debris collecting space should be provided which has an access for inspection and cleaning, fitted with a sealed double-cased door (see Fig. 14.4).

AD J1/2/3
sec. 1; 1.5
sec. 2; 2.11

Flues – linings to chimneys

There is a general recommendation that chimneys should be factory-made insulated chimneys or should be constructed of masonry. Masonry chimneys serving Type 1 appliances may be built of refractory material without a lining or may be lined with one of the following:

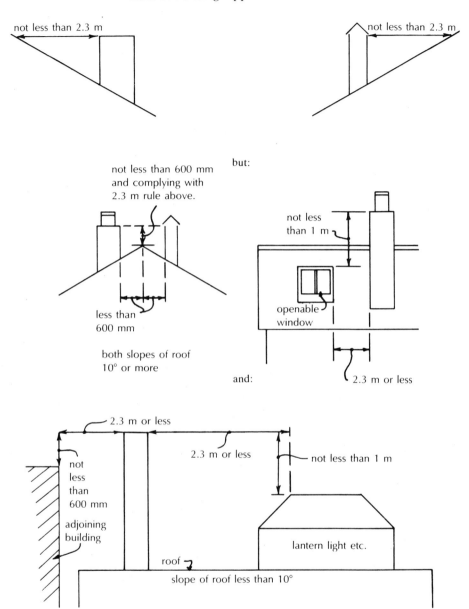

Fig. 14.2 Flue outlets – Type 1 appliances.

- Clay flue linings with rebated or socketed joints to BS 1181: 1989 *Specification for clay flue linings and flue terminals*.
- Clay pipes and fittings to BS 65: 1991 *Specification for vitrified clay pipes, fittings, joints and ducts*, socketed, imperforate and acid resistant.
- Flue linings made from kiln-burnt aggregate and high alumina cement with rebated or socketed joints or steel collars around the joint.

flue in chimney

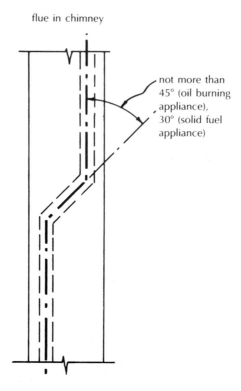

not more than
45° (oil burning
appliance),
30° (solid fuel
appliance)

Fig. 14.3 Bend in flues – Type 1 appliances.

AD J1/2/3
sec. 2; 2.12,
2.13
sec. 4; 4.7

AD J1/2/3
sec. 4
4.7(b)

The linings should be jointed and pointed with fire-proof mortar and built into the chimney with sockets uppermost. This prevents condensate from running out of the joints where it might adversely affect any caulking material. The space between the linings and the masonry should be filled with weak mortar or insulating concrete. For oil-burning appliances, if the temperature of the flue gases is unlikely to exceed 260°C (under the worst operating conditions) then the provisions for gas burning appliances (see below) may be followed instead of the above.

Factory-made insulated chimneys

These chimneys are prefabricated in a factory and are erected on site in sections.

For an appliance burning solid fuel the chimney should be constructed and tested to BS 4543 *Factory-made insulated chimneys*, Part 1: 1990 *Methods of test for factory-made insulated chimneys* and Part 2: *Specification for chimneys with stainless steel flue linings for use with solid fuel fired appliances*. For oil-fired appliances the chimney should comply with BS 4543 *Factory-made insulated chimneys*, Part 1: 1990 *Methods of test* and Part 3: 1990 *Specification for chimneys with stainless steel flue lining for use with oil-fired appliances*.

The chimney should also be installed in accordance with BS 6461 *Installation of chimneys and flues for domestic appliances burning solid fuel (including*

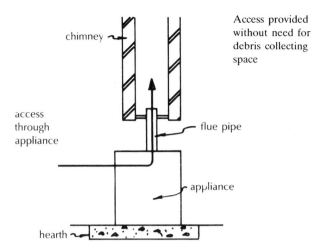

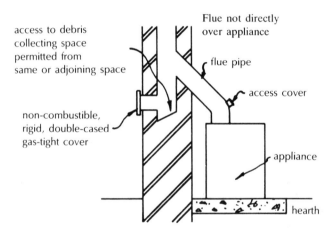

Fig. 14.4 Access for inspection and cleaning.

wood and peat), Part 2: 1984 *Code of practice for factory-made insulated chimneys for internal applications*; or to manufacturers' instructions.

AD J1/2/3
sec. 2; 2.16
sec. 4; 4.8

Flue pipes

Flue pipes serving Type 1 appliances should not pass through any roof space, and should only be used to connect an appliance to a chimney. Whether or not the space between the roof covering and the ceiling finish in a flat roof constitutes a roof space is open to conjecture. The obvious intention is that flue pipes should be as short as possible and should only be used to connect an appliance to a proper chimney.

A horizontal connection is permitted to connect a back outlet appliance to a chimney but this should not exceed 150 mm in length (see Fig. 14.5).

AD J1/2/3
sec. 2
2.6

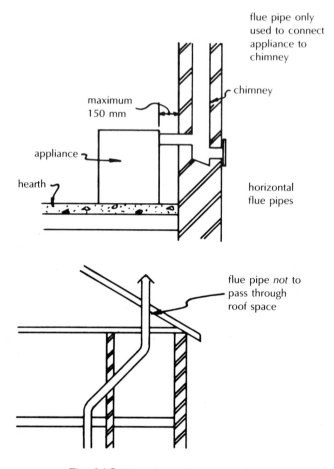

Fig. 14.5 Flue pipes – Type 1 appliances.

Flue pipes should be constructed of:

- Cast iron, in accordance with BS 41: 1973 (1981) *Specification for cast iron spigot and socket flue or smoke pipes and fittings.*
- Mild steel at least 3 mm thick.
- Stainless steel at least 1 mm thick as described in BS 1449 *Steel plate, sheet and strip*, Part 2: 1983 *Specification for stainless and heat-resisting steel plate, sheet and strip*, for Grade 316 S11, 316 S13, 316 S16, 316 S31, 316 S33 or equivalent Euronorm 88–71 designation.
- Vitreous enamelled steel complying with BS 6999: 1989 *Specification for vitreous enamelled low carbon steel flue pipes, other components and accessories for solid fuel burning appliances with a maximum rated output of 45 kW.*

**AD J1/2/3
sec. 2; 2.7,
2.8
sec. 4; 4.7**

Where spigot and socket flue pipes are used the sockets should be placed uppermost.

It should be noted that where the flue gas temperature of a Type 1 oil-burning appliance does not exceed 260°C then it is permissible to connect the appliance to a flue pipe or chimney as described below for Type 2 appliances.

Discharge of products of combustion – Type 2 appliances

Section 3 of AD J1/2/3 deals specifically with gas burning appliances with a rated input up to 60 kW as follows:

- Cooking appliances (ovens, hotplates, grills, etc.).
- Balanced-flued appliances (boilers, convector heaters, water heaters, etc.).
- Decorative log and solid fuel fire effect appliances.
- Individual, natural draught, open-flued appliances (boilers, back boilers, etc.). **AD J1/2/3** sec. 3

Fuel effect fires are a special case since ceramic fuel is heated by a live flame. Where these have been tested by an approved authority they may be installed in accordance with the manufacturers' instructions. If this is not the case, they should either comply with BS 5871: Parts 1 to 3: 1991 or Section 2 of AD J1/2/3 (i.e. solid fuel burning appliances). **AD J1/2/3** sec. 3; 3.1
  Apart from gas cooking appliances (which discharge their products of combustion into the air of the room in which they are situated), Type 2 appliances should discharge into balanced flues, flue linings in masonry chimneys, chimneys constructed of refractory materials without a lining (chimney walls), factory-made insulated chimneys, flue pipes or flexible flue liners.

Balanced flues

The recommendations listed above (see p. 14.5) for Type 1 appliances also apply to Type 2 appliances with one exception:

- Where an appliance terminal is wholly or partly beneath an opening (openable window, ventilator, etc.) no part of the terminal should be within 300 mm vertically of the bottom of the opening. **AD J1/2/3** sec. 3; 3.7

  Additionally any appliance situated in a bathroom, shower room or private garage should be a balanced-flued appliance. **AD J1/2/3** sec. 3; 3.2

Flues – general requirements

Flues should comply with the following rules:

- SIZE. The recommendations for the cross-sectional measurements and areas of flues to Type 2 appliances are:
  (a) The cross-sectional area of a flue serving a gas fire should be at least 12000 mm for round flues and 16500 mm for rectangular flues. Additionally, the flues should have a minimum dimension of 90 mm.
  (b) The cross-sectional area of a flue serving any other appliance should not be less than the area of the appliance outlet (see Fig. 14.6).

AD J1/2/3 sec. 3; 3.5
These rules do not apply to balanced flued or solid fuel effect appliances.

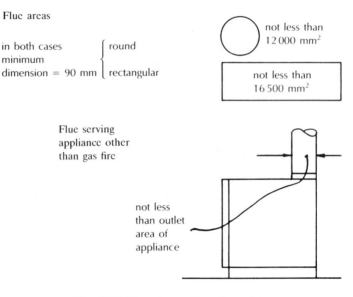

Flue areas

in both cases minimum dimension = 90 mm $\begin{cases} \text{round} \\ \text{rectangular} \end{cases}$

not less than 12 000 mm^2

not less than 16 500 mm^2

Flue serving appliance other than gas fire

not less than outlet area of appliance

Fig. 14.6 Flue sizes – Type 2 appliances.

- OPENINGS INTO FLUES. The only openings permitted into a flue serving a Type 2 appliance are:

AD J1/2/3
sec. 1; 1.5
sec. 2; 2.11

   (a) An opening for inspection or cleaning, fitted with a gas-tight cover of non-combustible material; *or,*
   (b) a draught diverter, draught stabiliser or explosion door.

- COMMUNICATIONS WITH FLUES. No flue should communicate with more than one room or internal space in a building. However, an opening for cleaning or inspection is permitted as for Type 1 appliances (see p. 14.8).

AD J1/2/3
sec. 1; 1.6

- OUTLETS OF FLUES. Any flue outlet should be:

   (a) fitted with a flue terminal if any dimension measured across the axis of the flue outlet is less than 175 mm; *and,*
   (b) placed so that an air current may freely pass over it at all times; *and,*
   (c) placed so that no part of the outlet is within 600 mm of any opening into the building (see Fig. 14.7).

AD J1/2/3
sec. 3; 3.8

- BENDS IN FLUES. If possible flues should be vertical. However, if it is necessary to form a bend in a flue it should not be greater than 45° to the vertical.

AD J1/2/3
sec. 3; 3.6

Flues – linings to chimneys

AD J1/2/3
sec. 2
Part B
2.12

There is a general recommendation that chimneys should either be assembled from factory-made insulated components or they should be constructed of masonry.

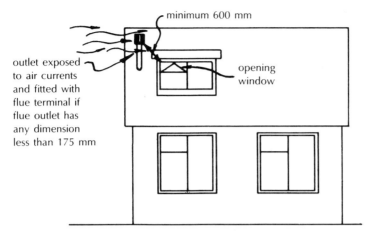

Fig. 4.7 Flue outlets – Type 2 appliances.

Masonry chimneys serving Type 2 appliances should be:

- Lined with any of the materials specified for Type 1 appliances (see p. 14.9). The linings should be jointed and pointed with cement mortar and built into the chimney with sockets or rebates uppermost. The space between the linings and the masonry should be filled with weak mortar or insulating concrete; *or*,
- Constructed of flue blocks without a lining. Flue blocks should comply with BS 1289 *Flue blocks and masonry terminals for gas appliances*, Part 1: 1986 *Specification for precast concrete flue blocks and terminals* and Part 2: 1989 *Specification for clay flue blocks and terminals.*

AD J1/2/3
sec. 3; 3.11,
3.12

It is also permissible to line a chimney with a flexible flue liner if:

- The liner complies with BS 715: 1989 *Specification for metal flue pipes, fittings, terminals and accessories for gas-fired appliances with a rated input not exceeding 60 kW.*
- The chimney was built before 1 February 1966; *or*,
- The chimney is already lined or is constructed of flue blocks as described above.

AD J1/2/3
sec. 3; 3.13

Where an appliance is connected to a chimney that is not lined or constructed of flue blocks as described in AD J1/2/3 (see above) then the chimney should be connected with a debris collection space with cleaning access. The collection space should extend at least 250 mm below the lowest point of entry of the appliance flue into the chimney and should have a minimum volume of 0.012 m^3.

AD J1/2/3
sec. 3; 3.14

Factory-made insulated chimneys

Any of the factory-made insulated chimneys described above (see p. 14.10) for Type 1 appliances may be used.

AD J1/2/3
sec. 3; 3.16

Flue pipes

AD J1/2/3
sec. 3; 3.9

Flue pipes for Type 2 appliances should be constructed of:

- Any of the materials specified for Type 1 appliances (see p. 14.12).
- Asbestos cement as described in BS 567: 1989 *Specification for asbestos-cement flue pipes and fittings, light quality*, or BS 835: 1989 *Specification for asbestos-cement flue pipes and fittings, heavy quality*.
- Sheet metal to BS 715: 1989 *Specification for metal flue pipes fittings, terminals and accessories for gas-fired appliances with a rated input not exceeding 60 kW*.
- Any other material which is suitable for its intended purpose.

Flue pipes should be fitted with the sockets uppermost.

Protection of building against fire and heat

Regs Sch. 1
J3

The construction of fireplaces and chimneys and the installation of heat producing appliances and flue-pipes must be carried out so as to reduce to a reasonable level the risk of the building catching fire in consequence of their use.

The requirements of this paragraph may be met by ensuring that hearths, fireplaces, chimneys and flue pipes:

- Are of sufficient size.
- Are constructed of suitable materials.
- Are suitably isolated from any adjacent combustible materials.

Protection of building against fire and heat – Type 1 appliances

Hearths for Type 1 appliances
Constructional hearths should be provided where a Type 1 appliance is to be installed. They should be constructed of solid non-combustible material at least 125 mm thick (including the thickness of any non-combustible floor under the hearth).

Constructional hearths built in connection with a fireplace recess should:

(a) Extend within the recess to the back and jambs of the recess; *and,*
(b) project at least 500 mm in front of the jamb; *and,*
(c) extend outside the recess to at least 150 mm beyond each side of the opening.

If not built in connection with a fireplace recess, the plan dimensions of the hearth should be such as to accommodate a square of at least 840 mm.

The recommendation that the hearth should project 500 mm in front of the jambs for a fireplace recess is to reduce the danger of fire from cinders, etc.

AD J1/2/3
sec. 2; 2.18,
2.19

AD J1/2/3
sec. 1
Part C
1.24 & 1.25

Combustible material should not be placed under a constructional hearth for a Type 1 appliance within a vertical distance of 250 mm from the upper surface of the hearth, unless there is an airspace of at least 50 mm between the combustible material and the underside of the hearth. Timber fillets supporting a hearth where it adjoins the floor are exempted from this rule (see Fig. 14.8).

Minimum plan dimensions:

with fireplace recess

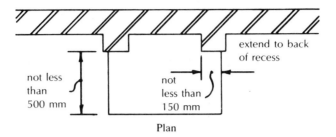

Plan

without fireplace recess

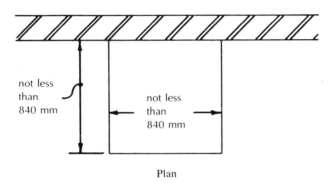

Plan

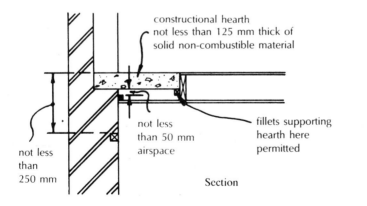

Section

Position of timber fillets

Fig. 14.8 Hearths for Type 1 appliances.

Fireplace recesses

Fireplace recesses should be constructed of solid non-combustible material and should have a jamb on each side at least 200 mm thick, a solid back wall at least 200 mm thick, or a cavity wall back with each leaf at least 100 mm thick. These thicknesses are to run the full height of the recess. However, if a fireplace recess is in an external wall the back may be a solid wall of not less than 100 mm thickness. Similarly, if part of a wall acts as the back of two recesses on opposite sides of the wall, it may be a solid wall not less than 100 mm thick. It is assumed that this latter exemption does not apply to a wall separating buildings or dwellings within a building since the requirements for chimney walls (see below) specify a minimum thickness of 200 mm in these circumstances (see Fig. 14.9).

AD J1/2/3
sec. 2; 2.20

Walls adjoining Type 1 appliances

AD J1/2/3
sec. 2; 2.21

Where a constructional hearth for an appliance is not situated in a fireplace recess, any wall or partition within a distance of 150 mm from the edge of the hearth should be constructed of solid non-combustible material at least 75 mm thick to a height of at least 1.2 m above the upper surface of the hearth (see Fig. 14.10).

Chimneys for Type 1 appliances – general requirements

Generally, if a chimney is built of masonry and is lined as described above any flue in that chimney should be:

- Surrounded and separated from any other flue in that chimney by at least 100 mm thickness of solid masonry material, excluding the thickness of any flue lining material.
- Separated by at least 200 mm of solid masonry material from another compartment of the same building, another building or another dwelling. (It is not clear how a cavity or compartment wall would be regarded, however.)
- Separated by at least 100 mm of solid masonry material from the outside air or from another part of the same building (but not a part which is a dwelling or a separate compartment). (See Fig. 14.11.)

AD J1/2/3
sec. 2; 2.14

Chimneys for solid fuel appliances should be designed to withstand a temperature of at least 1100°C without suffering structural changes which would impair the performance or stability of the chimney.

Chimneys for Type 1 appliances – proximity of combustible material

Combustible materials should not be placed nearer to a flue than 200 mm. Where the thickness of solid non-combustible material surrounding a flue in a chimney is less than 200 mm, no combustible material other than a floor board, skirting board, dado rail, picture rail, mantelshelf or architrave, should be placed within 40 mm of the outer surface of the chimney or fireplace recess.

Recesses built in masonry

Fireplace recesses generally

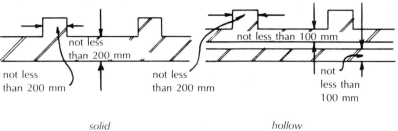

not less than 200 mm

not less than 100 mm

not less than 200 mm

not less than 200 mm

not less than 100 mm

solid

hollow

Recesses in external wall

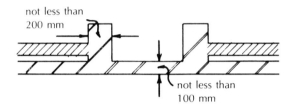

not less than 200 mm

not less than 100 mm

Back-to-back recesses

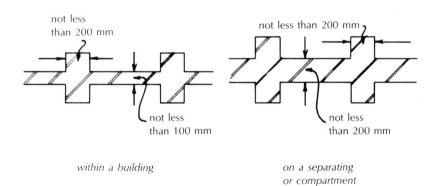

not less than 200 mm

not less than 100 mm

not less than 200 mm

not less than 200 mm

within a building

on a separating or compartment wall

Fig. 14.9 Fireplace recess – Type 1 appliances.

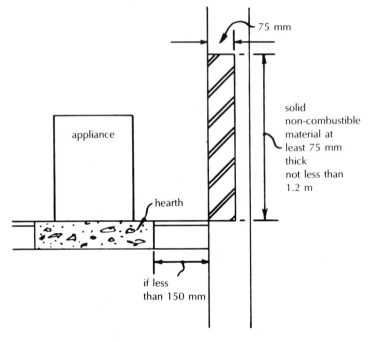

Fig. 14.10 Walls adjoining Type 1 appliances.

AD J1/2/3
sec. 2; 2.15
No metal fastening in contact with combustible material should be placed within 50 mm of the flue (see Fig. 14.12).

Chimneys for Type 1 appliances – factory-made insulated chimneys

A chimney serving a Type 1 appliance may consist of a factory-made insulated chimney having:

- No part of the chimney passing through or attached to a part of the building forming a separate compartment, unless it is surrounded in non-combustible material having at least half the fire resistance required for the compartment wall or floor (see Chapter 7, Fire).
- No combustible material nearer to the outer surface of the chimney than the distance (X) used for the test procedures specified in BS 4543 *Factory-made insulated chimneys*, Part 1: 1990 *Methods of test*.
- AD J1/2/3
sec. 2; 2.17
sec. 4; 4.9 A removable casing of suitable imperforate material enclosing any part of the chimney within a cupboard, storage space or roof space with no combustible material enclosed within the casing, and the distance between the inside of the casing and the outside of the chimney not less than the distance (X) specified above.

Placing and shielding of flue pipes for Type 1 appliances

A flue pipe adjacent to a wall or partition should be set at a minimum distance from any combustible material forming part of the wall or partition of at least:

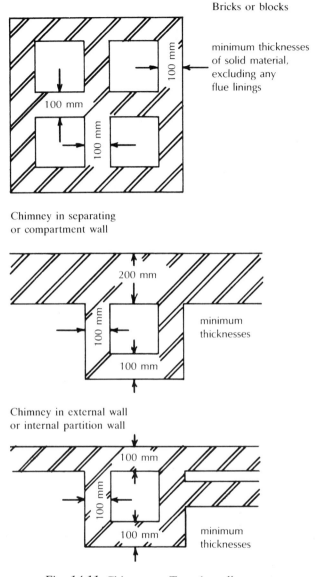

Fig. 14.11 Chimneys – Type 1 appliances.

(a) 3 times its external diameter; *or,*

(b) $1\frac{1}{2}$ times its external diameter if the combustible material is protected by a shield of non-combustible material which is placed so that there is an airspace of at least 12.5 mm between the shield and the combustible material. The non-combustible shield should have a width equal to at least 3 times the external diameter of the flue pipe (see Fig. 14.13).

AD J1/2/3
sec. 2; 2.9

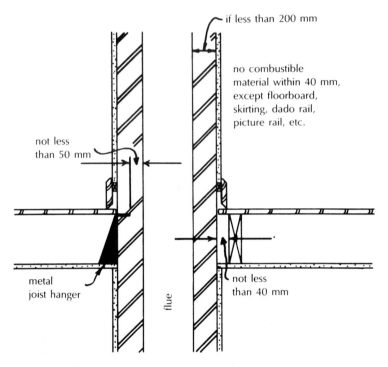

Fig. 14.12 Proximity of combustible material – Type 1 applicances

Positioning of Type 1 appliances

Where a Type 1 appliance is installed directly upon or over a constructional hearth, combustible material should not be laid closer to the base of the appliance than:

(a) *at the front*, 300 mm if the appliance is an open fire or stove which can, when opened, be operated as an open fire, or 225 mm in any other case;
(b) *at the back and sides*, 150 mm or as necessary to satisfy paragraph 2.22(b) AD J1/2/3 which relates to distance from hearth to walls (see below).

The distances fixed by this provision depend on the likelihood of danger from falling cinders or radiation. Where combustible material is not laid upon or over the constructional hearth, then the above distances should be the minimum distances from appliance base to edges of hearth.

Additionally, if any part of the back or sides of the appliance lies within 150 mm horizontally of the wall, then the wall should be of solid non-combustible construction at least 75 mm thick from floor level to a level of 300 mm above the top of the appliance.

If, however, any part of the back or sides of the appliance lies within 50 mm of the wall, then the wall should be of solid non-combustible construction at least 200 mm thick from floor level to a level of 300 mm above the top of the appliance (see Fig. 14.14).

**AD J1/2/3
sec. 2; 2.22**

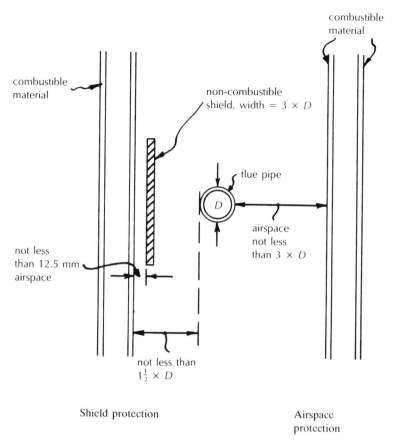

combustible
material

combustible
material

non-combustible
shield, width = 3 × *D*

flue pipe

D

airspace
not less
than 3 × *D*

not less
than 12.5 mm
airspace

not less than
$1\frac{1}{2}$ × *D*

Shield protection

Airspace
protection

Fig. 14.13 Protection of flue pipes next to combustible materials
– Type 1 appliances.

Special provisions relating to Type 1 appliances

Some oil-burning appliances are allowed to comply with lesser requirements
due to their low operating temperatures.

The exceptions permitted are as follows:

- A Type 1 appliance which has a hearth temperature of not more than 100°C
  and is placed on an imperforate rigid seating of non-combustible material
  does not need to comply with the general rules regarding the construction
  of hearths for Type 1 appliances (see paragraphs 2.18 and 2.19 of AD J1/2/3
  and pp.14.16 and 14.17).
- A Type 1 appliance which is constructed in such a manner that the surface
  temperature of the appliance's side and back panels does not exceed 100°C
  does not need to comply with the general rules regarding the construction
  of walls and partitions adjoining Type 1 appliances (see Paragraph 2.21 of
  AD J1/2/3 and pp. 14.18 to 14.20).

**AD J1/2/3
sec. 4; 4.10,
4.11**

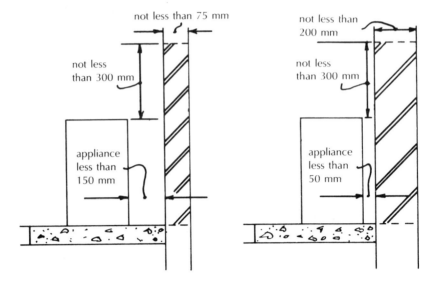

Sections

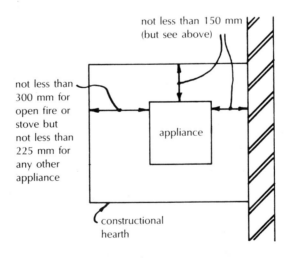

Plan

Fig. 14.14 Positioning of Type 1 appliances.

Protection of building against fire and heat – Type 2 appliances

The provisions described below apply to all Type 2 gas burning appliances with a rated input up to 60 kW except decorative log or solid fuel fire-effect appliances. These should comply with section 2 of AD J1/2/3.

AD J1/2/3
sec. 3; 3.1

Hearths for Type 2 appliances

With the exception of a gas-fired back boiler a Type 2 appliance should be placed over a solid non-combustible hearth, at least 12 mm thick, which,

- Extends at least 150 mm beyond the back and sides of the appliance.
- Extends forward at least 225 mm horizontally from any flame or incandescent material within the appliance.

Such a hearth is not required:

- If the appliance is installed so that no part of any flame or incandescent material is less than 225 mm above the floor.
- Where the appliance satisfies the requirements of the appropriate parts of BS 5258 *Safety of domestic gas appliances*, or BS 5386 *Gas burning appliances* for installation without a hearth.

For a back boiler, the hearth should be constructed of solid non-combustible material:

- At least 125 mm thick; *or*,
- At least 25 mm thick on 25 mm non-combustible supports.

It should extend at least 150 mm beyond the back and sides of the appliance and extend forward at least 225 mm beyond the front of the appliance.

These recommendations for gas appliance hearths are much less rigorous than those for solid fuel or oil-burning appliances, since there is no danger from falling fuel (see Fig. 14.15).

AD J1/2/3
sec. 3; 3.17,
3.18, 3.19

Shielding of structure adjoining Type 2 appliances

The back, top and sides of the appliance (including any associated draught diverter), should be separated from any combustible part of the building, by *either*,

(a) a shield of non-combustible material at least 25 mm thick, *or*,
(b) at least 75 mm airspace.

This recommendation again need not be fulfilled if the appliance complies with the relevant recommendations of the appropriate parts of BS 5258 *Safety of domestic gas appliances* or BS 5386 *Gas burning appliances* (see Fig. 14.16).

AD J1/2/3
sec. 3
3.20

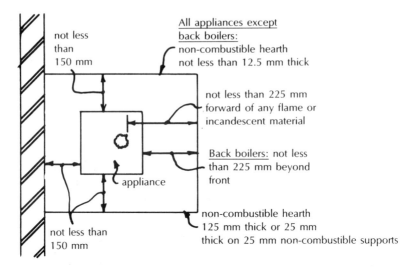

not less than 150 mm

All appliances except back boilers: non-combustible hearth not less than 12.5 mm thick

not less than 225 mm forward of any flame or incandescent material

Back boilers: not less than 225 mm beyond front

appliance

not less than 150 mm

non-combustible hearth 125 mm thick or 25 mm thick on 25 mm non-combustible supports

Plan

Unless:

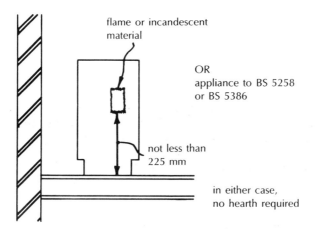

flame or incandescent material

OR
appliance to BS 5258 or BS 5386

not less than 225 mm

in either case, no hearth required

Section

Fig. 14.15 Hearths for Type 2 appliances.

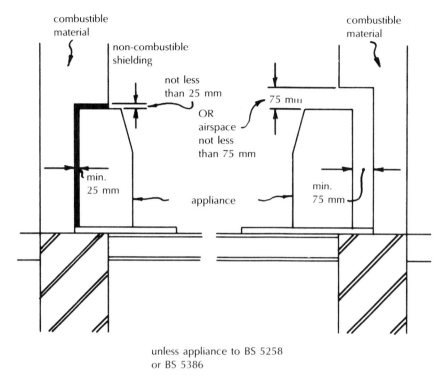

Fig. 14.16 Shielding of structure adjacent to Type 2 appliances.

Chimneys for Type 2 appliances

A chimney serving a Type 2 appliance, which is constructed of brickwork or blockwork, should have walls which are at least 25 mm thick. Additionally, it is permissible to pass any chimney serving a Type 2 appliance through a compartment wall or floor provided that the chimney walls have at least the fire resistance recommended by Approved Document B (see p. 7.78) for that wall or floor (see Fig. 14.17).

AD J1/2/3
sec. 3; 3.15
(as amended)

Placing and shielding of flue pipes for Type 2 appliances

Flue pipes serving Type 2 appliances should be placed so that:

- Every part of the flue pipe is at least 25 mm distant from any combustible material.
- Where the flue pipe passes through a roof, floor or wall formed of combustible materials, it is enclosed in a sleeve of non-combustible material and there is at least 25 mm airspace between the flue and the sleeve.
- Where the flue pipe passes through a compartment wall or compartment floor it is cased with non-combustible material having at least half the fire resistance required for the compartment wall or floor.

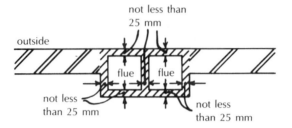

not less than
25 mm

outside

flue   flue

not less
than 25 mm

not less
than 25 mm

another part of same building
(but not separate compartment)

Chimney in external wall

another compartment, another building
or another dwelling

flue   flue

not less
than 25 mm

not less
than 25 mm

at least fire-resistance
required for
compartment
or
separating wall

another part of same building
(but not separate compartment)

Chimney on separating or compartment wall

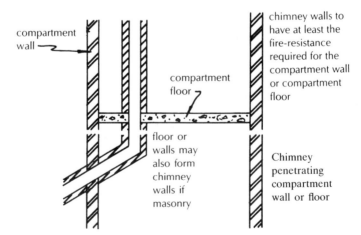

compartment
wall

compartment
floor

floor or
walls may
also form
chimney
walls if
masonry

chimney walls to
have at least the
fire-resistance
required for the
compartment wall
or compartment
floor

Chimney
penetrating
compartment
wall or floor

Fig. 14.17 Chimneys for Type 2 appliances.

Where a double-walled flue pipe is used the 25mm airspace should be measured from the outside of the inner pipe (see Fig. 14.18).

AD J1/2/3
sec. 3; 3.10

The recommendations contained in AD J1/2/3 may also be met by complying with the relevant recommendations of the National Standards and Codes of Practice listed in Appendix 2 (see pages A2.1 to A2.15).

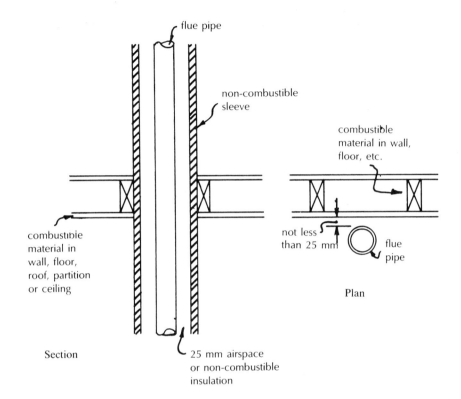

Fig. 14.18 Placing and shielding of flue pipes – Type 2 appliances.

Chapter 15

Protection from falling, collision and impact

Introduction

The control of stairways, ramps and guards has always formed an important part of Building Regulations since stairways represent, in many cases, the only way out of a building in the event of fire and there is a need to make buildings accessible to disabled people. Additionally, provisions are necessary to protect people from the risk of falling when they use exposed areas such as landings, balconies and accessible roofs.

The 1998 edition of Approved Document K introduces new provisions governing:

- The safe use of vehicle loading bays,
- Measures to reduce the risk of collisions with open windows, skylights and ventilators, *and*
- Measures designed to reduce the risk of injury when using various types of sliding or powered doors and gates,

mainly in order to ensure compliance with the Workplace (Health, Safety and Welfare) Regulations 1992.

Therefore, compliance with the revised regulations in Part K prevents action being taken against the occupier under the Workplace Regulations when the building is eventually in use. This involves considering aspects of design which will affect the way a building is used and applies only to workplaces. Although dwellings are excluded from the changes, in mixed use developments the requirements for the non-domestic part of the use would apply to any shared parts of the building (such as common access staircases and corridors). With flats the situation may be less clear for although certain sections of Part K do not apply to dwellings, it is still necessary for people such as cleaners, wardens and caretakers to work in the common parts. Therefore, the requirements of the Workplace Regulations may still apply even though the Building Regulations do not. Additionally, it is now necessary to provide safe access to areas used exclusively for maintenance purposes.

Stairways, ramps and ladders

Stairways, ramps and ladders which form part of the building must be designed, constructed and installed so that people may move safely between

Regs Sch. 1
K1

levels, in or about the building. Regulation K1 applies to all areas of a building which need to be accessed, including those used only for maintenance.

It should be noted that compliance with Regulation K1 prevents action being taken against the occupier of a building under regulation 17 of the Workplace (Health, Safety and Welfare) Regulations 1992 when the building is eventually in use. (Regulation 17 relates specifically to permanent stairs, ladders and ramps on routes used by people in places of work and includes access to areas used for maintenance.)

Application

The provisions contained in AD K1 regarding stairs, ramps and ladders only apply if there is a change in level of:

AD K1

- More than 600 mm in dwellings; *or,*
- Two or more risers in other buildings (the difference in levels in this case will depend on the recommended height for the risers), or 380 mm where there is no stair.

Since stairs, etc. must provide safe access for people, an acceptable level of safety in a building will be dictated by the circumstances. Therefore, the standard in a dwelling will be lower than that recommended for a public building because there are likely to be less people in the dwelling and they will be familiar with the stairs. Similarly, the standard of access to maintenance areas will be lower than that recommended for normal use to reflect the greater care expected of those gaining access.

AD K1
Performance

AD K1
0.2

Outside stairways and ramps, (e.g. entrance steps), are covered by the regulations if they form part of the building (obviously, the proximity of the steps to the building and the way they are associated with it will dictate whether or not they are part of the building, in most cases). Therefore, steps in paths leading to the building would not be covered by Part K. These access routes may, of course, need to comply with other parts of the regulations if:

- They form part of a means of escape in case of fire (see Approved Document B: Fire safety).

AD K1
0.3

- They are intended for use by disabled people (see Approved Document M: Access and facilities for disabled people).

In general, normal access routes in assembly buildings (e.g. sports stadia, theatres, cinemas etc.), should follow the guidance in AD K. Where special consideration needs to be given to guarding spectator areas or there are steps in gangways serving these areas then it may be preferable to follow more specialised guidance contained in the following:

- BS 5588 *Fire precautions in the design, construction and use of buildings*, Part 6:1991 *Code of practice for places of assembly* – for new assembly buildings.
- *Guide to Fire Precautions in Existing Places of Entertainment and Like Premises*, Home Office 1990 – for existing assembly buildings.

AD K1
0.4

- *Guide to Safety at Sports Grounds*, The Stationery Office 1997 – for stands at sports grounds.

Interpretation

A number of definitions are given which apply generally throughout AD K:

CONTAINMENT – a barrier to prevent people from falling from a floor to the storey below.

FLIGHT – the section of a ramp or stair running between landings with a continuous slope or series of steps.

RAMP – a slope which is steeper than 1 in 20 intended to enable pedestrians or wheelchair users to get from one floor level to another.

A number of definitions are given which are specific to stairways and ladders:

STAIR – steps and landings designed to enable pedestrians to get to different levels.

PRIVATE STAIRS – stairs in or serving only one dwelling.

INSTITUTIONAL AND ASSEMBLY STAIRS – stairs serving places where a substantial number of people gather.

OTHER STAIRS – stairs serving all other buildings apart from those referred to above.

GOING – the distance measured in plan across the tread less any overlap with the next tread above (see Fig. 15.1).

RISE – the vertical distance between the top surfaces of two consecutive treads (see Fig. 15.1).

PITCH AND PITCH LINE – these terms are illustrated in AD K1 but are not defined; *however,*

PITCH LINE – may be defined as a notional line connecting the nosings of all treads in a flight, including the nosing of the landing or ramp at the top of the flight. The line is taken so as to form the greatest possible angle with the horizontal, subject to the special recommendations for tapered treads (see below).

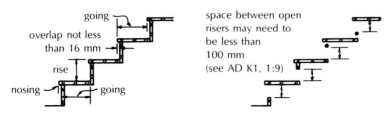

Fig. 15.1 Rise and going.

PITCH – may be defined as the angle between the pitch line and the horizontal (see Fig. 15.2).

ALTERNATING TREAD STAIRS – stairs with paddle shaped treads where the wider portion alternates from one side to the other on each consecutive tread.

HELICAL STAIRS – stairs which form a helix round a central void.

SPIRAL STAIRS – stairs which form a helix round a central column.

LADDER – a series of rungs or narrow treads used as a means of access from one level to another which is normally used by a person facing the ladder.

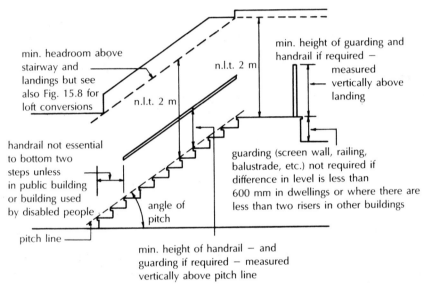

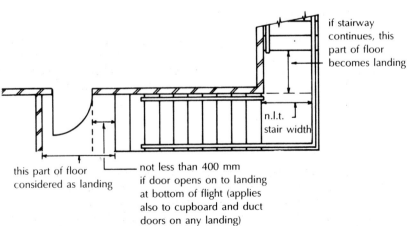

Fig. 15.2 General requirements.

TAPERED TREAD – a tread with a nosing which is not parallel to the nosing of the tread or landing above it.

AD K1
0.5 &
sec. 1
1.2, 1.3

General recommendations for stairways and ramps

Landings

As a general rule a landing should be provided at the top and bottom of every flight or ramp. Where a stairway or ramp is continuous, part of the floor of the building may count as a landing. The going and width of the landing should not be less than the width of the flight or ramp.

AD K1
sec. 1
1.15

Landings should be level and free from permanent obstructions. (This would allow, for example, the placing of a temporary barrier such as a child's safety gate between a landing and a flight.) A door is permitted to swing across a landing at the bottom of a flight or ramp but only if it leaves an area 400 mm wide across the full width of the flight or ramp. This rule also applies to cupboard and duct doors, but in this case they are permitted to open over any landing (including a landing at the top of a flight). Approved Document B contains details of restrictions on the use of cupboards situated in means of escape staircases.

AD K1
sec. 2
2.6

AD K1
sec. 1
1.16, 1.17
sec. 2
2.6

A landing of firm ground or paving at the top or bottom of an external flight or ramp may slope at a gradient of not more than 1 in 20 (see Fig. 15.2).

Handrails

Stairs and ramps should have a handrail on at least one side. Where the width of the stair or ramp is 1m or more, then a handrail should be fixed on both sides.

Handrails are not required:

- beside the bottom two steps in a stairway unless it is in a public building or is intended for use by disabled people (see Approved Document M and Chapter 17).
- where the rise of a ramp is 600 mm or less.

Handrails should provide firm support and grip and be fixed at a height of between 900 mm and 1 m vertically above the pitch line or floor. They can also form guarding where the heights can be matched. Further details regarding handrails for disabled people can be found in Chapter 17. (See also Fig. 15.2.)

AD K1
sec. 2
1.27 &
sec. 2
2.5

Headroom

Clear headroom of 2 m should be provided over the whole width of any stairway, ramp or landing. There are reduced dimensions for headroom over stairs in loft conversions (see p. 15.9, stairs to loft conversions).

Headroom is measured vertically from the pitch line, or where there is no pitch line, from the top surface of any ramp, floor or landing (see Fig. 15.2).

AD K1
sec. 1
1.10 &
sec. 2
2.2

Width

AD K1
sec. 1
1.11 &
sec. 2
2.3

AD K contains no recommendations for minimum stair or ramp widths; however there may be width recommendations in other Approved Documents. Reference should be made to AD B: Fire safety (Chapter 7) and AD M: Access and facilities for disabled people (Chapter 17), where minimum width guidance can be found.

Stairway recommendations

Rules applying to all stairways

AD K1
sec. 1
1.14

- In any stairway there should not be more than thirty-six rises in consecutive flights, unless there is a change in the direction of travel of at least 30° (see Fig. 15.3).

AD K1
sec. 1
1.5

- For any step the sum of twice its rise plus its going ($2R + G$) should not be more than 700 mm nor less than 550 mm. This rule is subject to variation at tapered steps, for which there are special rules.

AD K1
sec. 1
1.1

- The rise of any step should generally be constant throughout its length and all steps in a flight should have the same rise and going.

AD K1
sec. 1
1.8

- Open risers are permitted in a stairway but for safety the treads should overlap each other by at least 16 mm.
- Each tread in a stairway should be level.

Tapered treads should comply with the following rules:

- The minimum going at any part of a tread within the width of a stairway should not be less than 50 mm.
- The going should be measured:

  (i) if the stairway is less than 1m wide, at the centre point of the length or deemed length of a tread; *and,*

AD K1
sec. 1
1.18

  (ii) if the stairway is 1 m or more wide, at points 270 mm from each end of the length or deemed length of a tread. (When referring to a set of consecutive tapered treads of different lengths, the term 'deemed length' means the length of the shortest tread. This term is not used in AD K1.) (See Fig. 15.4.)

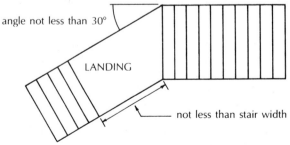

angle not less than 30°

LANDING

not less than stair width

change of direction required if more than 36 risers in consecutive flights

Fig. 15.3 Length of flights.

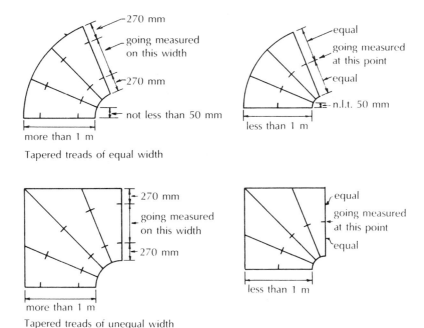

Fig. 15.4 Tapered treads.

- All consecutive tapered treads in a flight should have the same taper. **AD K1**
sec. 1
1.19
- Where stairs contain straight and tapered treads the goings of the tapered treads should not be less than those of the straight flight. **AD K1**
sec. 1
1.20
- In order to prevent small children from becoming trapped between the treads of open riser staircases there should be no opening in a riser of such size as to allow the passage of a sphere of 100 mm diameter. This rule applies to all stairways which are likely to be used by children under the age of five years. (See Fig. 15.1.) **AD K1**
sec. 1
1.9

Rules applying to private stairways

There are a number of recommendations in AD K which control the steepness, rise and going of private stairs as follows:

- The height of any rise should not be more than 220 mm.
- The going of any step should generally not be less than 220 mm (but see the rules relating to tapered treads above).
- The pitch should not be more than 42°. This means that it is not possible to combine a maximum rise with a minimum going. **AD K1**
sec. 1
1.3, 1.4

Figure 15.5 illustrates the practical limits for rise and going. This figure also incorporates the $(2R + G)$ relationship mentioned above.

The rules governing rise and going for private stairs will also satisfy the recommendations contained in AD M for access for disabled people.

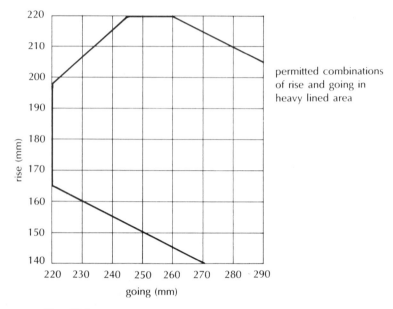

permitted combinations
of rise and going in
heavy lined area

Fig. 15.5 Practical limits for rise and going – private stairways.

Rules applying to institutional and assembly stairs

Institutional buildings are usually occupied by young children, old people or people with physical or mental disabilities. It is necessary, therefore, that staircases should be of slacker pitch than in other types of buildings and further recommendations regarding rise and going may be found in AD M. Additionally, they may need to comply with width recommendations in AD B or AD M. Assembly buildings contain large numbers of people and similar considerations will apply.

- The height of any rise should not be more than 180 mm.
- Subject to the rules governing tapered treads, the going of any step should generally not be less than 280 mm. This may be reduced to 250 mm if the floor area of the building served by the stairway is less than 100 m².

AD K1
sec. 1
1.3, 1.4

Figure 15.6 illustrates the practical limits for rise and going. This figure also incorporates the $(2R + G)$ relationship mentioned above.

AD K1
sec. 1
1.6

AD K1
sec. 1
1.13

AD K1
sec. 1
1.12

- In order to maintain sightlines for spectators in assembly buildings, gangways are permitted to be pitched at up to 35°.
- There should not be more than sixteen risers in a single flight where a stairway serves an area used for assembly purposes. (See also Rules applying to other stairs, below.)
- The width of a stairway in a public building which is wider than 1800 mm should be sub-divided with handrails or other suitable means so that the sub-divisions do not exceed 1800 mm in width.

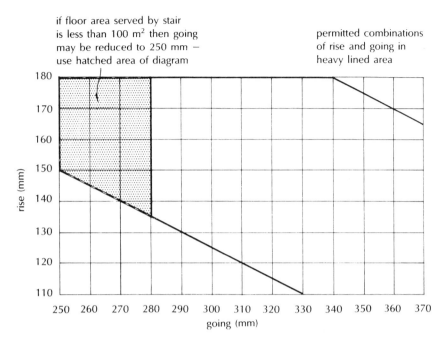

Fig. 15.6 Practical limits for rise and going – institutional and assembly buildings.

Rules applying to other stairs

These recommendations apply to all stairs other than those in private dwellings, institutional or assembly buildings. Many of the buildings covered by these recommendations will also need to be accessible to disabled people and should, therefore, conform to the guidance given in AD M.

- The height of any rise should not be more than 190 mm.
- The going of any step should generally not be less than 250 mm (but see the rules relating to tapered treads above).

AD K1
sec. 1
1.3, 1.4

Figure 15.7 illustrates the practical limits for rise and going. This figure also incorporates the $(2R + G)$ relationship mentioned above.

- There should not be more than sixteen risers in a single flight where a stairway serves an area used for shop purposes.

AD K1
sec. 1
1.13

Stairs to loft conversions

When carrying out loft conversions it is often extremely difficult to fit a conventional staircase without substantial alteration to the existing structure. This often results in serious loss of the space which the conversion is intended to provide. Approved Document K contains a number of alternative recommendations which are intended to assist in the better use of space where a conventional staircase would prove difficult to install.

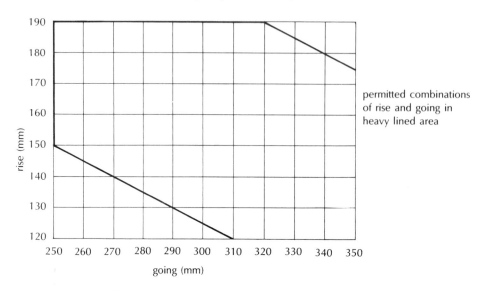

Fig. 15.7 Practical limits for rise and going – other buildings.

Headroom

AD K1
sec. 1
1.10
Where there is insufficient height to achieve the recommended 2 m headroom over a stairway, 1.9 m at the centre of the stair is acceptable reducing to 1.8 m at the side (see Fig. 15.8).

Ladders

Fixed ladders may be installed for access in a loft conversion where there is insufficient room to install a conventional staircase without alteration to the existing space if they conform to the following:

- There should be fixed handrails on both sides.
- The ladder should only serve one habitable room.
AD K1
sec. 1
1.25
- Retractable ladders are not acceptable for means of escape in case of fire.

The definition of habitable room is not included in AD K and it varies from one Approved Document to another. The only safe conclusion which can be drawn is that bathrooms or WCs are not habitable rooms. Therefore, it would seem that a ladder could be used to access one habitable room and a bathroom or WC, but see Alternating tread stairs below.

It would also seem to be the case that retractable ladders can be used where the roof of a bungalow is converted to provide a two-storey house. These houses do not need to have means of escape to a stairway since escape from a suitable first floor window is acceptable in AD B. (See AD B, Section 1, 1.1 and Chapter 7 above).

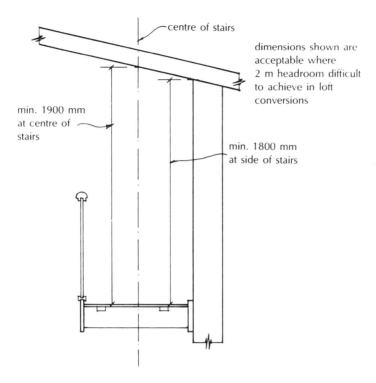

Fig. 15.8 Reduced headroom over stairs – loft conversions.

Spiral and helical stairs

Generally stairs designed in accordance with BS 5395 *Stairs, ladders and walkways*, Part 2: 1984 *Code of practice for the design of helical and spiral stairs* will satisfy the recommendations of AD K. It is permissible to provide stairs with lesser goings for conversion work if space is limited, but the stair should only serve one habitable room (and, perhaps a bathroom or WC). The degree of variation from the norm is, however, not specified.

AD K1
sec. 1
1.21

Alternating tread stairs

When first introduced a few years ago, these stairs caused quite a controversy since the treads are not of uniform width and the staircase is steeper than a conventional flight. They rely on a certain degree of familiarity on the part of the user since it is necessary to start the ascent or descent on the correct foot or disaster will ensue!

Alternating tread stairs should only be installed for loft conversion work where insufficient space is available to accommodate a conventional staircase and they should:

- Be in one or more straight flights.
- Provide access to only one habitable room plus bathroom or WC provided it is not the only WC in the dwelling.

- Be fitted with handrails on both sides.
- Contain treads which have slip-resistant surfaces.
- Have uniform steps with parallel nosings.
- Have a minimum going of 220 mm and a maximum rise of 220 mm when measured over the wider part of the tread.

AD K1
sec. 1
1.22 to 1.24
- Conform to the recommendations regarding maximum gap sizes for open riser stairs.

A typical design for an alternating tread stair is shown in Fig. 15.9.

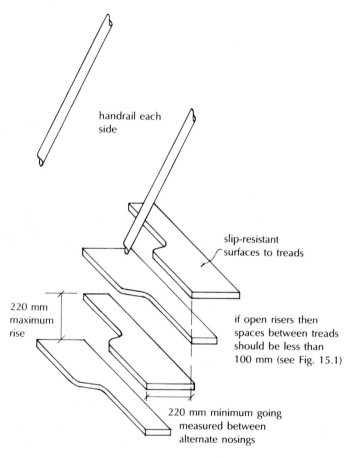

handrail each side

slip-resistant surfaces to treads

220 mm maximum rise

if open risers then spaces between treads should be less than 100 mm (see Fig. 15.1)

220 mm minimum going measured between alternate nosings

Fig. 15.9 Alternating tread stairs.

Ramp recommendations

AD K1
sec. 2
2.1
In addition to the general recommendations outlined above, Section 2 of AD K contains the following particular guidance for ramps:

AD K1
sec. 2
2.4
- The slope should not be greater than 1 in 12.
- There should be no permanent obstructions placed across any ramp.

Guarding of stairways, ramps and landings

Guarding should generally be provided at the sides of every flight, ramp or landing. However, guarding need not be provided in dwellings where there is a drop of 600 mm or less, or where there are fewer than two steps in other buildings.

AD K1
sec. 1, 1.28
& sec. 2, 2.7

The rules which apply to prevent small children from being trapped in open riser staircases also apply to guarding; that is, there should be no opening of such size as to allow the passage of a sphere of 100 mm diameter. This relates to the guarding on any staircase except that which is in a building which is unlikely to be used by children under the age of five years. It may be difficult in certain circumstances to decide where this exemption should apply. For example, some public houses have a policy of not admitting young children whereas others actually encourage them (accompanied by their parents, of course!). Guarding should also be designed so that it cannot easily be climbed by small children. (This might preclude the use of horizontal 'ranch' style balustrading, for example.)

AD K1
sec. 1, 1.29

Certain minimum heights for the guarding to flights, ramps and landings are given in AD K2/3, Section 3, Diagram 11. These are illustrated in Fig. 15.10 below and it should be noted that the guarding should be strong enough to resist at least the horizontal forces given in BS 6399: Part 1:1996.

AD K1
sec. 1, 1.30

Access to maintenance areas

It is important to establish how frequently access will be required to areas needing maintenance. If this is likely to be more than once per month then permanent stairs or ladders (such as those suggested in AD K1 for private stairs in dwellings) may need to be provided. Alternatively, the requirement may be satisfied by following the guidance in BS 5395: Part 3: 1985 *Code of practice for the design of industrial type stairs, permanent ladders and walkways*. Less frequent access to maintenance areas may be achieved using portable ladders.

AD K1
sec. 1, 1.31
& 1.32

Alternative approach to stairway design

It is permissible to use other sources of guidance when designing stairs.

- BS 5395 *Stairs, ladders and walkways*, Part 1: 1977 *Code of practice for the design of straight stairs* contains recommendations which will meet the steepness requirements of AD K.

AD K1
sec. 1
1.7

- Wood stairs designed in accordance with BS 585, Part 1: 1989 will offer reasonable safety to users.

AD K1
sec. 1
1.20

- Stairs, ladders or walkways in industrial buildings should follow the recommendations of BS 5395, Part 3: 1985 *Code of practice for the design of industrial stairs, permanent ladders and walkways* or BS 4211: 1987 *Specification for ladders for permanent access to chimneys, other high structures, silos and bins*.

AD K1
sec. 1
1.26

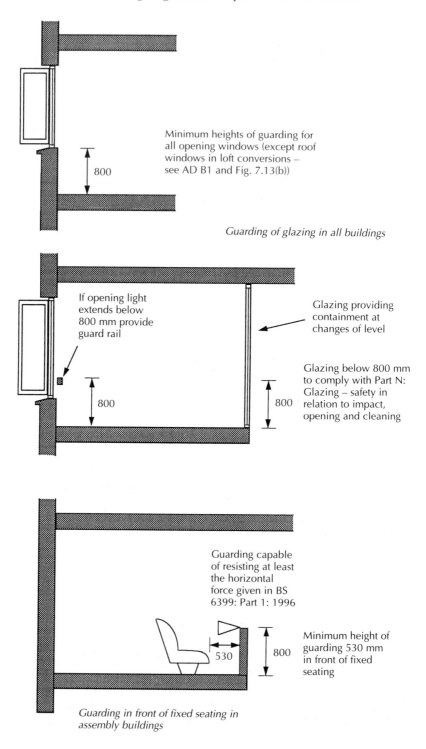

Fig. 15.10 Minimum height of guarding in all buildings.

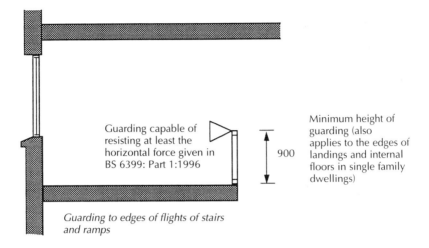

Guarding capable of resisting at least the horizontal force given in BS 6399: Part 1:1996

Minimum height of guarding (also applies to the edges of landings and internal floors in single family dwellings)

900

Guarding to edges of flights of stairs and ramps

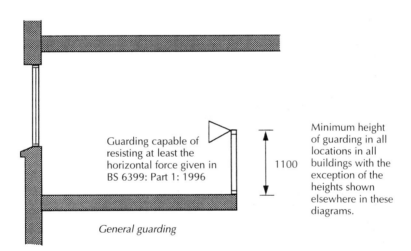

Guarding capable of resisting at least the horizontal force given in BS 6399: Part 1: 1996

Minimum height of guarding in all locations in all buildings with the exception of the heights shown elsewhere in these diagrams.

1100

General guarding

Fig. 15.10 continued.

Protection from falling

The following areas of the building must be provided with barriers to protect people in or about the building from falling:

Regs Sch. 1
K2

- Stairways and ramps which form part of the building, *and*
- Floors, balconies, and any roof to which people normally have access, *and*
- Light wells, basements or similar sunken areas which are connected to the building.

Regulation K2 applies to all areas of a building which need to be accessed, including those used only for maintenance.

It should be noted that compliance with Regulation K2 prevents action being taken against the occupier of a building under Regulation 13 of the Workplace (Health, Safety and Welfare) Regulations 1992 when the building is eventually in use. (Regulation 13 relates to requirements designed to protect people from the risk of falling a distance likely to cause personal injury.)

AD K2/3

The recommendations contained in AD K2 regarding the provision of pedestrian barriers to provide protection from falling only apply if there is a change in level of:

- More than 600 mm in dwellings; *or*,
- More than the height of two risers in other buildings (the difference in levels in this case will depend on the recommended height for the risers), or 380 mm where there is no stair.

As in the case of stairs (see p. 15.2) circumstances will usually dictate what constitutes an acceptable level of safety for the guarding in a building. Therefore, the standard in a dwelling may be lower than that recommended for a public building because there are likely to be less people in the dwelling and they will be familiar with its layout, etc. Similarly, the standard of access to maintenance areas will be lower than that recommended for normal use to reflect the greater care expected of those gaining access.

AD K2/3
Performance

AD K2/3

The provisions for guarding contained in AD K1 are extended by Approved Document K2/3 to cover the edges of any part of:

- A balcony, gallery, roof (including rooflights or other openings), floor (including the edge below any window openings) or other place to which people have access.
- Any light well, basement or sunken area adjoining a building.
- A vehicle park (but not, of course, on any vehicle access ramps).

AD K2/3
sec. 3
3.1

Guarding is not needed where it would obstruct normal use (for example, at the edges of loading bays).

Guarding recommendations

Where guarding is provided to meet the requirements of AD K2, then it should:

- Have a minimum height as shown in Fig. 15.10.
- Be capable of resisting the horizontal force given in BS 6399: Part 1: 1996.

AD K2/3
sec. 3
3.2

- Have any glazed part designed in accordance with the recommendations of AD N; Glazing – safety in relation to impact, opening and cleaning (see Chapter 18).
- Have no opening of such size as to permit the passage of a 100 mm diameter sphere if it is in a building which is likely to be used by children under 5 years of age.

AD K2/3
sec. 3
3.3

Guarding can consist of a wall, balustrade, parapet or similar barrier but should be designed so that it cannot be easily climbed by small children (i.e. horizontal rails should be avoided) if they are likely to be present in the building.

Guarding recommendations for maintenance areas

It is important to establish how frequently access will be required to areas needing maintenance. If this is likely to be more than once per month then guarding provisions such as those suggested for dwellings (see Fig. 15.10) will be satisfactory. For less frequent access it may only be necessary to provide temporary guarding and in many circumstances the posting of warning notices may be sufficient to satisfy the requirement. Building Regulations, of course, do not cover the design and installation of temporary guarding or warning signs although they are covered by the Construction (Design and Manage- **AD K2/3** ment) Regulations 1994. (For information on signs see the Health and Safety **sec. 3** (Signs and Signals) Regulations 1996.) **3.4 to 3.6**

Vehicle barriers and loading bays

Vehicle ramps and any levels in a building to which vehicles have access are required to have barriers in order to protect people in or about the building. Additionally, vehicle loading bays must either be constructed in such a way as to protect people in them from collision with vehicles or they must contain **Regs Sch. 1** features which achieve the same end. **K3**

Vehicle barriers

If the perimeter of any roof, ramp or floor to which vehicles have access, forms **AD K2/3** part of a building, it should have barriers to protect it, provided that it is level **sec. 3** with or above any adjacent floor, ground or vehicular route. **3.7**

Vehicle barriers can be formed by walls, parapets, balustrading or similar **sec. 3** obstructions and should be at least the heights shown in Fig. 15.11. Barriers **3.8** should be capable of resisting the horizontal forces as set out in BS 6399 *Loading for buildings*, Part 1: 1984 *Code of practice for dead and imposed loads*.

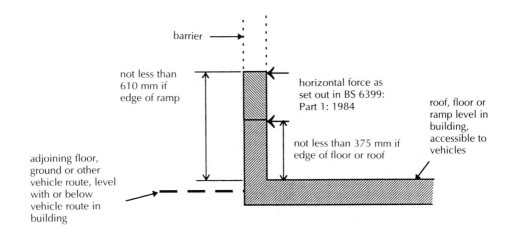

Fig. 15.11 Vehicle barriers – any building.

Loading bays

Loading bays for less than three vehicles should be provided with:

- One exit point (e.g. steps) from the lower level, ideally near to the centre of the rear wall; *or*
- A refuge into which people can go if in danger of being struck or crushed by a vehicle.

Larger loading bays (i.e. for three or more vehicles) should be provided with:

AD K2/3
sec. 3
3.9

- One exit point at each side, *or*
- A refuge.

See Fig. 15.12 for details.

Protection from collision with open windows, skylights or ventilators

Regulation K4 requires that provision be made to prevent people who are moving in or about a building from colliding with open windows, skylights or ventilators. This requirement does not apply to dwellings but it does apply to all parts of other buildings including, in a limited form, to areas of the building used exclusively for maintenance purposes.

Regs Sch. 1
K4

Again, compliance with this regulation prevents action being taken against the occupier of a building under Regulation 15(2) of the Workplace (Health, Safety and Welfare) Regulations 1992 (Regulation 15(2) also relates to projecting windows, skylights and ventilators) when the building is eventually in use.

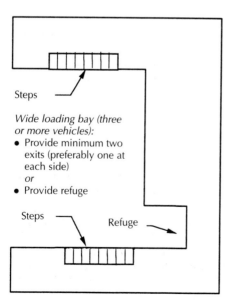

Steps

Wide loading bay (three or more vehicles):
- Provide minimum two exits (preferably one at each side)
  or
- Provide refuge

Steps   Refuge

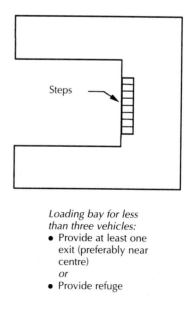

Steps

Loading bay for less than three vehicles:
- Provide at least one exit (preferably near centre)
  or
- Provide refuge

Fig. 15.12 Vehicle loading bays.

Since it is desirable to leave windows, etc. open for ventilation purposes, this should be possible without causing danger to people who might collide with them. Generally, this can be achieved by:

- Providing windows, skylights and ventilators so that projecting parts are kept away from people moving in and about the building; *or*
- Providing features which guide people away from these projections.

In certain special cases (e.g. in spaces used only for maintenance purposes) it is reasonable to expect the exercise of greater care by those gaining access, therefore less demanding provisions than those for normal access could satisfy the regulation.

AD K4 –
Performance

Avoiding projecting parts

AD K4 makes it clear that the requirements of Regulation K4 do not apply to the opening parts of windows, skylights or ventilators:

(a) Which are at least 2 m above ground or floor level; *or*
(b) Which do not project more than about 100 mm internally or externally into spaces in or about the building where people are likely to be present.

Since the siting of opening lights above 2 m may not always be possible, an acceptable solution might be to fit projecting lights with restraint straps designed to restrict the projection to about 100 mm. The restraint should only be removed for maintenance and cleaning purposes. An ordinary casement stay would probably not be considered suitable for this purpose. Any such solution should, of course, be checked for acceptability with the building control authority (local authority or approved inspector).

Alternative solutions given in AD K4 also include:

(a) Marking the projection with a feature such as a barrier or rail about 1100 mm high to prevent people walking into it; *or*
(b) Guiding people away from the projection by providing ground or floor surfaces with strong tactile differences or suitable landscaping features.

AD K4
sec. 4
4.1 & 4.2

These solutions are shown in Fig. 15.13 below.

Maintenance areas

Where such areas are used infrequently, projecting parts of windows, skylights and ventilators should be made easier to see by being clearly marked, in order to satisfy Regulation K4.

AD K4
sec. 4
4.4

It should be noted that there are other provisions in the Building Regulations which relate to safety on common circulation routes in buildings. In particular, reference should be made to Approved Document B – Fire safety (see p. 7.22 for guidance on the clear widths of escape routes), and Approved Document M – Access and facilities for disabled people (see p. 17.7 for guidance on the avoidance of hazards on circulation routes).

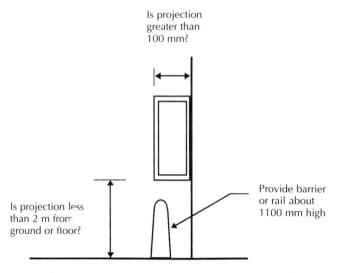

Is projection
greater than
100 mm?

Is projection less
than 2 m from
ground or floor?

Provide barrier
or rail about
1100 mm high

Fig. 15.13(a) Marking projections – barriers.

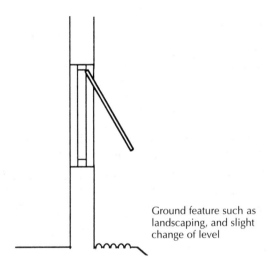

Ground feature such as
landscaping, and slight
change of level

Fig. 15.13(b) Marking projections – ground features.

Safe use of doors

In order that they may be safely used, provisions must be made to prevent doors or gates:

(a) Which slide or open upward, from falling onto any person, *and*
(b) Which are powered, from trapping anyone.

Furthermore, powered doors and gates must be openable in the event of a power failure and all swing doors and gates are required to be designed to allow a clear view of the space on either side.

Regulation K5 does not apply to dwellings, or to any door or gate which is part of a lift. Compliance with this regulation prevents action being taken against the occupier of a building under Regulation 18 (requirements for doors and gates) of the Workplace (Health, Safety and Welfare) Regulations 1992 when the building is eventually in use.

Regs Sch. 1
K5

Door and gate safety features

Typically, the following safety features will satisfy the requirements of Regulation K5 by preventing the opening and closing of doors and gates from presenting a safety hazard:

(a) Vision panels should be provided in doors and gates on traffic routes and in those which can be pushed open from either side (i.e. double swing doors) unless they can be seen over (e.g. lower than about 900 mm for a person in a wheelchair). For example, this means providing such doors and gates with a glazed panel giving a zone of visibility between 900 mm and 1500 mm from floor level (see also Fig. 17.3 on p. 17.9).

(b) Sliding doors and gates should be prevented from leaving the end of the suspension track by a stop or other suitable means. Additionally, if the suspension system fails or the rollers leave the track they should be prevented from falling by means of a retaining rail.

(c) Upward opening doors and gates should be designed to prevent them from falling and causing injury.

(d) If doors and gates are power operated the following safety features are examples of those which could be incorporated:
  • devices to prevent injury to people from being struck or trapped (e.g. doors with pressure sensitive edges)
  • readily accessible and identifiable stop switches
  • manual or automatic opening provisions in the event of power failure if this is necessary for health and safety reasons.

Reference should also be made to other provisions in the Building Regulations which relate to the design and use of doors in buildings. In particular, see Approved Document B – Fire safety (see page 7.58 above for guidance on the provision of doors on escape routes) and Approved Document M – Access and facilities for disabled people (see pages 17.8 to 17.11 below for guidance on the design of internal and external doors).

AD K5
5.1 & 5.2

Chapter 16

Conservation of fuel and power

Introduction

Part L of Schedule 1 to the 1991 Regulations (as amended) is concerned with the conservation of fuel and power. This is supported by Approved Document L which covers the following topics:

- Resistance to the passage of heat,
- Heating system controls,
- Insulation of hot water and heating services,
- Controls over the design and installation of artificial lighting systems in buildings other than dwellings.

National Building Regulation standards for insulation were first introduced in 1965 and were progressively raised during the 1970s in response to steep increases in the World price for oil. Specific proposals for conservation of fuel and power were not introduced until 1979 and the 1994 revisions to ADL represented perhaps the most extensive changes seen to date.

The prime aim of Part L is now seen to be concerned with environmental issues including the reduction of CO_2 emissions and the Government is committed to reducing CO_2 emissions to 20% below the 1990 levels by 2010. Since buildings contribute approximately half of all CO_2 emissions, the contribution from the building sector is likely to be significant. The Government is currently reviewing the way in which the scope of the Regulations can be widened to:

- Improve the as-built standards of new construction work,
- Progressively improve the energy efficiency of existing buildings,
- Make buildings more efficient to operate.

Conservation of fuel and power

Paragraph L1 of Schedule 1 requires that reasonable provision must be made for the conservation of fuel and power in buildings. This must be achieved by:

Regs Sch. 1
L1

(a) putting a limit on the amount of heat which is lost through the building fabric,
(b) making sure that the space heating and hot water systems are properly controlled,

(c) preventing unnecessary heat loss from hot water vessels and hot water service pipework, and
(d) limiting the amount of heat which is lost in space heating systems from hot water pipes and hot air ducts.

These requirements apply to dwellings, and also to any other buildings which have floor areas in excess of 30 m².

Additionally, one further requirement applies to certain buildings (but not including dwellings), where artificial lighting is to be provided to more than 100 m² of floor area. The lighting systems in such buildings must be designed and constructed so that:

- they use a reasonable amount of fuel and power, and
- reasonable provision is made for their control.

Approved Document L contains general and specific guidance on how the above requirements may be met.

AD L1
0.1, 0.2
In general terms, it is important to provide energy efficient measures so that heat losses through walls, roofs, floors, windows and doors, etc., are kept to a minimum, although it is permissible to take into account solar heat gains and the savings which can be made from installing more efficient heating systems.

Heat may also be lost by air leakage around ill-fitting doors and windows, so these should be adequately draught stripped.

Space heating and hot water systems can be made to function more efficiently if controls are placed on their duration of operation and operating temperatures. Hot water vessels, pipes and heating ducts should be adequately insulated unless the heat output from them is designed to contribute to the efficient heating of the building.

With regard to the need to provide energy efficient lighting systems, this can be achieved by installing suitable low energy lighting sources and by controlling the operation of the system by manual or automatic means so that the maximum benefit may be derived from natural light.

The specific recommendations of AD L1 are considered in the remainder of this chapter.

Interpretation

AD L1
0.13a
The introduction to AD L contains special definitions which apply throughout this document.

- EXPOSED ELEMENT means an element of a building which is exposed to the outside air (and this definition includes suspended floors whether or not the underfloor is vented, and floors which are in contact with the ground).
- SEMI-EXPOSED ELEMENT means an element which separates a heated part of the building from an unheated part which is exposed to the outside air and which does not comply with the recommendations for limitation of heat loss contained in AD L (see Fig. 16.1). It should be noted that the term ELEMENT is not defined in AD L. However, from the context of the document it can be assumed to mean a wall, floor, roof, window, door or other opening.

AD L1
0.13b

- U-VALUE means the thermal transmittance coefficient, which is the rate of heat transfer in watts through one square metre of a structure when the combined radiant and air temperatures on each side of the structure differ by 1 kelvin (i.e. 1°C). This is stated in watts per square metre of fabric per kelvin (W/m²K). It should be noted when calculating U-values, that certain elements of the construction, such as timber joists, structural and other types of framing, mortar joints and window frames act as thermal bridges and should be allowed for in U-value calculations. Typical calculations showing how to make these allowances are described on page 16.26 below where it can be seen that the effects of thermal bridging may be ignored if the thermal resistance between the bridged material and the bridging material is less than 0.1 m²K/W (e.g. as occurs between normal mortar joints and brickwork, but *not* between normal mortar joints and lightweight blockwork). AD L1
0.8, 0.11

- THERMAL CONDUCTIVITY is the amount of heat per unit area, conducted in unit time through a slab of material of unit thickness, per degree of temperature difference. It is expressed in watts per metre of thickness of material per degree kelvin (W/mK) and is usually denoted by the Greek letter λ. AD L1
0.8

  Certified test results of thermal conductivities (i.e. λ – values in W/mK) and thermal transmittances (i.e. U-values in W/m²K) for particular products should be obtained from individual manufactures. If this proves to be difficult the values contained in the tables in AD L (which are reproduced below) may be used instead. U-values may be calculated provided that suitable allowances have been made for the effects of thermal bridging as is the case with the tables mentioned above. AD L1
0.9, 0.10

- STANDARD ASSESSMENT PROCEDURE (SAP). This is the Government's approved method for arriving at an energy rating for a dwelling based on the calculated annual energy cost for space and water heating. Under Regulation 14A (see page 3.7), all new dwellings (whether created by new construction or by change of use) are required to be given a SAP Energy Rating and this must be supplied to the local authority or approved inspector. AD L1
0.14, to 0.20

  The rating is expressed on a scale of 1 to 100 where the higher numbers indicate the better standard and although there is no obligation to achieve any particular rating it will be seen from the detailed guidance given below that SAP ratings of 60 or less normally indicate that higher insulation standards should be applied. In fact, if a SAP rating of 80 or more is achieved then, depending on the floor area of the dwelling, this may in itself be a way of demonstrating compliance.

  In theory, any competent person may carry out SAP calculations or certify that they have been done correctly, provided that their competency is accepted by the local authority and this should be checked by applicants before submitting calculations. In this context, National Home Energy Rating (NHER) registered assessors who operate under the scheme run by the National Energy Foundation, are deemed to have been authorised by the Secretary of State as assessors for undertaking SAP calculations and should be acceptable to local authorities for that purpose.

  It should be noted that calculations certifying compliance with Part L from 'Approved Persons' (see Section 16 (9) of the Building Act 1984 on page 4.14 above) are also acceptable to local authorities. Such a person must be registered under a recognised scheme approved by the Secretary of State and since no such scheme exists at present, there are no approved persons.

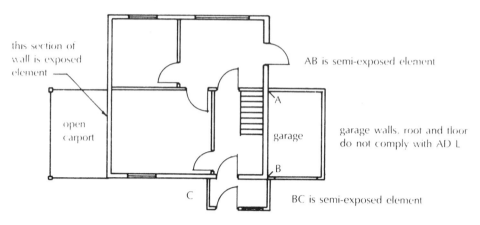

all external walls apart from AB and BC
are exposed elements

this section of
wall is exposed
element

AB is semi-exposed element

open
carport

garage

garage walls, roof and floor
do not comply with AD L

BC is semi-exposed element

unheated porch walls, roof and
floor do not comply with AD L

Dwelling

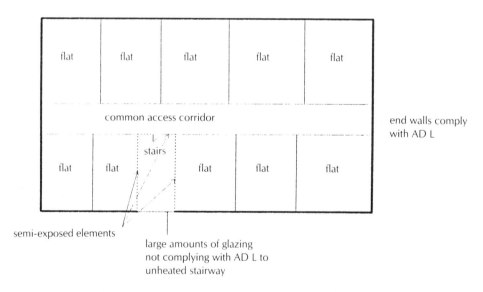

flat   flat   flat   flat   flat

common access corridor

end walls comply
with AD L

flat   flat

stairs

flat   flat   flat

semi-exposed elements

large amounts of glazing
not complying with AD L to
unheated stairway

Flats

Fig. 16.1 Exposed and semi-exposed elements.

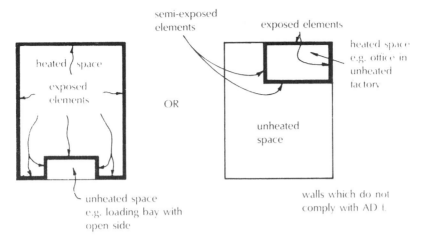

Fig. 16.1 continued.

General considerations

Rules for measurement of areas

Areas of the external enclosing elements of the building such as walls, floors and roofs should comply with the following rules:

- measure between internal finished surfaces,
- include any projecting elements such as bay windows,
- for roofs, measure in the plane of the insulation, and
- for floors, include non-usable spaces such as service ducts and stair wells.

AD L1
0.12

Large complex buildings

It is permissible to consider the different parts of large complex buildings separately when applying the measures for conservation of fuel and power contained in AD L.

AD L1
0.6

Buildings with low level heating needs

Those buildings which are not normally heated to any great extent (and would therefore have minimal heat losses) do not need to be insulated. A low level of heating is defined by reference to the output of the building's space heating system. This should not exceed:

AD L1
0.4, 0.5

- 50 watts per square metre of floor area for industrial and storage buildings,
- 25 watts per square metre of floor area for all other buildings (but not including dwellings).

Sometimes, speculative industrial and commercial developments are constructed where the final use of the building is not known. In these circumstances full insulation and sealing of the building fabric would be necessary.

Dwelling house extensions

AD L1
0.3

Small extensions to dwellings (i.e. not exceeding $10\,m^2$ floor area) are considered to have reasonable provision for the conservation of fuel and power if they are constructed in a similar manner to the existing dwelling.

Problems associated with increased thermal insulation

AD L1
0.7

It has been known for a number of years that higher insulation levels have increased the risk of interstitial condensation occurring in the building fabric. Also, greater thermal stresses have been placed on exposed materials such as bituminous roofing felts causing these to fail well before the completion of their normal design life and thus permitting rain penetration to occur.

By correct design and specification in addition to good workmanship and site supervision most of these problems may be overcome. To this end the Building Research Establishment has produced a report BR 262, entitled *Thermal insulation: avoiding risks* (1994 edition), which contains practical guidance (although it does not have Approved Document status). Reference should also be made to the National House Building Council's publication *Thermal insulation and ventilation Good Practice Guide* 1991 and to Approved Document F: Ventilation (see Chapter 11), especially F2: Condensation in roofs.

Conservation of fuel and power – specific guidance

Approved Document L treats dwellings separately from other buildings when giving recommendations as to how heat losses through the building fabric may be limited. Nevertheless, one technique for assessing adequate thermal performance is provided which is common to all buildings (although it does contain slight variations for dwellings). This is termed the *elemental method* in AD L and it provides a yardstick by which a number of other techniques may be judged. In addition to the elemental method various other techniques are described in AD L which give designers a wide choice of options for demonstrating compliance with the functional requirements. These include:

AD L
1.1 and 2.1

- For dwellings: the *Target U-value* method and the *Energy Rating* method; and,
- For buildings other than dwellings: a *Calculation* method and an *Energy Use* method.

These techniques are described in detail below.

Elemental method

AD L1
1.2 and 2.2

In this method certain maximum U-values, and window, door and rooflight areas are specified, for different types of buildings. These are summarised in

Tables 16.1 and 16.4 below and the values are illustrated in Fig. 16.2 for dwellings and Fig. 16.3 for all other building types.

Elemental method – specified insulation thickness

In order to satisfy the requirements of regulation L1 (and effectively, achieve the U-values shown in Table 16.1) a method is provided in Appendix A of AD L for choosing the required thickness of the principal insulating material in particular forms of construction. This method is based on the fact that in most wall, floor and roof constructions the total insulation value largely depends on one part of the construction (e.g. the cavity fill material in an external wall). This is because all good insulating materials have low thermal conductivity (i.e. a measure of the rate at which heat will pass through a material). Therefore, the thermal conductivity of a material may be used to assess its insulation value and the thermal conductivities of all the materials in the wall, floor or roof may be used to assess the insulation value of the total construction.

AD L
Appendix A

Table 16.1 Recommended U-values for all buildings.

Element	U-values (W/m²K)				Notes
	Dwellings		Other residential	Other buildings	
	SAP 60 or less	SAP over 60	—	—	
Pitched roofs	0.2	0.25	0.25	0.25	Applies to pitched roofs with a loft space. Roofs pitched at 70° or over may have wall U-value.
Flat roofs	0.2	0.35	0.35	0.45	Also applies to pitched roofs without loft space where insulation follows slope of roofs. Roofs pitched at 70° or over may have wall U-value
Exposed walls	0.45	0.45	0.45	0.45	
Exposed floors and ground floors	0.35	0.45	0.45	0.45	
Semi-exposed walls and floors	0.6	0.6	0.6	0.6	
Windows, doors and rooflights	3.0	3.3	3.3	3.3	Includes personnel doors in buildings other than dwellings
Vehicle access and similar large doors	—	—	0.7	0.7	

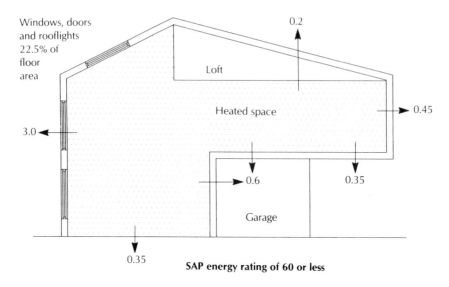

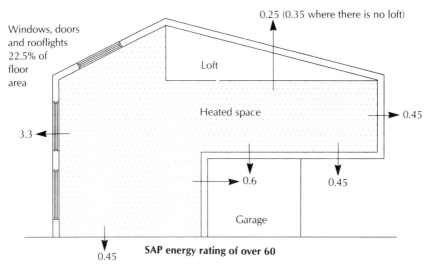

Fig. 16.2 Permitted U-values and opening areas – dwellings.

Using the tables from Appendix A of AD L (which are reproduced below) it is possible to select the minimum thickness of insulation recommended in a number of roof, wall and floor constructions and also make allowances for the contribution of the rest of the construction towards the overall insulation value. Therefore:

- for roofs, use Tables A1 to A4,
- for walls, use Tables A5 to A8, and
- for floors, use Tables A9 to A11 or use the perimeter to area ratio method as described on page 16.23 below.

Windows and personnel doors

residential buildings: 30% of exposed wall area
places of assembly [includes offices & shops]: 40% of exposed wall area [excludes display windows]
industrial and storage buildings: 15% of exposed wall area
rooflights: 20% of exposed roof area

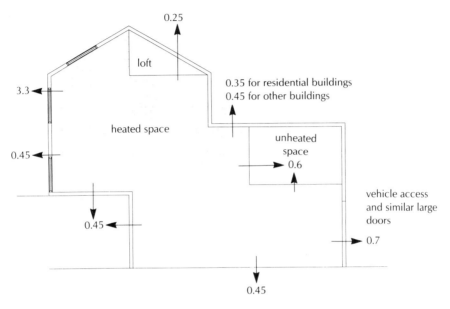

Fig. 16.3 Permitted U-values and opening areas – buildings other than dwellings.

The elemental method is applicable to all building types. However, for dwellings, it is first necessary to estimate the anticipated SAP rating for the chosen design since more stringent U-values should be used from the tables if the SAP rating is 60 or less. It should be noted that SAP ratings do *not* have to be assessed for buildings other than dwellings.

**AD L
1.2**

To help in this assessment of the anticipated SAP rating, the dwelling in Fig. 16.4 may be used as an example to show how the SAP rating may vary in response to changing items in the design. These are listed in Table 16.2 below.

Table 16.2 should only be used as a guide since the sensitivity to the various items is not absolute but depends on the values of the other items. The effect of changing the area of the windows, for example, depends on the relative size of the wall U-value and the window U-value. What is apparent from the table is the relatively high sensitivity to floor area, boiler type and choice of fuel.

To further assist the reader when using the elemental method for dwellings, a flow chart is illustrated in Fig. 16.5 which shows the importance of early assessment of the SAP rating and the necessity for reviewing this during the design process as design decisions are taken.

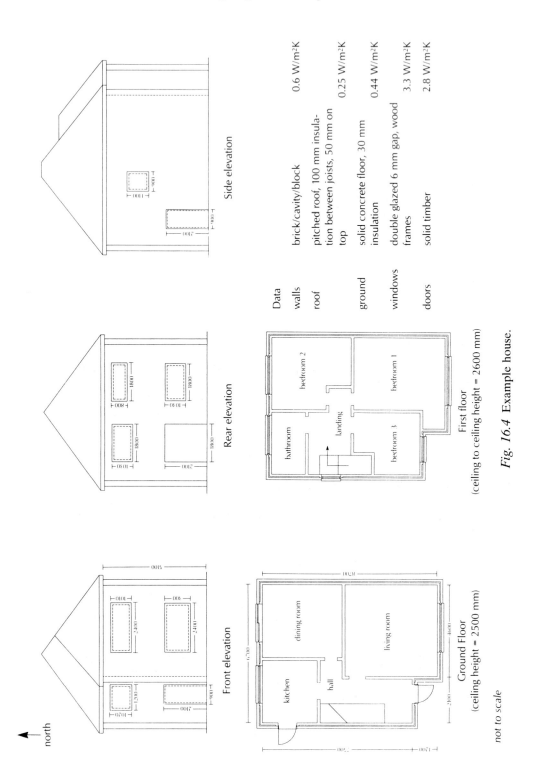

Data		
walls	brick/cavity/block	0.6 W/m²K
roof	pitched roof, 100 mm insulation between joists, 50 mm on top	0.25 W/m²K
ground	solid concrete floor, 30 mm insulation	0.44 W/m²K
windows	double glazed 6 mm gap, wood frames	3.3 W/m²K
doors	solid timber	2.8 W/m²K

Side elevation

Rear elevation

First floor
(ceiling to ceiling height = 2600 mm)

Front elevation

Ground Floor
(ceiling height = 2500 mm)

north

not to scale

Fig. 16.4 Example house.

Table 16.2 SAP sensitivity to design changes – detached dwelling.

Item changed	*Change in SAP*
Floor area changed by 10%	4 to 5
Floor perimeter changed by 10%	1 to 2
Wall area changed by 10%	2 to 3
Change U-values by 10%	2 to 3
Change window area by 10%	1 to 2
Change room height by 10%	1 to 2
Change heating system from electric storage heaters to low thermal capacity oil or mains gas boiler	+15
Change heating system from electric storage radiators to oil or mains gas condensing boiler	+25

Elemental method – examples of the use of standard tables

AD L
Appendix A

Flat roof to office building

Referring to Table A1 in Appendix A of AD L and Fig. 16.6.

- The thermal conductivity of the insulating material (e.g. mineral wool quilt, 0.04 W/mK) is chosen in column F.
- The design U-value of 0.45 is chosen in column A (i.e. row 6).
- Move horizontally across row 6 to column F and read off the insulation base thickness of 97 mm.

This thickness of insulation may be used in the roof or it may be more economical to take account of the remaining parts of the construction and thereby reduce the insulation thickness as follows:

- Assess each of the parts of the construction which appear in Table A4 of Appendix A of AD L and sum up the reductions:

  (i) Roof space (row 12, column F)   6 mm
  (ii) Plasterboard (row 8, column F)   3 mm

                                          Total reduction   9 mm

- Therefore final insulation thickness = 97 – 9 = 88 mm

Since mineral wool quilt comes in standard thicknesses, it is likely that the nearest to 88 mm is, in fact, 100 mm thick. This rather makes a mockery of the use of the reduction factors in this case but they may prove more useful in other forms of roof construction.

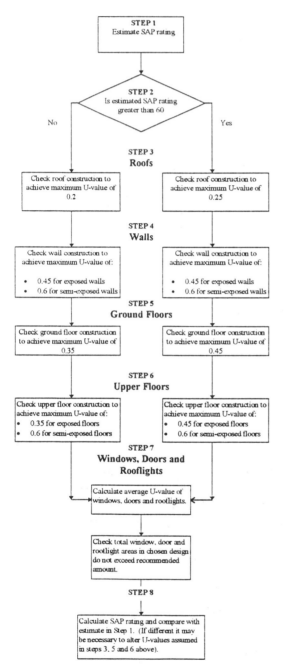

Notes

Use examples of dwellings in AD L Appendix G or Table 16.2 above to gain rough idea of expected SAP rating.

SAP ratings greater than 60 mean that less stringent U-values may be used for the walls, floors, roof, windows, doors and rooflights of the dwelling.

To assess thickness and types of insulation needed, use method described on pages 16.11 to 16.15 below. The method uses Tables A1 to A4 which are reproduced from AD L1 Appendix A.

To assess thickness and type of insulation needed, use method described on pages 16.15 to 16.18 below. The method uses Tables A5 to A8 which are reproduced from AD L1 Appendix A.

Either

To assess thickness and type of insulation needed, use method described on pages 16.16 to 16.22 below. The method uses Tables A9 to A11 which are reproduced from AD L1 Appendix A.

Or

Use the perimeter to area ratio method described on page 16.23.

To assess thickness and type of insulation needed, use method described on pages 16.24 to 16.26 below. The method uses Tables A12 to A14 which are reproduced from AD L1 Appendix A.

Use Table 2 from ADL to establish indicative U-values for door and window types in chosen design (or use certified manufacturer's test data if available). See page 16.33 for method of calculating average U-value.

Average U-value for windows, doors and rooflights obtained from previous step is compared against ADL Table 3 values to give recommended maximum area. Note that this is expressed as a percentage of the total floor area of the dwelling.

Follow the procedures shown in Appendix G of AD L or employ a suitably qualified person (see page 16.3 above) to do the calculation.

Fig. 16.5 Elemental method for dwellings – flowchart.

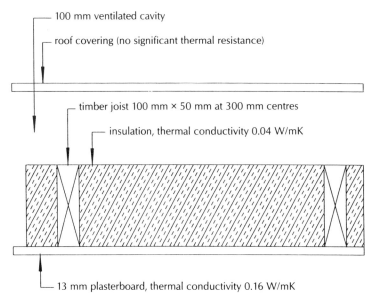

100 mm ventilated cavity

roof covering (no significant thermal resistance)

timber joist 100 mm × 50 mm at 300 mm centres

insulation, thermal conductivity 0.04 W/mK

13 mm plasterboard, thermal conductivity 0.16 W/mK

Fig. 16.6 Flat roof to office building – design U-value = 0.45 W/m²K.

AD L, Appendix A

Table **A1** **Base thickness of insulation between ceiling joists or rafters**

		Thermal conductivity of insulant (W/mK)						
		0.02	0.025	0.03	0.035	0.04	0.045	0.05
	Design U-value (W/m²K)	Base thickness of insulating material (mm)						
	A	B	C	D	E	F	G	H
1	0.20	167	209	251	293	335	376	418
2	0.25	114	142	170	199	227	256	284
3	0.30	86	107	129	150	171	193	214
4	0.35	69	86	103	120	137	154	172
5	0.40	57	71	86	100	114	128	143
6	0.45	49	61	73	85	97	110	122

Note: Tables A1 and A2 are derived for roofs with the proportion of timber at 8%, corresponding to 48 mm wide timbers at 600 mm centres, excluding noggings. For other proportions of timber the U-value can be calculated using the procedure in Appendix B of ADL. See also page 16.30 below.

AD L, Appendix A

Table A2   **Base thickness of insulation between and over joists or rafters**

	Design U-value (W/m²K)	Thermal conductivity of insulant (W/mK)						
		0.02	0.025	0.03	0.035	0.04	0.045	0.05
		Base thickness of insulating material (mm)						
	A	B	C	D	E	F	G	H
1	0.20	126	145	166	187	209	232	254
2	0.25	106	120	136	152	169	187	204
3	0.30	86	104	116	129	143	157	171
4	0.35	69	86	102	112	124	135	147
5	0.40	57	71	86	100	109	119	129
6	0.45	49	61	73	85	97	107	115

Note: Tables A1 and A2 are derived for roofs with the proportion of timber at 8%, corresponding to 48 mm wide timbers at 600 mm centres, excluding noggings. For other proportions of timber the U-value can be calculated using the procedure in Appendix B of ADL. See also page 16.30 below.

AD L, Appendix A

Table A3   **Base thickness for continuous insulation**

	Design U-value (W/m²K)	Thermal conductivity of insulant (W/mK)						
		0.02	0.025	0.03	0.035	0.04	0.045	0.05
		Base thickness of insulating material (mm)						
	A	B	C	D	E	F	G	H
1	0.20	97	122	146	170	194	219	243
2	0.25	77	97	116	135	154	174	193
3	0.30	64	80	96	112	128	144	160
4	0.35	54	68	82	95	109	122	136
5	0.40	47	59	71	83	94	106	118
6	0.45	42	52	62	73	83	94	104

AD L, Appendix A

Table **A4** **Allowable reductions in thickness for common roof components**

	Concrete slab density (kg/m³)	Thermal conductivity of insulant (W/mK)						
		0.02	0.025	0.03	0.035	0.04	0.045	0.05
		Reduction in base thickness of insulating material (mm) for each 100 mm of concrete slab						
	A	B	C	D	E	F	G	H
1	600	11	13	16	18	21	24	26
2	800	9	11	13	15	17	20	22
3	1100	6	7	9	10	12	13	15
4	1300	5	6	7	8	9	10	11
5	1700	3	3	4	5	5	6	7
6	2100	2	2	2	3	3	4	4

	Other materials and components	Reduction in base thickness of insulating material (mm)						
7	10 mm plasterboard	1	2	2	2	3	3	3
8	13 mm plasterboard	2	2	2	3	3	4	4
9	13 mm sarking board	2	2	3	3	4	4	5
10	12 mm Calcium Silicate liner board	1	2	2	2	3	3	4
11	Roof space (pitched)	4	5	5	6	7	8	9
12	Roof space (flat)	3	4	5	6	6	7	8
13	19 mm roof tiles	0	1	1	1	1	1	1
14	19 mm asphalt (or 3 layers of felt)	1	1	1	1	2	2	2
15	50 mm screed	2	3	4	4	5	5	6

Cavity wall to dwelling

Referring to Table A5 in Appendix A of AD L and Fig. 16.7 below:

- The thermal conductivity of the insulating material (e.g. expanded polystyrene slab, 0.035 W/mK) is chosen in column E.
- The design U-value of 0.45 is chosen in column A (i.e. row 4).
- Move horizontally across row 4 to column E and read off the insulation base thickness of 71 mm.

This thickness of insulation may be used in the wall or it may be more economical to take account of the remaining parts of the construction and thereby reduce the insulation thickness as follows:

Assess each of the parts of the construction which appear in Tables A6 and A7 of Appendix A of AD L and sum up the reductions:

From Table A6

(i)   Cavity (row 1, column E)   6 mm
(ii)   102 mm brick outer leaf (row 2, column E)   4 mm
(iii)   13 mm plasterboard (row 6, column E)   3 mm

From Table A7

(iv)   Concrete block inner leaf, density 600 kg/m (row 1,
      column E). Since the table is based on 100 mm
      thicknesses of blockwork it is necessary to adjust for
      the 150 mm blockwork used in this example.
      Therefore, value from row 1, column E of Table A7
      = 15 mm × 1.5 = 23 mm (when rounded up)   **23 mm**

                                      Total reduction   36 mm

• Therefore final insulation thickness = 71 – 36 = 35 mm

Ground floors

If a ground floor is sufficiently large it may be possible to achieve the recommended U-value of 0.35 W/m^{2}K or 0.45 W/m^{2}K without adding insulation material.

AD L
Appendix C
Diagram C1 from Appendix C of AD L is reproduced below and gives the range of floor dimensions for which insulation is recommended. When using the diagram the floor dimensions may be taken as referring to the whole building even if it consists of terraced or semi-detached properties. Alternatively, designers may use the floor dimensions of each individual property in a block at their own discretion.

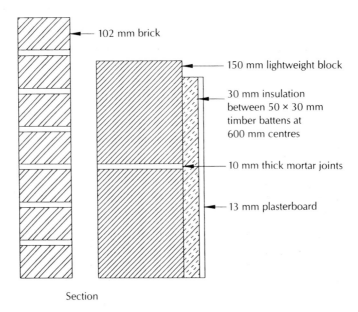

102 mm brick

150 mm lightweight block

30 mm insulation between 50 × 30 mm timber battens at 600 mm centres

10 mm thick mortar joints

13 mm plasterboard

Section

Fig. 16.7 Cavity wall to dwelling – design U-value = 0.45 W/m^2.

AD L, Appendix A

Table **A5** **Base thickness of insulation layer**

		Thermal conductivity of insulant (W/mK)						
		0.02	0.025	0.03	0.035	0.04	0.045	0.05
	Design U-value (W/m²K)	Base thickness of insulating material (mm)						
	A	B	C	D	E	F	G	H
1	0.30	63	79	95	110	126	142	158
2	0.35	54	67	80	94	107	120	134
3	0.40	46	58	70	81	93	104	116
4	0.45	41	51	61	71	82	92	102
5	0.60	30	37	45	52	59	67	74

AD L, Appendix A

Table **A6** **Allowable reductions in base thickness for common components**

		Thermal conductivity of insulant (W/mK)						
		0.02	0.025	0.03	0.035	0.04	0.045	0.05
	Component	Reduction in base thickness of insulating material (mm)						
	A	B	C	D	E	F	G	H
1	Cavity (25 mm min.)	4	5	5	6	7	8	9
2	Outer leaf brick	2	3	4	4	5	6	6
3	13 mm plaster	1	1	1	1	1	1	1
4	13 mm lightweight plaster	2	2	2	3	3	4	4
5	10 mm plasterboard	1	2	2	2	3	3	3
6	13 mm plasterboard	2	2	2	3	3	4	4
7	Airspace behind plasterboard dry-lining	2	3	3	4	4	5	6
8	9 mm sheathing ply	1	2	2	2	3	3	3
9	20 mm cement render	1	1	1	1	2	2	2
10	13 mm tile hanging	0	0	0	1	1	1	1

AD L, Appendix A

Table **A7**   **Allowable reduction in base thickness for concrete components**

	Density (kg/m)	Thermal conductivity of insulant (W/mK)						
		0.02	0.025	0.03	0.035	0.04	0.045	0.05
		Reduction in base thickness of insulation (mm) for each 100 mm of concrete						
	A	B	C	D	E	F	G	H
Concrete inner leaf								
1	600	9	11	13	15	17	20	22
2	800	7	9	11	13	15	17	19
3	1000	6	8	9	11	12	14	15
4	1200	5	6	7	9	10	11	12
5	1400	4	5	6	7	8	9	9
6	1600	3	4	4	5	6	7	7
Concrete outer leaf or single leaf wall								
7	600	8	10	13	15	17	19	21
8	800	7	8	10	12	14	15	17
9	1000	6	7	8	10	11	12	14
10	1200	4	6	7	8	9	10	11
11	1400	3	4	5	6	7	8	9
12	1600	3	3	4	5	5	6	7
13	1800	2	3	3	4	4	5	5
14	2000	2	2	2	3	3	4	4
15	2400	1	1	2	2	2	2	3

AD L, Appendix A

Table **A8**   **Allowable reduction in base thickness for insulated timber frame walls**

	Thermal conductivity of insulation within frame (W/mK)	Thermal conductivity of insulant (W/mK)						
		0.02	0.025	0.03	0.035	0.04	0.045	0.05
		Reduction in base thickness of insulation for each 100 mm of frame (mm)						
	A	B	C	D	E	F	G	H
1	0.035	42	53	63	74	84	95	105
2	0.040	38	48	58	67	77	87	96

Note: The table is derived for walls for which the proportion of timber is 12%, which corresponds to 48 mm wide studs at 400 mm centres. For other proportions of timber the U-value can be calculated using the procedure in Appendix B.

AD L, Appendix C

Diagram C1   Floor dimensions for which insulation is required

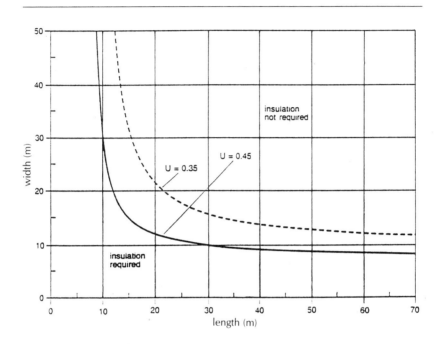

Where it is considered that floor insulation is needed, Tables A9 to A11 of Appendix A of AD L show how this may be estimated for solid and suspended ground floors.

As an example, take a detached dwelling with a SAP rating of 75, which has internal dimensions of 6 m by 8 m. Assuming a solid ground-bearing floor and insulation with thermal conductivity of 0.035 W/mK (e.g. expanded polystyrene slab) with a design U-value of 0.45 W/m²K,

- Calculate the ratio P/A, where P is the perimeter length of the dwelling in metres and A is the floor area in square metres. Therefore P/A = 28/48 = 0.6
- For a U-value of 0.45 W/mK and P/A of 0.6 go to row 23 of Table A9
- Move horizontally to column E (thermal conductivity of 0.035 W/mK) and read off the insulation thickness of 35 mm.

This is the thickness recommended and there are no reductions possible (in contrast with the tables for roofs and walls mentioned above).

The assessment of floor U-values

Alternative methods for determining the U-values of insulated and uninsulated ground floors are given in Appendix C of AD L, paragraphs C4 to C9.

AD L
Appendix A

AD L
Appendix C

AD L, Appendix A

Table A9   **Insulation thickness for solid floors in contact with the ground**

	P/A*	0.02	0.025	0.03	0.035	0.04	0.045	0.05
		A	B	C	D	E	F	G	H
		Insulation thickness (mm) for: U-value of 0.25 W/m²K							
		Thermal conductivity of insulant (W/mK)							
1	1.00	62	77	93	108	124	139	155
2	0.90	61	76	91	107	122	137	152
3	0.80	60	75	90	105	119	134	149
4	0.70	58	73	87	102	116	131	145
5	0.60	56	70	84	98	111	125	139
6	0.50	52	66	79	92	105	118	131
7	0.40	47	59	71	83	95	107	119
8	0.30	39	49	59	69	79	88	98
9	0.20	24	30	36	42	48	54	60

U-value of 0.35 W/m²K

	P/A*	0.02	0.025	0.03	0.035	0.04	0.045	0.05
10	1.00	39	49	58	68	78	88	97
11	0.90	38	48	57	67	76	86	95
12	0.80	37	46	55	65	74	83	92
13	0.70	35	44	53	62	70	79	88
14	0.60	33	41	49	58	66	74	82
15	0.50	30	37	44	52	59	67	74
16	0.40	25	31	37	43	49	55	61
17	0.30	16	21	25	29	33	37	41
18	0.20	1	1	1	2	2	2	2

U-value of 0.45 W/m²K

	P/A*	0.02	0.025	0.03	0.035	0.04	0.045	0.05
19	1.00	26	33	39	46	53	59	66
20	0.90	25	32	38	44	51	57	63
21	0.80	24	30	36	42	48	54	60
22	0.70	22	28	34	39	45	51	56
23	0.60	20	25	30	35	40	45	50
24	0.50	17	21	25	30	34	38	42
25	0.40	12	15	18	21	24	27	30
26	0.30	4	5	6	6	7	8	9
27	<0.27	0	0	0	0	0	0	0

Note: *P/A is the ratio of floor perimeter (m) to floor area (m²).

AD L, Appendix A

Table **A10**   **Insulation thickness for suspended timber ground floors**

	P/A*	Insulation thickness (mm) for: U-value of 0.25 W/m²K						
		Thermal conductivity of insulant (W/mK)						
		0.02	0.025	0.03	0.035	0.04	0.045	0.05
	A	B	C	D	E	F	G	H
1	1.00	95	110	126	140	155	170	184
2	0.90	93	109	124	138	153	167	181
3	0.80	91	106	121	136	150	164	178
4	0.70	88	103	118	132	145	159	173
5	0.60	85	99	113	126	139	153	166
6	0.50	79	92	106	118	131	143	156
7	0.40	71	83	95	107	118	129	140
8	0.30	57	68	78	88	97	106	116
9	0.20	33	39	46	52	58	64	69
		U-value of 0.35 W/m²K						
10	1.00	57	67	77	87	96	106	115
11	0.90	55	66	75	85	94	103	112
12	0.80	53	63	73	82	91	100	109
13	0.70	51	60	69	78	87	95	104
14	0.60	47	56	64	73	81	89	97
15	0.50	42	50	57	65	72	79	87
16	0.40	34	41	47	53	60	66	72
17	0.30	22	26	31	35	39	43	47
18	0.20	1	1	2	2	2	2	3
		U-value of 0.45 W/m²K						
19	1.00	37	44	51	57	64	70	77
20	0.90	35	42	49	55	62	68	74
21	0.80	33	40	46	53	59	65	70
22	0.70	31	37	43	49	54	60	65
23	0.60	27	33	38	43	49	54	58
24	0.50	22	27	32	36	40	44	49
25	0.40	15	18	22	25	28	31	34
26	0.30	4	5	6	7	8	9	10
27	<0.27	0	0	0	0	0	0	0

Notes: *P/A is the ratio of floor perimeter (m) to floor area (m²).

The table is derived for suspended timber floors for which the proportion of timber is 12%, which corresponds to 48 mm wide timbers at 400 mm centres.

For other proportions of timber the U-value can be calculated using the procedure in Appendix B.

AD L, Appendix A

Table **A11**   **Insulation thickness for suspended concrete beam and block ground floors**

				Insulation thickness (mm) for:				
				U-value of 0.25 W/m²K				
		Thermal conductivity of insulant (W/mK)						
	P/A*	0.02	0.025	0.03	0.035	0.04	0.045	0.05
	A	B	C	D	E	F	G	H
1	1.00	60	75	90	104	119	134	149
2	0.90	59	73	88	103	118	132	147
3	0.80	58	72	86	101	115	130	144
4	0.70	56	70	84	98	112	126	140
5	0.60	54	67	80	94	107	121	134
6	0.50	50	63	75	88	101	113	126
7	0.40	45	57	68	79	91	102	113
8	0.30	37	46	56	65	74	84	93
9	0.20	22	27	33	38	43	49	54
				U-value of 0.35 W/m²K				
10	1.00	37	46	55	64	74	83	92
11	0.90	36	45	54	63	72	81	90
12	0.80	35	43	52	61	69	78	87
13	0.70	33	41	50	58	66	74	83
14	0.60	31	38	46	54	61	69	77
15	0.50	27	34	41	48	55	62	69
16	0.40	22	28	34	39	45	50	56
17	0.30	14	18	21	25	29	32	36
18	0.20	0	0	0	0	0	0	0
				U-value of 0.45 W/m²K				
19	1.00	24	30	36	42	48	54	60
20	0.90	23	29	35	41	46	52	58
21	0.80	22	28	33	39	44	50	55
22	0.70	20	25	31	36	41	46	51
23	0.60	18	23	27	32	36	41	45
24	0.50	15	18	22	26	29	33	37
25	0.40	10	12	15	17	19	22	24
26	0.30	2	2	2	3	3	4	4
27	<0.28	0	0	0	0	0	0	0

Note: *P/A is the ratio of floor perimeter (m) to floor area (m²).

For ground floors the U-value depends upon the length of exposed edge. The length of exposed edge is a function of the shape of the floor, its dimensions and the built form of the dwelling. The full calculation must take all these factors into account and involves the use of sophisticated mathematical functions.

U-values of floors without insulation

The U-value of an uninsulated floor depends on the ratio of its perimeter to its area (P/A). An approximate formula for the floor U-value has been developed by Anderson at BRE and is

AD L Appendix C C4

$$U_0 = 0.05 + 1.65 \times (P/A) - 0.6(P/A)^2 \qquad \text{(equation 1)}$$

where P = exposed perimeter length (m) (ignoring unheated spaces such as porches and garages although the wall between the unheated space and heated building is included in the total perimeter length) and A = floor slab area (m^2).

AD L Appendix C C6

Table 16.3 below is based on this formula and is more extensive than that contained in AD L Appendix C. The U-value of an uninsulated floor may be obtained directly from the table once the P/A ratio has been determined.

AD L Appendix C C7

It should be noted that for buildings consisting of a number of dwellings, such as blocks of flats, semi-detached houses and terraces, the dimensions used relate to the area under the whole building.

Table 16.3 U-values of uninsulated floors.

Ratio P/A	U_0	Ratio P/A	U_0
0.05	0.13	0.55	0.78
0.10	0.21	0.60	0.82
0.15	0.28	0.65	0.87
0.20	0.36	0.70	0.91
0.25	0.43	0.75	0.95
0.30	0.49	0.80	0.99
0.35	0.55	0.85	1.02
0.40	0.61	0.90	1.05
0.45	0.67	0.95	1.08
0.50	0.73	1.00	1.10

U-values of insulated floors

The U-value of a floor with insulation can be calculated from the equation below:

$$U_{ins} = \cfrac{1}{\cfrac{1}{U_o} + R_{ins}} \qquad \text{(equation 2)}$$

AD L Appendix C C8

U_o is the U-value of an uninsulated floor of the same dimensions (as found from Table 16.3 or by using the formula). R_{ins} is the thermal resistance of the insulation (m^2K/W). For suspended floors, U_{ins} includes the thermal resistance of the structural deck (up to a value of $0.2\,m^2K/W$). Therefore, R_{ins} should only be calculated to include the resistance of any added insulation and any resistance from the structural deck over and above $0.2\,m^2K/W$.

AD L
Appendix C
C8

Example U-value calculation for ground floor

Taking the dwelling shown in Fig. 16.4 as an example:

Perimeter (P) $= 6.7 + 8.7 + 4.6 + 1.2 + 2.1 + 7.5 = 30.8\,m$
Area (A) $= (7.5 \times 6.7) + (4.6 \times 1.2) = 55.77\,m^2$

$$\frac{\text{Perimeter}}{\text{area}} = \frac{30.8}{55.77} = 0.55\,m^{-1}$$

From the Table 16.3, U-value $= 0.78\,W/m^2K$ or by equation:

$$U_o = 0.05 + 1.65 \times (P/A) - 0.6(P/A)^2 = 0.05 + 1.65 \times 0.55 - 0.6 \times 0.55^2$$
$$= 0.776\,W/m^2K$$

Since this value is greater than that recommended by AD L, it is necessary to insulate the floor. Assuming a SAP rating of over 60, then the U-value should not exceed $0.45\,W/m^2$. If this value is inserted in equation 2 and it is rearranged, the thermal resistance required for the floor insulation may be obtained:

$$R_{ins} = 1/0.45 - 1/0.776 = 2.222 - 1.288 = 0.934\,m^2K/W$$

An insulation material such as expanded polystyrene slab has a thermal conductivity (K) of $0.035\,W/mK$ and the thickness required can be found from:

Thickness (t) $= R_{ins} \times K = 0.934 \times 0.035 = 0.032\,m$ or $32\,mm$.

Further information on floor U-values may be obtained from:

AD L
Appendix C
C10

- BRE Information Paper IP 3/90 *The U-value of ground floors: application to building regulations*,
- BRE Information Paper IP 7/93 *The U-value of solid ground floors with edge insulation*, and
- BRE Information Paper IP 14/94 *U-values for basements*.

Upper (exposed and semi-exposed) floors

AD
Appendix
A

Upper floors (whether exposed or semi-exposed) may be designed under the elemental method by using Tables A12 to A14 of Appendix A of AD L which are reproduced below. The procedure to be followed is similar to that described above for roofs. The base thickness of the chosen insulation material is selected from either Table A12 (for floors of timber joist construction) or Table A13 (for floors of concrete construction). The base thickness is then reduced by the amounts given in Table A14 for the plasterboard ceiling, the floor boarding or the screed as appropriate.

AD L, Appendix A

Table **A12**   Exposed and semi-exposed upper floors of timber construction

Design U-value (W/m²K)	Thermal conductivity of insulant (W/mK)						
	0.02	0.025	0.03	0.035	0.04	0.045	0.05
	Base thickness of insulation between joists to achieve design U-values						
A	B	C	D	E	F	G	H
Exposed floor							
1 0.35	61	76	92	107	122	146	162
2 0.45	42	53	63	74	84	95	106
Semi-exposed floor							
3 0.60	25	32	38	44	50	57	63

Note: Table A12 is derived for floors with the proportion of timber at 12% which corresponds to 48 mm wide timbers at 400 mm centres. For other proportions of timber the U-value can be calculated using the procedure in Appendix E.

AD L, Appendix A

Table **A13**   Exposed and semi-exposed upper floors of concrete construction

Design U-value (W/m²K)	Thermal conductivity of insulant (W/mK)						
	0.02	0.025	0.03	0.035	0.04	0.045	0.05
	Base thickness of insulation to achieve design U-values						
A	B	C	D	E	F	G	H
Exposed floor							
1 0.35	52	65	78	91	104	117	130
1 0.45	39	49	59	69	79	89	98
Semi-exposed floor							
3 0.60	26	33	39	46	52	59	65

AD L, Appendix A

Table A14   **Exposed and semi-exposed upper floors: allowable reductions in base thickness for common components**

		Thermal conductivity of insulant (W/mK)						
		0.02	0.025	0.03	0.035	0.04	0.045	0.05
	Component	Reduction in base thickness of insulating material (mm)						
	A	B	C	D	E	F	G	H
1	10 mm plasterboard	1	2	2	2	3	3	3
2	19 mm timber flooring	3	3	4	5	5	6	7
3	50 mm screed	2	3	4	4	5	5	6

The proportional area calculation method

AD L
Appendix B It will be apparent to most designers when using the tables in Appendix A, that they cover only a limited range of constructional forms. Where there is a desire to vary from the standard forms or where innovative designs are being contemplated it will be necessary to carry out a U-value calculation in order to take full advantage of the thermal properties of the construction.

Many structures contain repeating thermal bridges such as insulation between timber studding or mortar joints in blockwork and these need to be allowed for when calculating the U-value. The example calculation procedure given in Appendix B of AD L should be employed. This is based on the calculation method described in the CIBSE Design Guide A3, which should be consulted for more detailed guidance. It should be noted that the tables given in Appendix A are based on this method of calculation. The thermal conductivity of a particular material may be obtained from the supplier or manufacturer or the CIBSE Guide mentioned above. Average values of thermal conductivity for some commonly used building materials are given in Table A15 of Appendix A of AD L and this is reproduced below.

For each leaf that contains more than one material the calculation is as follows:

• Calculate the resistance of any non-bridged homogeneous leaves,
• Calculate the resistance through each section containing the different materials,
• Combine these resistances in proportion to their areas to give an area weighted average for that leaf.

Evaluate the total resistance by adding the resistances found as above to the resistance of any homogeneous leaves.

AD L, Appendix A

Table **A15**   **Thermal conductivity of some common building materials**

Material	Density (kg/m³)	Thermal conductivity (W/mK)
Walls (external and internal)		
Brickwork (outer leaf)	1700	0.84
Brickwork (inner leaf)	1700	0.62
Cast concrete (dense)	2100	1.40
Cast concrete (lightweight)	1200	0.38
Concrete block (heavyweight)	2300	1.63
Concrete block (medium weight)	1400	0.51
Concrete block (lightweight)	600	0.19
Normal mortar	1750	0.8
Fibreboard	300	0.06
Plasterboard	950	0.16
Tile hanging	1900	0.84
Timber	650	0.14
Surface finishes		
External rendering	1300	0.50
Plaster (dense)	1300	0.50
Plaster (lightweight)	600	0.16
Calcium Silicate board	875	0.17
Roofs		
Aerated concrete slab	500	0.16
Asphalt	1700	0.50
Felt/bitumen layers	1700	0.50
Screed	1200	0.41
Stone chippings	1800	0.96
Tile	1900	0.84
Wood wool slab	500	0.10
Floors		
Cast concrete	2000	1.13
Metal tray	7800	50.00
Screed	1200	0.41
Timber flooring	650	0.14
Wood blocks	650	0.14
Insulation		
Expanded polystyrene (EPS) slab	25	0.035
Mineral wool quilt	12	0.040
Mineral wool slab	25	0.035
Phenolic foam board	30	0.020
Polyurethane board	30	0.025

Note: If available, certified test values should be used in preference to those in the table.

Exposed walls – example U-value calculation

Referring to Fig. 16.8:

Data:

102 mm brickwork, thermal conductivity 0.84 W/mK
50 mm cavity, thermal resistance 0.18 m^{2}K/W
150 mm blockwork (440 mm × 215 mm), thermal conductivity
  0.12 W/mK
10 mm mortar joints, thermal conductivity 0.8 W/mK
30 mm insulation, thermal conductivity 0.04 W/mK
50 × 30 mm timber studs at 600 mm centres, thermal conductivity
  0.14 W/mK
13 mm plasterboard, thermal conductivity 0.16 W/mK

In this example there are two layers with cold bridges:

(a) the blockwork with its mortar joints;
(b) the insulation and timber studs.

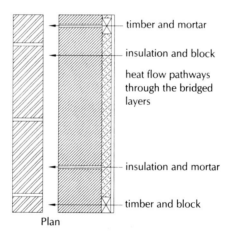

Plan

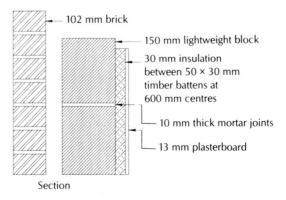

Section

Fig. 16.8 Example wall calculation – dwelling.

The mortar joints in the outer leaf do not give rise to cold bridging as they have a thermal conductivity similar to that of the bricks.

There are four different combinations of heat flow pathways through the blockwork and insulation:

- block/insulation,
- block/timber,
- mortar/insulation, and
- mortar/timber.

The calculation is as follows:

1. Resistance of inner leaf

Resistance of non-bridged layers,

Resistance of half the cavity		$=$	0.09 m^{2}K/W
Resistance of plasterboard	$= 0.013/0.16$	$=$	0.08 m^{2}K/W
Resistance of inside surface	R_{si}	$=$	0.12 m^{2}K/W
Total resistance of non-bridged layers	R_{nb}	$=$	0.29 m^{2}K/W

2. Resistance of bridged layers

Material resistances (thickness/conductivity)

Resistance of block, R_b		$= 0.150/0.12$	$=$	1.25 m^{2}K/W
Resistance of mortar, R_m		$= 0.150/0.8$	$=$	0.19 m^{2}K/W
Resistance of insulation, R_{ins}		$= 0.030/0.04$	$=$	0.75 m^{2}K/W
Resistance of timber, R_t		$= 0.030/0.14$	$=$	0.21 m^{2}K/W

Resistance of heat flow paths

Resistance of block/insulation
$$Rb,in = R_b + R_{ins} + R_{nb} \quad = \quad 1.25 + 0.75 + 0.29 \quad = \quad 2.29 \text{ m}^2\text{K/W}$$
Resistance of block/timber
$$R_{b,t} = R_b + R_t + R_{nb} \quad = \quad 1.25 + 0.21 + 0.29 \quad = \quad 1.75 \text{ m}^2\text{K/W}$$
Resistance of mortar/insulation
$$R_{m,ins} = R_m + R_{ins} + R_{nb} \quad = \quad 0.19 + 0.75 + 0.29 \quad = \quad 1.23 \text{ m}^2\text{K/W}$$
Resistance of mortar/timber
$$Rm,t = Rm + Rt + Rnb \quad = \quad 0.19 + 0.21 + 0.29 \quad = \quad 0.69 \text{ m}^2\text{K/W}$$

Face area of materials, as a fraction of total face area

block $\quad F_b = \dfrac{440 \times 215}{450 \times 225} \qquad = 0.934$

mortar: $\quad F_m = 1 - F_b = 1 - 0.934 = 0.066$

insulation: $F_{ins} = \dfrac{550}{600} \qquad = 0.917$

timber: $\quad F_t = 1 - F_{ins} = 1 - 0.917 = 0.083$

Fraction of face area of heat flow paths

block/insulation: $\quad F_{b,ins} = F_b \times F_{ins} = 0.934 \times 0.917 = 0.856$
block/timber: $\quad F_{b,t} \quad = F_b \times F_t \quad = 0.934 \times 0.083 = 0.078$

mortar/insulation: $F_{min,ins} = F_m \times F_{ins} = 0.066 \times 0.917 = 0.061$
mortar/timber: $\quad F_{m,t} \quad = F_m \times F_t \quad = 0.066 \times 0.083 = 0.005$

Sum of resistances in parallel

The four heat flow paths are resistances in parallel, hence the total resistance of the bridged layers is the sum of resistances in parallel, which in this case is given by:

$$\frac{1}{R_{\text{inner leaf}}} = \frac{F_{b,\text{ins}}}{R_{b,\text{ins}}} + \frac{R_{b,t}}{R_{b,t}} + \frac{F_{m,\text{ins}}}{R_{m,\text{ins}}} + \frac{F_{m,t}}{R_{m,t}}$$

which gives:

$$\frac{1}{R_{\text{inner leaf}}} = \frac{0.856}{2.29} + \frac{0.078}{1.75} + \frac{0.061}{1.23} + \frac{0.005}{0.69} \qquad = 0.475$$

Resistance of the inner leaf $R_{\text{inner leaf}}$ $1/0.475$ = 2.1 m^{2}K/W

3. Resistance of outer leaf

Resistance of outside surface R_{so}		= 0.06 m^{2}K/W
Resistance of brick outer leaf	= 0.102/0.84	= 0.12 m^{2}K/W
Resistance of half the cavity Rcav		= 0.09 m^{2}K/W

Resistance of the outer leaf R$_{\text{outer leaf}}$ **= 0.27 m^{2}K/W**

4. Total resistance of the wall

The total resistance of the wall is the sum of the resistances of the outer and inner leaves.

$R_{\text{total}} = R_{\text{inner leaf}} + R_{\text{outer leaf}}$ = 2.1 + 0.27 = 2.37 m^{2}K/W

5. U-value of the wall

$$U = \frac{1}{R_{\text{total}}} = \frac{1}{2.37} \qquad = 0.42 \, \text{W/m}^2\text{K}$$

Exposed roof – example U-value calculation

Referring to Fig. 16.9:

Data:

An outer covering with no significant thermal resistance.
100 mm ventilated cavity, thermal resistance 0.16 m^{2}K/W.
100 mm insulation, thermal conductivity 0.04 W/mK, between
 timber joists.
100 × 50 mm timber joists at 300 mm centres, thermal conductivity
 0.14 W/mK.
13 mm plasterboard, thermal conductivity 0.16 W/mK.

In this case there is only one thermally bridged layer, the insulation bridged by the timber joists. The calculation is thus:

Inside layers

Resistance through section containing timber joists

Resistance of inside surface R_{si}		= 0.10 m^{2}K/W
Resistance of plasterboard	= 0.013/0.16	= 0.08 m^{2}K/W

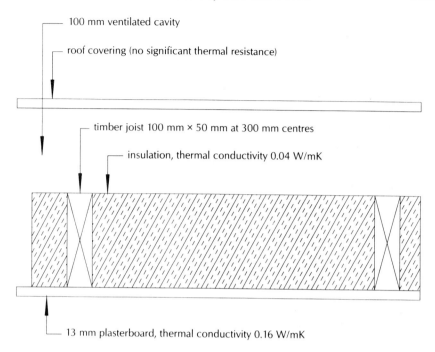

100 mm ventilated cavity

roof covering (no significant thermal resistance)

timber joist 100 mm × 50 mm at 300 mm centres

insulation, thermal conductivity 0.04 W/mK

13 mm plasterboard, thermal conductivity 0.16 W/mK

Fig. 16.9 Example roof calculation – office.

Resistance of timber joists	= 0.1/0.14	= 0.71 m²K/W
Half resistance of cavity		= 0.08 m²K/W

Resistance of section containing timber joists R_j $\qquad$ **= 0.97 m²K/W**

Resistance through section containing insulation

Resistance of inside surface R_{si}		= 0.10 m²K/W
Resistance of plasterboard	= 0.013/0.16	= 0.08 m²K/W
Resistance of insulation	= 0.1/0.04	= 2.5 m²K/W
Half resistance of cavity		= 0.08 m²K/W

Resistance of section containing insulation R_{ins} $\qquad$ **= 2.76 m²K/W**

Combining R_j with R_{ins}

Fractional area of joists:

$$F_j = \frac{\text{thickness of joists}}{\text{joist centres}} = \frac{50}{300} = 0.167$$

Fractional area of insulation:

$$F_{ins} = (1 - F_j) = (1 - 0.167) = 0.833$$

Resistance of inner layers is given by:

$$R_{inner} = \cfrac{1}{\cfrac{F_j}{R_j} + \cfrac{F_{ins}}{R_{ins}}} = \cfrac{1}{\cfrac{0.167}{0.97} + \cfrac{0.833}{2.76}} \qquad = 2.11\,\text{m}^2\text{K/W}$$

Resistance of outer layer

Half resistance of cavity	$= \quad 0.08 \text{ m}^2\text{K/W}$
Resistance of outside surface R_{so}	$= \quad 0.04 \text{ m}^2\text{K/W}$
Resistance of outer layer	$= \quad \textbf{0.12 m}^2\textbf{K/W}$

Total resistance of roof

The total resistance of the roof is the sum of the resistances of the inner and outer layers.

$$R_{inner} + R_{outer} = 2.11 + 0.12 \qquad\qquad = \quad 2.22 \text{ m}^2\text{K/W}$$

$$\text{U-value of roof} = \frac{1}{\text{total resistance}} = \frac{1}{2.22} \qquad = \quad 0.45 \text{ W/m}^2\text{K}$$

Elemental method – allowances for windows, doors and rooflights

The requirement of paragraph L1 of Schedule 1 for conservation of fuel and power will be met:

1. If the average U-value of the windows, doors and rooflights does not exceed the figure given in Table 16.1 above for each building type; and,
2. The area allowances given in Table 16.4 are not exceeded.

AD L 1.2

It should be noted that for dwellings, the maximum U-value recommended will vary with the SAP rating. This means a maximum of 3.0 W/m²K for a SAP of 60 or less and this may be increased to 3.3 W/m²K where the SAP exceeds 60. The area allowance is based on the total floor area of the dwelling (i.e. first floor plus ground floor in a two-storey dwelling) which should not exceed 22.5%.

AD L 1.4

AD L 2.3

For buildings other than dwellings there is only one maximum average U-value of 3.3 W/m²K recommended. Additionally, the area allowances for windows and doors are based on the total exposed wall area, and for rooflights, on the total roof area.

AD L 1.9 and 2.9

AD L 1.6 and 2.6

In order to select appropriate U-values for the windows, doors and rooflights in a building, it is preferable to use certified manufacturers' data. Where this is unavailable, Table 2 of AD L (which is reproduced below) provides indicative U-values for a range of commonly used components. Single-glazed panels in external doors are acceptable provided that they do not cause the average U-value of all the openings to exceed the area limit.

AD L 1.8 and 2.8

When selecting and installing sealed double-glazed windows care should be taken, since there is a risk of condensation forming between the panes. Information on how to avoid this problem may be obtained from the BRE Report BR 262 *Thermal insulation: avoiding risks*.

Permitted variation in the areas of openings

AD L 1.10 and 2.10

The areas of windows, doors and rooflights given above may be varied provided that they are compensated for by changes in the average U-value of the openings. The graphs shown in Fig. 16.10 and Fig. 16.11 indicate the variations that are possible and are based on Table 3 for dwellings and Table

AD L, Section 1

Table **2**　**Indicative U-values** (W/m^2K) **for windows, doors and rooflights**

Item	Type of frame							
	Wood		Metal		Thermal break		PVC-U	
Air gap in sealed unit (mm)	6	12	6	12	6	12	6	12
Window, double-glazed	3.3	3.0	4.2	3.8	3.6	3.3	3.3	3.0
Window, double-glazed, low-E	2.9	2.4	3.7	3.2	3.1	2.6	2.9	2.4
Window, double-glazed, Argon fill	3.1	2.9	4.0	3.7	3.4	3.2	3.1	2.9
Window, double-glazed, low-E, Argon fill	2.6	2.2	3.4	2.9	2.8	2.4	2.6	2.2
Window, triple-glazed	2.6	2.4	3.4	3.2	2.9	2.6	2.6	2.4
Door, half double-glazed	3.1	3.0	3.6	3.4	3.3	3.2	3.1	3.0
Door, fully double-glazed	3.3	3.0	4.2	3.8	3.6	3.3	3.3	3.0
Rooflights, double-glazed at less than 70° from horizontal	3.6	3.4	4.6	4.4	4.0	3.8	3.6	3.4
Windows and doors, single-glazed	4.7		5.8		5.3		4.7	
Door, solid timber panel or similar	3.0		—		—		—	
Door, half single-glazed, half timber or similar	3.7		—		—		—	

8 for buildings other than dwellings from AD L. These are reproduced below.

Calculation of average U-value for windows doors and rooflights

Table 16.5 shows the U-values and areas for the windows and doors of the detached house in Fig. 16.4 above.
AD L
Appendix E

The average U-value for the windows and doors is given by:

$$U = \frac{\text{Total rate of heat loss through the openings}}{\text{Total area of window and door openings}}$$

From table 16.5 above

$$U = \frac{10.58 + 55.06}{3.78 + 16.686} = 3.2 \, W/m^2K$$

Therefore the area of openings as a percentage of the dwelling's floor area:

$$= \frac{\text{Total area of openings}}{\text{Total floor area}} = \frac{3.78 + 16.686}{55.77 + 55.77} = \frac{20.466}{111.54} = \frac{18\% \text{ of the}}{\text{floor area}}$$

Table 16.4 Recommended area allowance for windows, doors and rooflights for all buildings.

Building type	Windows and doors* (As % of exposed wall area)	Rooflights (As % of roof area)	Notes
Residential	30%	20%	Applies to residential buildings (*not* dwellings) where people reside whether temporarily or permanently, e.g. institutional buildings, hotels and boarding houses
Shops, offices and assembly	40%	20%	When carrying out calculations for conservation of fuel and power purposes, there is no need to include areas such as display windows, shop entrance doors and similar glazing for these building types
Industrial and storage	15%	20%	
Dwellings	22.5% of total *floor* area for windows, doors and rooflights combined		

* There are no restrictions over the total area of vehicle access doors in buildings other than dwellings.

From graph in Fig. 16.10 or Table 3 in AD L it can be seen that for an average U-value of 3.2 W/m²K the area of openings should not exceed 21% for a SAP of 60 or less or 23.5% for a SAP of over 60. Therefore the dwelling will satisfy the requirement whatever the SAP rating.

Extensions to buildings

Reference has been made on page 16.6 regarding the application of AD L to small extensions to dwellings. For extensions larger than $10\,m^2$ to any building,

Table 16.5 Door and window openings for dwelling shown in Fig. 16.4.

Element	Area	U-value	Rate of heat loss per degree
Windows	16.686	3.3	55.06
Doors	3.78	2.8	10.58
Totals	20.466		65.64

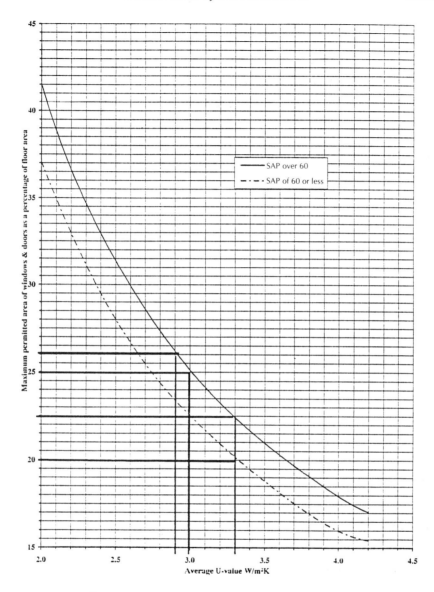

Fig. 16.10 Percentage opening areas – dwellings.

the recommendations vary slightly from the standard expected of new buildings as follows:

- For dwellings, it is appropriate to use the U-values related to a SAP rating of over 60. This is not mentioned in the Approved Document, however the Department of the Environment issued a guidance note to this effect after the publication of AD L.
- Windows, doors and personnel doors containing single-glazed panels may be assumed to have a U-value of $3.3\,W/m^2K$ if they are enclosed by unheated, draught-proof conservatories or porches.

AD L
1.7 and 2.7

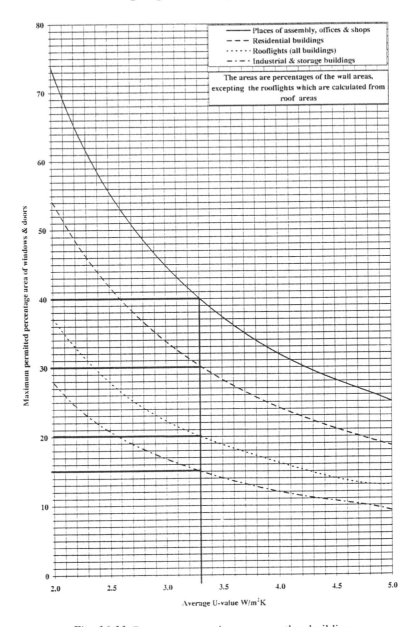

Fig. 16.11 Percentage opening areas – other buildings.

AD L
1.5 and 2.5

- The average U-values of the openings in extensions should not exceed 3.3 W/m²K. When establishing if the areas of the openings fall within the allowances shown in Table 16.4, it is permissible to apply the allowance by considering either the extension itself or the extension and the existing building together.

When a conservatory is attached to and built as part of a new dwelling, it can be a source of heat gain if placed on a south facing elevation, and research is

AD L, Section 1

Table 3 **Permitted variation in the area of windows and doors for dwellings**

Average U-value (W/m²K)	Maximum permitted area of windows and doors as a percentage of floor area for SAP Energy Ratings of:	
	60 or less	over 60
2.0	37.0%	41.5%
2.1	35.0%	39.0%
2.2	33.0%	36.5%
2.3	31.0%	34.5%
2.4	29.5%	33.0%
2.5	28.0%	31.5%
2.6	26.5%	30.0%
2.7	25.5%	28.5%
2.8	24.5%	27.5%
2.9	23.5%	26.0%
3.0	**22.5%**	25.0%
3.1	21.5%	24.0%
3.2	21.0%	23.5%
3.3	20.0%	**22.5%**
3.4	19.5%	21.5%
3.5	19.0%	21.0%
3.6	18.0%	20.5%
3.7	17.5%	19.5%
3.8	17.0%	19.0%
3.9	16.5%	18.5%
4.0	16.0%	18.0%
4.1	15.5%	17.5%
4.2	15.5%	17.0%

Note: The data in this table is derived assuming a constant heat loss through the elevations amounting to the loss when the basic allowance for openings of 22.5% of floor area is provided and the standard U-values given in Table 16.1 apply. It is also assumed for the purposes of this table that there are no rooflights.

being carried out to assess this. AD L does not, as yet, recognise the positive role that can be played by conservatories, but it does allow them to be accepted as draught lobbies (see above).

A conservatory is defined as having at least 75% of its roof and 50% of its walls constructed from translucent material (see also the definition in Appendix A of AD B, Fire, on page 7.5 above and the recommendations regarding safety glazing in AD N in Chapter 18). If it is not heated, it should be treated as an integral part of the dwelling unless it is separated from it by:

AD L
1.42 to 1.45

AD L, Section 2

Table **8**   **Permitted variation in the areas of windows, doors and rooflights for buildings other than dwellings**

Average U-value	Residential buildings percentage of wall area	Places of assembly, offices and shops percentage of wall area	Industrial and storage buildings percentage of wall area	Rooflights (all) percentage of roof area
2.0	55	74	28	37
2.1	52	69	26	35
2.2	49	65	24	33
2.3	46	62	23	31
2.4	44	58	22	29
2.5	42	56	21	28
2.6	40	53	20	27
2.7	38	51	19	25
2.8	36	49	18	24
2.9	35	47	17	23
3.0	34	45	17	22
3.1	32	43	16	22
3.2	31	41	16	21
3.3	**30**	**40**	**15**	**20**
3.4	29	39	14	19
3.5	28	37	14	19
3.6	27	36	14	18
3.7	26	35	13	18
3.8	26	34	13	17
3.9	25	33	12	17
4.0	24	32	12	16
4.1	23	31	12	16
4.2	23	30	11	15
4.3	22	30	11	15
4.4	22	29	11	14
4.5	21	28	11	14
4.6	21	27	10	14
4.7	20	27	10	13
4.8	20	26	10	13
4.9	19	26	10	13
5.0	19	25	9	13

Note: The data in this table is derived assuming a constant heat loss through the exposed wall or roof area as appropriate. The constant heat loss amounts to the loss through the wall or roof component plus the loss through the basic area allowance of windows, personnel doors or rooflights respectively as calculated using the U-values in Table 16.1.

- Walls and floors which are insulated to at least the standard recommended for semi-exposed walls and floors; and/or,
- Windows and doors which are draught-stripped and have U-values as recommended for the other exposed windows and doors in the dwelling.

If a fixed heating installation is installed in a conservatory it should have its own temperature and on/off controls.

The target U-Value method of compliance for dwellings

The requirements of paragraph L1 of Schedule 1 will be met if the average U-values are calculated and do not exceed the following targets:

AD L 1.12 to 1.15

- *For dwellings with SAP Energy Rating of 60 or less*

$$\text{Target U-value} = \frac{\text{total floor area} \times 0.57}{\text{total area of exposed elements}} + 0.36\,\text{W/m}^2\text{K}$$

- *For dwellings with SAP Energy Rating over 60*

$$\text{Target U-value} = \frac{\text{total floor area}}{\text{total area of exposed elements}} + 0.4\,\text{W/m}^2\text{K}$$

(*Note:* remember to include the area of the exposed fabric, including ground floor, windows, doors and rooflights which are exposed to the outside air when calculating the total in the above equations. Semi-exposed elements may be left out of the above calculation if they are insulated to the standard set out in Table 16.1.)

Allowing for the benefits of solar gain and more efficient heating systems

When demonstrating if a particular building achieves the target U-value, the benefits of solar gain and more efficient heating systems can be taken into account.

Solar gains
The Target U-Value is based on the assumption that the glazing is equally distributed on the north and south elevations. The procedure allows the area of glazing used in the calculation to be reduced as a way of accounting for solar gains. Where the area of glazing facing south exceeds that facing north by $\pm 30°$, then the window area can be taken to be the total window area less 40% of the difference between the area of glazing facing south $\pm 30°$ and the area of glazing facing north $\pm 30°$.

AD L 1.16

High efficiency heating systems
The target U-value is based on the assumption that the central heating system is a standard gas boiler or oil fired hot water central heating system with a seasonal efficiency of at least 72%. A more efficient heating system allows for an increase in the target U-value by up to 10%, e.g. a 10% increase where a condensing gas boiler is fitted.

AD L 1.17 and 1.18

Example of target U-value method

AD L
Appendix F For the detached house shown in Fig. 16.4, the rate of heat loss through the exposed elements is calculated in Table 16.6 below. The design U-values are shown in the table and it can be seen that the walls are of a lower standard than that recommended for the elemental method, whereas the roof and ground floor are slightly better than that recommended. The SAP Energy rating is assumed to be greater than 60.

Table 16.6 Calculation of rate of heat loss.

Element	Area (m²)	U-value (W/m²K)	Rate of heat loss (W/K)
Ground floor	55.77	0.44	24.54
Roof	55.77	0.25	13.94
Exposed walls	136.614	0.6	81.97
Windows	16.686	3.3	55.06
Doors	3.78	2.8	10.58

Total area of exposed elements = 268.62 m²   Total rate of heat loss per degree = 186.09 W/K

Therefore the target is calculated thus:

$$\text{Target U-value} = \frac{\text{total floor area} \times 0.64}{\text{total exposed elements}} + 0.4 \text{ W/m}^2\text{K}$$

$$\text{Target U-value} = \frac{111.54 \times 0.64}{268.62} + 0.4 = 0.66 \text{ W/m}^2\text{K}$$

The average U-value is given by:

$$\text{Average U-value [U]} = \frac{\text{Total rate of heat loss per degree}}{\text{Total area of exposed elements}} = \frac{186.09}{268.62}$$

$$= 0.69 \text{ W/m}^2\text{K}$$

- The average U-value is more than the target U-value therefore the proposed dwelling does not comply and the following modifications should be investigated:
- Improving the thermal resistance of some element(s).
- Taking solar gain into account.
- Utilising a higher performance heating system.

Improving the U-value of the walls

The walls are to be upgraded to give a U-value of 0.45 W/m²K,

$$\text{Average U-value [U]} = \frac{\text{Total rate of heat loss per degree}}{\text{Total area of exposed elements}} = \frac{165.61}{268.62}$$

$$= 0.62 \text{ W/m}^2\text{K}$$

This is less than the target, hence compliance.

Using a higher performance heating system

The installation of a condensing boiler will allow the target U-value to be relaxed by 10% as follows:

Target U-value = $0.66 + (0.66 \times 0.1) = 0.73$ W/m^{2}K

The average U-value calculated above is now less than the revised target U-value, hence compliance.

The proposed dwelling will comply if either the wall U-value is upgraded or a condensing boiler is installed.

The energy rating method of compliance for dwellings

This method allows any valid energy conservation measure to be used in the design of the dwelling.

AD L 1.19

A newly created dwelling (or each dwelling in a block of flats or converted building) will automatically meet the requirement of paragraph 1 of Schedule 1 for conservation of fuel and power if the SAP energy rating, when related to its floor area, is at least equal to the value given in Table 4 of AD L, which is reproduced below.

The SAP is an energy cost rating based on the annual energy cost required for space and water heating (under typical occupancy conditions). Details of how it is calculated are given in Appendix G of AD L, including a blank worksheet and the tables necessary to complete it. Experience has shown that when using this worksheet, although the procedure is not difficult, it is tedious and time-consuming and there is much scope for error. Alternatively, there are many excellent computer software packages on the market today which simplify the process considerably and make mistakes much less likely. Additionally, it is usually a simple matter to re-run the program with different methods of construction or boiler types, etc. should an unacceptable SAP rating emerge from the first run-through.

AD L Appendix G

U-values which are higher than those recommended under the elemental method are possible when using the Target U-value and Energy Rating methods. This could result in surface condensation forming in certain circumstances. For this reason it may be desirable to limit the U-values as follows:

AD L 1.20 and 1.21

AD L

Table **4**   SAP Energy Ratings to demonstrate compliance

Dwelling floor area (m²)	SAP Energy Rating
80 or less	80
more than 80 up to 90	81
more than 90 up to 100	82
more than 100 up to 110	83
more than 110 up to 120	84
more than 120	85

- exposed wall and floor U-values not exceeding 0.7 W/m^{2}K
- roof U-value not exceeding 0.35 W/m^{2}K.

Furthermore, when using the Energy Rating method lintel, cill and jamb details should follow the guidance shown on pages 16.44 to 16.47 below.

The calculation method for buildings other than dwellings

AD L 2.12
When using the elemental method above, for buildings other than dwellings, it is necessary to keep within the permitted limits shown for window, door and rooflight areas, and the standard U-values for the exposed and semi-exposed elements shown in the various tables and diagrams. Since the purpose of Part L of Schedule 1 to the 1991 Regulations is the conservation of fuel and power, it should be possible to satisfy the requirements without rigid adherence to specified areas or U-values provided that the proposed building is not less efficient thermally than a building that was constructed using the specified areas and U-values.

The *calculation method* allows greater flexibility between the areas of windows, doors and rooflights and/or insulation levels of the exposed and semi-exposed elements of the building envelope.

AD L 2.13
The essential feature of the calculation method is that the rate of heat loss from the proposed building should be shown by calculation to be no greater than that from a standard building of similar size and shape which is designed to comply with the elemental method. The following conditions apply when using the calculation method:

AD L 2.14
- If the proposed building has smaller openings than the maximum recommended by the elemental method these smaller areas should be assumed in the standard building when comparing heat losses.

AD L 2.15
- If the proposed building has a floor U-value of better than 0.45 W/m^{2}K then this better value should be assumed in the standard building.

AD L 2.20
- U-values which are higher than those recommended under the elemental method are possible when using the calculation method. This could result in surface condensation forming in certain circumstances. For this reason it may be desirable to limit the U-values as follows:

  (a) For residential buildings
  — exposed wall and floor U-values not exceeding 0.7 W/m^{2}K
  — roof U-value not exceeding 0.45 W/m^{2}K
  (b) For all other buildings – exposed walls, roofs and floors not exceeding 0.7 W/m^{2}K.

An example to demonstrate the use of the calculation method is shown below.

Window/U-value trade-off for office – example of the calculation method

AD L Appendix H
A three-storey detached office building has internal dimensions of 40 m × 10 m × 9 m. It is proposed to insulate the exposed floor, walls and roof to a U-value of 0.6. W/m^{2}K. 80% double-glazed windows and 20% double-glazed rooflights are required. The windows are to have metal frames with thermal breaks and the double-glazing is to be low-E coated with a 12 mm air gap. The

rooflights are to have metal thermal break frames, with a 12 mm air gap. Confirm that the proposed building complies with the requirements of Part L. Referring to Fig. 16.12 below:

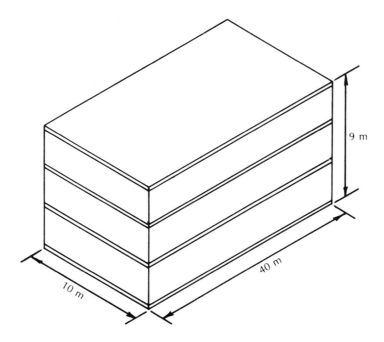

Fig. 16.12 Window/rooflight trade-off – example of the calculation method.

For the proposed building:

- Area of exposed walls (including openings) $= (40 + 40 + 10 + 10) \times 9$
  $= 900 \text{ m}^2$
- Area of windows $= 80\% \times 900$ $= 720 \text{ m}^2$
- Area of personnel doors $= 8 \text{ m}^2$
- Area of exposed wall at U-value of 0.6 $= 900 - (720 + 8)$
  $= 172 \text{ m}^2$
- Area of roof and ground floor $= 40 \times 10$ $= 400 \text{ m}^2$
- Area of rooflights $= 20\% \times 400$ $= 80 \text{ m}^2$
- Area of roof at U-value of $0.6 = 400 - 80$ $= 320 \text{ m}^2$
- Rate of heat loss per degree is calculated in the table below:

Element	Area (m^2)	U-value (W/m^2K)	Rate of heat loss (W/K)
Exposed walls	172	0.6	103
Windows	720	2.6	1872
Roof	320	0.6	192
Rooflights	80	3.8	304
Personnel doors	8	3.3	26
Ground floor	400	0.6	240
Total rate of heat loss			2737

For the standard building:

- Area of exposed walls (including openings) $= (40 + 40 + 10 + 10) \times 9$
    $= 900 \text{ m}^2$
- Area of windows $= 40\% \times 900$ $= 360 \text{ m}^2$
- Area of exposed walls at U-value of 0.45 $= 900 - 360$
    $= 540 \text{ m}^2$
- Area of roof and ground floor $= 40 \times 10$ $= 400 \text{ m}^2$
- Area of rooflights $= 20\% \times 400$ $= 80 \text{ m}^2$
- Area of roof at U-value of 0.45 $= 400 - 80$ $= 320 \text{ m}^2$
- Rate of heat loss per degree is calculated in the table below:

Element	Area (m^2)	U-value (W/m^2K)	Rate of heat loss (W/K)
Exposed walls	540	0.45	243
Windows and doors	360	3.3	1188
Roof	320	0.45	144
Rooflights	80	3.3	264
Ground floor	400	0.45	180
Total rate of heat loss			2019

The total rate of heat loss per degree from the proposed building exceeds that from the standard building, therefore the proposal does not comply with the requirements of Part L and the design should be re-appraised, perhaps by altering the U-values or changing the window and rooflight areas, until a satisfactory solution can be found.

Energy use method for buildings other than dwellings

AD L
2.17 to 2.19

This procedure allows complete freedom of design and permits the use of any valid energy conservation measure. Therefore, full account may be taken of useful heat gains from solar radiation through the fabric, industrial processes, artificial lighting and any other form of heat gain to which the building is subject.

An energy use calculation should be carried out to show that the annual energy use of the proposed building (after taking account of useful heat gains) is no greater than the calculated energy use of a similar building designed in accordance with the elemental method described above.

Where a building is to be naturally ventilated, compliance may be demonstrated by following the guidance in CIBSE *Building Energy Code 1981, Part 2a (worksheets 1a to 1e)*.

Thermal bridging around openings in walls

AD L
1.22 and
1.23, 2.21
and 2.22

Excessive additional heat losses and local condensation can occur due to the closing of cavities around openings in walls for windows and doors, etc. Particular attention should be paid to the design of lintels, jambs and cills so that the effects of thermal bridging can be minimised. Fig. 16.13 illustrates typical details which would satisfy the functional requirements of L1.

Insulating blockwork   **Insulated internal lining**   **Full or partial cavity fill**

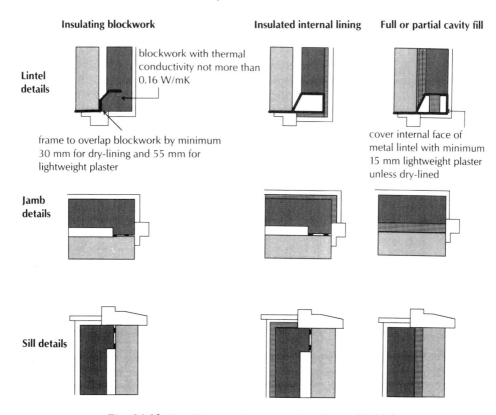

Fig. 16.13 Detailing openings to reduce thermal bridging.

Alternatively, it is possible to show by calculation that the edge details around openings are satisfactory, by using the procedure described in Appendix D of AD L. This makes it possible to establish:

**AD L
1.24 and
2.23,
Appendix D**

- Whether or not there is a risk of condensation occurring at the edges of openings; and/or,
- Whether any possible heat losses are significant in these positions.

Calculation procedure for minimum thermal resistance path

The calculation procedure has the following steps:

**AD L
Appendix D**

1. Identify the minimum thermal resistance path.
2. Establish the thermal conductivity of each material in the path.
3. Divide the length of each part of the thermal resistance path in metres by its thermal conductivity to obtain its thermal resistance.
4. Add together the individual thermal resistances to obtain the total thermal resistance (R_{min}) for the path identified in step 1.

The minimum thermal resistance Rmin, is then compared with the following performance criteria to see whether corrective action is necessary:

- For negligible risk of surface condensation and mould growth at the edges of openings, R_{min} should be at least $0.20\,m^2K/W$ (rounded to two decimal places). If the value of R_{min} is less than this then the design should be modified to reduce the risk of surface condensation or more rigorous calculation methods should be employed.
- For heat losses to be insignificant at the edges of openings, R_{min} should be at least $0.45\,m^2K/W$ (rounded to two decimal places). If the value of R_{min} is less than this then it will be necessary to take these additional heat losses into account. The edge detail should be redesigned (perhaps using an insulated cavity closer, for example) and the calculation redone until it can be shown that R_{min} is at least $0.45\,m^2K/W$.

It is also possible to show compliance in other ways and these will depend on the overall approach which has been adopted to the thermal design. For example, if the **Target U-value method** has been used for a dwelling, then it is possible to increase the average U-value by the following amount:

$$\frac{0.3 \times \text{total length of relevant opening surrounds}}{\text{total exposed surface area}} = (W/m^2K)$$

The overall design should then be checked to ensure that the modified average U-value does not exceed the Target U-value.

If the **calculation procedure** has been adopted for buildings other than dwellings, then the rate of heat loss from the proposed building could be increased by:

$$0.3 \times \text{total length of relevant opening surrounds}$$

The modified rate of heat loss from the proposed building should then be checked against the calculated rate of heat loss from the notional building to ensure that it is not exceeded.

It should be noted that the procedures shown above need to be modified where the edge detail contains thin layers of material not exceeding 4 mm in thickness (such as steel lintels). Further calculations are necessary in these cases and details of these may be obtained from BRE Information Paper 12/94 *Assessing condensation risk and heat loss at thermal bridges around openings*.

The following example illustrated in Fig. 16.14 shows the basic calculation method for a window jamb.

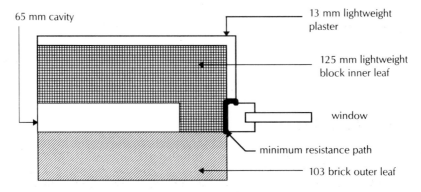

Fig. 16.14 Example calculation for minimum resistance path at edge of opening.

Step 1. The minimum resistance path is shown by the thick black line.
Step 2. The thermal conductivities for each of the materials in the path are shown in column 2 of Table 16.7 below.
Step 3. Obtain the thermal resistance of each part of the path by dividing the path length by its thermal conductivity (see columns 3 and 4 in Table 16.7).
Step 4. Add together the individual thermal resistances to obtain the minimum resistance Rmin which from the table is shown as $0.42 \, m^2 K/W$.

Table 16.7 Calculation of minimum thermal resistance path at edge of openings.

(1) Material	(2) Thermal conductivity (W/mK)	(3) Path length (m)	(4) Thermal resistance R (m²K/W)
Lightweight block inner leaf	0.19	0.065	0.34
Lightweight plaster	0.16	0.013	0.08
		Minimum resistance R_{min} =	0.42

This value shows that the risk of condensation and mould growth is negligible. However, the value is less than $0.45 \, m^2 K/W$ which means that the heat loss at the edge of the opening is significant and it is necessary to modify the design as shown in Table 16.8 below by the inclusion of an insulated cavity closer. The resulting value for R_{min} of $1.94 \, m^2 K/W$ is more than adequate.

Table 16.8 Calculation of minimum thermal resistance path including insulated cavity closer.

(1) Material	(2) Thermal conductivity (W/mK)	(3) Path length (m)	(4) Thermal resistance R (m²K/W)
Insulated cavity closer	0.035	0.065	1.86
Lightweight plaster	0.16	0.013	0.08
		Minimum resistance R_{min} =	1.94

Limitation of air leakage

One of the ways in which energy is wasted in buildings is by unintentional infiltration of cold outside air through the building envelope. This leakage should be limited, as far as is practicable, and the following measures would help to satisfy the requirements of regulation L1:

AD L
1.25 and
2.24

- In dry-lined construction, the gaps between the lining and the masonry construction should be sealed at the edges of window and door openings, and at the junction with walls, floors and ceilings. This is normally achieved by using continuous bands of fixing plaster;
- In timber-frame construction, vapour control membranes should be sealed;
- Openable parts of windows, doors and rooflights should be draught-stripped;
- Loft hatches and access hatches in unheated floors should be provided with draught-seals;
- Service pipe penetrations into hollow construction or voids should be sealed at the point of penetration, and boxing for concealed services should be sealed at floor and ceiling level.

These recommendations are summarised in Fig. 16.15 below and further advice on measures to limit infiltration is also obtainable from:

AD L
1.26 and
2.25

- BRE Report BR 262 *Thermal insulation: avoiding the risks*; and,
- *Thermal insulation and ventilation Good Practice Guide 1991*, published by the NHBC and the Energy Efficiency Office; and,
- In office buildings, BRE Report BR 265 *Minimising air infiltration in office buildings*.

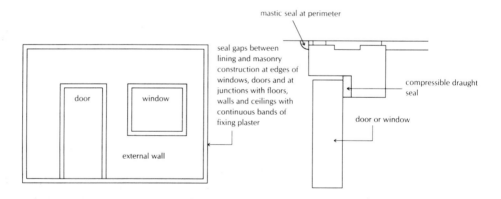

a) Dry-lined construction   b) Draught-stripping of windows and doors

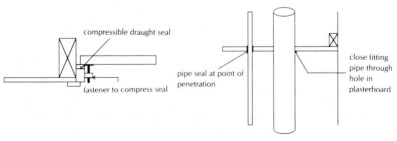

c) Draught-sealing of access hatches   d) Sealing to service pipe penetrations

Fig. 16.15 Limiting infiltration of cold air in buildings.

Space heating and hot water storage system controls

Energy savings can also be made if space heating and hot water storage systems are fitted with controls which adjust the energy input to the system to suit the normal usage of the building.

Under the terms of AD L this means providing space heating or hot water storage systems in buildings with automatic means for controlling the operation and output of the space heating system and the temperature of the stored water.

Application

AD L covers control systems in all types of buildings, the only exception being for those systems which control commercial or industrial processes. The guidance given in AD L for space heating systems in dwellings is restricted to the more common varieties of heating system and excludes heating provided by separate gas, electric and solid fuel fires or room heaters with integral controls.

AD L
2.26

AD L
1.27

Space heating controls

AD L describes three types of space heating system control which apply to all buildings:

AD L
1.28

- Temperature control;
- Timing control; *and*
- Boiler control.

Temperature control

In dwellings, it is normal for there to be two heating zones (relating to separate living and sleeping areas, for example) where different temperatures are required, however, a single zone may be appropriate in single storey, open plan flats or bedsitters. Therefore, the temperature control system chosen should be able to cope with up to two zones and could consist of any suitable temperature sensing devices such as room thermostats and/or individual thermostatic radiator valves. This form of zone control would be suitable for hot water central heating systems, fan controlled electric storage heaters and electric panel heaters. Zone controls would not be suitable for ducted warm air systems or flap controlled electric storage heaters, however thermostatic control should still be provided to these systems.

In buildings other than dwellings, temperature control may be effected by using room thermostats or thermostatic radiator valves or any other form of temperature sensing devices on each part of the space heating system which is designed to be separately controlled. Additionally, space heating systems which use hot water should be provided with an external temperature sensing device placed outside the building connected to a weather compensator controller. This allows for the temperature of the water flowing in the heating circuit to be regulated in response to changes in weather conditions, see Fig. 16.16.

AD L
1.29 to 1.31

AD L
2.27

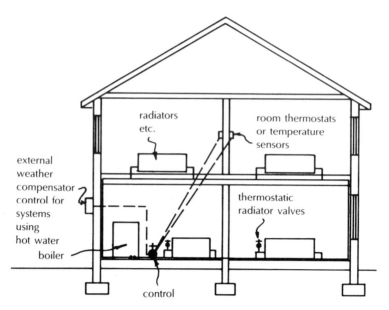

Fig. 16.16 Control of room temperatures.

Timing control

Energy savings can be made if space heating systems are capable of being switched on and off in accordance with their normal pattern of use.

In dwellings, a simple programmable clock may be installed. Most of these devices permit two heating periods per day. AD L now recommends that some form of timer should be provided on gas or oil fired heating systems and on systems with solid fuel boilers operated by a forced-draught fan. Clearly, it would be pointless to install a timer on a solid fuel boiler which operated by means of natural draught so this form is excluded from the recommendation.

**AD L
1.32**

For buildings other than dwellings which are used intermittently, such as offices, shops and factories where shift working is not carried out, timing controls are necessary to shut down the heating system at certain times (at night or weekends) and to switch it on again prior to recommencement of normal use.

**AD L
2.28 and
2.29**

The system of control used will depend on the size and type of building and the output of the heating system as follows:

- For space heating systems with an output of not more than 100 kW, a clock control which can be set manually to give start and stop times may be used.
- For space heating systems with an output in excess of 100 kW, a more complex system of control is recommended which will take into account the rate at which the building will react when heating is shut off and restarted. This arrangement (known as 'optimum start control') allows variable start times for each of the heating systems in the building.

If heating systems are completely shut off during periods when a building is unoccupied, frost and excessive humidity or condensation may cause damage to the structure, services or contents of the building. Therefore, controls may be provided which will allow certain minimum temperatures to be maintained.

Boiler control

In most dwellings with gas or oil fired hot water central heating the heat source is a single boiler. As we have seen from above, control is usually by room thermostats or separate thermostatic radiator valves. These devices should switch the boiler off when there is no call for heat, however systems controlled only by thermostatic radiator valves may need to be fitted with flow control or other similar devices to prevent unnecessary boiler cycling.

AD L
1.33

In buildings other than dwellings multiple boiler installations are common. Since boilers run most efficiently when they are at, or near full output, controls become essential when two or more boilers are required to support the same heat demands, if this exceeds 100 kW.

AD L
2.30

The form of control provided (sequence control) should be able to detect variations in heat demands and so start, stop or modulate the boilers as required in order to optimise the conservation of fuel and power.

Control of hot water storage

The recommendations apply to all types of buildings and are designed to ensure that stored hot water is efficiently heated, delivered at the correct temperature and provided only when there is a demand. In order to achieve this the following guidance is given in AD L:

AD L
1.34, 1.35
2.31 and
2.32

- The water storage vessel should contain a heat exchanger with sufficient capacity for effective control. This can be achieved by installing a vessel which complies with BS 1566: 1984 (1990) *Copper indirect cylinders for domestic purposes* or BS 3198: 1981 (with amendments prior to June 1994) *Specification for copper hot water combination units for domestic purposes* or their equivalent. In these documents particular attention should be given to the sections dealing with the surface areas and pipe diameters of the heat exchangers.
- A thermostat should be provided to shut off the heat supply when the chosen water temperature is reached. Where the water is heated in conjunction with a hot water central heating system the water storage thermostat should be interconnected with the room thermostat(s) so that the boiler is switched off when no heat is required.
- A time switch should be provided which can either be part of the central heating system or a separate local device, in order to allow the heat supply to be shut off when there is no hot water demand.

Quite clearly, these recommendations would not be suitable for a solid fuel fired hot water system where the heat supply cannot be switched on and off and where the hot water cylinder is not providing the slumber load. In this case a thermostatically controlled valve should be provided.

Alternative recommendations for heating and hot water control

Provided that the above recommendations regarding temperature, timing and boiler control are met then the guidance given in following standards may be adopted:

For dwellings

AD L
1.36

- BS 5449: 1990 *specification for forced circulation hot water central heating systems for domestic premises*;
- BS 5864: 1989 *specification for installation in domestic premises of gas-fire ducted air-heaters of rated output not exceeding 60 kW*;
- Any other authoritative design specifications recognised by the heating fuel supply company.

For buildings other than dwellings

AD L
2.33

- BS 6880: 1988 *Code of practice for low temperature hot water heating systems of output greater than 45 kW*;
- CIBSE Applications Manual AM1: 1985 *Automatic controls and their implications for systems design*.

(See also details of unvented hot water storage system controls in Chapter 12.)

Insulation of heating vessels, pipes and ducts

In general, heating vessels, pipes and ducts should be insulated. This applies especially to pipes where they pass through unheated spaces such as roofs, under floors or in attached garages where it may be necessary to increase the insulation thicknesses indicated below for protection against freezing. (Further guidance on suitable protection in these circumstances may be obtained from the BRE Report BR 262 *Thermal insulation: avoiding risks*.)

AD L
1.41 and
2.41

This general rule does not apply:

- Where the pipes or ducts are intended to provide useful heating to a room or space, or

AD L
2.34

- Where the vessels, pipes and ducts form part of an industrial or commercial process.

Insulation of hot water storage vessels

AD L
1.37, 2.35
and 2.36

The guidance given in AD L is based on the insulation of a 120 litre storage vessel (i.e. 450 mm diameter by 900 mm high). The insulation should be factory applied and designed to restrict standing heat losses to 1 watt per litre or less using the test method described in BS 1566 Part 1, Appendix B.4 or equivalent.

This will also apply to vessels of other sizes if the material and its thickness are equivalent to that provided for a 120 litre vessel. In practical terms this means providing vessels with a 35 mm thick, factory applied coating of PU-foam with a minimum density of 30 kg/m^3 and of zero ozone depletion potential.

Additionally, for buildings other than dwellings it is permissible to use vessels which comply with BS 853: 1990 *Specification for calorifiers and storage vessels for central heating and hot water supply* or its equivalent. In this case it is recommended that at least 50 mm of material should be provided with a thermal conductivity of 0.045 W/mK (or other materials which give the same performance) and it may be necessary in certain cases to provide an outer metal casing to protect the insulation.

Chapter 12 above deals with unvented hot water storage systems. It should be noted that the safety fittings and pipework which are needed for their safe operation will also need additional insulation to control heat losses. Clearly, this should be provided without impeding safe operation and the visibility needed to observe warning discharges.

**AD L
1.38 and
2.37**

Insulation of pipes and ducts

Pipes should be insulated with a material which:

**AD L
1.39 and
2.38**

- Has a thermal conductivity not exceeding 0.045 W/mK and a thickness at least equal to the outside diameter of the pipe up to a maximum of 40 mm; *or*,
- Satisfies the requirements of BS 5422; 1990 *Methods for specifying thermal insulation materials on pipes, ductwork and equipment in the temperature range of –40°C to + 700°C.*

For warm air ducts the insulation material should meet the recommendations of BS 5422: 1990 as defined for pipes above.

An appreciable amount of heat may also be lost from the lengths of hot pipework connected immediately adjacent to the hot water storage vessel. Therefore, in buildings other than dwellings where it is permissible to install a hot water storage vessel which complies with BS 853 or its equivalent (see above page 16.52). The hot pipes connected to such a vessel (including the vent pipe, and the primary and secondary flow and return pipes if fitted) should be insulated as described immediately above.

**AD L
2.40**

Similar recommendations apply to all building types employing vessels which comply with BS 1566 and BS 3198. In these cases the hot pipes (including the vent pipe, and the primary flow and return to the heat exchanger if fitted) should be insulated up to at least 1 metre from their connection point with the vessel or until they become concealed in the building fabric, etc. This insulation should give a performance of not less than that which would be provided by a material having a thermal conductivity of 0.045 W/mK and a thickness of 15 mm.

**AD L
1.40 and
2.39**

Lighting

Where artificial lighting is to be provided to more than 100 m^2 of floor area in buildings other than dwellings, the lighting systems in such buildings must be designed and constructed so that:

**Regs Sch. 1
L1(e)**

- they use a reasonable amount of fuel and power, and
- reasonable provision is made for their control.

These requirements do not apply to display lighting or emergency escape lighting.

**AD L
2.43**

Interpretation

The following definitions apply to the provision of lighting in buildings other than dwellings:

AD L
2.42

- EMERGENCY ESCAPE LIGHTING – part of the general emergency lighting designed to provide illumination for the safety of people leaving an area in a building or attempting to terminate a dangerous process before leaving.
- DISPLAY LIGHTING – lighting used to highlight displays of merchandise or exhibits.
- CIRCUIT WATTS – power consumed by lamps, their control gear and power correction factor equipment.
- SWITCH – includes dimmer switches. Generally, dimming should be achieved by reduction of the energy supply and not its diversion.
- SWITCHING – includes dimming.

Efficacy of lamps

AD L
2.44 and
2.48

Two methods for providing efficient lighting installations are recommended in AD L as follows:

1. By using high efficacy lamps so that at least 95% of the installed lighting capacity in circuit watts is provided from fittings which use lamps chosen from those listed in Table 9 of AD L, which is reproduced below; *or,*
2. By installing lighting fittings using lamps with an average initial (100 hour) efficacy of at least 50 lumens per circuit watt.

AD L
Appendix H

Therefore, in order to check if a proposed lighting installation complies with the recommendations listed above it will be necessary to schedule the lamps either:

- according to their type and circuit watts, in which case:

$$\frac{\text{circuit watts consumed by lamps listed in Table 9} \times 100}{\text{total circuit watts}} \geq 95\%$$

AD L

Table **9**   **Types of high efficacy lamps**

Light source	Types
High pressure sodium	
Metal halide	All types of ratings
Induction lighting	
Tubular fluorescent	All 25 mm diameter (T8) lamps provided with low-loss or high frequency control gear
Compact fluorescent	All ratings above 11 W

or,

- according to their circuit watts and lumen output, in which case:

$$\frac{\text{total lumen output of installation}}{\text{total circuit watts}} \geq 50 \text{ lumens/watt}$$

The values of the circuit watts per lamp and the lumen output per lamp may be obtained from manufacturer's data.

Control of lighting installations

Wherever possible, lighting controls should be provided so that maximum use is made of daylight and lights are switched off when spaces are unoccupied. This can be achieved by the use of automatic switching but it must be designed so that the building's occupants are not put at risk should the lights go out. AD L 2.45

Easily accessible local switches should be provided as follows: AD L 2.46 and 2.47

- Within 8 metres or 3 × height of the fitting above floor level, whichever is the greater;
- Located within each work area or at the boundaries between work areas and general circulation routes;
- Operated by deliberate action of the occupants either manually (e.g. by rocker switches, press buttons or pull cords) or by remote control (e.g. infra-red transmitters, sonic, ultra-sonic and telephone headset controllers);
- Operated by automatic switching systems such as those which switch off the lights when they sense the absence of occupants.

Lighting controls – alternative approaches

Although the recommendations listed above may be appropriate in most cases, AD L gives alternative guidance which may also be adopted as follows: AD L 2.49 to 2.51

- In offices and storage buildings, local switching which follows the general guidance above may be provided for each separate working area. This should be arranged to maximise daylight and could be incorporated with other controls such as time-switches and photo-electric switches.
- In other types of buildings time switching may be considered where there are operational areas with clearly timetabled occupational characteristics. Photo-electric switching may also be appropriate in certain circumstances.
- Lighting controls designed in accordance with the recommendations of the CIBSE *Code for interior lighting* 1994, may also meet the requirements for conservation of fuel and power provided that the designs achieved perform at least as well as those derived using the guidance in AD L.

Material alterations and changes of use – the application of Part L

Reference to Chapter 2 above will confirm that certain works to existing buildings may result in the need to upgrade their thermal performance. AD L provides general guidance on the extent to which this upgrading may be necessary.

Material alterations

It will be seen from page 2.12 above, that in general terms, alterations to a building should not result in a new or greater contravention of the regulations as a result of the alterations. With this in mind, Part L of Schedule 1 should be interpreted as follows and the recommendations apply equally to dwellings and other buildings:

AD L, 1.46 to 1.48 and 2.52 to 2.54

- Roof insulation should be provided to the standard recommended for new construction only if the roof structure is being substantially replaced.
- Floor insulation should be provided if the ground floor structure is being substantially replaced. Again, this should be to the standard for new buildings in rooms which are heated.
- Where it is proposed to substantially replace a complete external wall, insulation should be provided to a reasonable thickness and it should incorporate the sealing measures described above (see page 16.47).
- If work is carried out on space heating and hot water systems which can be classified as a material alteration, then the controls and insulation measures recommended for new installations should be incorporated (see pages 16.49 to 16.53 above).

Material change of use

AD L, 1.49 to 1.50 and 2.55 to 2.56

Material change of use is defined on pages 2.13 and 2.14 above where it will be seen that Part L is applicable in all cases. AD L gives the following guidance on the extent to which it is necessary to thermally upgrade a building when its use is changed:

- By following the guidance for roofs, floors, walls, and space heating and hot water systems as outlined above for material alterations.
- Where a roof is not being replaced it should still be possible to upgrade the insulation if the loft is accessible. The U-value of the roof should be improved to a maximum of $0.35 \, \text{W/m}^2\text{K}$ if the existing insulation provides a value which is worse than $0.45 \, \text{W/m}^2\text{K}$.
- The insulation of exposed and semi-exposed walls should be improved if the internal surfaces are being renovated over a substantial area. If this is achieved by using an insulated dry-lining system, then it should be sealed in accordance with the recommendations described on page 16.47 above.
- Where replacement windows are being fitted they should be draught-stripped and should achieve a U-value not greater than $3.3. \, \text{W/m}^2\text{K}$. AD L recognises that this recommendation may be inappropriate in conservation work and in other situations where it is desirable to retain existing window designs.
- In buildings other than dwellings, if the lighting systems are to be substantially replaced then the new installations should follow the guidance on pages 16.53 to 16.55 above.

Chapter 17

Access and facilities for disabled people

Introduction

The law concerning access for disabled people to buildings has a relatively short history. The first provisions were contained in the Chronically Sick and Disabled Persons Act 1970. These provisions were mostly advisory and were only applied if it was reasonably practicable to do so. There were no enforcement powers contained in the Act and it proved to be rather ineffective. It was clear that some form of legislation with 'teeth' was required.

Therefore, it was considered that the Building Regulations were the most suitable medium for any future legislation. This resulted in the fourth amendment to the 1976 Regulations which introduced Part T, *Facilities for disabled people*, in August 1985. Since then the former Part T has been recast in the current format as Part M (supported by Approved Document M) and has been the subject of three further revisions to extend its scope and coverage.

The most recent amendments to Part M (and Approved Document M) mean that requirements M1 to M3 apply to new dwellings as well as to other types of buildings. The new regulations come into force on 25 October 1999, however, the Government has taken precautions to prevent floods of applications arriving with local authorities or approved inspectors just before the deadline in order to circumvent the new requirements. Therefore, plans of new dwellings will need to be approved by local authorities, (or plans certificates from approved inspectors will need to be given to local authorities), before 1 June 1999, since the new requirements will apply to any submissions placed after that date.

When dealing with access requirements for disabled people reference should also be made to the Disability Discrimination Act 1995 (see Chapter 5).

Interpretation

DISABLED PEOPLE – this is a narrowly defined term in regulation M1 which applies to those people with:

(a) A physical impairment which limits their ability to walk or makes them dependent on a wheelchair for mobility, *or*
(b) Impaired sight or hearing.

A number of other terms are defined in Approved Document M, which apply throughout the document as follows:

ACCESS – approach or entry.

ACCESSIBLE – access is facilitated for disabled people to buildings or parts of buildings.

AD M 0.10

BUILDING – in addition to dwellings (see page 17.24 below) the rules in AD M apply to the following buildings:

- Shops, offices, factories and warehouses.
- Schools, other educational establishments and student residential accommodation in traditional halls of residence.
- Institutions.
- Premises which admit the public whether on immediate payment, subscription, fee or otherwise.

The rules apply to the whole building and equally to any parts of the building which comprise separate individual premises.

SUITABLE – means of access and facilities which are designed for the use of disabled people.

PRINCIPAL ENTRANCE STOREY – the storey of a building which contains the main entrance or entrances.

AD M 0.10

Sometimes it may be necessary to provide an alternative, accessible entrance into the building (see below). In this case the storey containing the alternative entrance would be the principal entrance storey.

Application

Part M of Schedule 1 to the 1991 Regulations applies to:

AD M 0.1

(a) New buildings (including dwellings) or buildings which have been demolished to leave only the external walls standing.

AD M 0.2

In buildings, other than dwellings, which have been substantially reconstructed, it may prove impractical to provide an accessible entrance suitable for disabled people. Where this is the case AD M still recommends that the other requirements of Part M be applied.

AD M 0.3

(b) extensions to existing buildings (but not to dwellings, since extensions to these are excluded from Part M) if the extension has a ground storey.

AD M 0.4 & 0.5

The extension must comply with the requirements of Part M but there is no obligation to bring the existing building up to the standards. On the other hand the extension must not adversely affect the existing building with regard to the provision of Part M requirements which may already exist. Where the extension is accessed through the existing building then it would be unreasonable to expect a higher standard of access than that provided in the existing building. However, if the extension is capable of being independently approached and entered from the boundary of the site, then it should be treated as if it were a new building.

(c) those external features which are needed to provide access from the edge of the site to the building and from any car parking within the site.

AD M
0.7

Part M does not apply to the following:

- Material alterations – although facilities which existed before the alterations were carried out must not be made worse as a consequence of the alterations. For example, existing sanitary conveniences provided for disabled people may be moved to another, equally accessible, location in the building, but they may not be removed.
- Any part of a building used solely for inspection, maintenance or repair of the building or its services or fittings.
- *Any* extension to an existing dwelling, and an extension to any other building which does not have a ground storey (e.g. where a new floor is added to an existing building).
- Additionally, Part M4 – Audience or spectator seating – does not apply to dwellings.

Regs Sch. 1
M1 to M4

Buildings other than dwellings

The main provisions

Reasonable provision must be made in buildings for:

- Disabled people to gain access to and use their facilities.
- Sanitary conveniences suitable for disabled people in any building where sanitary conveniences are provided.
- A reasonable number of wheelchair spaces, where the building contains audience or spectator seating.

Regs Sch. 1
M2
Regs Sch. 1
M3
Regs Sch. 1
M4

Therefore, in buildings other than dwellings, it should be reasonably safe and convenient for disabled people to:

- Approach the principal entrance (or other entrances permitted by AD M) to the building from the edge of the site or from car parking within the site.
- Gain access into the building.
- Gain access within the building.
- Use the facilities which are provided in the building (including sanitary conveniences and accommodation for disabled people within audience or spectator seating).
- Enjoy the use of aids to communication if they have impaired hearing or sight in auditoria, meeting rooms, reception areas and ticket offices.

AD M
Performance
and 0.8i)

It should be noted that the provision of access and facilities for disabled people in buildings is not only for those who work there, but also for visitors to the building.

Means of access

The approach to the building

AD M
Section 1

Disabled people should be able to reach the principal entrance into the building or any other entrances which are provided (see Access to the building, below).

Access should be provided from the entrance into the site curtilege or from any car parking which is provided for disabled people within the building site.

Disabled people should also be able to get from one building to another on the site.

The following recommendations are given in AD M regarding approach to the building:

AD M
1.1 to 1.11

- It should be level where possible and certainly no steeper than 1 in 20 since disabled people have difficulty negotiating changes in level and people with impaired vision may be unaware of abrupt level changes.
- It should have a surface width of at least 1200 mm in order to provide adequate space for wheelchairs or helpers and for people passing in the opposite direction.
- Tactile warnings should be provided on pedestrian routes for people with impaired vision where the route crosses a vehicular carriageway or at the top of steps. Examples of typical tactile pavings are shown in Fig. 17.1.
- If it needs to be steeper than 1 in 20 then a proper ramped approach should be provided. (See Ramped access below and Fig. 17.2.)
- Dropped kerbs should be provided for wheelchair users at carriageway crossings.
- Many ambulent disabled people find it easier to negotiate steps than ramps, therefore, where possible easy-going steps should complement a ramped approach. (See Complementary steps below and Fig. 17.2.)
- Care should be taken to avoid hazards to people with impaired sight when they use access routes close to the building. (See Avoidance of hazards below.)

Access to the building

In general, buildings should be designed so that there is a convenient access into the building suitable for disabled people. This rule applies equally to visitors or staff whether they arrive on foot or in a wheelchair.

If it is necessary to provide separate entrances for visitors and customers or staff, each entrance should be suitable and accessible.

In certain cases it may not be possible to make the main entrance accessible due to space restriction and congestion or sloping ground. Sometimes, car parking spaces are provided in areas where access to the principal entrance is not possible. In these cases an additional accessible entrance may be provided, but this should also be for general use and should give suitable internal access to the principal entrance.

AD M
1.30, 1.31

Ramped access

Steep ramps are difficult for wheelchair users to negotiate and they may add to the difficulties of an ambulent disabled person by increasing their unsteadiness in adverse weather conditions or by becoming slippery.

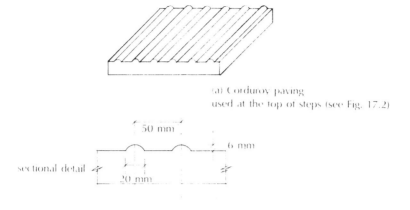

(a) Corduroy paving
used at the top of steps (see Fig. 17.2)

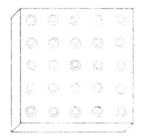

sectional detail

paving slabs 300 mm, 400 mm
or 500 mm square

(b) Modified blister paving
used at carriageway crossings

sectional detail

Fig. 17.1 Tactile pavings.

Equally, long ramps create difficulties since disabled people and their helpers may need to make frequent stops to regain their strength or breath, or to ease pain. Also, adequate space is needed for people passing and for the negotiation of door openings.

AD M
1.12 to 1.19

The recommendations for a ramped approach to the building are illustrated in Fig.17.2 and may be summarised as follows.

Ramps should:

- Have a non-slip surface.
- Ideally, be not steeper than 1 in 20. However, a 1 in 15 ramp is permitted if the individual flight is no longer than 10 m. Similarly a 1 in 12 ramp is allowed where the flight does not exceed 5 m in length.
- Have a landing at top and bottom at least 1200 mm long, clear of any door swing.
- Have intermediate landings at least 1500 mm long, clear of any door swing.

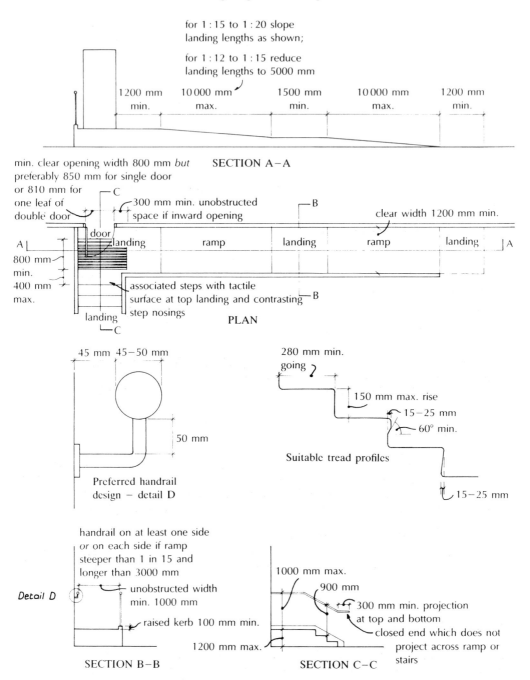

Fig. 17.2 Access to the building.

- Have on any open side of a flight or landing a raised kerb at least 100 mm high. This helps to avoid the risk to wheelchair users of their feet catching under or in an open balustrade.
- Have a surface width of at least 1200 mm.
- Have handrails on both sides of the flight or landing if the ramp is more than 2 m long, in order to provide support. Handrails should be designed to be easily gripped and should be well supported.   **AD M 1.19**

Complementary steps

Many of the design factors which apply to ramps also apply to steps. However, there are also a few additional factors.

Sudden changes of level marked by steps may create dangers for people with impaired sight especially at the head of a flight. Individual steps, whether on their own or in a flight, should also be apparent.

Some ambulent disabled people may have stiffness in their hip or knee joints or may need to wear calipers. Here there is a danger that they may catch their feet under nosings or treads on the staircase.

Clearly, people with physical weakness on one side or the other or who have sight impairments will need to place their feet squarely onto the treads.

Therefore, complementary steps should:

- Have a landing with a tactile surface at the top of the flight to give advance warning of the change in level. This should be the 'corduroy' type of paving shown in Fig. 17.1 and it should extend at least 400 mm beyond each side of the stairs.
- Have step nosings which are clearly visible by the use of contrasting brightness.
- Not rise more than 1200 mm between landings.
- Have landings at top and bottom of the flight and intermediate landings at least 1200 mm long, clear of any door swing.   **AD M 1.20 to 1.24**
- Have uniform rises no greater than 150 mm high and uniform goings at least 280 mm long (measured 270 mm in from the narrow edge of the flight for tapered steps).
- Have a clear width of at least 1000 mm.
- Have a suitable handrail on each side if comprising more than one riser and a suitable tread profile as illustrated in Fig. 17.2. It should be noted that open risers are not recommended.   **AD M 1.25 to 1.26**

Avoidance of hazards

Where a circulation route passes close to a building care should be taken to avoid projections which might be a hazard to people with sight impairments. This is most commonly caused by doors and windows which open outwards across paths. Opening lights and outward opening doors should be guarded if the path is adjacent to the building.   **AD M 1.27 to 1.29**

Alternatively, the path can be separated from the building by a slightly raised edging with a strong tactile surface such as cobbles, for example.

Entrance doors and lobbies

A wheelchair user will need extra space when negotiating an entrance lobby to allow for assistance and to avoid others who may be passing in the opposite direction. It should also be possible to move clear of one door when opening the next.

AD M
1.40, 1.41

For this reason entrance doors and lobbies need to be built to certain minimum dimensions.

Figure 17.3 shows the design principles which should be followed.

Entrance doors should have an absolute minimum clear width of 800 mm. Ideally, the minimum clear width should be that provided by a 1000 mm single leaf external doorset or by one leaf of an 1800 mm double leaf doorset i.e. 850 mm clear or 810 mm clear – see Table 2, BS 4787: Part 1: 1980 (1985).

Whichever type of door is used it is important to allow sufficient room for the door to be opened by a person in a wheelchair. Therefore, the space into which the door opens should be unobstructed on the door handle side for at least 300 mm.

AD M
1.32 to 1.35

Disabled people cannot normally react quickly to avoid collisions when a door is suddenly opened. Therefore they should be able to see people who are approaching a door from the other side and should be seen themselves. This can be achieved by providing a glazed panel in the door which gives a zone of visibility between 900 mm and 1500 mm from the floor. (See Fig. 17.3.)

Revolving doors

Small revolving doors are not negotiable by wheelchair users and may create entry and exit difficulties for people with sight impairments or ambulatory problems. Therefore, revolving doors should always be accompanied by a conventional, accessible entrance door.

AD M
1.36 to 1.39

Some public buildings, supermarkets and the like are fitted with large revolving doors which may be suitable for wheelchair users. They should be capable of accommodating several people at once and should revolve very slowly. They should also be fitted with mechanisms which slow them down still further or stop them as soon as any resistance is felt.

Access within the building

Once inside the building a disabled person must be able to reach and use the facilities provided.

Different building types contain unique facilities to which it may be reasonable to provide access for disabled people.

AD M does not attempt to provide exhaustive guidance on all the facilities which may be relevant and the choice may appear somewhat arbitrary. However, a number of aims are stated:

- People with hearing impairments should be able to play a full part in such things as conferences and committee meetings.

AD M
3.1 to 3.3

- Common facilities such as waiting rooms, canteens or cloakrooms should be located in an accessible storey of the building.

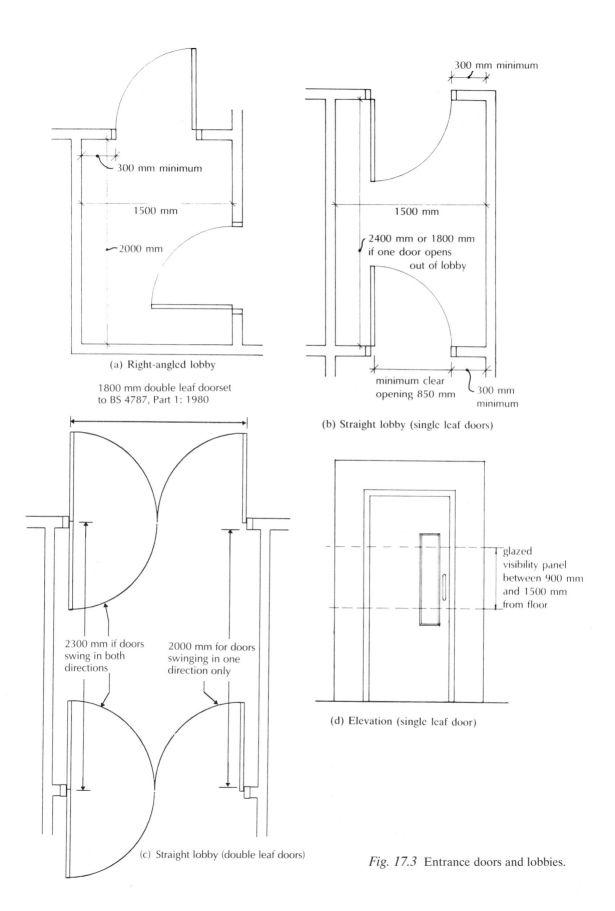

(a) Right-angled lobby

300 mm minimum

1500 mm

2000 mm

1800 mm double leaf doorset
to BS 4787, Part 1: 1980

300 mm minimum

1500 mm

2400 mm or 1800 mm
if one door opens
out of lobby

minimum clear
opening 850 mm

300 mm
minimum

(b) Straight lobby (single leaf doors)

2300 mm if doors
swing in both
directions

2000 mm for doors
swinging in one
direction only

(c) Straight lobby (double leaf doors)

glazed
visibility panel
between 900 mm
and 1500 mm
from floor

(d) Elevation (single leaf door)

Fig. 17.3 Entrance doors and lobbies.

Specific guidance is given regarding:

- Restaurants and bars.
- Hotel and motel bedrooms.
- Changing facilities (including dressing cubicles and shower compartments).
- Aids to communication for people with impaired hearing.

These recommendations are in addition to the requirements for sanitary conveniences and audience or spectator seating.

AD M
2.1, 2.2

Therefore it is necessary to provide sufficient space for wheelchair manoeuvre and convenient ways of travelling from one level to another. People with impaired hearing or sight should be catered for by providing features which enable them to find their way around the building safely and conveniently.

Corridors, passageways, internal doors and internal lobbies can present problems for disabled people unless care is taken in their design and adequate space is provided to enable a wheelchair to be manoeuvred and for other people to pass.

Corridors and passageways

AD M
2.5, 2.6

Corridors and passageways generally should have a clear width of at least 1200 mm. However, this figure may be reduced to 1000 mm where lift access is not provided to the corridor or it is situated in an extension approached through an existing building.

With internal doors the important factor is the minimum clear opening width.

AD M quotes an absolute minimum size of 750 mm. However, the use of 900 mm single leaf internal doorsets or 1800 mm double leaf internal doorsets complying with Table 1 of BS 4787 will give clear opening sizes of 770 mm for the single leaf doorset and 820 mm for one leaf of the double doorset.

The 300 mm space requirement adjacent to the leading edge of the door mentioned above (see p. 17.8) also applies to internal doors. In some circumstances, e.g. when leaving a fellow guest's hotel bedroom, it may be reasonable to assume that assistance will be on hand and therefore the 300 mm space may not be needed.

AD M
2.3, 2.4

Each door across an accessible corridor should be provided with a glazed vision panel giving a zone of visibility between 900 mm and 1500 mm from the floor.

It should be noted that BS 4787 does not permit a fire resisting single leaf doorset to exceed 900 mm in overall width.

Internal lobbies

AD M
2.7, 2.8

Internal lobbies may be smaller than principal entrance lobbies since fewer people are likely to use them at the same time. However there should still be room for a wheelchair user to move clear of one door before opening the next.

Internal lobbies should comply with the minimum dimensions shown in Fig. 17.4.

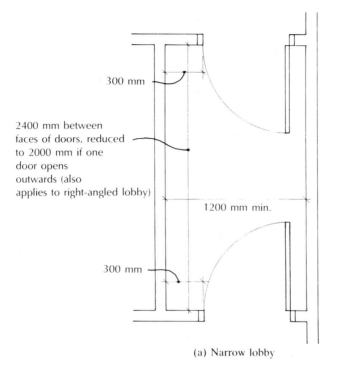

300 mm

2400 mm between faces of doors, reduced to 2000 mm if one door opens outwards (also applies to right-angled lobby)

1200 mm min.

300 mm

(a) Narrow lobby

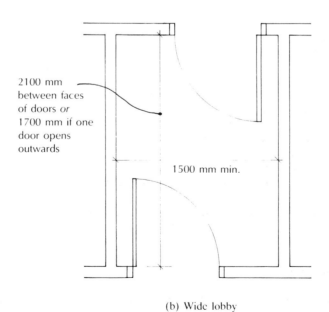

2100 mm between faces of doors *or* 1700 mm if one door opens outwards

1500 mm min.

(b) Wide lobby

Fig. 17.4 Internal lobbies.

Vertical means of access

In the majority of buildings the most suitable form of vertical access for a wheelchair user is a lift. Lifts are expensive and take up usable space, therefore, it may be unreasonable to expect a lift to be provided in all instances especially in smaller buildings. Accordingly, lifts should be provided to serve any floor above or below the principal entrance storey where the following nett floor areas are exceeded:

AD M
2.9 to 2.13

- Two-storey buildings – 280 m^2 of nett floor area.
- Buildings exceeding two storeys – 200 m^2 of nett floor area.

The figures are derived by adding together the areas of all the parts of a storey which use the same entrance from the street or an indoor mall, even if they are in different parts of the same storey or are used for different purposes. Thus the figures given are for each storey above or below the principal entrance storey. In calculating the figures given above it is permissible to exclude the area of the vertical circulation, sanitary accommodation or maintenance areas in the storey. It is, of course, essential to provide means of access from the lift to the rest of the storey.

Where passenger lifts are not provided a stair suitable for ambulent disabled people should be installed. This should also be suitable for people with impaired vision.

Lift design

AD M
2.14

Lifts for wheelchair users need to be large enough to allow access and egress, and should not be cramped internally.

Controls should be within reach and since disabled people need more time to enter and leave the lift car, suitable delay systems should be provided to lessen the risk of contact with the closing doors. For people with sensory impairments it may be necessary to provide visual and vocal floor indication.

A suitable lift design is shown in Fig. 17.5. Its main features are:

- An unobstructed, accessible landing space at least 1500 mm square in front of the lift doors.
- A door or doors with a clear opening width of 800 mm.
- A car with minimum dimensions of 1100 mm wide by 1400 mm deep.
- Landing and car controls between 900 mm and 1400 mm from landing or car floor levels and at least 400 mm from the front wall.
- Tactile floor level indication on each landing next to the lift call button and, where the lift serves more than three floors, on or adjacent to the lift buttons in the lift car to confirm the floor selected.
- Visual and vocal floor indication where the lift serves more than three floors.
- A signalling system which gives five seconds warning that the lift is about to stop at a floor and once stopped, a minimum of five seconds before the doors begin to close after being fully open.

It is necessary to ensure that the door controls can be overridden in order that lift passengers do not get caught in the closing doors. In the past it has

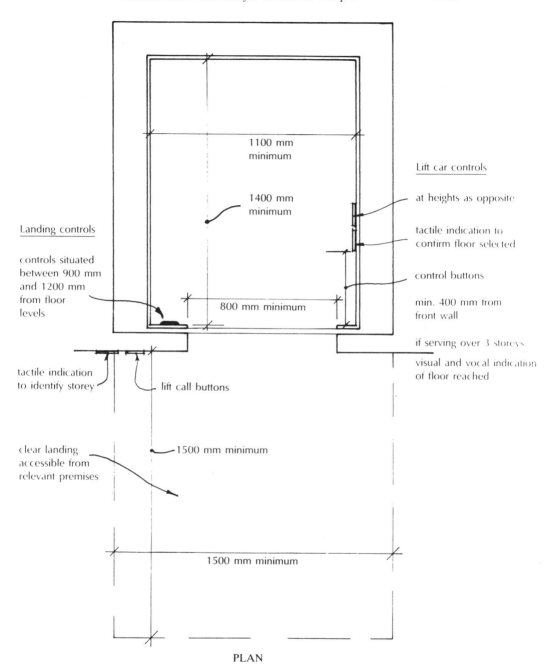

Lift car controls

at heights as opposite

tactile indication to confirm floor selected

control buttons

min. 400 mm from front wall

if serving over 3 storeys

visual and vocal indication of floor reached

Landing controls

controls situated between 900 mm and 1200 mm from floor levels

tactile indication to identify storey

lift call buttons

clear landing accessible from relevant premises

1100 mm minimum

1400 mm minimum

800 mm minimum

1500 mm minimum

1500 mm minimum

PLAN

Fig. 17.5 Lift (suitable for disabled people).

been common to incorporate a door edge pressure system which causes the doors to re-open when resistance is encountered. This type of system is not suitable for disabled people since it could cause them to become unbalanced. A door re-opening activator which uses a photo-eye or infra-red detector is satisfactory providing the door remains fully open for at least three seconds. BS 5655 *Lifts and service lifts* contains details of some suitable lifts systems in Parts 1, 2 and 5: 1989 and Part 7: 1983.

Stairlifts

**AD M
2.15, 2.16**

Some buildings contain small areas with unique facilities such as staff rest and training rooms or small galleried libraries which are often on upper or lower floors. It is reasonable to expect access for wheelchair users to such areas but it may not be practical to provide a full passenger lift.

Storeys which have a nett floor area exceeding $100\,m^2$ containing unique facilities, may be accessed by a wheelchair stairlift provided it complies with BS 5776: 1996 *Specification for powered stairlifts*.

Platform lifts

**AD M
2.17, 2.18**

The provision of ramps to effect level changes inside buildings can sometimes have serious planning implications since they tend to take up large areas of floor space. A platform lift which complies with BS 6440: 1983 *Code of practice for powered lifting platforms for use by disabled people* may be provided in lieu of a ramp but not at the expense of a suitable staircase.

Internal stairways

Where a lift is not provided in a building then a stairway suitable for ambulent disabled people or people with impaired sight should be provided.

With certain exceptions, the design for external stairways shown in Fig. 17.2 is also suitable for internal stairways.

Internal stairs are subject to more constraints in the form of ceiling heights and space restrictions. However, it is not considered reasonable to require the provision of tactile warnings at the start of level changes. Nevertheless, stair nosings should still be distinguishable for the benefit of people with impaired sight.

The principal variations permitted are:

**AD M
2.19 to 2.22**

● Uniform risers should not exceed 170 mm in height.
● Uniform goings should not be less than 250 mm in length (measured 270 mm in from the narrow edge of the flight for tapered steps).
● Tactile warnings of level changes are not necessary.
● The maximum rise of a flight between landings should not exceed 1800 mm.

This last figure is somewhat flexible since it depends to a large extent on site constraints such as landing levels, storey heights or the space required for the extra staircase length. If the recommendations contained in Approved Document K regarding numbers of risers are followed, then this would be acceptable in exceptional cases.

Internal ramps

Internal ramps should follow the recommendations which are specified for external ramps (see p. 17.4 and Fig. 17.2).

AD M
2.23, 2.24

Use of facilities within the building

Access to restaurant and bar facilities

Disabled people should be able to visit and use restaurants and bars whether accompanied or alone. The choice of waiter or self-service facilities should be available and there should be suitable access to seating areas. Changes of floor levels in seating areas are permissible if kept to a reasonable scale and should be accessible by ambulent disabled people.

Therefore, the full range of services offered should be accessible, including bars and self-service counters and at least half the area where seating is provided. Sometimes, the nature of the service provided in a restaurant varies. For example, some areas may be self-service and some may be waitress service. In these cases at least half the area of each should be accessible even if they are in different storeys.

AD M
3.4 to 3.6

Access to hotel and motel guest bedrooms

It is essential that a certain proportion of guest bedrooms in hotels or motels should be suitable for wheelchair users. This means that the bedroom should be accessible and should contain sufficient space to allow wheelchair manoeuvre within it and into en-suite bathroom facilities, if provided.

Since it is usual for guests to visit each other in their bedrooms when on holiday or attending residential conferences for example, it is reasonable to limit the disabled facilities to the provision of an accessible doorway. This assumes that an able-bodied guest will open and close the door.

The following main provisions apply:

AD M
3.7 to 3.9

- Each guest bedroom entrance door should follow the recommendations for internal doors given on p. 17.10.
- One guest bedroom out of every 20 should be accessible in terms of layout, dimensions and facilities for a wheelchair user, i.e. if 21 guest bedrooms were provided then two of them would have to be suitable for a wheelchair user (see Fig. 17.6 for details).
- All other guest bedrooms should have entrance doors with a minimum clear opening width of 750 mm. It is permissible to omit the 300 mm space at the side of the door.

Shower and changing facilities

Many disabled people enjoy swimming and other recreational activities but find it difficult to participate due to the lack of suitable changing and showering facilities. Adequate space is required to manoeuvre the wheelchair and to transfer onto a seat. Seats, shower heads, taps, clothes hooks and mirrors all need to be mounted at suitable heights. Figure 17.7 gives details of suitable arrangements for shower and changing cubicles.

AD M
3.10, 3.11

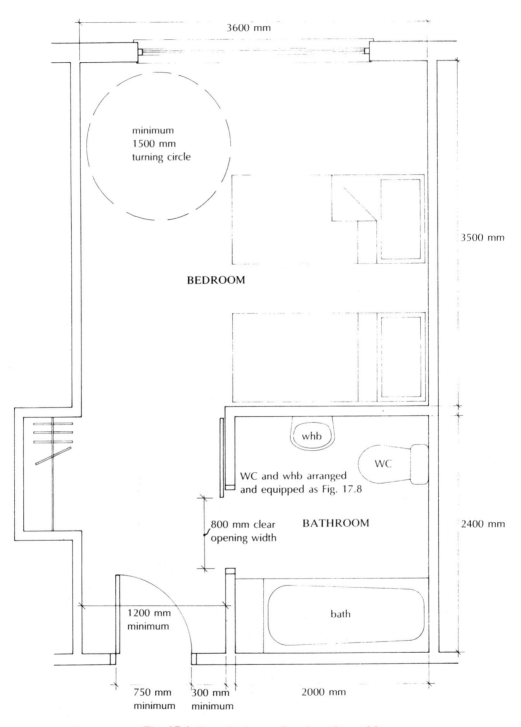

Fig. 17.6 Guest bedroom (hotels and motels).

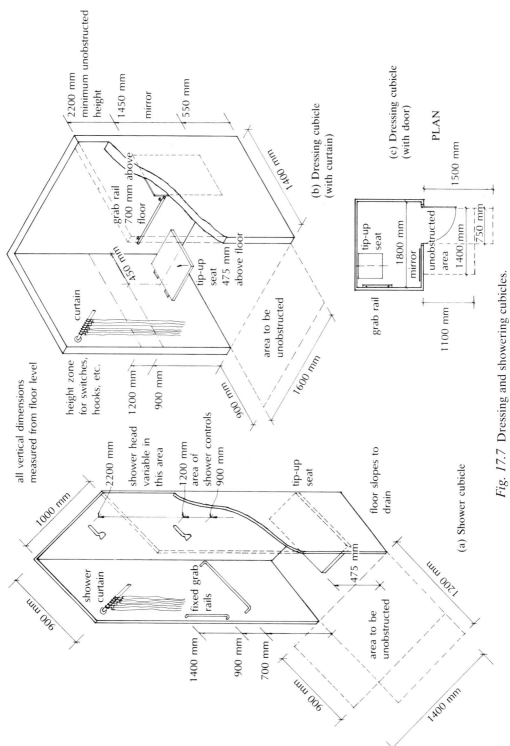

Fig. 17.7 Dressing and showering cubicles.

Communication aids

When people attend public performances or conferences it is reasonable to expect them to be able to enjoy the proceedings. People with hearing difficulties may need to receive an amplified signal some 20 dB above the level received by a person with 'normal' hearing. The amplification system should be able to suppress reverberation, and audience or other noises.

A particular problem is in glazed ticket or booking offices where it may be very difficult to hear the attendant.

AD M
3.12 to 3.17

There are two systems commonly in use at present – the loop induction system and the infra-red system. AD M gives a limited description of the principles of each and points out some of their disadvantages. No British Standard references are given and it is left to building owners to decide which system best suits the layout and use of their buildings. Obviously, there will be a need to consult specialist companies if it is necessary to install an amplification system.

AD M recommends that aids to communication should be provided at booking and ticket offices where a glazed screen separates customer and vendor.

Further, large reception areas, auditoria and meeting rooms where the floor area exceeds $100 \, \text{m}^2$ are also covered.

The systems installed should incorporate features which allow the person wearing the hearing aid to receive the amplified sound without loss or distortion due to bad acoustics or extraneous noise.

It is felt that this guidance is sufficiently vague to be of little use to building owners, designers and building control authorities and will only lead to variations in interpretation and consequent delays in passing of plans and approval of work on site.

Sanitary conveniences

Section 4 of AD M3 sets out the requirements for the provision of sanitary conveniences suitable for both wheelchair users and ambulant disabled people.

AD M
4.1

The general principle is that where sanitary conveniences would normally be provided for able-bodied people then disabled people should also be catered for with suitable accommodation.

Some of the recommendations under the heading Staff in AD M3 would appear to be of general application and have been interpreted as such in the following discussion. (These can be identified by the margin reference AD M3, 4.15 and 4.17.)

Design of sanitary conveniences for wheelchair users

The following general design principles should be considered:

AD M
4.2 to 4.8

● Unisex and/or integral conveniences may be provided under the terms of AD M3, unisex facilities being separate and self-contained whereas integral facilities are contained within the separate accommodation for men and women.

- The scale of provision will depend on the size and nature of the building and on the ease of access to the facility.
- A more flexible arrangement can be provided in large buildings by the inclusion of both unisex and integral toilet facilities.
- Travel distances should reflect the fact that some disabled people need to reach a WC quickly.
- It should not be necessary to travel more than one storey to reach a WC.
- Where more than one WC compartment is provided in a building then both left- and right-hand transfer layouts should be provided.

Individual compartments should be designed for ease of access and use and should provide:

- Adequate space for wheelchair manoeuvre and for the presence of a helper to assist with frontal, lateral, diagonal or backward transfer onto the WC.
- Hand washing and drying facilities which may be reached from the WC before transfer back to the wheelchair.
- Both unisex and integral WC compartments should be similar in layout and content.

The AD appears to show a marked preference for unisex WC accommodation and lists the following advantages over integral facilities:

- The approach to it is separated from other sanitary accommodation.
- It is more easily identified.
- There is more chance of it being available when needed.
- Assistance is permitted by helpers of either sex. (This would not be possible with integral facilities since existing custom prevents access by a helper whose sex is different from those for whom the provision is made.)
- Less space is needed overall, since duplication of facilities is necessary with integral accommodation for the same level of provision.

Provision of sanitary accommodation for wheelchair users

It is necessary to provide sanitary conveniences for all of a building's users whether they be visitors, customers or staff.

Disabled visitors and customers are more likely to be accompanied by a member of the opposite sex, therefore unisex facilities should be provided for them. **AD M 4.9**

On the other hand, staff may be less likely to need assistance. If help is needed then it is more likely that a member of the same sex will provide it.

Therefore, for staff, the facilities may be either unisex or integral. There may be separate provision for both sexes on alternative floors if the building is provided with lifts and the sanitary conveniences are in areas with unrestricted access. In this case it should not be necessary to travel more than a cumulative horizontal distance of 40 m from a work station to the WC. **AD M 4.13 to 4.17**

Visitors to hotels and motels will often find that guest bedrooms have en-suite sanitary accommodation. If this is the case, then this should also be the arrangement for bedrooms suitable for disabled people. Where en-suite facilities are not provided in the general sanitary arrangement then unisex facilities should be provided within easy reach. There is still, of course, the **AD M 4.10 to 4.12** necessity to provide additional sanitary accommodation for staff and for daytime or non-resident visitors.

Figure 17.8 illustrates a typical layout for a WC compartment suitable for wheelchair users in either unisex or integral facilities. It is suitable in any of the situations referred to above.

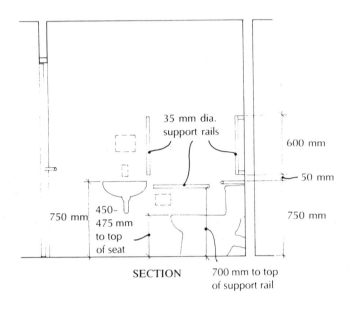

SECTION

PLAN

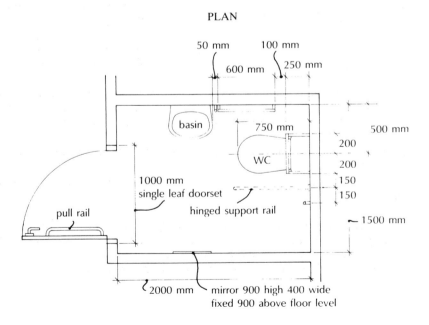

Fig. 17.8 WC for wheelchair users.

Provision of sanitary accommodation for ambulant disabled people

It is permissible for certain small buildings to have stair access only (see p. 17.12).

In these cases suitable sanitary conveniences for wheelchair users should be provided in the principal entrance storey unless this contains only vertical circulation areas or the principal entrance. AD M 4.18 to 4.20

Any sanitary accommodation containing WC compartments situated on floors without lift access should contain at least one WC compartment designed for use by ambulant disabled people (i.e. people with a limited ability to walk or to support themselves). This should be in addition to any sanitary conveniences provided for wheelchair users in the principal entrance storey.

A WC compartment suitable for ambulant disabled people is illustrated in Fig. 17.9.

Audience or spectator seating

Where fixed audience or spectator seating is provided in theatres, cinemas, concert halls, sports stadia and similar buildings, then reasonable provision must be made to accommodate disabled people.

A sufficient number of wheelchair spaces should be provided which are accessible and provide a clear view of the event. The spaces may be kept clear at all times or may contain seating which can be removed easily for each occasion. AD M 5.1 to 5.6

Wheelchair spaces should be:

- At least 900 mm wide.
- At least 1400 mm deep.
- Dispersed throughout the theatre or stadium so that disabled people have a choice of sitting next to able-bodied or disabled companions.

There should be provided at least six spaces or one for every 100 fixed audience or spectator seats available to the public, whichever figure is the greater.

In a large stadium AD M4 recommends that a smaller proportion of wheelchair spaces could be provided. Unfortunately, no further guidance is given as to what constitutes a large stadium or a smaller proportion of wheelchair spaces.

Figure 17.10 shows typical wheelchair space layouts for both theatres and sports stadia.

Additionally, guidance on access for disabled people to sports stadia may be obtained from:

Guide to Safety at Sports Ground, published by the Stationery Office, 1997.
Designing for Spectators with Disabilities, published by the Football Stadia Advisory Council, 1993 (Available from the Sports Council).
Access for disabled people, English Sports Council Guidance Notes.

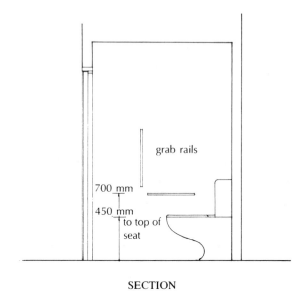

SECTION

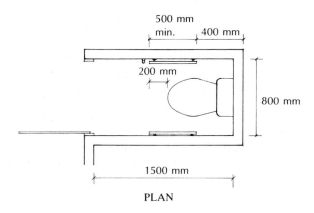

PLAN

Fig. 17.9 WC for ambulent disabled people.

Other legislation and technical guidance

Access and facilities in schools and other educational buildings

AD M 0.9 Where work is to be carried out in schools or other educational establishments, the DfEE publication, Design Note 18, 1984 *Access for Disabled Persons to Educational Buildings* will satisfy the requirements of paragraphs M2, M3 and M4. The relevant paragraphs are: 2.1/2/4/6; 3.1; 4.1/2/4/6 and 5.1. However, it may also be necessary to incorporate some of the general design

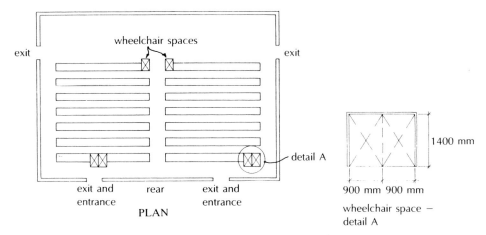

A   Theatre — notional disposition of wheelchair spaces

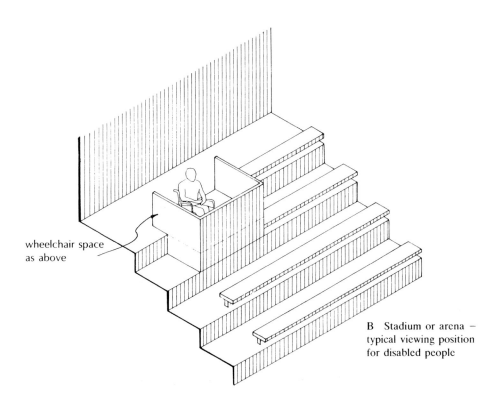

B   Stadium or arena — typical viewing position for disabled people

Fig. 17.10 Audience or spectator seating.

considerations in Design Note 18 into educational buildings, since regulation M2 also includes provisions for people with impaired hearing or sight.

Means of escape in case of fire

It is clear that Part M is limited to the provision of access into and around a building, and the use of that building. Therefore it does not extend to provisions for means of escape for disabled people if a fire occurs. For this, reference should be made to Approved Document B, Fire safety, and BS 5588, Part 8: *Code of practice for means of escape for disabled people* (see also p. 7.63).

Disability Discrimination (Employment) Regulations 1996

Under Regulation 8 of the Disability Discrimination (Employment) Regulations, an employer who installs facilities for disabled people which comply with the requirements of Part M (and continue substantially to meet those requirements in use), cannot be required to alter such facilities.

Workplace (Health, Safety and Welfare) Regulations 1992

It should be noted that compliance with Regulation M2, in conjunction with Part K, prevents action being taken against the occupier of a building under Regulation 17 of the Workplace (Health, Safety and Welfare) Regulations 1992 when the building is eventually in use. (Regulation 17 relates specifically to permanent stairs, ladders and ramps on routes used by people in places of work).

Access and facilities in dwellings

Important Note: These provisions come into force on 25 October 1999 and do not apply to dwellings which are already under construction on that date, provided that the construction work began in accordance with the provisions of the Regulations concerning notification and approval of plans, etc. However, the provisions effectively came into force on 1 June 1999, since any plans of new dwellings which were passed by a local authority (or any plans certificates which were given to a local authority by an approved inspector) on or after that date will be judged against the new requirements.

Interpretation

The following definitions of terms used in Approved Document M apply only to the provisions concerning dwellings:

**AD M
0.8 ii)**

DWELLINGS – in addition to dwelling-houses, flats and maisonettes, the term includes purpose-built student living accommodation (but not traditional halls of residence which only provide bedrooms and are not equipped as self-contained accommodation).

COMMON – serving more than a single dwelling.

AD M
0.11

HABITABLE – when used to define the principal storey in a dwelling, means a room intended to be used for dwelling purposes. This includes a kitchen but not a utility room or bathroom.

MAISONETTE – a self-contained dwelling which occupies more than one storey in a building, but which is not a dwelling-house.

POINT OF ACCESS – the place where a person would alight from a vehicle when visiting a dwelling, before they approached the dwelling. This could be inside or outside the plot boundary.

PRINCIPAL ENTRANCE – the entrance which would normally be used by a visitor who was not familiar with the dwelling. In a block of flats it would be the common entrance.

PLOT GRADIENT – the slope measured between the point of access and the finished floor level of the dwelling.

STEEPLY SLOPING PLOT – a plot gradient which exceeds 1 in 15.

The main provisions

Reasonable provision must be made:

- For disabled people to gain access to and use the facilities in the dwelling.
- For sanitary conveniences suitable for disabled people in the entrance storey of the dwelling.

Regs Sch. 1
M2

Regs Sch. 1
M3

The 'entrance storey' is the storey which contains the principal entrance to the dwelling (see above for definition) and in the design of some dwellings the entrance storey may not contain any habitable rooms. If this is the case, then the option is given of placing the sanitary conveniences in either the entrance storey or a principal storey. Regulation M3 defines 'principal storey' as '*the storey nearest to the entrance storey which contains a habitable room, or where there are two such storeys equally near, either such storey*'.

Although this legal terminology may seem confusing, the intention of the regulation is simply to require sanitary conveniences suitable for disabled people to be placed in the entrance storey of the dwelling, *if this contains any habitable rooms*. Otherwise the sanitary conveniences may be placed in either the entrance storey or the nearest storey to the principal entrance which contains habitable rooms.

Therefore, in dwellings, it should be reasonably safe and convenient for disabled people to:

- Approach the principal (or other suitable alternative) entrance to the dwelling from the edge of the site or from car parking within the site.
- Gain access into the building.
- Gain access within the building.
- Use sanitary accommodation located in the nearest storey containing habitable rooms to the principal entrance.

AD M
Performance
and 0.8ii)

It should be noted that the provision of access and facilities for disabled people in dwellings is to enable them to visit new dwellings and use the principal storey. The intention is that disabled occupants will be able to cope better with reducing mobility and will be able to remain living in their own homes longer than would otherwise be the case. On the other hand, the provisions are not intended to facilitate fully independent living for all disabled people.

Means of access

AD M
6.1 to 6.10

In general terms, a disabled person should be able to gain access into a dwelling from the point of leaving a vehicle (which is parked either within or outside the plot boundary).

In most cases it should be possible to provide a safe and convenient level or ramped approach, thereby permitting wheelchair users to gain access to the dwelling. Clearly, there will be situations on steeply sloping plots (i.e. where the plot gradient exceeds 1 in 15), where it will only be practicable to provide a stepped approach. Where this is the case the approach should be suitable for an ambulant disabled person using a walking aid.

The approach to the dwelling

The choice of a suitable approach to the dwelling from the point of access to the plot, will be influenced by the topography and available area of the plot, and by the distance to the dwelling from the point of access. Account may also need to be taken of local planning requirements, especially for new developments in conservation areas. Developers are advised to discuss the access requirements of Part M with their local planning authority, in conjunction with their building control supervisor (local authority or approved inspector) at an early stage in the design process, to avoid later conflicts.

It may be possible to reduce the effect of a steeply sloping plot by means of a suitable driveway. This could allow for the parking space within the plot boundary to be at a sufficiently high level to permit a level or ramped approach from the parking space to the dwelling.

The surface material of the approach to the dwelling should be firm enough to support a wheelchair and user, and smooth enough to allow satisfactory manoeuvre. Ambulant disabled people using walking aids also need to be considered. Therefore, loose surfacing materials such as gravel or shingle are unlikely to be satisfactory and the approach should be sufficiently wide (in addition to the width of the parking space) to allow safe and convenient passage.

In practical design terms the provisions can be summarised in the following paragraphs.

Provide a suitable approach:

AD M
6.11, 6.12

- From a reasonably level point of access (i.e. from the vehicle parking position) to the dwelling entrance (i.e. the principal entrance or a suitable alternative entrance if it is not possible to access the principal entrance).
- With crossfalls which do not exceed 1 in 40.
- Which may consist in whole or in part of a vehicle driveway.

A level approach will have:

- A gradient not exceeding 1 in 20.
- A firm and even surface.
- A minimum width of 900 mm.

AD M
6.13

A ramped approach will be needed where the overall plot gradient exceeds 1 in 20 but does not exceed 1 in 15. In this case the ramped approach should have:

AD M
6.14, 6.15

- A firm and even surface.
- Minimum unobstructed flight widths of 900 mm.
- Individual flights no longer than 10 m in length with a maximum gradient of 1 in 15 (although gradients not exceeding 1 in 12 are allowed where the individual flight does not exceed 5 m in length).
- Top, bottom and if necessary, intermediate landings at least 1200 mm long, clear of any door or gate swinging across it.

A stepped approach will be needed where the overall plot gradient exceeds 1 in 15.
In this case the stepped approach should have:

AD M
6.16 to 6.18

- Minimum unobstructed flight widths of 900 mm.
- A maximum rise of 1800 mm between landings.
- Top, bottom and if necessary, intermediate landings at least 900 mm long.
- A suitable tread profile as illustrated in Fig. 17.2.
- Uniform risers between 75 and 150 mm high and goings at least 280 mm long (measured 270 mm in from the narrow edge of the tread for tapered steps).
- Where the flight consists of three or more risers, a suitable handrail should be provided on one side. The handrail should have a profile which can be gripped, be positioned between 850 and 1000 mm above the pitch line of the flight and project at least 300 mm beyond the top and bottom nosings (see Fig. 17.11).

Access into the dwelling

In general, the entrance into a dwelling or a block of flats from outside should be provided with an accessible threshold irrespective of whether the approach to the entrance is level, ramped or stepped. Exceptionally, if the approach is stepped and, for practical reasons a step into the dwelling is unavoidable, it should not exceed 150 mm in height.

AD M
6.19 to 6.23

There has been considerable difficulty in deciding on design details for accessible thresholds, (since there is a need to comply with Part C, see Chapter 8), and it was one of the factors which delayed the implementation of the new approved document guidance. Ultimately, AD M was published without the guidance but with a comment that design considerations for accessible thresholds would be published separately. At the time of writing, the guidance was still unavailable.

The entrance door to an individual dwelling and/or a block of flats should be wide enough to accommodate a person using a wheelchair. This require-

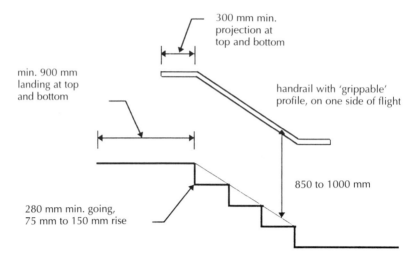

Fig. 17.11 Stepped approach to dwelling.

ment can be satisfied if such a door has a minimum clear opening width of 775 mm.

Access within the dwelling

AD M
7.1 to 7.5

The requirements of regulations M2 and M3 mean that access must be facilitated to habitable rooms and to a WC (which may be in a bathroom) in the entrance storey or the principal storey of the dwelling, as appropriate.

Where it is not possible to make the principal entrance accessible and an alternative is provided instead, the route to the remainder of the entrance storey from the alternative entrance must be carefully considered, especially if the route passes through other rooms. Therefore, corridors and passageways should be wide enough to allow convenient access for a person in a wheelchair, whilst at the same time allowing for manoeuvre past local obstructions such as radiators and other fixtures.

Doors to rooms need to be wide enough to cater for both head-on approach and right angled approach from a person in a wheelchair. The rules for access within the entrance or principal storey of a dwelling are illustrated in Fig. 17.12 below.

Vertical circulation

AD M
7.6 and 7.7

Steps within the entrance storey of a dwelling should be avoided wherever possible. Sometimes, such as in the case of severely sloping plots, it may not be possible to avoid putting a change of level involving steps, in the entrance storey. In these circumstances a stair should be provided which is wide enough to be negotiated by an ambulant disabled person with assistance and with

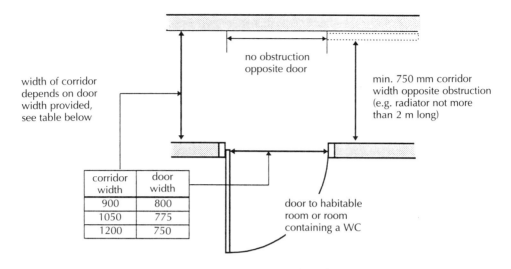

Right-angled approach to door

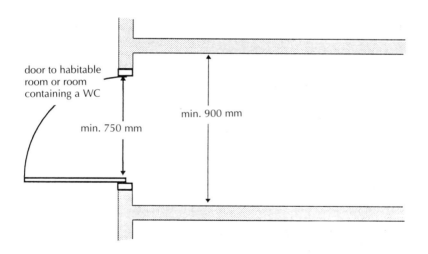

Head-on approach to door

Fig. 17.12 Corridor and door widths in dwellings.

handrails on both sides. Therefore, any stair provided in the entrance storey which gives access to habitable rooms should have:

- A minimum clear width of 900 mm;
- A continuous handrail on each side, and on any intermediate landings, where the flight consists of three or more risers; *and*
- Rise and going in accordance with the guidance for private stairs in Approved Document K (see Chapter 15).

Access to socket outlets and switches in dwellings

AD M
8.1 to 8.3

Ambulant disabled people and people who use wheelchairs are less mobile and likely to have more limited reach than able-bodied people. Therefore switches and socket outlets for such things as electrical appliances, lighting, television aerials, telephone jack points, etc., should be mounted at suitable heights so that they can be easily reached. Essentially, this means locating sockets and switches in habitable rooms between 450 mm and 1200 mm from finished floor level.

Common access stairs in blocks of flats

AD M
9.1 to 9.5

It should be possible for a disabled person to visit an occupant, on any storey in a building containing flats. The most suitable means of vertical access for a disabled person is a lift; however, AD M recognises that lifts are not always provided (and does not, at present, recommend that lifts should be installed). Therefore, where there are no passenger lifts, the common access stairs should be suitable for use by ambulant disabled people, (as well as being suitable for people with impaired sight) and be designed to have:

- Step nosings which are clearly visible by the use of contrasting brightness;
- Top and bottom landings which follow the guidance contained in Part K1 of Approved Document K (see Chapter 15);
- Uniform risers not exceeding 170 mm in height;
- Uniform goings not less than 250 mm in length (measured 270 mm in from the narrow edge of the flight for tapered steps);
- A suitable handrail on each side of flights and landings if comprising more than one riser;
- A suitable tread nosing profile (e.g. as illustrated in Fig. 17.2 but with 170 mm maximum rise and 250 mm minimum going); *and*
- Risers which are not open.

Some of this guidance is illustrated in Fig. 17.13.

Passenger lifts in blocks of flats

AD M
9.4, 9.6 and
9.7

If passenger lift access is to be provided to flats above the entrance storey in a building, it should be suitable for both unaccompanied wheelchair users and people with sensory impairments. It should also contain suitable delay sytems to enable disabled people more time to enter and leave the car and lessen the risk of contact with the closing doors.

A suitable passenger lift should have:

- A minimum load capacity of 400 kg.
- An unobstructed, accessible landing space at least 1500 mm square in front of the lift doors.
- A door or doors with a clear opening width of 800 mm.
- A car with minimum dimensions of 900 mm wide by 1250 mm deep (although other dimensions may be suitable if it can be demonstrated by test evidence or experience in use, etc. that they are suitable for an unaccompanied wheelchair user).

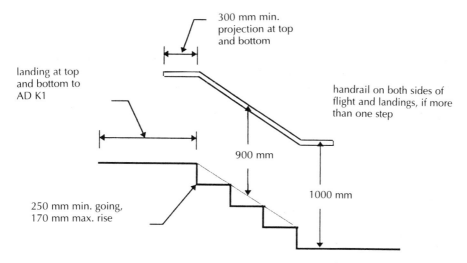

Fig. 17.13 Common access stairs in blocks of flats.

- Landing and car controls between 900 mm and 1400 mm from landing or car floor levels and at least 400 mm from the front wall.
- Tactile floor level indication on each landing next to the lift call button to identify the storey in question, and on or adjacent to the lift buttons in the lift car to confirm the floor selected.
- Visual and audible floor indication where the lift serves more than three floors.
- A signalling system which gives a visual warning that the lift is about to stop at a floor and once stopped, a minimum of 5 seconds before the doors begin to close after being fully open.

It is necessary to ensure that the door controls can be overriden in order that lift passengers do not get caught in the closing doors. This should be a suitable electronic system (such as a photo-eye or infrared detector), but not a door edge pressure system which might cause a disabled person to lose their balance. The door should remain fully open for at least 3 seconds.

Provision of accessible sanitary conveniences in dwellings

For dwellings which contain more than one storey, sanitary conveniences suitable for disabled people should be provided in the entrance storey, *if this contains any habitable rooms.* Otherwise the sanitary conveniences may be placed in either the entrance storey or the nearest storey to the principal entrance which contains habitable rooms. There should be no need to negotiate a stairway to reach the WC from the habitable rooms in that storey. (Obviously, for single storey dwellings and individual flats the sanitary conveniences can only be provided in the entrance storey).

Additionally, the following provisions apply to the design and location of the sanitary accommodation:

AD M
10.1 to 10.3

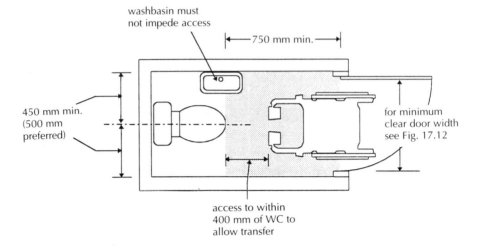

Typical WC compartment with frontal access

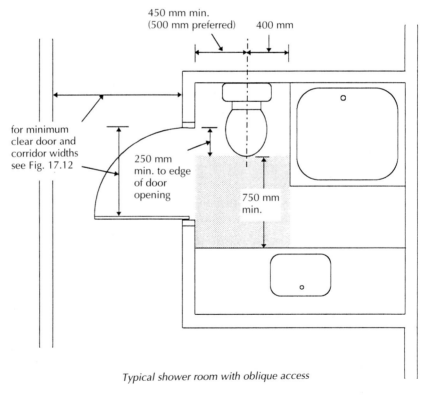

Typical shower room with oblique access

Fig. 17.14 Accessible sanitary conveniences in dwellings.

- The WC may be located in a bathroom if there is one available in the relevant storey of the dwelling.
- It is accepted in AD M that it may not always be practical for a wheelchair to be fully accommodated inside the WC compartment.
- The WC door should open outwards and should be positioned so as to allow access to the WC for a person in a wheelchair.
- The WC door should have a minimum width as shown in Fig 17.12 (and be wider if possible, so as to allow easier access and manoeuvring by wheelchair users).
- There should be sufficient space in the WC compartment to allow wheelchair users to access the WC.
- The position of the washbasin should not impede access to the WC.

Typical, suitable WC compartments are illustrated in Fig. 17.14.

Chapter 18

Glazing

Introduction

This chapter deals with safety issues associated with glazing in terms of protecting people against the risks of impact and making sure that glazed elements may be opened, closed and cleaned safely.

The 1998 edition of Approved Document N introduces new provisions governing the safe use and cleaning of glazed elements mainly in order to ensure compliance with the Workplace (Health, Safety and Welfare) Regulations 1992. Therefore, compliance with the revised regulations in Part N prevents action being taken against the occupier under the Workplace Regulations when the building is eventually in use. This involves considering aspects of design which will affect the way a building is used and applies only to workplaces.

Although dwellings are excluded from the changes, in mixed use developments the requirements for the non-domestic part of the use would apply to any shared parts of the building (such as common access staircases and corridors). With flats the situation may be less clear for although certain sections of Part N do not apply to dwellings, it is still necessary for people such as cleaners, wardens and caretakers to work in the common parts. Therefore, the requirements of the Workplace Regulations may still apply even though the Building Regulations do not.

Application

Part N of Schedule 1 to the 1991 Regulations applies to all new glazing used in the erection, extension or material alteration of a building.

It does not apply to exempt buildings (except small conservatories or porches in Class VII of Schedule 2, see page 2.11), or to replacement glazing although this may be the subject of consumer protection legislation in the future.

Requirement N1 applies to all building types whereas requirements N2, N3 and N4 apply to all building types *except* dwellings (i.e. dwelling-houses and flats).

Protection against impact

People are likely to come into contact with glazing in certain critical locations, when they are moving in or about a building. Accordingly, the glazing must:

Regs Sch. 1
N1

(a) if broken on impact, break so that the dangers of injury are minimised (i.e. break safely); *or,*
(b) resist impact without breaking; *or,*
(c) be protected or shielded from impact.

Critical locations under the terms of paragraph N1 are shown in Fig. 18.1. Two main areas may be identified where an accident may result in cutting or piercing injuries:

- Doors and door side panels between finished floor level and 1500 mm above.
- Internal and external walls and partitions between finished floor level and 800 mm above.

AD N1
0.2, 0.3, 1.1

In doors and door side panels the main risk is in the area of door handles and push plates especially since doors are prone to stick. Also it is possible that an initial impact above waist level may result in a fall through the glass. In low level glazing away from doors the main risks are to children.

Possible solutions to N1

AD N1
1.2

Approved Document N lists a number of solutions which can be adopted in order to minimise the risks of injury in these critical areas as follows:

(a) If breakage occurs the glazing should break safely.

The concept of safe breakage is taken from BS 6206: 1981 *Specification for impact performance requirements for flat safety glass and safety plastics for use in buildings:* clause 5.3.
    A test is carried out using a leather bag filled with lead shot. This is swung pendulum-fashion to impact a sheet of safety glazing material and the results are noted.
    The test material is required to remain unbroken or it may break safely as defined in one of the following ways:

(i) Cracks and fissures are allowed to develop provided it is not possible to pass a sphere of 76 mm diameter through any openings and any detached particles are of limited size; *or,*
(ii) Disintegration is allowed to occur provided the particles are of small size; *or,*

AD N1
1.3

(iii) Breakage is allowed to occur provided the pieces are not sharp or pointed.

In essence a glazing material in a critical location will be satisfactory if it can be classified under the requirements of BS 6206 as Class C (i.e. it remains unbroken or breaks safely when impacted from a height of 305 mm).

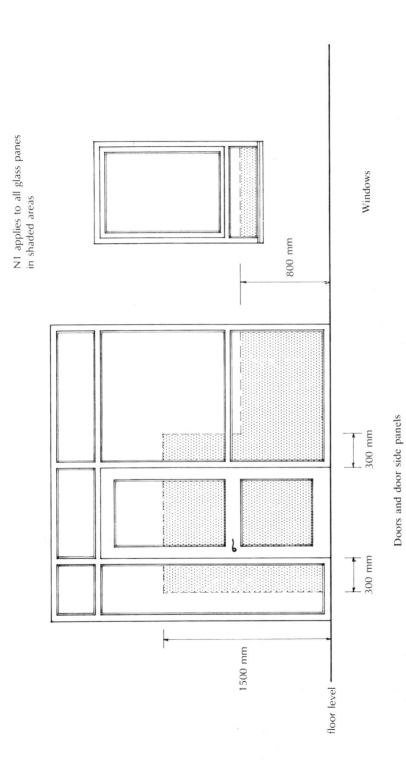

Fig. 18.1 Internal and external walls – critical locations.

Additionally, if it is installed in a door or door side panel and has a pane width which exceeds 900 mm, then it should meet the requirements of Class B of BS 6206 (i.e. it should remain unbroken or break safely when impacted from heights of 305 mm and 457 mm).

(b) The glazing should be robust.

AD N1
1.4
Robustness refers to the strength of the glazing material. Some materials such as glass blocks or polycarbonates are inherently strong. Annealed glass gains its strength through increased thickness and AD N1 describes the use of this material for large glazed areas forming fronts in shops, showrooms, factories, offices or public buildings. The dimensions of these glazed areas and their related glass thicknesses are shown in Table 18.1.

Table 18.1 Annealed glass – thickness/dimension limits.

| Height (mm) | | Length (mm) | | Thickness (mm) |
From	To	From	To	
0	1100	0	1100	8
1100	2250	1100	2250	10
2250	3000	2250	4500	12
3000	any	4500	any	15

Note
Annealed glass sizes and thicknesses for use in large areas to shopfronts, showrooms, offices, factories and public buildings.

(c) The glazing should be in small panes.

This relates to the use of a single pane or one of a number of panes within glazing bars, in either case having a smaller dimension not exceeding 250 mm and an area not greater than 0.5 m². Annealed glass in small panes should not be less than 6 mm in thickness although it is possible to install traditional copper or leaded lights using 4mm glass provided that fire resistance is not a factor. (See Fig. 18.2.)

(d) The glazing should be permanently protected.

Permanent protection means that the glazing should be installed behind a permanent screen which:

• prevents a sphere of 75 mm diameter touching the glazing,
• is itself robust; *and*,
• is difficult to climb in cases where the glazing forms part of protection from falling.

AD N1
1.7, 1.8
Where permanent screen protection is provided then the glazing itself does not need to comply with requirement N1. (See Fig. 18.2.)

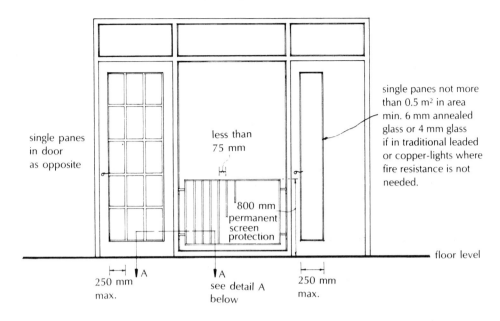

single panes in door as opposite

less than 75 mm

single panes not more than 0.5 m² in area min. 6 mm annealed glass or 4 mm glass if in traditional leaded or copper-lights where fire resistance is not needed.

800 mm permanent screen protection

floor level

250 mm max.

A
see detail A
below

250 mm max.

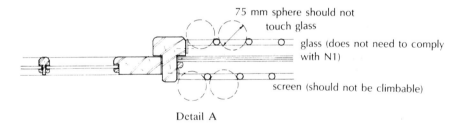

75 mm sphere should not touch glass

glass (does not need to comply with N1)

screen (should not be climbable)

Detail A

Fig. 18.2 Small panes and permanent screen protection.

Manifestation of glazing

If there is a risk that people may come into contact with large, uninterrupted areas of transparent glazing while moving in or about a building, then paragraph N2 of Schedule 1 requires that such areas must incorporate features which make the glazing apparent. As mentioned above, this requirement does not apply to dwellings.

Regs Sch. 1 N2

The risk of collision and consequent injury is most serious where parts of a building or its surroundings are separated by transparent glazing and the impression is given that direct access is possible through the area without interruption. In these critical locations (i.e. internal or external walls of shops, showrooms, offices, factories, public or other non-domestic buildings) it is necessary to adopt some means of making the glazing more apparent.

This is termed 'manifestation' of the glazing in AD N2 and it may take the form of patterns, company logos, broken or solid lines, etc. marked on the glazing at appropriate heights and intervals. This is illustrated in Fig. 18.3.

AD N2 0.6 to 0.8 2.1 to 2.4

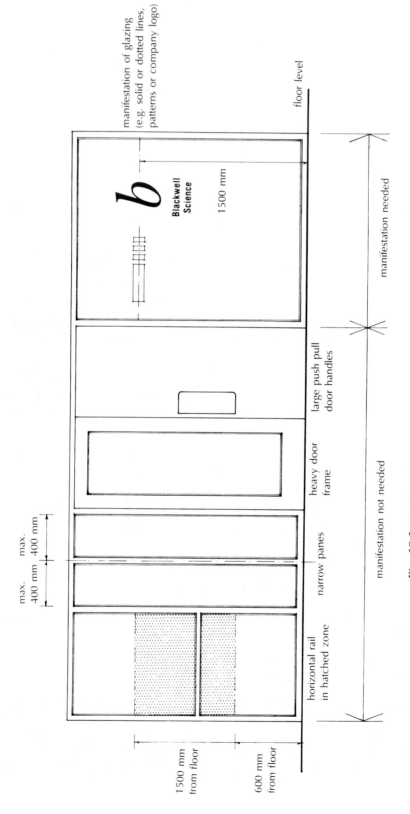

Fig. 18.3 Manifestation of large glazed areas (internal or external).

It is, of course, possible to indicate the presence of glazing by other means. Such features as mullions, transoms, door frames or large push or pull handles can be effectively used and AD N2 acknowledges that where these features are incorporated in the design then permanent manifestation will not be necessary.

Some examples of these features are shown in Fig. 18.3.

AD N2
2.5, 2.6

Safe use of windows, skylights and ventilators

Windows, skylights and ventilators must be constructed or equipped so that they can be opened, closed or adjusted safely, if they are so positioned as to be operable by people in or about the building. This requirement does not apply to dwellings.

Regs Sch. 1
N3

Compliance with Requirement N3 prevents action being taken against the occupier of a building under Regulation 15(1) of the Workplace (Health, Safety and Welfare) Regulations 1992 when the building is eventually in use. (Regulation 15(1) relates to requirements for opening, closing or adjusting windows, skylights and ventilators).

In order to meet the performance standard for safe operation of windows, skylights and ventilators **controls** should typically be located as follows:

 (i) Not more than 1.9 m above the floor or other stable surface where there is unobstructed access (ignoring small recesses, such as window reveals).
(ii) At a lower level where there is an obstruction (e.g. 1.7 m from floor level if there is an obstruction 600 mm deep (including any recess) and not more than 900 mm high).

If controls cannot be positioned at a safe distance from a permanent stable surface, then it may be necessary to install either manual or electrical remote controls.

Above ground level there may be a danger of the operator or other person falling through a window. Where this is the case suitable opening limiters should be fitted or the window should be guarded as described in Approved Document K (see Chapter 15).

AD N3
3.1 to 3.3

Safe access for cleaning glazed surfaces

Provision must be made for glazed surfaces to be safely accessible for cleaning. This includes:

● Windows and skylights; *and*
● Translucent walls, ceilings or roofs.

Regs Sch. 1
N4

Regulation N4 does not apply to dwellings, or to any of the above transparent or translucent elements if their surfaces are not intended to be cleaned. In this case, compliance with Requirement N4 prevents action being taken against the occupier of a building under Regulation 16 of the Workplace (Health, Safety and Welfare) Regulations 1992 when the building is eventually in use. (Regulation 16 relates to requirements for cleaning glazed surfaces in buildings.)

In the context of Regulation N4, it will be necessary to make provision for safe means of access for cleaning *both* sides of any glazed surfaces which are positioned so that there is a danger of falling more than 2 m. Furthermore, glazed surfaces which cannot be cleaned safely by a person standing on the ground, a floor or other permanent stable surface will need to be catered for in other ways.

The following arrangements (illustrated in Figs 18.4 and 18.5) are typical examples of how it may be possible to satisfy Regulation N4:

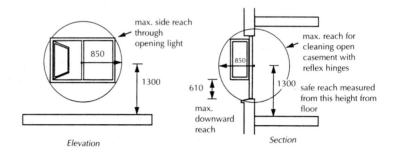

Fig. 18.4 Typical safe reaches for cleaning windows.

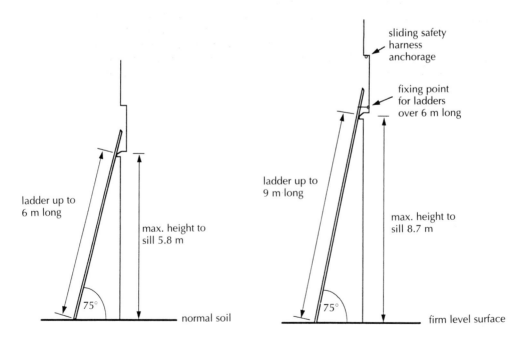

Fig. 18.5 Ladder access for cleaning windows. (***Note:*** Since designers will need to relate the length of access ladders to window sill heights, the maximum sill heights from ground level are shown for both 6 m and 9 m ladders.)

- Install windows of a suitable design and size so that they can be cleaned from inside the building. Reversible type windows should be capable of being fixed in the reversed position for cleaning purposes (see Fig. 18.4). For additional information on windows see BS 8213: Part 1 *Windows, doors and rooflights,* and Approved Document K (see Chapter 15) for guidance on minimum sill heights.
- Use portable ladders up to 9 m long if there is an adequate area of firm, level ground situated in a safe place for siting the ladders. Ladders up to 6 m long may be sited on normal soil. (For ladders over 6 m long provide permanent tying or fixing points.) (See Fig. 18.5).
- Provide catwalks at least 400 mm wide with either 1100 mm high guarding or provision for anchorages for sliding safety harnesses.
- Provide access equipment, e.g. suspended cradles or travelling ladders with safety harness attachments.
- Provide adequate anchorage points for safety harnesses or abseiling hooks.
- In exceptional circumstances where other means of access cannot be used, provide suitably located space for scaffold towers.

**AD N4
4.1 and 4.2**

Further references to glazing

Attention is drawn to the following Approved Documents where further information regarding glazing may be found:

- Approved Document B: *Fire safety* – guidance on fire resisting glazing and the reaction of glass to fire (see Chapter 7).
- Approved Document K: *Protection from falling, collision and impact* – guidance on glazing which forms part of protection from falling from one level to another, and which needs to provide containment as well as limiting the risk of injuries through contact. Recommendations are given concerning the heights of guarding and the means for achieving containment.

III
Appendices

Appendix 1

Local Acts of Parliament

The following list of Local Acts of Parliament is reproduced by kind permission of Tarmac Professional Services – Special Services, which is a corporate approved inspector under the Building Regulations. It has been researched and compiled by TPS Special Services and any enquiries about it or about the approved inspector system of building control in general, may be made by contacting Mr K.W.G. Blount, Honorary Secretary, the Association of Corporate Approved Inspectors, c/o TPS Special Services, Waterlinks House, Richard Street, Birmingham, B7 4AA (*Tel*: 0121 333 2811).

The sections marked with an asterisk (*) in column 3 of the list are applicable where there are matters to be satisfied by a developer either as part of a submission under Building Regulations or which otherwise need to be addressed in parallel with that submission.

County Acts usually, but not inevitably, apply across a whole county. References to district authorities in the following list however merely illustrate those which may have been consulted, and do not necessarily constitute a complete list of districts within that county. However, where it is known that a particular section applies only in a particular district this is indicated.

The DETR is developing proposals to repeal redundant provisions of Local Acts but this will take some time to achieve. This list was last updated on 18 September 1996.

(1) *Local Act*	(2) *Relevant sections*	(3) *Apply?*
County of Avon Act 1982 • Bath City Council • Bristol City Council	s.7 Parking places, safety requirements s.35 Hot springs, excavations in certain areas of Bath	*
Berkshire Act 1986 • Newbury District Council	s.28 Safety of stands s.32 Access for fire brigade s.36 Parking places, safety requirements s.37 Fire precautions in large storage buildings s.38 Fire precautions in high buildings	 * * * *
• Reading Borough Council		

(1) *Local Act*	(2) *Relevant sections*	(3) *Apply?*
Bournemouth Borough Council Act 1985	s.15 Access for fire brigade	*
	s.16 Parking places, safety requirements	*
	s.17 Fire precautions in certain large buildings	*
	s.18 Fire precautions in high buildings	*
	s.19 Amending s.72 Building Act 1984	*
Cheshire County Council Act 1980	s.48 Parking places, safety requirements	*
● Chester City Council	s.50 Access for fire brigade	*
● Warrington BC	s.49 Fireman switches	
	s.54 Means of escape, safety requirements	
County of Cleveland Act 1987		
● Stockton on Tees Borough Council	s.5 Access for fire fighting	*
	s.6 Parking places, safety requirements	*
	s.15 Safety of stands	
● Middlesborough Borough Council		
Clwyd Act 1985		
● Borough of Rhuddlan	s.19 Parking places, safety requirements	*
● Colwyn District Council	s.20 Access for fire brigade	*
Cornwall Act 1984		
● Caradon DC		
● North Cornwall		
Croydon Corporation Act 1960	s.93/94 Buildings of excess cubic capacity	*
	s.95 Buildings used for trade and for dwellings	
Cumbria Act 1982		
● Barrow Borough Council	s.23 Parking places, safety requirements	*
● Carlisle City Council	s.25 Access for fire brigade	*
	s.28 Means of escape from certain buildings	
Derbyshire Act 1981		
● Borough of High Peak	s.16 Safety of stands	*
	s.23 Access for fire brigade	*
	s.24 MoE from certain buildings	*
	s.28 Parking places; safety requirements	*
	s.25 Fireman switches	
Dyfed Act 1987		
● South Pembrokeshire District Council	s.46 Safety of stands	
	s.47 Parking places; safety requirements	*
	s.51 Access for fire brigade	*
● Carmarthen District Council		
East Ham Corporation Act 1957		
● Newham London Borough Council	s.38 Separate drainage systems	*
	s.54 Separate access to tenements	*
	s.61 Access for fire brigade	*

(1) *Local Act*	(2) *Relevant sections*	(3) *Apply?*
East Sussex Act 1981 ● Hastings Borough Council	s.34 Fireman switches s.35 Access for fire brigade	 *
Essex Acts 1952 & 1958 (GLC areas formerly in Essex)		
Essex Act 1987 ● Uttlesford District Council	s.13 Access for fire brigade	*
Exeter Act 1987		
Greater Manchester Act 1981 ● Trafford Metropolitan 　Borough Council	s.58 Safety of stands	
	s.61 Parking places, safety requirements s.62 Fireman switches s.63 Access for fire brigade s.64 Fire precautions in high buildings s.65 Fire precautions in large storage buildings s.66 Fire and safety precautions in public and other buildings	* * * * *
● Manchester City Council		
Hampshire Act 1983 ● Southampton City Council	s.11 Parking places, safety requirements s.12 Access for fire brigade s.13 Fire precautions in certain large buildings	* * *
Hastings Act 1988		
Hereford City Council Act 1985	s.17 Parking places; safety requirements s.18 Access for fire brigade	* *
Humberside Act 1982 ● Gt.Grimsby BC	s.12 Parking places, safety requirements s.13 Fireman switches s.14 Access for the fire brigade s.15 Means of escape in certain buildings s.17 Temporary structures, byelaws	* *
Isle of Wight Act 1980 ● (Part VI)	s.32 Access for fire brigade s.31 Fireman switches s.30 Parking places; safety requirements	* *
Kent 1958 (GLC areas formerly in Kent)		
County of Kent Act 1981 ● Canterbury City Council ● Rochester City Council	s.51 Parking places, safety requirements s.52 Fireman switches s.53 Access for fire brigade s.78 Annulment of plans approvals	* * *
Lancashire Act 1984	s.31 Access for fire brigade	*

(1) *Local Act*	(2) *Relevant sections*	(3) *Apply?*
Leicester Act 1985 ● Leicester City Council	s.21 Safety of stands s.49 Parking places, safety requirements s.50 Access for fire brigade s.52 Fire precautions in high buildings s.53 Fire precautions in large storage buildings	 * * * *
● North West Leicestershire District Council	s.54 Means of escape, safety requirements	
London Building Acts 1930–1939	s.20 Buildings of excess height or cubic capacity **N.B.:** By definition (BA 1984 s.88 & Sch.3), these Acts do not constitute Local Acts but otherwise have a similar effect (see also Chapter 5).	*
● Corporation of London ● City of Westminster London Borough Council ● Royal Borough of Kensington & Chelsea		
County of Merseyside Act 1980 ● Liverpool City Council	s.20 Safety of stands s.48/49 Means of escape from fire s.50 Parking places, safety requirements s.51 Fire and safety precautions in public and other buildings s.52 Fire precautions in high buildings s.53 Fire precautions in large storage buildings s.54 Fireman switches s.55 Access for fire brigade	 * * * * *
● Borough of Wirral		
Middlesex Act 1956 (GLC areas formerly in Middlesex)	s.33 Access for fire brigade	*
Mid Glamorgan County Council Act 1987 ● Merthyr Tydfil ● Taff-Ely BC	s.9 Access for fire brigade	*
Nottinghamshire Act 1985 ● City of Nottingham		
Plymouth Act 1987		
Poole Act 1986	s.10 Parking places, safety requirements s.11 Access for fire brigade s.14 Fire precautions in certain large buildings s.15 Fire precautions in high buildings	* * *
County of South Glamorgan Act 1976 ● Cardiff City Metropolitan District Council	s.27 Safety of stands s.48/50 Underground parking places s.51 Means of escape for certain buildings s.52 Fireman switches s.53 Precautions against fire in high buildings s.54 Byelaws for temporary structures.	 * *

(1) Local Act	(2) Relevant sections	(3) Apply?
South Yorkshire Act 1980		

● City of Sheffield
  Metropolitan District
  Council
	s.53 Parking places, safety precautions	*

● Barnsley Metropolitan Borough Council	s.54 Fireman switches	
● Rotherham Metropolitan Borough Council	s.55 Access for fire brigade	*
● Doncaster Metropolitan Borough Council		
Staffordshire Act 1983		
● Staffordshire Moorlands District Council	s.25 Parking places, safety precautions s.26 Access for fire brigade	* *
Surrey Act 1985		
● Guildford Borough Council	s.18 Parking places, safety requirements	*
● Spelthorne Borough Council	s.19 Fire precautions in large storage buildings s.20 Access for fire brigade	* *
Tyne & Wear 1980	s.24 Access for fire brigade	*
West Glamorgan Act 1987	s.43 Parking places, safety precautions	*
● City of Swansea		
● Neath Borough Council		
West Midlands Act 1980		
● Birmingham City Council	s.39 Safety of stands s.44 Parking places, safety requirements s.45 Fireman switches s.46 Access for fire brigade s.49 Means of escape from certain buildings	 * *
West Yorkshire Act 1980		
● Kirklees Metropolitan District Council	s.9 Culverting water courses s.50 Separate drainage systems	* *
● Bradford City Council	s.51 Fireman switches	
Worcester City Council Act 1985		

Appendix 2

National Standards and Codes of Practice

Chapter 6 – Structural stability

The Building Regulations 1991, Schedule 1
A1/2
BS 12: 1989 *Specification for Portland cement.*
BS 187: 1978 *Specification for calcium silicate (sandlime and flintlime) bricks*
   Amendment slip number 1: AMD 5427.
BS 449: *Specification for the use of structural steel in building*, Part 2: 1969 *Metric units*
   Amendment slip number 1: AMD 416
                2: AMD 523
                3: AMD 661
                4: AMD 1135
                5: AMD 1787
                6: AMD 4576
                7: AMD 5698
                8: AMD 6255.
BS 882: 1983 *Specification for aggregates from natural sources for concrete*
   Amendment slip number 1: AMD 5150.
BS 1243: 1978 *Specification for metal ties for cavity wall construction*
   Amendment slip number 1: AMD 3651
                2: AMD 4024.
BS 1297: 1987 *Specification for tongued and grooved softwood flooring.*
BS 3921: 1985 *Specification for clay bricks.*
BS 4471: 1987 *Specification for sizes of sawn and processed softwood.*
BS 4978: 1988 *Specification for softwood grades for structural use.*
BS 5268: *Structural use of timber,*
Part 2: 1991 *Code of practice for permissible stress design, materials and workmanship,*
Part 3: 1985 *Code of practice for trussed rafter roofs*
   Amendment slip number 1: AMD 5931,
Part 6: *Code of practice for timber framed walls*, Section 6.1: 1988 *Dwellings not exceeding three storeys.*
BS 5328: *Concrete,*
Part 1: 1991 *Guide to specifying concrete*
Part 2: 1991 *Methods for specifying concrete mixes.*
BS 5390: 1976 *Code of practice for stone masonry*

Amendment slip number 1: AMD 4272.

BS 5628: *Code of practice for use of masonry,*

Part 1: 1978 *Structural use of unreinforced masonry*

Amendment slip number 1: AMD 2747

2: AMD 3445

3: AMD 4800

4: AMD 5736

Part 3: 1985 *Materials and components, design and workmanship*

Amendment slip number 1: AMD 4974

BS 5950: *Structural use of steelwork in building,*

Part 1: 1990 *Code of practice for design in simple and continuous construction: hot rolled sections,*

Part 2: 1992 *Specification for materials, fabrication and erection: hot rolled sections,*

Part 3: *Design in composite construction,* Section 3.1: 1990 *Code of practice for design of simple and continuous composite beams,*

Part 4: 1982 *Code of practice for design of floors with profiled steel sheeting,*

Part 5: 1987 *Code of practice for design of cold rolled sections*

Amendment slip number 1: AMD 5957.

BS 6073: *Precast concrete masonry units,*

Part 1: 1981 *Specification for precast concrete masonry units*

Amendment slip number 1: AMD 3944

2: AMD 4462.

BS 6399: *Loading for buildings,*

Part 1: 1984 *Code of practice for dead and imposed loads*

Amendment slip number 1: AMD 4949

2: AMD 5881

3: AMD 6031,

Part 3: 1988 *Code of practice for imposed roof loads*

Amendment slip number 1: AMD 6033.

BS 6649: 1985 *Specification for clay and calcium silicate modular bricks.*

BS 6750: 1986 *Specification for modular co-ordination in building.*

BS 8004: 1986 *Code of practice for foundations.*

BS 8110: *Structural use of concrete,*

Part 1: 1985 *Code of practice for design and construction*

Amendment slip number 1: AMD 5917

2: AMD 6276,

Part 2: 1985 *Code of practice for special circumstances*

Amendment slip number 1: AMD 5914,

Part 3: 1985 *Design charts for single reinforced beams, doubly reinforced beams and rectangular columns*

Amendment slip number 1: AMD 5918.

BS 8200: 1985 *Code of practice for design of non-loadbearing external vertical enclosure of buildings.*

BS 8298: 1989 *Code of practice for design and installation of natural stone cladding and lining.*

CP3: Chapter V: *Loading,*

Part 2: 1972 *Wind loads*

Amendment slip number 1: AMD 4952

2: AMD 5152

3: AMD 5343

4: AMD 6028.

CP 118: 1969 *The structural use of aluminium*
Amendment slip number 1: AMD 1129.

A3
BS 5628: *Code of practice for use of masonry,*
Part 1: 1978 *Structural use of unreinforced masonry*
Amendment slip number 1: AMD 2747
2: AMD 3445
3: AMD 4800
4: AMD 5736.
BS 5950: *Structural use of steelwork in building,*
Part 1: 1990 *Code of practice for design in simple and continuous construction: hot rolled sections.*
BS 8110: *Structural use of concrete,*
Part 1: 1985 *Code of practice for design and construction*
Amendment slip number 1: AMD 5917
2: AMD 6276,
Part 2: 1985 *Code of practice for special circumstances*
Amendment slip number 1: AMD 5914.

Chapter 7 – Fire

The Building Regulations 1991, Schedule 1
Introduction
BS 5588: *Fire precautions in the design, construction and use of buildings,*
Part 10: 1991 *Code of practice for enclosed shopping complexes.*

B1. Sections 1–5
BS 5266: *Emergency lighting,*
Part 1: 1988 *Code of practice for the emergency lighting of premises other than cinemas and certain other specified premises used for entertainment.*
BS 5306: *Fire extinguishing installations and equipment on premises,*
Part 2: 1990 *Specification for sprinkler systems.*
BS 5395: *Stairs, ladders and walkways,*
Part 2: 1984 *Code of practice for the design of helical and spiral stairs*
Amendment slip number 1: AMD 6076.
BS 5446: *Specification for components of automatic fire alarm systems for residential purposes:*
Part 1: 1990 *Point-type smoke detectors.*
BS 5499: *Fire safety signs, notices and graphic symbols,*
Part 1: 1990 *Specification for fire safety signs.*
BS 5588: *Fire precautions in the design, construction and use of buildings,*
Part 1: 1990 *Code of practice for residential buildings,*
Part 2: 1985 *Code of practice for shops*
Amendment slip number 1: AMD 5555
2: AMD 6239
3: AMD 6478,
Part 3: 1983 *Code of practice for office buildings*
Amendment slip number 1: AMD 5556
2: AMD 5825
3: AMD 6160,
Part 4: 1978 *Code of practice for smoke control in protected escape routes using pressurization*

Amendment slip number 1: AMD 5377,

Part 5: 1991 *Code of practice for firefighting stairs and lifts,*

Part 6: 1991 *Code of practice for assembly buildings,*

Part 8: 1988 *Code of practice for means of escape for disabled people,*

Part 9: 1989 *Code of practice for ventilation and air conditioning ductwork,*

Part 10: 1991 *Code of practice for enclosed shopping complexes.*

BS 5720: 1979 *Code of practice for mechanical ventilation and air conditioning in buildings.*

BS 5839: *Fire detection and alarm systems for buildings,*

Part 1: 1988 *Code of practice for system design, installation and servicing.*

BS 5906: 1980 *Code of practice for storage and onsite treatment of solid waste from buildings.*

BS 6387: 1983 *Specification for performance requirements for cables required to maintain circuit integrity under fire conditions*

Amendment slip number 1: AMD 4989

2: AMD 5615.

CP 1007: 1955 *Maintained lighting for cinemas.*

B2. Section 6

BS 476: *Fire tests on building materials and structures,*

Part 6: 1981 *Method of test for fire propagation for products,*

Part 6: 1989 *Method of test for fire propagation for products,*

Part 7: 1971 *Surface spread of flame test for materials,*

Part 7: 1987 *Method for classification of the surface spread of flame of products*

Amendment slip number 1: AMD 6249.

B3. Sections 7–11

BS 4514: 1983 *Specification for unplasticized PVC soil and ventilating pipes, fittings and accessories*

Amendment slip number 1: AMD 4517

2: AMD 5584.

BS 5255: 1989 *Specification for thermoplastics waste pipe and fittings.*

BS 5306: *Fire extinguishing installations and equipment on premises,*

Part 2: 1990 *Specification for sprinkler systems.*

BS 5588: *Fire precautions in the design, construction and use of buildings,*

Part 5: 1991 *Code of practice for firefighting stairs and lifts,*

Part 9: 1989 *Code of practice for ventilation and air conditioning ductwork,*

Part 10: 1991 *Code of practice for enclosed shopping complexes.*

B4. Sections 12–14

BS 476: *Fire tests on building materials and structures,*

Part 3: 1958 *External fire exposure roof tests,*

Part 6: 1981 *Method of test for fire propagation for products,*

Part 6: 1989 *Method of test for fire propagation for products.*

BS 5306: *Fire extinguishing installations and equipment on premises,*

Part 2: 1990 *Sprinkler systems.*

B5. Sections 15–18

BS 5306: *Fire extinguishing installations and equipment on premises,*

Part 1: 1976 (1988) *Hydrant systems, hose reels and foam inlets*

Amendment slip number 1: AMD 4649

2: AMD 5756,

Part 2: 1990 *Specification for sprinkler systems.*

BS 5588: *Fire precautions in the design, construction and use of buildings*,
Part 5: 1991 *Code of practice for firefighting stairs and lifts*,
Part 10: 1991 *Code of practice for enclosed shopping complexes*.

Appendix A
BS 476: *Fire tests on building materials and structures*,
Part 3: 1958 *External fire exposure roof tests*,
Part 4: 1970 (1984) *Non-combustibility test for materials*
   Amendment slip number 1: AMD 2483
                       2: AMD 4390,
Part 6: 1981 *Method of test for fire propagation for products*,
Part 6: 1989 *Method of test for fire propagation for products*,
Part 7: 1971 *Surface spread of flame tests for materials*,
Part 7: 1987 *Method for classification of the surface spread of flame of products*
   Amendment slip number 1: AMD 6249,
Part 8: 1972 *Test methods and criteria for the fire resistance of elements of building construction*
   Amendment slip number 1: AMD 1873
                       2: AMD 3816
                       3: AMD 4822,
Part 11: 1982 *Method for assessing the heat emission from building products*,
Part 20: 1987 *Method for determination of the fire resistance of elements of construction (general principles)*
   Amendment slip number 1: AMD 6487.
Part 21: 1987 *Methods for determination of the fire resistance of loadbearing elements of construction*,
Part 22: 1987 *Methods for determination of the fire resistance of non-loadbearing elements of construction*,
Part 23: 1987 *Methods for determination of the contribution of components to the fire resistance of a structure*,
Part 24: 1987 *Method for determination of the fire resistance of ventilation ducts.*
BS 747: 1977: *Specification for roofing felts*
   Amendment slip number 1: AMD 3775
                       2: AMD 4609
                       3: AMD 5101.
BS 2782: 1970 *Methods of testing plastics*,
Part 1: *Thermal properties*, Method 120A *Determination of the Vicat softening temperature for thermoplastics*,
Part 5: *Miscellaneous methods*, Method 508A.
BS 5306: *Fire extinguishing installations and equipment on premises*,
Part 2: 1990 *Specification for sprinkler systems.*
BS 5438: 1976 *Methods of test for flammability of vertically oriented textile fabrics and fabric assemblies subjected to a small igniting flame*, Test 2: 1989.
BS 5867: *Specification for fabrics for curtains and drapes*,
Part 2: 1980 *Flammability requirements*
   Amendment slip number 1: AMD 4319.
BS 6073: *Precast concrete masonry units*,
Part 1: 1981 *Specification for precast concrete masonry units*
   Amendment slip number 1: AMD 3944
                       2: AMD 4462.

BS 6336: 1982 *Guide to development and presentation of fire tests and their use in hazard assessment.*
CP 144: *Roof coverings,*
Part 3: 1970. *Build-up bitumen felt*
Amendment slip number 1: AMD 2527
2: AMD 5229.
PD 6520: 1988 *Guide to fire test methods for building materials and elements of construction.*

Appendix B
BS 476: *Fire tests on building materials and structures,*
Part 8: 1972 *Test methods and criteria for the fire resistance of elements of building construction*
Amendment slip number 1: AMD 1873
2: AMD 3816
3: AMD 4822,
Part 22: 1987 *Methods for determination of the fire resistance of non-load-bearing elements of construction,*
Part 31: *Methods for measuring smoke penetration through doorsets and shutter assemblies,* Section 31.1: 1983 *Measurement under ambient temperature conditions.*
BS 5499: *Fire safety signs, notices and graphic symbols,*
Part 1: 1990 *Specification for fire safety signs.*
BS 5588 *Fire precautions in the design, construction and use of buildings,*
Part 4: 1978 *Code of practice for smoke control in protected escape routes using pressurization*
Amendment slip number 1: AMD 5377.

Chapter 8 – Materials, workmanship, site preparation and moisture exclusion

The Building Regulations 1991, Regulation 7
BS EN ISO 9000: *Quality management and quality assurance standards.*
BS EN ISO 9001: 1994 *Quality Systems, Model for quality assurance in design, development, production, installation and servicing.*
BS EN ISO 9002: 1994 *Quality Systems, Model for quality assurance in production, installation and servicing.*
BS 8000: *Workmanship on building sites,*
Part 1: 1989 *Code of practice for excavation and filling,*
Part 2: *Code of practice for concrete work,* Section 2.1: 1990 *Mixing and transporting concrete,* Amendment AMD 9324, February 1997, Section 2.2: 1990 *Sitework with in situ and precast concrete,*
Part 3: 1989 *Code of practice for masonry*
Amendment AMD 6195, May 1990,
Part 4: 1989 *Code of practice for waterproofing,*
Part 5: 1990 *Code of practice for carpentry, joinery and general fixings,*
Part 6: 1990 *Code of practice for slating and tiling of roofs and claddings,*
Part 7: 1990 *Code of practice for glazing,*
Part 8: 1989 *Code of practice for plasterboard partitions and dry linings*
Part 9: 1989 *Code of practice for cement/sand floor screeds and concrete floor toppings,*
Part 10: 1995 *Code of practice for plastering and rendering*
Amendment AMD 9271, November 1996,
Part 11: *Code of practice for wall and floor tiling,* Section 11.1: 1989 *Ceramic*

tiles, Terrazzo tiles and mosaics, Section 11.2: 1990 *Natural stone tiles*, Amendment AMD 8623, August 1995,

Part 12: 1989 *Code of practice for decorative wallcoverings and painting*,

Part 13: 1989 *Code of practice for above ground drainage and sanitary appliances*,

Part 14: 1989 *Code of practice for below ground drainage*,

Part 15: 1990 *Code of practice for hot and cold water services (domestic scale)*,

Part 16: 1997 *Code of practice for sealing joints in buildings using sealants*.

The Building Regulations 1991, Schedule 1
C1/2/3
BS 5930: 1981 *Code of practice for site investigation.*
DD 175: 1988 *Code of practice for the identification of potentially contaminated land and its investigation.*

C4
BS 1282: 1975 *Guide to the choice, use and application of wood preservatives.*
BS 5247: *Code of practice for sheet roof and wall coverings,*
Part 14: 1975 *Corrugated asbestos-cement*
Amendment slip number 1: AMD 2821
2: AMD 3502.
BS 5262: 1976 *Code of practice. External rendered finishes*
Amendment slip number 1: AMD 2103
2: AMD 6246.
BS 5328: *Concrete,*
Part 1: 1990 *Guide to specifying concrete*
Part 2: 1990 *Method for specifying concrete mixes.*
BS 5390: 1976 (1984) *Code of practice for stone masonry*
Amendment slip number 1: AMD 4272.
BS 5617: 1985 *Specification for urea-formaldehyde (UF) foam systems suitable for thermal insulation of cavity walls with masonry or concrete inner and outer leaves.*
BS 5618: 1985 *Code of practice for thermal insulation of cavity walls (with masonry or concrete inner and outer leaves) by filling with urea-formaldehyde (UF) foam systems.*
Amendment slip number 1: AMD 6262.
BS 5628 *Code of practice for use of masonry,*
Part 3: 1985 *Materials and components, design and workmanship.*
BS 6232: *Thermal insulation of cavity walls by filling with blown man-made mineral fibre,*
Part 1: 1982 *Specification for the performance of installation systems*
Amendment slip number 1: AMD 5428,
Part 2: 1982 *Code of practice for installation of blown man-made mineral fibre in cavity walls with masonry and/or concrete leaves.*
BS 6915: 1988 *Specification for design and construction of fully supported lead sheet roof and wall coverings.*
BS 8102: 1990 *Code of practice for protection of structures against water from the ground.*
BS 8200: 1985 *Code of practice for design of non-loadbearing external vertical enclosures of buildings.*
BS 8208: *Guide to assessment of suitability of external cavity walls for filling with thermal insulants,*

Part 1: 1985 *Existing traditional cavity construction.*
BS 8215: 1991 *Code of practice for design and installation of damp-proof courses in masonry construction.*
BS 8298: *Code of practice for design and installation of natural stone cladding and lining.*
CP 102: 1973 *Code of practice for protection of buildings against water from the ground*
Amendment slip number 1: AMD 1151
                       2: AMD 2196
                       3: AMD 2470.

CP 143 *Code of practice for sheet roof and wall coverings,*
Part 1: 1958 *Aluminium, corrugated and troughed,*
Part 5: 1964 *Zinc,*
Part 10: 1973 *Galvanised corrugated steel,*
Part 12: 1970 *Copper*
Amendment slip number 1: AMD 863
                       2: AMD 5193,
Part 15: 1973 (1986) *Aluminium*
Amendment slip number 1: AMD 4473,
Part 16: 1974 *Semi-rigid asbestos bitumen sheet.*
CP 297: 1972 *Precast concrete cladding (non-loadbearing).*
DD 93: 1984 *Methods for assessing exposure to wind-driven rain.*

Chapter 9 – Toxic substances

The Building Regulations 1991, Schedule 1
D1
BS 5617: 1985 *Specification for urea-formaldehyde (UF) foam systems suitable for thermal insulation of cavity walls with masonry or concrete inner and outer leaves.*
BS 5618: 1985 *Code of practice for thermal insulation of cavity walls (with masonry or concrete inner and outer leaves) by filling with urea-formaldehyde (UF) foam systems.*
Amendment slip number 1: AMD 6262.
BS 8208: *Guide to assessment of suitability of external cavity walls for filling with thermal insulants,*
Part 1: 1985 *Existing traditional cavity construction.*
Amendment slip number 1: AMD 4996.

Chapter 10 – Sound insulation

The Building Regulations 1991, Schedule 1
E1/2/3
BS 1142: 1989 *Specification for fibre boards.*
BS 2750: *Measurement of sound insulation in buildings and of building elements,*
Part 1: 1980 *Recommendations for laboratories,*
Part 3: 1980 *Laboratory measurement of airborne sound insulation of building elements,*
Part 4: 1980 *Field measurement of airborne sound insulation between rooms,*
Part 6: 1980 *Laboratory measurement of impact sound insulation of floors,*
Part 7: 1980 *Field measurements of impact sound insulation of floors.*
BS 5628: *Code of practice for use of masonry,*

Part: 3 1985 *Materials and components, design and workmanship.*
BS 5821: *Methods for rating the sound insulation in building elements,*
Part 1: 1984 *Method for rating the airborne sound insulation in buildings and interior building elements,*
Part 2: 1984 *Method for rating the impact sound insulation.*

Chapter 11 – Ventilation

The Building Regulations 1991, Schedule 1
F1 and F2
BS 4434: 1989 *Specification for safety aspects in the design, construction and installation of refrigerating appliances and systems.*
BS 5250: 1989 *Code of practice for the control of condensation in buildings.*
BS 5440: Part 1: 1990 *Specification for installation of flues.*
BS 5720: 1979 *Code of practice for mechanical ventilation and air-conditioning in buildings.*
BS 5925: 1991 *Code of practice for ventilation principles and designing for natural ventilation.*

Chapter 12 – Hygiene

The Building Regulations 1991, Schedule 1
G1
BS 6465 *Sanitary installations,*
Part 1: 1984 *Code of practice for scale of provision, selection and installation of sanitary appliances.*

G3
BS 3955: 1986 *Specification for electrical controls for household and similar general purposes.*
    Amendment slip number 1: AMD 5940.
BS 4201: 1979 (1984) *Specification for thermostats for gas-burning appliances*
    Amendment slip number 1: AMD 4531
                          2: AMD 6268.
BS 6283 *Safety and control devices for use in hot water systems,*
Part 2: 1991 *Specification for temperature relief valves for pressures from 1 bar to 10 bar,*
Part 3: 1991 *Specification for combined temperature and pressure relief valves for pressures from 1 bar to 10 bar.*
BS 6700: 1987 *Specification for design, installation, testing and maintenance of services supplying water for domestic use within buildings and their curtilages.*
BS 7206: 1990 *Specification for unvented hot water storage units and packages.*

Chapter 13 – Drainage and waste disposal

The Building Regulations 1991, Schedule 1
H1
BS 65: 1991 *Specification for vitrified clay pipes, fittings and ducts, also flexible mechanical joints for use solely with surface water pipes and fittings.*
BSEN 295: *Vitrified clay pipes and fittings and pipe joints for drains and sewers,*

Part 1: 1991 *Test requirements,*
Part 2: 1991 *Quality control and sampling,*
Part 3: 1991 *Test methods.*
BS 416 *Discharge and ventilating pipes and fittings, sand-cast or spun in cast iron,*
Part 1: 1990 *Specification for spigot and socket systems,*
Part 2: 1990 *Specification for socketless systems.*
BS 437: 1978 *Specification for cast iron spigot and socket drain pipes and fittings*
Amendment slip number 1: 5877.
BS 864 *Capillary and compression tube fittings of copper and copper alloy,*
Part 2: 1983 *Specification for capillary and compression fittings for copper tubes*
Amendment slip number 1: AMD 5097
2: AMD 5651.
BS 882: 1983 *Specification for aggregates from natural sources for concrete*
Amendment slip number 1: AMD 5150.
BS 2871 *Copper and copper alloys. Tubes,*
Part 1: 1971 *Copper tubes for water, gas and sanitation*
Amendment slip number 1: AMD 1422
2: AMD 2203.
BS 3656: 1981 (1990) *Specification for asbestos-cement pipes, joints and fittings for sewerage and drainage.*
Amendment slip number 1: AMD 5531.
BS 3868: 1973 *Prefabricated drainage stack units: galvanized steel.*
BS 3921: 1985 *Specification for clay bricks.*
BS 3943: 1979 *Specification for plastics waste traps*
Amendment slip number 1: AMD 3206
2: AMD 4191
3: AMD 4692.
BS 4514: 1983 *Specification for unplasticized PVC soil and ventilating pipes, fittings and accessories*
Amendment slip number 1: AMD 4517
2: AMD 5584.
BS 4660: 1989 *Specification for unplasticized polyvinyl chloride (PVC-U) pipes and plastics fittings of nominal sizes 110 and 160 for below ground drainage and sewerage.*
BS 5254: 1976 *Specification for polypropylene waste pipe and fittings (external diameter 34.6 mm, 41.0 mm and 54.1 mm)*
Amendment slip number 1: AMD 3588
2: AMD 4438.
BS 5255: 1989 *Specification for thermoplastics waste pipe and fittings.*
BS 5481: 1977 (1989) *Specification for unplasticized PVC pipe and fittings for gravity sewers*
Amendment slip number 1: AMD 3631
2: AMD 4436.
BS 5572: 1978 *Code of practice for sanitary pipework*
Amendment slip number 1: AMD 3613
2: AMD 4202.
BS 5911 *Precast concrete pipes, fittings and ancillary products.*
Part 2: 1982 *Specification for inspection chambers and street gullies*
Amendment slip number 1: AMD 5146.

Part 100: 1988 *Specification for unreinforced and reinforced pipes and fittings with flexible joints*
    Amendment slip number 1: AMD 6269.
Part 101: 1988 *Specification for glass composite concrete (GCC) pipes and fittings with flexible joints,*
Part 120: 1989 *Specification for reinforced jacking pipes with flexible joints.*
Part 200: 1989 *Specification for unreinforced and reinforced manholes and soakaways of circular cross section.*
BS 6087: 1990 *Specification for flexible joints for grey or ductile cast iron drain pipes and fittings (BS 437) and for discharge and ventilating pipes and fittings (BS 416)*
    Amendment slip number 1: AMD 6357.
BS 7158: 1989 *Specification for plastics inspection chambers and drains.*
BS 8110 *Structural use of concrete,*
Part 1: 1985 *Code of practice for design and construction*
    Amendment slip number 1: AMD 5917
                              2: AMD 6276.
BS 8301: 1985 *Code of practice for building drainage,*
    Amendment slip number 1: AMD 5904.

H2
BS 5328 *Concrete,*
Part 1: 1991 *Guide to specifying concrete,*
Part 2: 1991 *Methods specifying concrete mixes,*
Part 3: 1990 *Specification for the procedures to be used in producing and transporting concrete*
    Amendment slip number 1: AMD 6927,
Part 4: 1990 *Specification for the procedures to be used in sampling, testing and assessing compliance of concrete.*
BS 6297: 1983 *Code of practice for design and installation of small sewage treatment works and cesspools.*

H3
BS 6367: 1983 *Code of practice for drainage of roofs and paved areas*
    Amendment slip number 1: AMD 4444.
BS 8301: 1985 *Code of practice for building drainage*
    Amendment slip number 1: AMD 5904 2: AMD 6580.

H4
BS 5906: 1980 (1987) *Code of practice for the storage and on-site treatment of solid waste from buildings.*

Chapter 14 – Heat producing appliances

The Building Regulations 1991, Schedule 1
J1/2/3
BS 41: 1973 (1981) *Cast iron spigot and socket flue or smoke pipes and fittings.*
BS 65: 1991 *Specification for vitrified clay pipes, fittings and ducts, also flexible mechanical joints for use solely with surface water pipes and fittings.*
BS 476 *Fire tests on building materials and structures,*
Part 4: 1970 (1984) *Non-combustibility tests for materials*
    Amendment slip number 1: AMD 2483
                              2: AMD 4390.

BS 567: 1973 (1984) *Specification for asbestos-cement flue pipes and fittings, light quality,*
> Amendment slip number 1: AMD 5963.

BS 715: 1989 *Specification for metal flue pipes, fittings, terminals and accessories for gas-fired appliances with a rated input not exceeding 60 kW,*
> Amendment slip number 1: AMD 6615
> 2: AMD 6335.

BS 835: 1973 (1989) *Specification for asbestos-cement flue pipes and fittings, heavy quality,*
> Amendment slip number 1: AMD 5964.

BS 1181: 1989 *Specification for clay flue linings and flue terminals.*

BS 1289: *Flue blocks and masonry terminals for gas appliances,*

Part 1: 1986 *Specification for precast concrete flue blocks and terminals,*

Part 2: 1989 *Specification for clay flue blocks and terminals.*

BS 1449 *Steel plate, sheet and strip,*

Part 2: 1983 *Specification for stainless and heat resisting steel plate, sheet and strip,*
> Amendment slip number 1: AMD 4807
> 2: AMD 6646.

BS 4543 *Factory-made insulated chimneys,*

Part 1: 1990 *Methods of test,*

Part 2: 1990 *Specification for chimneys with stainless steel fire linings for use with solid fuel fired appliances.*

Part 3: 1990 *Specification for chimneys with stainless steel fire linings for use with oil fired appliances.*

BS 5258 *Safety of domestic gas appliances,*

Part 1: 1986 *Specification for central heating boilers and circulators,*

Part 4: 1987 *Specification for fanned-circulation ducted-air heaters,*

Part 5: 1989 *Specification for gas fires*
> Amendment slip number 1: AMD 4076
> 2: AMD 4745,

Part 7: 1977 *Storage water heaters,*

Part 8: 1980 *Combined appliances: gas fire/back boiler,*

Part 12: 1990 *Specification for decorative fuel effect gas appliances (2nd and 3rd family gases).*

Part 13: 1986 *Specification for convector heaters.*

BS 5386 *Specification for gas burning appliances,*

Part 1: 1976 *Gas burning appliances for instantaneous production of hot water for domestic use*
> Amendment slip number 1: AMD 2990
> 2: AMD 5832,

Part 2: 1981 (1986) *Mini water heaters (2nd and 3rd family gases),*

Part 3: 1980 *Domestic cooking appliances burning gas*
> Amendment slip number 1: AMD 4162
> 2: AMD 4405
> 3: AMD 4878
> 4: AMD 5220
> 5: AMD 6642,

Part 4: 1991 *Built-in domestic cooking appliances.*

BS 5410 *Code of practice for oil firing,*

Part 1: 1977 *Installations up to 44 kW output capacity for space heating and hot water supply purposes*

Amendment slip number 1: AMD 3637.

BS 5440 *Installation of flues and ventilation for gas appliances or rated input not exceeding 60 kW (1st, 2nd and 3rd family gases),*
Part 1: 1990 *Specification for installation of flues.*
Part 2: 1989 *Specification for installation of ventilation for gas appliances.*

BS 5546: 1990 *Specification for installation of gas hot water supplies for domestic purposes (1st, 2nd and 3rd family gases)*
Amendment slip number 1: AMD 6656.

BS 5864: 1989 *Specification for installation in domestic premises of gas-fired ducted air heaters of rated output not exceeding 60 kW.*

BS 5871: 1980 (1983) *Specification for installation of gas fires, convector heater fire/back boilers and decorative fuel effect gas appliances,*
Part 3: 1991 *Decorative fuel effect gas appliances of heat input not exceeding 15 kW (2nd and 3rd family gases)*
Amendment slip number 1: AMD 7033.

BS 6172: 1990 *Specification for installation of domestic gas cooking appliances (1st, 2nd and 3rd family gases).*

BS 6173: 1990 *Specification for installation of gas fired catering appliances for use in all types of catering establishments (1st, 2nd and 3rd family gases).*

BS 6461: *Installation of chimneys and flues for domestic appliances burning solid fuel (including wood and peat),*
Part 2: 1984 *Code of practice for factory-made insulated chimneys for internal applications.*

BS 6798: 1987 *Specification for installation of gas-fired hot water boilers of rated input not exceeding 60 kW.*

BS 6999: 1989 *Specification for vitreous enamelled low carbon steel flue pipes, other components and accessories for solid fuel burning appliances with a maximum rated output of 45 kW.*

BS 8303: 1986 *Code of practice for installation of domestic heating and cooking appliances burning solid mineral fuels*
Amendment slip number 1: AMD 5723.

Chapter 15 – Stairways, ramps and guards

The Building Regulations 1991, Schedule 1
K1
BS 585: *Wood stairs,*
Part 1: 1989 *Specification for stairs with closed risers for domestic use, including straight and winder flights and quarter and half landings.*

BS 4211: 1987 *Specification for ladders for permanent access to chimneys, other high structures, silos and bins.*

BS 5395: *Stairs, ladders and walkways,*
Part 1: 1977 *Code of practice for stairs*
　　Amendment slip number 1: AMD 3355
　　　　　　　　　　　 2: AMD 4450,
Part 2: 1984 *Code of practice for the design of helical and spiral stairs,*
Part 3: 1985 *Code of practice for the design of industrial type stairs, permanent ladders and walkways.*

BS 5588: *Fire precautions in the design, construction and use of buildings,*
Part 6: 1991 *Code of practice for places of assembly.*

K2/3
BS 6399: *Loading for buildings,*

Part 1: 1996 *Code of practice for dead and imposed loads*
BS 6180: 1995 *Code of practice for protective barriers in and about buildings.*

Chapter 16 – Conservation of fuel and power

The Building Regulations 1991, Schedule 1
L1
BS 699: 1984 (1990) with amendments prior to June 1994 *Specification for copper direct cylinders for domestic purposes.*
BS 853: 1990 *Specification for calorifiers and storage vessels for central heating and hot water supply.*
BS 1566: 1984 (1990) *Copper indirect cylinders for domestic purposes.*
BS 3198: 1981 with amendments prior to June 1994 *Specification for copper hot water combination units for domestic purposes.*
BS 5422: 1990 *Methods for specifying thermal insulation materials on pipes, ductwork and equipment in the temperature range of –40°C to +700°C.*
BS 5449: 1990 *Specification for forced circulation hot water central heating systems for domestic premises.*
BS 5864: 1989 *Specification for installation in domestic premises of gas-fired ducted air-heaters of rated output not exceeding 60 kW.*
BS 6880: 1988 *Code of practice for low temperature hot water heating systems of output greater than 45 kW.*

Chapter 17 – Access and facilities for disabled people

The Building Regulations 1991, Schedule 1
M2
BS 4787: *Internal and external wood doorsets, door leaves and frames,*
Part 1: 1980 (1985) *Specification for dimensional requirements.*
BS 5655: *Lifts and service lifts,*
Part 1: 1986 *Safety rules for the construction and installation of electric lifts*
Amendment slip number 1: AMD 5840 (Part 1 to be replaced by BS EN 81-1, when published),
Part 2: 1988 *Safety rules for the construction and installation of hydraulic lifts*
Amendment slip number 1: AMD 6220 (Part 2 to be replaced by BS EN 81-2, when published),
Part 5: 1989 *Specifications for dimensions for standard lift arrangements,*
Part 7: 1983 *Specification for manual control devices, indicators and additional fittings*
Amendment slip number 1: AMD 4912.
BS 5776: 1996 *Specification for powered stairlifts*
BS 6440: 1983 *Code of practice for powered lifting platforms for use by disabled persons* (Amendment due 1999).

Chapter 18 – Safety glazing

The Building Regulations 1991, Schedule 1
N1
BS 6206: 1981 *Specification for impact performance requirements for flat safety glass and safety plastics for use in buildings.*

N4

BS 8213: Part 1: 1991 Windows, doors and rooflights.

Part 1. *Code of practice for safety in use and during cleaning of windows and doors (including guidance on cleaning materials and methods).*

Index